ADVANCES IN CHEMICAL PHYSICS

VOLUME LXVII

ADVANCES IN CHEMICAL PHYSICS—VOLUME LXVII

I. Prigogine and Stuart A. Rice—Editors

AB INITIO METHODS IN QUANTUM CHEMISTRY—Part I

Edited by

K. P. LAWLEY

Department of Chemistry
Edinburgh University

A WILEY–INTERSCIENCE PUBLICATION

JOHN WILEY & SONS

CHICHESTER · NEW YORK · BRISBANE · TORONTO · SINGAPORE

Library of Congress Cataloging-in-Publication Data:

Ab initio methods in quantum chemistry.

  (Advances in chemical physics; v. 67)
  'A Wiley–Interscience publication.'
  Includes index.
  1. Quantum chemistry.   I. Lawley, K. P.   II. Series.
QD453.A27   [QD462.5]   541.2'8   87–9168

ISBN 0 471 90900 9

British Library Cataloguing in Publication Data:

Lawley, K. P.
  Ab initio methods in quantum chemistry I.
  —(Advances in chemical physics, ISSN
  0065–2385)
  1. Quantum chemistry
  I. Title   II. Series
  541.2'8   QD462

ISBN 0 471 90900 9

Printed and bound in Great Britain

INTRODUCTION

Few of us can any longer keep up with the flood of scientific literature, even in specialized subfields. Any attempt to do more, and be broadly educated with respect to a large domain of science, has the appearance of tilting at windmills. Yet the synthesis of ideas drawn from different subjects into new, powerful, general concepts is as valuable as ever, and the desire to remain educated persists in all scientists. This series, *Advances in Chemical Physics*, is devoted to helping the reader obtain general information about a wide variety of topics in chemical physics, which field we interpret very broadly. Our intent is to have experts present comprehensive analyses of subjects of interest and to encourage the expression of individual points of view. We hope that this approach to the presentation of an overview of a subject will both stimulate new research and serve as a personalized learning text for beginners in a field.

ILYA PRIGOGINE
STUART A. RICE

CONTRIBUTORS TO VOLUME LXVII

R. AHLRICHS, Institut für Physikalische Chemie und Elektrochemie, Lehrstuhl für Theoretische Chemie, Universität Karlsruhe, 7500 Karlsruhe, West Germany

R. D. AMOS, University Chemical Laboratory, Cambridge CB2 1EW, UK

K. BALASUBRAMANIAN, Department of Chemistry, Arizona State University, Tempe, Arizona 85287, USA

F. BERNARDI, Istituto Chimico G. Ciamician, Universita di Bologna, 40136 Bologna, Italy

P. J. BRUNA, Lehrstuhl für Theoretische Chemie, Universität Bonn, D-5300 Bonn 1, West Germany

P. DURAND, Laboratoire de Physique Quantique, Unité Associée au CNRS no. 505, Université Paul Sabatier, 31602 Toulouse Cedex, France

R. O. JONES, Institut für Festkörperforshung der Kernforschungsanlage Jülich, D-5170 Jülich, Federal Republic of Germany

J.-P. MALRIEU, Laboratoire de Physique Quantique, Unité Associée au CNRS no. 505, Université Paul Sabatier, 31602 Toulouse Cedex, France

S. D. PEYERIMHOFF, Lehrstuhl für Theoretische Chemie, Universität Bonn, D-5300 Bonn 1, West Germany

K. S. PITZER, Department of Chemistry and Lawrence Berkeley Laboratory, University of California, Berkeley, California 94720, USA

M. A. ROBB, Department of Chemistry, Kings College, London WC2R 2LS, UK

P. SCHARF, Institut für Physikalische Chemie und Elektrochemie, Lehrstuhl für Theoretische Chemie, Universität Karlsruhe, 7500 Karlsruhe, West Germany

H. B. SCHLEGEL, Department of Chemistry, Wayne State University, Detroit, Michigan 48202, USA

S. WILSON, Theoretical Chemistry Department, Oxford OX1 3TG, UK

CONTENTS

Ab Initio Methods in Quantum Chemistry—I
Edited by K. P. Lawley
© 1987 John Wiley & Sons Ltd.

EXCITED-STATE POTENTIALS

P. J. BRUNA* and S. D. PEYERIMHOFF

Lehrstuhl für Theoretische Chemie, Universität Bonn, Wegelerstrasse 12, D-5300 Bonn 1, West Germany

CONTENTS

*Present address: Department of Chemistry, University of New Brunswick, Fredericton, New Brunswick, Canada.

I. INTRODUCTION

Quantum-mechanical *ab initio* calculations for small molecular systems are widely used these days as an instrument in studying problems in various fields of chemistry and molecular physics[1-5]. Most studies deal with ground-state phenomena, i.e. the structure and properties of compounds, thermal reaction pathways and dynamical behavior based on this information. There has been a noticeable increase in excited-state studies in recent years, however, in particular in connection with problems in molecular spectroscopy, in ionization processes or in the detailed study of photochemical reactions, such as photodissociation, energy-transfer and charge-exchange $A^+ + B \rightarrow A + B^+$ reactions. The calculations are especially powerful for small molecules (for example, for systems up to 50 electrons and six atoms other than hydrogen), and hence numerous applications are found in particular in the area of atmospheric and interstellar chemistry and in the study of combustion processes. In these fields it is often found that experimental and theoretical studies are undertaken in close conjunction and that the two yield complementary data which, taken together, are able to clarify a process. In other instances it is not uncommon that for short-lived species the values obtained from calculations are so far the only ones available.

Basic to the understanding of excited-state processes is knowledge of the excited states themselves, i.e. their character (assignment of spin and spatial symmetry), their absolute location on the energy scale and their change in stability as a function of nuclear displacements, plus knowledge of their lifetimes. The first properties, namely character and potential energy surface, are obtained in a straightforward manner from the quantum-mechanical calculation of an individual state, while lifetime determination is generally more complex and depends not only on the state itself but also on its interaction with other states. Generally one differentiates between radiative lifetimes (involving dipole-allowed or dipole-forbidden processes) and those determined by intersystem crossings.

The determination of potential energy surfaces by quantum-mechanical *ab initio* methods has a number of advantages over the determination of the same quantity by experimental measurements:

1. In the theoretical study, a compound is characterized only by its number of electrons plus the charges and locations of its constituent nuclei. Hence calculations can be carried out for systems irrespective of their thermodynamic stability, i.e. for stable molecules as well as for explosives or very short-lived radicals, independent of their possible experimentally unpleasant features such as being poisonous, environmentally harmful or contaminating to the equipment. Furthermore, positive or energetically stable negatively charged species can be handled with the same ease, and recently methods have been evaluated so that short-lived negative-ion

(resonance) states that lie energetically above the neutral ground state can be treated in a similar manner to the other species.

2. The same theoretical method can be employed for the entire wavelength region from inner-shell ionization or excitation all the way to the far-infrared, whereas different experimental equipment is generally needed for ultraviolet and unfrared spectroscopy, for example.

3. The calculations are in principle the same for all electronic states irrespective of their multiplicity or character (intra-valance shell or Rydberg transitions), in contrast to spectroscopic measurements which depend on selection rules between states, and often can only deduce those states for which intercombination with lower states is unlikely in an indirect manner, by the perturbation they cause in a spectral band system.

4. Calculations can be carried out for all possible nuclear conformations and hence allow one to determine the entire potential energy surface, while spectroscopic measurements generally sample only that portion of the surface which is in a favorable Franck–Condon area (i.e. in an area in which vibrational levels can be populated) and molecular-beam studies generally yield potential energy information only in an indirect way.

5. A special advantage of the theoretical method is the possibility of studying directly the interaction of potential surfaces, a factor that is essential for the molecular dynamics. The energy of the interacting surfaces is thereby obtained with not much more effort than is required to predict points for an isolated surface. The actual magnitudes of the interactions (i.e. the matrix element for radial or angular coupling between states of the same multiplicity, for example; or the matrix element for spin–orbit interaction between surfaces of different symmetry; or more complicated terms which couple electronic and nuclear motion as in the Renner–Teller or Jahn–Teller effects) have not so far been calculated very often, but are expected to attract more interest in the near future as the treatment of dynamic aspects becomes more common. The matrix element calculation is possible, as will be shown later.

6. The calculation of fine-structure effects originating from spin–orbit coupling is also now possible. In analogy to the potential surface determination, the advantage of the theoretical method is the possibility of obtaining these data for the entire region of nuclear conformers. Furthermore, if energy splittings are due to a combination of different phenomena (as for example Λ-doubling due to an interaction of rotation and spin–orbit effects), the measurements give the combined result to high accuracy, while the theoretical investigation is able to differentiate between the various contributions, which may be quite important for further understanding of a process.

7. Finally, it is clear that an analysis of the calculated wavefunctions or properties gives insight into the qualitative principles that govern certain

effects, and from this information qualitative rules can be derived on the basis of which trends in related systems can be predicted without actual calculations.

Considering this list of advantages it is not surprising that the quantum-chemical instrument has become an interesting and powerful tool. There is one serious drawback, however: the accuracy of this instrument with respect to energy levels is generally lower than in standard spectroscopic measurements. Depending on the theoretical effort and computation time spent on a problem, the results correspond to high- or low-resolution spectroscopy. Unfortunately, very often only an insider is able to judge the reliability of a calculation, whereas a scientist not so familiar with theoretical methods often takes the results at face value and can then be misled considerably, or takes the opposite point of view and does not trust the results at all. Neither of these extreme situations is constructive. Fortunately the theoretical procedures for treating excited states slowly converge to a level of treatment whose reliability can be judged to a large extent solely on the basis of a few parameters; as such calculations become more and more available, fairly reliable error limits can be attached to the theoretical results so that the uncertainties in relying on quantum-chemical data will be reduced.

In the following sections we will discuss some of the features mentioned above. The first (Section II) will address itself to the question of the overall accuracy that is generally achieved in excited-state potential surface calculations. The next (Section III) will present examples for species for which the theoretical predictions are sometimes the only data available challenging the experimentalists, or for which computed data have been around prior to measurements and are able to explain and predict general trends. Section IV will deal with the interaction of potential surfaces and the contribution of theoretical methods to this area. Section V will deal with the treatment of short-lived negative ions; Section VI will show calculated fine-structure effects in the potential energy curves of simple systems; and Section VII will give a selected but representative set of examples from our own work carried out in conjunction with experiments or examples that have led to a re-evaluation of measured data.

II. OVERALL ACCURACY OF EXCITED-STATE CALCULATIONS

A. Definition of Potential Energy Surface

The Schrödinger equation for a system of electrons and nuclei can be written as

$$[T_n(\mathbf{R}) + T_e(\mathbf{r}) + V_{nn}(\mathbf{R}) + V_{ne}(\mathbf{R}, \mathbf{r}) + V_{ee}(\mathbf{r})]\Psi(\mathbf{R}, \mathbf{r}) = E\Psi(\mathbf{R}, \mathbf{r}) \qquad (1)$$

where all terms with index n and coordinate $\mathbf{R}$ refer to the nuclei and those with e and **r** refer to the electrons. The Hamiltonian includes in addition to the kinetic energy T all electrostatic interactions V among electrons and nuclei; it can be extended by terms corresponding to other interactions such as spin–orbit terms H_{so}, spin–spin terms H_{ss}, etc. If the wavefunction is written as a product of electronic and nuclear terms

$$\Psi(\mathbf{R}, \mathbf{r}) = \Psi_e(\mathbf{r}, \mathbf{R})\chi(\mathbf{R}) \tag{2}$$

some cross-terms involving nuclear and electronic coordinates arise from the nuclear kinetic energy operator, namely

$$T_n(\mathbf{R})\Psi(\mathbf{R}, \mathbf{r}) = -\sum_K \frac{1}{2M_K} \nabla_K^2(\mathbf{R})\Psi_e(\mathbf{R}, \mathbf{r})\chi(\mathbf{R})$$

$$= -\tfrac{1}{2}\Psi_e(\mathbf{R}, \mathbf{r})\left(\sum_K \frac{1}{M_K}\nabla_K^2\chi(\mathbf{R})\right) - \tfrac{1}{2}\chi(\mathbf{R})\left(\sum_K \frac{1}{M_K}\nabla_K^2\Psi_e(\mathbf{R}, \mathbf{r})\right)$$

$$-\sum_K \frac{1}{M_K}[\nabla_K\Psi_e(\mathbf{R}, \mathbf{r})\nabla_K\chi(\mathbf{R})] \tag{3}$$

If the second and third terms involving derivatives of the electronic wavefunction Ψ_e with respect to nuclear displacements are neglected, the Schrödinger equation in Eq. (1) can be separated into two equations (the Born–Oppenheimer approximation):

$$[T_e(\mathbf{r}) + V_{ne}(R, \mathbf{r}) + V_{ee}(\mathbf{r})]\Psi_e(\mathbf{r}, R) = E_e(R)\Psi_e(\mathbf{r}, R) \tag{4}$$

$$[T_n(\mathbf{R}) + V_{nn}(\mathbf{R}) + E_e(R)]\chi(\mathbf{R}) = E\chi(\mathbf{R}) \tag{5}$$

The first depends on the electronic coordinates at a fixed position R of the nuclei and yields as solution the electronic wavefunction Ψ_e and electronic energy E_e for a given nuclear arrangement. The second describes nuclear motion in the potential

$$U(\mathbf{R}) = V_{nn}(\mathbf{R}) + E_e(\mathbf{R}) \tag{6}$$

Hence in order to obtain what is generally referred to as the potential energy surface $U(\mathbf{R})$, one has to solve the 'electronic' Schrödinger equation in Eq. (4) for a number of nuclear positions. The further term

$$V_{nn}(\mathbf{R}) = \sum_{K < K'} \frac{Z_K Z_{K'}}{\mathbf{R}_K - \mathbf{R}_{K'}} \tag{7}$$

is easily evaluated.

In the case in which further effects such as spin–orbit interaction, for example, are taken into account, the corresponding operator H_{so} is added to those in Eq. (4) and this contribution is also evaluated at a given value of the nuclear conformation R. The spin–orbit contribution to the energy then modifies E_e so that the modified value appears in the formula for the potential surface $U(\mathbf{R})$. In practice the spin–orbit contribution is often evaluated by perturbation theory so that its energy is simply added to the standard value of the electronic energy E_e.

In order to make direct comparisons with experimental data, the motion of the nuclei, in particular vibrations, must also be taken into account. This can be done in all those cases in which the neglect of the terms $\nabla_K^2(\mathbf{R})\Psi_e$ and $\nabla_K(\mathbf{R})\Psi_e$ is justified simply by solving Eq. (5) on the basis of the calculated potential energy surface. If the derivatives cannot be neglected because the electronic wavefunction changes drastically with nuclear displacements, they can be accounted for at least in an approximate manner in the following way. The total wavefunction $\Psi(\mathbf{R}, \mathbf{r})$ is expanded in terms of products over electronic states of the system

$$\Psi(\mathbf{R}, \mathbf{r}) = \sum_{\beta} \Psi_e^{\beta}(\mathbf{R}, \mathbf{r})\chi_{\beta}(\mathbf{R}) \tag{8}$$

If this expression is substituted into the Schrödinger equation in Eq. (1), multiplied by $\Psi_e^{\alpha}(\mathbf{R}, \mathbf{r})$ and integrated over the electronic coordinates r, one obtains

$$[T_n(\mathbf{R}) + U_{\alpha}(\mathbf{R}) - E]\chi_{\alpha}(\mathbf{R}) + \sum_{\beta}\sum_{K} -\frac{1}{2M_K}[\langle \Psi_e^{\alpha}|\nabla_K|\Psi_e^{\beta}\rangle\nabla_K$$
$$+ \langle \Psi_e^{\alpha}|\nabla_K^2|\Psi_e^{\beta}\rangle]\chi_{\beta}(\mathbf{R}) = 0 \tag{9}$$

Generally it is found that only a small number of electronic states interact strongly. For those the so-called non-adiabatic coupling matrix elements

$$\left\langle \Psi_e^{\alpha}\left|\frac{\partial}{\partial R_K}\right|\Psi_e^{\beta}\right\rangle \quad \text{and} \quad \left\langle \Psi_e^{\alpha}\left|\frac{\partial^2}{\partial R_K\partial R_{K'}}\right|\Psi_e^{\beta}\right\rangle \tag{10}$$

can be calculated,[6–8] as shown in a later section, and the small number of coupled equations of Eq. (9) can be solved directly or they can be transformed into a matrix representation and treated as a general eigenvalue problem.[6]

A prerequisite to the treatment of nuclear motion and a detailed comparison with experiment is the solution of the electronic Schrödinger equation, i.e. the determination of the potential energy surfaces, which is indeed still the time-consuming step in molecular calculations.

B. The Appropriate Atomic-orbital Basis Set

The overall accuracy of calculated potential energy surfaces depends in today's state-of-the-art calculations almost entirely on the atomic-orbital (AO) basis set employed. There are numerous AO basis sets in the literature,[9] but most of them have been tested only for ground-state calculations. Generally AO basis sets are optimized in order to give an optimal charge distribution for the atoms; then polarization functions are supplemented to describe the change in the atomic charge distribution under the influence of the molecular environment; and finally functions are added to account for correlation energy. The latter are only important in configuration-interaction (CI) type calculations, while the former show their influence already at the self-consistent field (SCF) level of treatment. Generally no stringent distinction is made between these two types of functions accounting for polarization and correlation.

There are various studies in the literature to determine the influence of polarization functions on ground-state properties, in particular on total energies, dissociation energies and bond lengths. Normally to insure acceptable accuracy the recommendation is to add at least one d function with Gaussian exponent between 0.4 (B) and 1.5 (F) for molecules containing first-row atoms[9,10] and between 0.15 (Al) and 0.7 (Cl) if the constituent atoms are in the row from Na to Ar. One of the more recent investigations[11] analyses the role of d functions in the sulfur compounds SO_2 and SO_3^{2-} and points out that a description of correlation in the 3p shell requires more compact d functions (Gaussian exponents around 8.0) than has normally been assumed; omission of those is shown to have a non-negligible effect on the calculated results for bond lengths, which according to the authors amounts to a decrease of as much as 0.03 Å in CaH, for example, if the 3p electrons are properly correlated. A systematic study of the relative importance of functions of d, f and g type for the F_2 ground-state potential curve[12] shows that in this case f functions are almost as important to account for external correlation as a second d function and should also be included if highly accurate results are desired.

All such considerations in the design of ground-state calculations are naturally also valid for excited-state treatments. Furthermore, since a reliable description of transition energies must account for the difference in correlation energy between various states, correlation orbitals might sometimes be even more critical in excited-state calculations than in those for the ground state. Typical examples discussed some time ago by Weiss[13] are $s^2p^n \to sp^{n+1}$ excitations and various transitions within the $s^l p^m d^n$ configurations. Fortunately, in a molecule in the neighborhood of its equilibrium, there seems to be a good number of orbitals available with various nodes in their charge distribution to make up jointly to some extent for the omission of adequate correlation orbitals: if on the other hand the potential surface is taken to its limits of separated compounds or atoms, the AO basis has to be optimized first to describe accurately those products.[13]

Finally, it is clear that in excited-state calculations 'spectroscopic' orbitals have also generally to be added to the standard AO basis sets, i.e. orbitals of higher quantum numbers that are occupied in the excited state but not in the ground state. A typical example is Rydberg orbitals, i.e. hydrogen-like large-orbit functions. Because of their diffuse nature, triplet- and corresponding singlet-state Rydberg orbitals are very alike and can often be described by a single function (or at least not more than two functions).[14] In smaller systems these can be placed in the center of the molecule rather than at each atom because of their united-atom nature; this is not possible, of course, if the potential surface is calculated up to the dissociation limits. Rydberg exponents are normally given in bibliographies for atomic functions[9,10] and because of their large radial expansion vary little from one atom to the next. Typical gaussian Rydberg exponents are 0.023, 0.020 and 0.015 for carbon 3s, 3p and 3d respectively and are identical to those recommended for sulfur 4s, 4p and 3d. They increase slightly for the more electronegative atoms; 0.028 (3s in N), 0.023 (2s in O) and 0.036 (3s in F), while the 3d penetrates so little into the inner molecular environment that its exponent of 0.015 is generally taken to be the same for all the atoms B to Cl. Other spectroscopic orbitals (with perhaps the exception of hydrogen 2s and 2p AO) are unfortunately not contained in standard AO basis set bibliographies, so that often orbital optimization is a prerequisite to a good excited-state potential surface calculation.

In what follows we will give some examples for additional excited-state basis sets from our own work in connection with molecular calculations. A typical example for optimization of spectroscopic orbitals is found for atoms on the far left and far right of the periodic system. A calculation for the charge-exchange reaction[15] $Na^+ + Mg \rightarrow Na + Mg^+$ requires in the asymptotic limits the occupation of Na in its $^2S(2p^63s)$, $^2P(2p^63p)$, $^2S(2p^64s)$ and ionic $^1S(2p^6)$ states and magnesium in the $^1S(2p^63s^2)$, $^{3,1}P(2p^63s3p)$, $^{3,1}S(2p^63s4s)$, $^{3,1}D(2p^63s3d)$, $^{3,1}P(2p^63s4p)$ and ionic $^2S(2p^63s)$ states. The optimized functional basis is given in Table I, obtained by employing correlated multi-reference single and double-excitation configuration interaction (MRD-CI) wavefunctions.

TABLE I

Optimized Gaussian AO basis set for excited-state calculations involving Na and Mg up to 3d and 4p single occupation.

	Na	Mg
Standard basis[16]		
(12s6p) in contraction	(6s2p)	(6s2p)
Additional functions		
3p	0.608	0.077
	0.024	0.027
3d	0.01533	0.0175
4s	0.00812	0.01168
4p	0.0063	0.00855
p pol	0.09	0.12
d pol	0.10	0.10

TABLE II

Calculated excited atomic states (in eV) by employing the optimized AO basis of Table I.[15]

	ΔE(calc)	ΔE(expt)		ΔE(calc)	ΔE(expt)
Na ^{2}S(3s)	0.0	0.0	Mg ^{1}S(3s^2)	0.0	0.0
^{2}P(3p)	2.00	2.103	3P(3s3p)	2.63	2.712
^{2}S(4s)	3.07	3.191	^{1}P(3s3p)	4.32	4.345
^{2}D(3d)	3.53	3.617	^{3}S(3s4s)	5.00	5.107
^{2}P(4p)	3.61	3.753	^{1}S(3s4s)	5.29	5.393
^{1}S ion	4.98	5.138	^{1}D(3s3d)	5.73	5.753
Electron affinity	0.507	0.546	3P(3s4p)	5.81	5.932
			^{3}D(3s3d)	5.93	5.946
			^{2}S ion	7.50	7.64

TABLE III

Optimized Gaussian AO basis set for excited-state calculations in ArH requiring occupation of H up to $n = 3$ and argon 3p^6 and 3p^{5}4s.

Ar basis		H basis	
Standard basis[17]	(12s9p) → [6s5p]	Standard basis (Ref. 19a)	(5s) → [4s]
Additional functions		Additional functions (Ref. 19b)	
4s (Ref. 18)	0.175 48	2s	0.984 130 ⎫
	0.030 28		0.037 634 ⎭
			0.016 60
4p (Ref. 18)	0.023 08	3s	0.220 40 ⎱
			0.056 60 ⎰
3d	0.015		
d pol	0.643 91		0.006 685
	0.169 87	4s	0.002 67
		2p	0.337 07 ⎱
			0.079 83 ⎰
			0.024 684
		3p	0.148 51 ⎱
			0.018 95 ⎰
			0.007 169
		4p	0.002 867
		3d (Ref. 19c)	0.036 356 ⎱
			0.010 769 ⎰
		p pol	0.75
Total	(14s10p3d) → [8s6p3d]	Total	(12s8p2d) → [9s6p1d]

The 3p representation in the magnesium $^{3,1}P$ states requires two basis functions; a good description of ^1P relative to ^3P and X^1S is only obtained after addition of the d polarization/correlation function. The results for the various states in Table II shows the basis set quality: deviations for all states are less than 0.15 eV, i.e. an error that can normally be tolerated. All calculated excitation energies are too small, which indicates that the ground-state correlation is described somewhat worse than that of the excited states. The standard (6s2p) AO basis alone yields an ionization potential of only 6.64 eV for Mg instead of the correct 7.64 eV.

Another example is a calculation on various excited states of ArH. In the asymptotic channels, excited argon atoms as well as excited hydrogen atoms appear. The lowest excited-state dissociation products at 10.2 eV are $Ar(^1S)$ + $H(n = 2)$ followed by three relatively close-lying limits corresponding to

TABLE IV

Comparison of energies (in eV) for some excited states of the positive ion Ar^+ relative to the ground state of neutral argon ^1S_g during AO basis set optimization. The standard Gaussian AO basis is the (12s9p) set given in Ref. 17 contracted to [6s5p]. Supplementary functions are always optimized for given states as indicated. Total SCF and estimated full CI energies (in hartree) are given for the ground state.

AO basis	Ar, ^1S_g $3s^23p^6$	$Ar^+, {}^2P_u$ $(3s^23p^5)$	$Ar^+, {}^2S_g$ $(3s3p^6)$	$Ar^+, {}^2D_g$ $(3s3p^44s)$	$Ar^+, {}^2D_g$ $(3s^23p^43d)$
Standard [6s5p]	−526.806 −526.857	15.18	–	–	–
Functions added					
d pol for Ar(^1S_g) exponent 0.736	−526.806 −526.971	15.26	13.53	18.75	–
s (0.0296) for Ar($3s^23p^54s$) plus (0.073) for $Ar^+(3s^23p^44s)$	−526.807 −526.972	15.27	14.36	18.27	–
d (0.0188) for Ar($3s^23p^53d$)	−526.807 −526.985	15.24	14.38	18.21	–
d (0.2) for $Ar^+(3s^23p^43d)$	−526.807 −526.985	15.46	13.87	18.20	20.05
d (0.04) for $Ar^+(3s^23p^43d)$	−526.807 −526.985	15.45	13.87	18.23	19.40
d (3.0) for Ar^+, $3s3p^6$ plus p (0.05) for $Ar^+(3s^23p^44p)$	−526.807 −526.995	15.44	13.11	18.09	18.81
contracted d exponents (3.0, 0.736) coefficients (0.062 76, 0.241 76) for final basis (8s6p4d)	−526.807 −526.994	15.42	13.20	18.10	19.05
Experiment	–	15.75	13.48	18.44	18.695

$Ar(^3P) + H(n = 1)$ at $11.6\,eV$, $Ar(^1P) + H(n = 1)$ at $11.8\,eV$ and $Ar(^1S) + H(n = 3)$ at $12.1\,eV$. Hence the basis requires spectroscopic 2s, 2p, 3s and 3p AOs for hydrogen and 4s for argon—and naturally an appropriate treatment of correlation for the excitation from the $3p^6$ in argon to the singlet and triplet excited $3p^5 4s$ configuration. The functions employed in addition to the general basis are summarized in Table III. The minimum that is required for a description of correlation is added in the form of d functions to argon and a p Gaussian to hydrogen, whereas all other functions must be looked upon as spectroscopic AOs. Results obtained with this basis are discussed later.

Further excited states of argon and its positive ion are needed in calculations designed to analyze the important states in atomic-beam $He^+ + Ar \rightarrow ArHe^+$ measurements,[20] since practically all excited states correspond to excited argon ions plus He in its ground state. Table IV presents the results of such a basis set optimization.[21] The standard McLean–Chandler[17] basis is supplemented by a number of functions always optimized to describe a particular state. It is obvious that the functions have practically no effect on the ground-state SCF energy and hence would not appear in any standard basis set optimization. Two s orbitals are required for a proper description of the 3s and 4s shells relative to $Ar(^1S_g)$, as seen from the large change in the $Ar^+ \, ^2S_g$ and ^2D_g energy upon introducing these functions; note that the state doubly occupying the 3s shell (^2P_u in Ar^+) is not affected. Similarly the 3d shell occupied in the ^2D_g state requires two functions for a proper description. At the same time, a d function is quite important to describe the correlation in the 3p shell, as expected. The absolute error in the first ionized state is still $0.28\,eV$,

TABLE V
Comparison of energies (in eV) for some excited states of the sulfur atom in various basis sets.

AO basis	^3P_g $3s^2p^4$	^1D_g $3s^2p^4$	^1S_g $3s^2p^4$	^3P_u $3sp^5$	^5D_u $3s^2p^3d$
(1) Standard[17] (128p) [6s4p] plus d (0.54)	0.0	1.34	2.82	10.02	(20.1)
(2) [6s4p] plus d (0.54) plus f (0.55)	0.0	1.24	2.86	10.14	(19.77)
(3) [6s4p] plus six contracted d functions[a] exponents (8.35, 2.21, 0.65) 0.27, 0.08, 0.0273	0.0	1.29	2.73	8.61	8.03
(4) [6s4p] plus six contracted d plus f (0.55)	0.0	1.19	2.78	8.745	8.28
Expt	0.0	1.12	2.72	8.93	8.39

[a]Taken from Ref. 11.

TABLE VI

Comparison of energies (in eV) for some excited states in the S^+ ion obtained with different AO basis sets.

AO basis[a]	^4S_u $3s^2p^3$	^2D_u $3s^2p^3$	^2P_u $3s^2p^3$	^4P_g $3sp^4$	^2D_g $3sp^4$	^2P_g $3sp^4$	^2S_g $3sp^4$	^4F_g $3s^2p^2d$
(1)	0.0	2.21	3.36	10.22	13.47	16.96	16.37	(22.56)
(2)	0.0	2.015	3.21	10.27	13.34	16.72	16.22	(21.96)
(3)	0.0	2.13	3.33	9.66	12.15	13.21	14.60	13.94
(4)	0.0	1.96	3.16	9.68	12.04	13.06	14.34	13.82
Expt	0.0	1.84	3.04	9.87	–	13.11	–	13.70

[a] As given in Table V.

which indicates that the correlation energy difference between Ar and Ar^+ has not been accounted for entirely. This would require f functions in the AO basis. On the other hand, the relative energies between argon $^2D_g(4s)$, ^2S_g and ^2P_u states are described with an accuracy better than 0.1 eV.

Further AO basis set studies have been made on sulfur- and chlorine-containing compounds. Tables V and VI show again that spectroscopic d orbitals must be present for a proper ^5D_u or ^4F_g description. While all three sulfur states with p^4 occupation are described fairly realistically in all basis sets chosen (improvement of 0.1 eV upon introduction of f functions), it is seen that the change from p^4 to p^5 occupation requires the use of additional d species (and possibly one further s function) to give a better account of the difference in correlation energy in these states. Similar conclusions can be drawn from comparison of the various excited S^+ states. Changes of the order of 0.1–0.2 eV occur upon addition of f functions, while again a large effect due to inclusion of d functions is seen on the excited states that involve a transition from p^3 to p^4 occupation. Finally, Table VII shows the relative location of the ionic ground state relative to the neutral system.

TABLE VII

Calculated ionization energy ($^4S_u - {}^3P_g$, in eV) for sulfur according to various treatments in different AO basis sets. The standard MRD-CI treatment assumes a doubly occupied core of K and L shells.

AO basis[a]/treatment	IP
AO basis (1), without core	9.58
AO basis (3)	9.74
AO basis (3), but d functions uncontracted	9.75
AO basis (2)	9.85
AO basis (4)	10.03
AO basis (4), but a total of two f	10.083
Expt (corrected for j-splitting)	10.333

[a] As given in Table V.

It is seen that the influence of the f functions on the calculated results is of the order of 0.25 eV, and if this can be generalized it is clear why standard AO basis sets generally underestimate the ionization potentials in molecules up to 0.5 eV. The contraction of the d functions has no effect at all in this calculation. As expected, the contribution of the second f function is quite small. The discrepancy with the measured value is still 0.2 eV, however. In this connection it might be worth while to investigate further the effect of correlating the core orbitals. On the other hand, even more functions have to be added in this case as proper correlation functions for the 2p shell (Gaussian d functions with exponents around 12,[22] for example, for Cl). Since our main interest is to deal with molecules in which the feature of a doubly occupied core is almost a necessary requirement in order to avoid the calculations becoming too extensive, one has simply to realize that errors of this order will appear if systems containing a different number of electrons with large changes in correlation energy are to be compared.

While in first- and second-row atoms the inclusion of f functions in the AO basis is desirable, as seen so far, this is found to be necessary in heavier atoms such as transition metals, for example. Reliable transition energies (i.e. errors smaller than 0.2 eV) are only found after inclusion of two f functions, where a tight component is necessary for the correlation description of the innermost $3s^2 3p^6$ subshell and the less compact f AO is to account for the essential correlation energy in the outermost 3d4s subshell. A detailed study of correlation effects in the (3s3p) subshell of Sch, TiH and VH has just been made[23] in connection with the electronic spectrum of these compounds, and the reader is referred to this article.

In dealing with heavier atoms, the spin–orbit effects become sizeable and the question arises as to which AO basis sets are best suited for such properties. A systematic study testing nine AO basis sets has been undertaken for the multiplet splitting of the $X^3\Sigma^-$ and $C^3\Pi$ states of SO as well as for the $X^2\Pi$ state of SO^+.[24] Without going into details (which can be found in the original reference), the salient features seem to be as follows: The accuracy of the spin–orbit matrix elements depends to a large extent on the representation of the valence shell and is influenced considerably less by the inner 1s or 2s shells. Hence small basis sets—smaller than STO-4G (Slater-type orbitals, four Gaussian), for example—are expected to give unreliable results for spin–orbit matrix elements, while those standard basis sets which yield good energies in molecular calculations can also be used efficiently for the calculation of spin–orbit properties. A further more limited study has been undertaken for the large ^3P_g splitting in the Se atom.[25] Various contractions of the standard[26] (14s11p5d) set have been tested and it has been found that a (11s9p) contraction can be made with practically no loss of energy, a 10s8p grouping will yield higher energies by approximately 0.001 hartree, while all other heavier contractions seem to be unacceptable on the basis of total energy. The similar trend has also been found in the evaluation of the spin–orbit splitting. The $\langle {}^3P|H_{so}|{}^3P \rangle$ matrix element derived in the more contracted (11s8p2d) basis (for technical restrictions in the spin–orbit program, the 5d have to be contracted to two functions) differs from that of the more flexible p space representation (11s9p2d) by $14\,\text{cm}^{-1}$ (out of a total of $830\,\text{cm}^{-1}$) in the standard non-relativistic calculation; the difference between the latter and the uncontracted 11p space is only $2\,\text{cm}^{-1}$, i.e. practically also without loss in accuracy. In a first-order treatment this means that the ${}^3P_2$–${}^3P_1$ splitting is underestimated by $28\,\text{cm}^{-1}$ in the 8p contraction relative to that of 9p; it increases to $32\,\text{cm}^{-1}$ if second-order effects (${}^3P$–${}^1D$ spin–orbit interaction) are also accounted for. A somewhat larger deviation of $45\,\text{cm}^{-1}$ (out of $1954\,\text{cm}^{-1}$ compared to the measured splitting of $1989.5\,\text{cm}^{-1}$) between the 8p and 9p contraction is obtained if the non-relativistic treatment is modified by some relativistic corrections to the kinetic energy and one-electron potential term,[27] and this finding indicates that the increased freedom in the p space is more important for the relativistic than for the non-relativistic results. Considering these (and other) results, there is an indication that for the treatment of spin–orbit effects (at least up to third-row members) no additional orbitals over those present for a good calculation of the total energy are necessary, but that contraction scheme which usually give only freedom to the outermost members of a group might have to be modified somewhat such that the total p space representation is rather flexible.

Before concluding this section, one type of species that has not been mentioned explicitly so far, the negative molecular ions, deserve some remarks. It is well known that stable negative ions generally have a more expanded charge distribution than the corresponding neutral species and hence require some semi-diffuse functions, normally listed as p functions[10] for first- and second-row atoms (Gaussian exponents in the range from 0.026 for Be and 0.074 for F, and between 0.017 for 3p in Al and 0.049 for 3p in Cl). In addition to this spectroscopic-type orbital, the description of correlation is very important. Calculations for carbon and silicon,[28,29] for example, show that SCF treatments grossly underestimate the electron affinity (EA); a standard (4s2p) basis plus polarization d function and a semi-diffuse p function yields only 0.49 eV (C) and 0.92 eV (Si) compared to the known electron affinity of 1.27 eV (C) and 1.39 eV (Si). The estimated full CI limit attributes 2.11 eV correlation energy to the neutral ${}^3P_g$ carbon and 2.47 eV to the ${}^4S_u$ C^- state and hence increases the EA value to 0.85 eV. Addition of another d-type function (exponent 0.15) adds only 0.04 eV to the SCF value of the EA but accounts for a larger portion of the absolute and differential correlation energy (-2.29 eV for ${}^3P_g$ and -2.89 eV for ${}^4S_u$) and hence yields an EA of 1.13 eV, which is still 0.15 eV below the measured value but far along the way from the SCF result. The corresponding data for silicon in the basis with the extra d function (exponent 0.05) are -1.95 eV correlation energy for Si and -2.27 eV for Si$^-$, which gives a calculated EA of 1.27 eV, 0.12 eV above the experiment. Addition of one f function (exponent 0.34) results in an EA (Si) of 1.32 eV, which is only 0.07 eV above the measured value. Hence an extra d correlation-type function should also be added to the negative-ion p function if accuracies better than 0.2 eV in the EA are desired. In this context it is interesting that silicon is one of the few atoms that possesses several stable negative-ion states, and calculated values[29] are 0.92 eV (^2D_u above ^4S_u) and 1.39 eV (^2P_u relative to ^4S_u) compared to 0.86 eV and 1.36 eV measured. This accuracy carries over to molecules[32] and predicts three stable states in SiH$^-$ ($^3\Sigma^-$, $^1\Delta$, $^1\Sigma^+$) and two in CSi$^-$ ($^2\Sigma^+$, $^2\Pi$), in contrast to C$_2^-$ which possesses a third such species ($^2\Sigma_u^+$). A further reduction in the calculated error in the EA can be obtained upon drastic AO basis set expansion. Raghavachari[30] has studied the EA of the atoms B through F in MP4 calculations (Moller–Plesset theory, to fourth order) by employing a triple-zeta representation of the valence shell plus semi-diffuse sp and either two d, or three d and one f, or four d and one f functions; the calculated EA of carbon was within 0.1 eV of the measured value. Feller and Davidson[31] studied the EAs of carbon and oxygen by employing a multi-reference CI calculation using an (8s5p4d2f1g) AO basis and obtained an electron affinity of 1.22 eV for carbon compared to the 1.27 eV found experimentally. Hence it seems clear that a large enough basis accounts for enough differential correlation to insure accuracies better than 0.05 eV. On the other hand, such large basis sets cannot be employed routinely for molecular calculations on standard mainframe computers because of economic reasons and hence it seems still a reasonable compromise to settle for a 0.1–0.2 eV accuracy, which seems to be attainable by simple addition of one pair of semi-diffuse p and d functions.

Finally, the description of negative-ion resonance states requires a number of diffuse functions for the description of the coupling to the continuum (free wavepacket). For the nitrogen $^2\Pi_g(\pi_u^4\pi_g)$ and $^2\Pi_u(\pi_u^3 3s^2)$ resonances, for example, a standard contracted [5s3p] set plus two polarization d functions (exponents 1.8846 and 0.5582) was expanded by three diffuse Rydberg-like molecule-centered p AOs (0.08, 0.02, 0.01), three of s type (0.09, 0.025, 0.015) and one d function (0.004). This basis has been found to be appropriate to describe the short-lived negative-ion states in N$_2^-$ around 2 eV and 11.5 eV by standard, although slightly modified, CI-type methods.[33] Further work to obtain more information on the basis set requirements for such problems of excited negative-ion states are under way.

In summary, then, the AO basis set required for excited-state calculations must generally be larger than that for ground states if similar accuracy is desired. On one hand, a larger degree of correlation must be accounted for in certain instances, which is possible through the use of additional correlation/polarization orbitals. On the other hand, spectroscopic orbitals, while not found in ground state configurations, are necessary to represent particular excited states in which they are occupied. All this is of particular importance if potential energy surfaces are calculated not only in the neighborhood of the lower- and upper-state minima but rather over a large area of nuclear conformations up to the dissociation limits. The straight forward use of acceptable ground-state AO sets (including some d polarization, for example) can produce misleading results for excited states: an incomplete account of the differential correlation energy will show up primarily in an error in the transition energy; omission of the spectroscopic orbital will not always be seen immediately since often other functions in the basis will try to represent this orbital in a more or less reasonable fashion— naturally with different success for different geometrical conformations so that the shape of the potential surface might be most affected by this error. Hence at the outset of excited-state calculations one has to construct carefully the AO basis with an eye toward the goal of which states one would like to describe. Fortunately, the treatment of fine-structure (multiplet splitting) effects does not seem to require basically different AOs than are employed for potential energy calculations.

C. Methods for Calculating Potential Surfaces

1. Methods Involving a Truncated Expansion

In order to study potential energy surfaces it is necessary to use theoretical approaches for the solution of the Schrödinger equation which give a balanced description for all states involved at various nuclear conformations, in particular for geometries in the neighborhood of the equilibrium as well as close to the dissociation limits. It is clear that only methods that account for correlation are at all acceptable for this purpose; the SCF solution would lead to an unrealistic ordering of states and distorted surfaces. Presently, all such methods applicable to larger systems are based on expansion techniques.

Generally the composition of electronic states varies along a dissociation path, and hence all theoretical methods (variational or perturbative) based on a truncated configuration expansion with a single-reference scheme only do not satisfy the requirement of a balanced description at different geometries. Sometimes such methods are useful to study spectroscopic properties near the equilibrium region of the ground state, but they are generally unable to describe consistently the excited states over a large range of nuclear conformations and are therefore not applicable to the theoretical study of photochemical or dynamic processes on the basis of *ab initio* hypersurfaces.

These deficiencies in a truncated configuration-type expansion disappear if a multi-reference scheme is adopted, whereby the reference space from which higher (normally single and double) excitation species are generated contains all such configurations necessary for an adequate zero-order description of the entire surface of a given electronic state. Such treatments do not reach the full CI energy, which would be the exact and therefore optimal solution of the non-relativistic Schrödinger equation in the Born–Oppenheimer approximation as given in Eq. 4, but they approximate it closely (depending on the computational expenditure) and, more importantly, are able to predict total energies to approximately the same degree of accuracy in all regions of the hypersurface; in other words, the goal of such multi-reference treatments is to predict the surfaces in a parallel manner to the exact solution at acceptable computational cost.

Even though comparisons with experiments have substantiated this claim, only recently have direct comparisons with explicit full CI calculations become possible. Handy and coworkers[34–37] have produced benchmarks with their full CI calculations for various small molecules (employing relatively restricted but realistic AO basis sets), which can be used to test a number of theoretical procedures. A typical example is the water molecule in its ground state, treated in a full CI at its equilibrium R_e, at $1.5R_e$ and at $2R_e$. Table VIII shows for comparison the results of a number of highly advocated theoretical schemes.

At the bottom of Table VIII one finds that the full CI space contains 256 473 configuration-state functions (CSFs)—or symmetry-adapted functions (SAFs) as they are sometimes called—in the given AO basis. The procedures based on perturbation methods—many-body perturbation theory (MBPT (4)) and Moller–Plesset theory (MP4)—up to fourth order account for a high percentage of correlation at equilibrium (99.33% in MBPT (4)) but deteriorate markedly with increased bond separation (only 94.72% at $2R_e$), which is of course not surprising since the coefficient of the leading configuration in the full CI expansion amounts to only 0.76 at $2R_e$ while it is 0.975 at R_e.[36] These results show very convincingly the shortcoming of such methods for generating potential energy surfaces. Lower-order treatments are even worse (a decrease from 95.16% to 77.28% at $2R_e$ for MBPT (3), for example) as shown in detail by Handy.[35,36] The same article also discusses the convergence pattern of MP calculations for various molecules; it seems to be always good at R_e but quite erratic at $2R_e$ if restricted Hartree–Fock (RHF) orbitals are taken, and extremely slow or not convergent at all if unrestricted Hartree–Fock (UHF) molecular orbitals (MOs) are employed.[35] The coupled-cluster (CC) or symmetry-adapted cluster (SAC) approaches also show considerable variation with change in bond length; the SAC-B version includes disjoint triply excited clusters as well as disjoint excited clusters to obtain a balanced description of the orbital and electron-pair cluster optimization when an SCF function is not a good description, and as such this method maintains approximately the same percentage of correlation correction for various geometries, even though on an absolute scale its performance is somewhat inferior to the previously discussed procedures. Coupled electron-pair appro-

TABLE VIII
Correlation energies (in percent) for the ground state of H_2O at three different geometries as obtained from various theoretical approaches.

Method	Ref.	R_e	$1.5R_e$	$2.0R_e$
MBPT(3)	37	95.16	88.09	72.28
MBPT(4), SDTQ		99.33	97.10	94.72
Moller–Plesset MP4	35	99.18	96.79	94.04
Symmetry-adapted cluster	38			
SAC-A		98.90	97.20	95.20
SAC-C		98.60	97.50	98.80
Coupled-cluster CC(SD)	37	98.79	97.35	96.99
CC(SD + T)		99.71	99.21	100.89
CEPA-2	39	97.89	96.24	–
CEPA-2V (variational)		97.89	96.26	–
CEPA-2V + singles		97.85	98.20	–
Single-reference CI	35			
SD (361)^a		95.47	91.15	83.96
SDTQ (17 678)		99.82	99.48	98.60
MR-CI(SD)	40			
CAS-CI(7) (70–7906)^b		98.69	99.03	99.39
CAS-CI(8) (328–22 644)		99.75	99.77	99.83
CAS-CI(9) (1436–52 5452)		99.95	99.92	99.95
MRD-CI	41			
(50–13 876 at R_e)		99.84	99.71	99.84
(47–14 227 at $1.5R_e$)		99.62^c		
(37–15 045 at $2R_e$)		(13504)		
Full CI^d (256 473)	34, 35	100.00	100.00	100.00
		(−0.148 028)	(−0.210 992)	(−0.310 067)

^{a}Numbers in parentheses give the size of the secular equation.
^{b}The first number in parentheses refers to the size of the reference space, the second to the order of the secular equation.
^{c}Obtained by employing $^1B_1$ excited-state MOs.
^{d}In parentheses are the absolute values for the correlation energies.

ximations (CEPA), variational or non-variational versions, also account for a slightly smaller percentage of correlation than the other methods. Finally, among the CI expansions, the simple single and double-excitation (SD) CI with respect to a single-reference configuration underestimates the exact data by 7.85 (R_e), 22.39 ($1.5R_e$) and 60.43 mhartree ($2R_e$) respectively and as such gives a very unbalanced description at the various geometries. Addition of triple excitations (T) increases the size of the CI expansion considerably from 361 to 3203 but shows still large errors, and only addition of all quadruples (Q) resulting in a secular equation size of 17 678 remedies the situation to some extent.

The situation is very difficult if a multi-reference basis is chosen. This has been done in a systematic manner in the MR-CI (SD) calculation of Shavitt and coworkers[40] in which the reference space consists of the so-called complete active space (CAS), i.e. all configuration-state functions that can be obtained by distributing the 10 electrons among the active orbitals in all possible ways consistent with the $^1A_1$ symmetry of the H_2O ground state. Such a procedure leads to 70, 328 and 1436 reference configurations seven, eight and nine orbitals respectively are chosen. The total CI spaces are then of order 7906, 22 644 and 52 452 as shown in Table VIII, and it is seen that the absolute percentage of the correlation energy accounted for is rather large and that it does not vary much for the three different geometries. The value closest to the full CI is obtained with 52 452 configurations (CSFs), but not that this space results from an extremely large reference set and already accounts for 20% of the total full CI space. The authors[40] have pointed out in their study that the final results will be of poor quality, however, if the number of active orbitals is less than seven (96.41%, 90.11% and 88.73% respectively at the various internuclear separations employing 12 reference configurations).

Our own approach is also of the multi-reference CI (MRD-CI) type,[42−44] but the methodology to chose the reference space is simply based on an energy criteria rather than on a systematic sampling of orbitals. All configurations that appear in the final CI expansion with a weight larger than a given value or that contribute in an m main reference CI more than a certain value to the energy (in H_2O more than 0.5 mhartree, for example) are chosen to constitute the set of reference configurations. This procedure is automatic in the way that at each geometry it finds as reference species those configurations which show the most important interaction independent of excitation class. In this manner the method does not necessarily generate all quadruple excitations, but it picks from the total space those which are most important energetically, and at the same time allows inclusion of higher excitation classes than of quadruple type. Furthermore, it keeps the size of the computation down to what is really necessary. It is thus designed to be essentially equivalent at various geometries, and this is supported by the results in Table VIII. The percentage of the correlation energy accounted for in the MRD-CI calculation is quite similar for all three internuclear distances. The absolute magnitude is the second best in the table and the secular equations and reference sets are smaller than for treatments giving comparable results (CAS eight-orbital MR-CI with 328 reference configurations and a secular equation order of 22 644; and the SDTQ single-reference CI calculation at R_e with 17 678 configurations). In particular, comparison of the MRD-CI and SDTQ results at $2R_e$ shows that, from the standpoint of total energies, it is more advantageous to select the strongly interacting configurations out of a total generated space (thereby keeping the dimension of the final secular equation manageable) rather than working with the larger space such as the SDTQ constrained to a more limited class of excitations. Finally, a change of the MO basis, as undertaken at R_e to employ excited-state $^1B_1$ orbitals, leads to quite similar results. Generally the

TABLE IX

Correlation energies (in percent) for the ground state of the BH radical as obtained from various theoretical approaches that correlate six or four valence electrons.

Method	Ref.	No core (6 VE)	With core (4 VE)
MBPT(4), SDTQ	37	95.06	94.23
Symmetry-adapted cluster	38		
SAC-A		98.20	–
SAC-B		98.30	–
Coupled-cluster CC(SD)	37	98.25	98.11
CC(SD + T)		98.86	98.77
CEPA-2	39	–	98.71
CEPA-2V		–	98.74
CEPA-2 + singles		–	99.54
Single-reference CI[a]	34		
CI(SD)		94.91(568)	95.34(228)
CI(SDTQ)		99.97(28698)	100.00(3036, full CI)
Full CI[a]	34	100.00(132686)	
MRD-CI[a]	45	99.95(17049)	–

[a] For the CI calculations the order of secular equation solved is always given in parentheses.

only difference appears in the size and composition of the reference set, which is often smaller when the set of parent molecular orbitals/natural orbitals (MOs/NOs) is chosen and hence the calculation is less expensive from the point of view of computation.

Similar comparisons are also available for the smaller system BH (Table IX) in its ground state, for which the full CI energy in a given basis is also known.[34]

In summary, then, the results show that all treatments of variational or perturbative nature based on a single-reference configuration generally cover less than 99% of the total correlation energy and are characterized by fluctuations in the absolute errors along the potential surface. They support the statement made at the beginning of this section that only multi-reference treatments will be able to give reliable results for potential energy surfaces of ground and, even more so, excited states. All the results to be discussed in later sections are obtained with the MRD-CI method.

2. Estimation of the Full Configuration-interaction Limit

In the last section we analyzed the behavior of the multi-reference methods when comparison with the exact full CI value was possible. For the majority of chemical problems, however, one has to deal with a larger number of electrons and many more basis functions as in the examples given, so that the full CI spaces are of the order of many million or billion configurations, which makes a direct determination of the full CI energy not possible by the present theoretical techniques, at least within a reasonable amount of computation time. Even though the MRD-CI method should by design—and through experience with experimental facts—be able to give potential surfaces roughly parallel to those of the exact full CI treatment, it would be interesting to obtain a rough estimate of the exact position of the full CI eigenvalue—if only for a better understanding or as a further check on internal consistency or the reliability of results. An estimated full CI value could then be used as a reference level to measure the quality of the directly calculated energies in a similar manner as has been done with the exactly calculated full CI energies in H_2O or BH.

An estimate can be obtained by employing various expressions as discussed elsewhere,[43,46,47] basically in the perturbative form

$$E(\text{full CI}) = E(\text{MRD-CI}) + E_{\text{error}} \quad (11)$$

whereby the full CI correction is of the form

$$E_{\text{error}} = \lambda[E(\text{MRD-CI}) - E(\text{ref})] \quad (12)$$

with various formulas for λ;[48-50] $E(\text{ref})$ is the energy obtained in a CI for only the reference configurations.

If relative energies are of interest, for example between two electronic states or between two different regions of the potential hypersurface, only the difference in the corrections is important, i.e.

$$\Delta E(\text{full CI}) = \Delta E(\text{MRD-CI}) + \Delta(E_{\text{error}})$$

Hence in order to have an efficient error compensation so that $\Delta E(\text{MRD-CI})$ is very close to the actual difference in full CI values, it is only necessary to insure that the corrections are within the same energy range, and normally as small as possible.

One of the simplest expressions is the generalized Davidson[48] formula, first suggested in our group and recently derived theoretically in two publications,[51,52] namely

$$E_{\text{error}} = \left(1 - \sum_{\text{ref}} c^2\right)[E(\text{MRD-CI}) - E(\text{ref})] \quad (13)$$

where $E(\text{ref})$ refers to the energy of the reference space only. Other approximation formulas[49,50] are numerically almost equivalent as soon as the number of electrons is larger than six and the accumulative weight of the reference configurations is above 90%. The magnitude of E_{error} can be manipulated if necessary, for example, by altering the reference set. The quantity $E(\text{ref})$ generally changes more abruptly than does $E(\text{MRD-CI})$ when additional configurations are added to the reference space, so that the second term in Eq. 13 decreases upon expansion of the reference set. At the same time, such an expansion increases the total weight $\sum_{\text{ref}} c^2$ in the final CI, which leads to a reduction of the first factor in Eq. (13). In other words, the simple inclusion of more reference configurations produces a reduction of the total error because the two factors involved in the correction formula (Eq. 13) are simultaneously made smaller. This is of course not surprising since such an expansion constitutes a step further on the way to the true full CI.

In cases in which in a given MO basis the accumulative weight $\sum_{ref} c^2$ is poor (i.e. below 90%), one also has the possibility of improving the quality of the calculation by employing the corresponding natural orbitals obtained in the first MRD-CI. The new one-electron NO basis has the property of improving the convergence of a given CI expansion and yielding a more compact

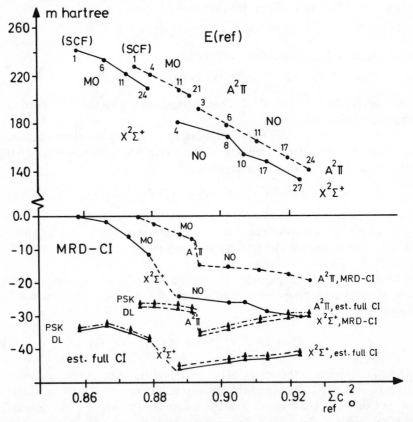

Fig. 1. Energies of the $X^2\Sigma^+$ and $A^2\prod$ states of CSi^- as a function of the size of the reference set (the number of SAFs is always indicated) whose contribution to the total wave function is $\sum_{ref} c_0^2$. The zero in energy is the energy obtained in a single-reference CI. Two different MO and NO basis sets are employed. The upper part shows the energies resulting from the reference set alone; the lower part contains the total MRD-CI and estimated full CI energy whereby the formulas of Davidson and Langhoff,[48] on the one hand, and Pople et al.,[49] on the other, are taken.

wavefunction (larger $\sum_{ref} c^2$). This methodology was applied earlier to the study of the relative stabilities of the $^2\Sigma$ and $^2\Pi$ states in the CN^+ radical and is exemplified in Fig. 1 for the two lowest states $X^2\Sigma^+$ and $A^2\Pi$ of the CSi^- negative ion.[29]

The relative position between these states is poorly described in the SCF approximation, which places the $^2\Pi$ state at -0.26 eV below the ground state $X^2\Sigma^+$ when compared with the best estimate of $+0.39$ eV for $^2\Pi$. A simple CI(SD) treatment gives the correct ordering, but the difference between them is erroneously only 0.1 eV. This finding is in accordance with a rather poor description of both states at this level, possessing $\sum_{ref} c^2$ values of only 85.84% and 87.52% respectively, which leads to a large full CI correction of 30–35 mhartree. In the most reliable prediction using the natural-orbital basis, the final contribution of the reference set is around 92.5% for both states so that the estimated error for each state is only around 10 mhartree and the difference in the correction ΔE_{corr} amounts to only 0.02 eV. In this connection it should be noted that a similar degree of internal consistency is also obtained for the same system in a less flexible AO basis without extra semi-diffuse correlation functions; hence even though the energy difference between both states has been calculated in each AO basis close to the respective full CI limits, the transition energy in the more limited basis is 0.63 eV compared to the more reliable value of 0.39 eV obtained in the more flexible AO basis. This example underlines the fact that for overall accuracy it is first and foremost the AO basis which is the determining factor, provided the degree of sophistication in the MRD-CI expansion is carried far enough so that its influence on the final result is less critical.

From a practical point of view, errors in transition energies are of the order of 0.2 eV if a standard AO and MRD-CI treatment is employed in the calculation, and similar errors can occur across potential surfaces. Standard treatments are thereby those which employ a double-zeta AO basis plus some polarization/correlation functions (normally one d function) and the necessary spectroscopic orbitals, and employ reference sets whose total contribution to the CI expansion is at least 90%. In order to insure higher accuracy, the AO set has to be increased in accordance with the discussion in Section II.B and the error limits in the MRD-CI procedure have to be evaluated in the manner outlined above.

III. SELECTED POTENTIAL ENERGY SURFACES

In the following section we will present a selection of calculated potential energy surfaces. Most of them have not yet been determined by experimental methods, or if so are at least only very fragmentary, and as such they will furnish a basis on which further and more detailed experimental or theoretical studies can be undertaken. Some of the species play a role in atmospheric chemistry and astrophysics or in combustion processes and are short-lived reaction intermediates which are difficult to handle by experimental procedures so that study by *ab initio* methods seems particularly appropriate. Others have not been observed at all so far and therefore the present data may serve as a guide to their observation, and some of the molecules may be taken as pieces of larger entities, such as species adsorbed on surfaces or as compounds embedded in solid-state structures or clusters. Since the *ab initio* computations require only knowledge of the number of electrons and the locations, numbers and charges of nuclei, they are equally applicable to neutral and positive or stable negative ions; for this reason charged species will be treated along with the neutral molecules in what follows.

A. Diatomic Molecules

In diatomic molecules there is only one geometrical variable and hence the potential surface reduces to a simple potential curve. Even though many of the states of C_2, N_2, O_2 or CO and NO are very well known,[53] much less information is available for the second- and higher-row analogs Si_2, Ge_2, P_2, As_2 or S_2, for example, and relatively little is known about diatomics with constituent atoms from different rows of the periodic table, such as CSi, SO or SiO, PO, etc. Even though the study of such 'exotic' molecules may not always seem to be attractive for a given species, the information on the various states is nevertheless important for at least two reasons: first, knowledge of the skeleton AB molecule is generally quite important for all such species related to it by formal hydrogen addition, i.e. for the various $H_n AB$ ($n = 1$ to 6) structures; and secondly, a critical analysis of the results for selected first- and second-row species will be very useful for further predictions of the diatomic behavior in many molecules not explicitly treated. Finally from a technical point of view, diatomic information is often required in somewhat less expensive calculations for larger systems, for example DIM (diatomic-in-molecule) approaches.

1. General Rules for Location and Shape of Potential Curves

For an analysis we will make use of the molecular-orbital diagram schematically drawn in Fig. 2 and the diatomics will be ordered according to their number of electrons. The bonding and antibonding properties of the MOs are well known, i.e. σ_g and π_u (or the equivalent in heteronuclear diatomics) are bonding, σ_u and π_g antibonding, while Rydberg species must normally be characterized as non-bonding because of their diffuse or large-orbit character. According to the Mulliken–Walsh model,[54–56] occupation of a bonding (antibonding) MO causes a decrease (increase) in equilibrium bond lengths, and on the basis of this rule the bond lengths of all excited states

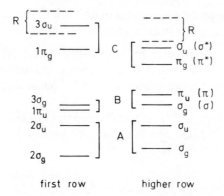

Fig. 2. Schematic ordering of orbitals according to their stability in first- and second-row diatomics.

relative to that of the ground state can easily be predicted qualitatively simply from the occupation scheme of the MO levels. Prediction of the relative energies of potential energy curves, on the other hand, is somewhat less obvious. Experience from a number of examples leads to the following observations: (a) in first-row diatomics, 2s–2p hybridization is considerable and as a consequence the $3\sigma_g$ MO is pushed upwards in a close energetic neighborhood or above $1\pi_u$—furthermore the splitting between bonding and antibonding $p\pi$ combinations is large and the corresponding bonding and antibonding character of $\pi_u(\pi)$ and $\pi_g(\pi^*)$ and $\sigma_u(\sigma^*)$ is strong; (b) in the second- and higher-row diatomics, the order of MO levels is 'normal' and because of the more diffuse charge distribution of 3p in comparison to 2p in first-row analogs the $\pi-\pi^*$ and $\sigma-\sigma^*$ splitting is smaller and the bonding/antibonding character is not as strong.

This analysis leads to the following rules:

1. The proximity of levels between blocks A and B (Fig. 2) in first-row diatomics leads to many low-lying states which involve excitations from $2\sigma_u$ into $1\pi_u$ and $3\sigma_g$ (unless all are doubly occupied); these states possess small equilibrium separations because of electron transfer from an antibonding to a bonding MO. They are much less favored in heavier diatomics.
2. The proximity and order of levels within block B favors σ occupation in second- (and higher-) row diatomics relative to that of π, which is the one more preferred in the first-row systems.
3. The larger energy gap between blocks B and C in first-row diatomics puts $\sigma \to \pi^*$ and $\pi \to \pi^*$ excitations at higher energies than in second- or higher-row analogs.
4. Because of the strong repulsive nature of the high-lying σ^*MO, at small internuclear separations there is often competition between occupying the

σ^* and Rydberg pσ orbitals, which in first-row diatomics is more often in favor of Rydberg states than in second-row molecules.

For heteronuclear diatomics, certain modifications in the orbital stability scheme of Fig. 2 are necessary in the standard manner depending on the electronegativity of the two partners or their charge distribution and AO orbital stability.

All these differences will be especially apparent in systems that partially occupy the π and σ MOs and therefore emphasis will be placed on these diatomics in what follows.

2. Diatomics with Seven to Ten Valence Electrons

Representative examples for diatomics with *seven* valence electrons studied by *ab initio* methods are C_2^+, Si_2^+, BN^+ and CSi^+. Potential curves for all of these can be found in the literature.[57-60] These ions have three electrons in the MOs of block B (Fig. 2) and all possess an $X^4\Sigma^-$ ground state according to

TABLE X

Calculated relative positions of various electronic states in the isovalent species C_2^+, BN^+, CSi^+ and Si_2^+ (all possessing seven valence electrons). Values in parentheses are vertical energy differences.

	Occupation					Relative energies (eV)			
State	σ_u	π	σ	π^*	σ^*	C_2^+	BN^+	CSi^+	Si_2^+
$^4\Sigma^-$ (g)	2	2	1	0	0	0.0	0.0	0.0	
(u)	1	2	2	0	0	2.47	3.40	$\{$ (3.53)	$\{$ (2.74)
(u)	2	1	1	1	0	(8.74)	–		$\{$ (4.59)
(u)	2	2	0	0	1	–	–		
Expt	$X^4\Sigma_g^- - {}^4\Sigma_u^-$					2.50			
$^4\Delta$ (u)	2	1	1	1	0	3.29	8.02	(3.83)	1.39(2.25)
$^4\Sigma^+$ (u)	2	1	1	1	0	3.33	7.12	(3.94)	1.47(2.36)
$^4\Pi$ (g)	1	3	1	0	0	1.39	1.73	(3.10)	$\{$ (2.67)
(g)	2	2	0	1	0	4.93	4.93	(4.38)	$\{$ (4.12)
(g)	2	1	1	0	1	–	–	–	
$^2\Pi$ (u)	2	3	0	0	0	0.72	0.85		1.32(1.37)
								$\{$ 1.31	
								$\{$ 1.76	
(u)	2	1	2	0	0	1.80	$\{$3.32		0.61(0.75)
(g)	1	3	1	0	0	3.02			(3.10)
Expt	$^2\Pi_g - {}^2\Pi_u$					2.29			
Calc						2.30			

theoretical work with a $\pi^2\sigma$ configuration; the same has been found for the neutral BC radical.[61,62] Information on excited states is contained in Table X and until recently these theoretical data were the only ones available for these systems. First of all it is seen that a $2\sigma_u \rightarrow \pi_g(^4\Sigma_u^-)$ and $2\sigma_u \rightarrow \pi(^{2,4}\Pi_g)$ transition is present at low energy in C_2^+ and BN^+ in accordance with rule 1. The first is presumably the one which has very recently been observed[58] in experiments using translational energy spectroscopy for C_2^+ and C_2H^+, with a measured peak at 2.50 eV. The only other low-lying states in the first-row atoms are those resulting from transitions within block B $(1\pi_u, 3\sigma_g)$ orbitals with $^2\Pi_u(\sigma_u^2\pi^3)$ being the lowest followed by a $\pi \rightarrow \sigma$ transition relative to the ground state resulting in $^2\Pi_u(\sigma_u^2\pi\sigma^2)$. This theoretical prediction has probably also been verified since it seems likely that the other recently measured intense peak[58] at 2.29 eV corresponds to the $^2\Pi_u-^2\Pi_g$ transition in the doublet manifold. Excitations into the π^* and σ^* MOs lead in the first-row diatomics to higher states according to Table X, in agreement with the preceding rule 3.

The fact that in second-row systems σ occupation is favored relative to π (rule 2) is best seen from Fig. 3. While in C_2^+ the order of $^2\Pi_u$ states is π_u^3 followed by $\pi_u\sigma_g^2$, it is the opposite in Si_2^+ in which the lowest state at 0.61 eV (Table X) possesses a $\pi_u\sigma_g^2$ occupation and the higher the π_u^3 population. The situation in CSi^+ is intermediate, as also seen from the figure, and the $^2\Pi$ potential curve[60] shows two minima corresponding to the two configurations. Furthermore it is seen from Table X that the transitions to π^* and σ^* in Si_2^+ ($^4\Delta$, $^4\Sigma^+$ and $^4\Pi$ states) are found at lower energy than in the first-row

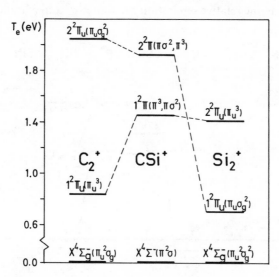

Fig. 3. Relative energies of various electronic states in the isovalent systems C_2^+, CSi^+ and Si_2^+.

homolog, in accordance with rule 3 above; furthermore there is more mixing between such configurations, as indicated in Table X, while in C_2^+ the $\pi \to \sigma^*$ and $\sigma \to \pi^*$ states are well separated. The latter is also expected to have some consequence for the experimental identification of such states: while the $^4\Sigma_u^-$ – $X^4\Sigma_g^-$ state in C_2^+ seems to have been identified,[59] the mixing of the two $^4\Sigma_u^-$ configurations in Si_2^+ will shift the minimum of $^4\Sigma_u^-$ to larger bond lengths (due to the dominance of the $\pi \to \pi^*$ excitation) and as such into an unfavorable Franck–Condon area in which measurements will be more complicated. Details of the BN^+ states[59] can also be analyzed on the basis of the charge distribution and dissociation limits but will not be pursued further in the present context.

Diatomics possessing *eight* valence electrons are the most versatile family with respect to character and multiplicity of their ground state. In this case four electrons have to be distributed among the block B MOs in Fig. 2, which leads to possible lowest states $^1\Sigma_g^+$ $(1\pi_u^4)$, $^3\Pi_u$ $(\pi_u^3\sigma_g)$ and $^3\Sigma_g^-$ $(\sigma_g^2\pi_u^2)$. Indeed, all three are observed as ground states, namely $X^1\Sigma^+$ for C_2 and CN^+, $X^3\Pi$ for BN, CSi and CP^+, and $X^3\Sigma^-$ for the species SiN^+, SiP^+, Si_2, Ge_2 and Pb_2, and this is in full accord with the rules outlined earlier.

In all molecules the $^3\Pi$ state is quite low in energy, about 0.1 eV above $X^1\Sigma^+$ in C_2 and CN^+ (Fig. 4) and very close to $X^3\Sigma_g^-$ in Si_2; only in the mixed compounds CP^+ and CSi with an $X^3\Pi$ ground state does a larger separation of 0.6–0.7 eV exist to the first excited $^3\Sigma^-$ state, a value comparable with the $^3\Sigma^-$–$^3\Pi$ separation of 0.71 eV known for C_2 and 0.94 eV predicted by calculations[63] for the CN^+ ion. One observes that, as long as the carbon atom is a constituent, the relative separation between $\pi^2\sigma^2$ and $\pi^3\sigma$ states is similar,

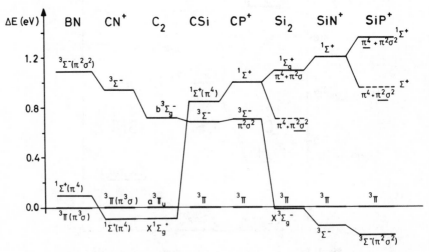

Fig. 4. Relative energies of various isovalent diatomics in their $^3\Pi$, $^1\Sigma^+$ and $^3\Sigma$ states.

fairly independent of the second atom, but it would have to be checked whether this is also true for higher-row diatomics such as combinations with Ge or As$^+$.

For illustration, the systems C_2,[64-66] BN,[67] SiP$^+$[68] and Si$_2$[69,70] have been chosen in Table XI; more details about the other species mentioned can be found in the literature for compounds containing second-row[46,69] and higher-row[71,72] atoms. In C_2 and BN the two $^1\Sigma^+$ states with π^4 and $\pi^2\sigma^2$ occupation are well separated by 1.8 eV and 3.1 eV respectively (indicating a stronger $\pi-\sigma$ separation in BN within block B), while in the second-row radicals Si$_2$ and SiP$^+$ a sizeable interaction of both configurations occurs in analogy to the $^2\Pi$ states in the diatomics with seven valence electrons discussed before.

TABLE XI

Calculated relative positions of various electronic states in the isovalent species C_2, BN, SiP$^+$ and Si$_2$ (all possessing eight valence electrons). Values for C_2 given in parentheses are experimental values.

| State | \multicolumn{5}{c}{Occupation} | | | | | \multicolumn{4}{c}{Relative energies (eV)} | | | |
	σ_u	π	σ	π^*	σ^*	C_2	BN	SiP$^+$	Si$_2$
$^3\Pi$ (u)	2	3	1	0	0	0.00, 0.00^a, (0.00) (1.32 Å)	0.00 1.32 Å	0.00 2.13 Å	0.00 2.20 Å
$^3\Sigma^-$ (g)	2	2	2	0	0	0.83, 0.74^a, (0.71) (1.37 Å)	1.09 1.48 Å	−0.23 2.48 Å	−0.02 2.29 Å
$^1\Sigma^+$ (g)	2	4	0	0	0	0.03, (−0.09) (1.24 Å)	0.10 1.28 Å	} 1.20	} 0.94
	2	2	2	0	0	1.84 –	3.29 1.43		
$^3\Sigma^+$ (u)	1	4	1	0	0	1.12, (1.65) (1.23 Å)	1.26 1.24 Å	Higher; interaction with other states	
(u)	2	3	0	1	0	6.06 –	5.20 1.47 Å	} 2.78	} 2.05
(u)	2	1	2	1	0	– –	–		
$^3\Pi$ (g)	1	3	2	0	0	2.50, 2.35^a, (2.39) (1.27 Å)	3.45 1.39 Å	Not studied, but higher	
$^5\Pi$ (g)	2	2	1	1	0	3.88 –	3.54	1.17 2.55 Å	1.12 2.5 Å
$^3\Pi$ (g)	2	2	1	1	0	5.54, (4.97)	4.79 1.53 Å	2.31	1.82

^{a}Refs. 65 and 66; the other values for C_2 are from Ref. 64.

Excitation from the $2\sigma_u$ again leads to low-energy states in C_2 and BN, in accordance with rule 1 stated above, but produces higher states in SiP^+ and Si_2; by contrast transitions into the C block π^* and σ^* are more favorable in the higher-row molecules. The trends in bond lengths are also seen quite nicely from Table XI and Fig. 5, with the smallest value for the configuration occupying the strongly bonding π_u^4 and larger values if the less bonding σ MO $^{3,1}\Pi$ and $^3\Sigma^-$, $^1\Delta$ is occupied instead; the bonding character is thereby described best as a change in orbital stability with internuclear separation, i.e. $\partial\varepsilon/\partial R > 0$ is bonding and π_u is more bonding than σ_g if the slope is steeper for π_u than for σ_g, $|\partial\varepsilon_\pi/\partial R| > |\partial\varepsilon_\sigma/\partial R|$. The decrease in bond length upon σ_u

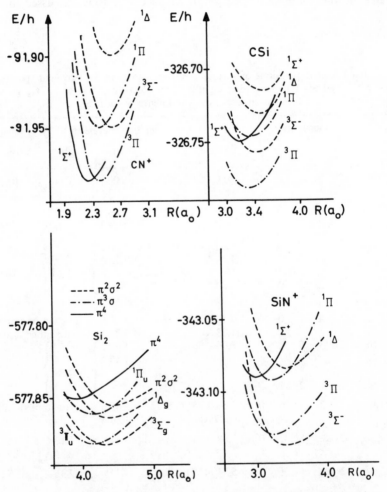

Fig. 5. Calculated potential energy curves for a number of electronic states in isovalent diatomics possessing eight valence electrons.

vacation and the corresponding increase upon π^* occupation is also evident. A $\pi \to \pi^*$ excitation relative to the low $^3\Pi$ state leads to the $^5\Pi$ state with a large equilibrium distance (and low vibrational frequency) which lies below the first dissociation limit already in the first-row diatomics and is even found among the lowest excited states in the silicon-containing species in Table XI. The lowest $^3\Pi_g$-type state in Si_2 and SiP^+ also results from this $\pi^2\sigma\pi^*$ configuration. The high-multiplicity states are more likely to be predicted reliably by computations, since their direct observation is normally not possible. One typical example is the $^5\Sigma_g^+$ state of N_2 for which 10 years ago[73] an MC-SCF (multi-configuration SCF) study found a very shallow minimum at large internuclear separation (with a small barrier to dissociation), and this has recently been supported by a detailed experimental study[74] of the perturbations is the high vibrational levels of the $^3\Pi_g$ state due to this $^5\Sigma_g^+$ state. Another such example is the $^5\Sigma_u^-$ state in O_2. Finally, on the more quantitative side, the calculated $^3\Pi_u$–$^3\Pi_g$ separation in C_2 of 2.35 eV compared to the measured 2.39 eV and the calculated 3.45 eV in BN[75] compared to the experimental value of 3.46 eV shows the power of the theoretical tool.

TABLE XII

Relative positions between various electronic states of the isovalent species CN, N_2^+, CP, SiN, SiP and P_2^+, all possessing nine valence electrons.[a]

State	Occupation							Relative energies (eV)					
	σ_u	π	σ	π^*	σ^*			CN	N_2^+	CP	SiN	SiP	P_2^+
$^2\Sigma^+$ (g)	2	4	1	0	0			0.0	0.0	0.0	0.0	0.0	0.0
$^2\Pi$ (u)	2	3	2	0	0		Expt	1.15	1.14	0.86	0.12	–	−0.27
							MRD-CI	1.05	–	0.75	0.19⎱ 0.22⎰	−0.07	−0.26
$^2\Sigma^+$ (u)	1	4	2	0	0		Expt	3.19	3.16	3.61	3.01	–	4.71
							MRD-CI	3.17	–	3.45	(2.77)	–	4.73
	2	3	1	1	0		Expt	7.33	8.01	–	–	–	2.90
							MRD-CI						2.93
$^4\Sigma^+$ (u)	2	3	1	1	0		Expt	4.49[c]	–	–	–	–	–
							MRD-CI	4.51	(4.68)[b]	2.27	2.30	1.58	1.54
$^4\Delta$ (u)	2	3	1	1	0		MRD-CI	5.77	–	3.22	3.00	2.26	2.46
$^4\Sigma^-$ (u)	2	3	1	1	0		MRD-CI	6.50	–	3.91	3.35	2.74	3.08
$^4\Pi$ (g)	2	2	2	1	0		MRD-CI	5.51	(5.68)[b]	3.30	2.59	1.58	1.57
							Expt	5.45[c]	–	–	–	–	–
$^6\Sigma^+$ (g)	2	2	1	2	0		MRD-CI	–	(6.18)[b]	4.0	3.9	2.5	2.26

[a]The experimental data are from Ref. 53. Calculated data on CN, SiN and SiP from Refs. 76 and 77; those for P_2^+ from 78; calculated data for CP as well as for all sextet states stem from Ref. 79.
[b]Results of Ref. 80.
[c]Results of Ref. 81.

In diatomics possessing *nine* valence electrons, one again finds a clear distinction between the second- and first-row compounds (Table XII). The five electrons to be distributed among the π and σ levels prefer the $\pi^4\sigma$ $^2\Sigma^+$ occupation in the first-row molecules CN and N_2^+, but the second-row counterparts SiP and P_2^+ are more stable in their maximum occupation of the σ MO, i.e. they prefer a $^2\Pi(\sigma^2\pi^3)$ ground state. The ground states of As_2^+ and GeAs would also be predicted to be $^2\Pi$ based on the general trends.

It is interesting that the various quartet states are totally unknown from experiment, with the only exception being recently published work on CN,[81] the result of which deviates by less than 0.1 eV from the corresponding MRD-CI predictions. The theoretical results for CP, SiN, SiP and P_2^+ clearly indicate that these quartet states are in the low-energy range, in CP and SiN even below the second excited state $B^2\Sigma^+$, and should be observable via perturbations in the doublet spectrum. Owing to the different potential curve characteristics (short bond length in the $B^2\Sigma^+$ $\sigma_u\pi^4\sigma^2$ state and larger distances in the quartet states, in particular $^4\Pi$), a crossing of doublet states with the high-spin states is expected. In this connection it must be pointed out that in the second-row compounds the $B^2\Sigma^+$ state must predominantly be characterized as $\pi^3\pi^*$ configuration (i.e. the same as for the first $^4\Sigma^+$ state) and hence its equilibrium bond length is increased relative to the ground state, in contrast to the first-row $B^2\Sigma^+$ states which possess shorter equilibrium distances. The first-row $B^2\Sigma^+$ $\sigma_u\pi^4\sigma^2$ configuration is sometimes visible as a shoulder or as a second minimum. A typical representative of such a second-row arrangement is seen in the potential curves for P_2^+ in Fig. 6. The almost parallel behavior between $B^2\Sigma_u^+$ and the states $^3\Sigma_u^+$ and $^4\Sigma_u^-$ is quite obvious. The $^4\Pi_g$ state shows a larger bond separation because the occupation of the more bonding π is exchanged for that of σ. The $^2\Sigma_u^+$ $(\sigma_u\pi^4\sigma)$ is shown to mix with higher states of the same symmetry. In all systems there also exist states of sextet multiplicity which correlate with the lowest dissociation products and either possess minima at large bond lengths as seen in P_2^+ or are in the main repulsive.

In *ten*-electron systems both π and σ MOs are fully occupied and therefore all systems possess a $^1\Sigma^+$ ground state. Examples are CO, N_2, CS, SiO, SiS and P_2 and most of these have been studied experimentally in considerable detail. Since they occupy all bonding MOs, they possess the highest stability of all the diatomics, and as a consequence the first excited states $\pi \to \pi^*$ or $\sigma \to \pi^*$ are relatively high in energy. The first $^3\Pi(\sigma \to \pi^*)$ in CO lies at 6.07 eV and $^3\Sigma_u^+(\pi \to \pi^*)$ at 6.92 eV, while the corresponding values in P_2 are 3.50 eV and 2.33 eV due to the smaller splitting between bonding and antibonding MOs as discussed in the previous section. Values in SiO or CS are somewhat in between at 4.21 and 3.43 eV (calc 3.37 eV) for $^3\Pi$ and 4.14 and 3.88 eV for $^3\Sigma_u^+$. States occupying the σ^* MO are very high in first-row diatomics and lie above the corresponding Rydberg states of the same symmetry, which are found around 10.7 eV in CO and 12.4 eV in N_2. In the second-row diatomics one

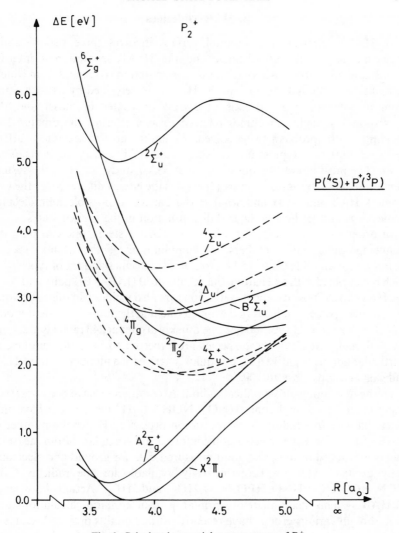

Fig. 6. Calculated potential energy curves of P_2^+.

observes some mixing between σ^* and corresponding Rydberg species and furthermore mixing of the $\pi^4\sigma\sigma^*$ and $\pi^3\sigma^2\pi^*$ states, as has been pointed out in various *ab initio* studies for the $E^1\Sigma^+$ state. Since many features of the potential curves are known from spectroscopic studies for these systems, they will not be treated any further in this context.

Diatomics with more than 10 electrons start to occupy the antibonding MOs and excited-state surfaces and differences between first- and higher-row systems can be analyzed in analogous fashion as in this section.

B. HAB Molecules

In HAB molecules the potential surface involves three variables, the molecular angle, the AB distance and the H–AB separation if internal coordinates are chosen. Alternatively a representation is often taken in which, in addition to the AB separation, the H atom is described by its separation from a point on the AB axis (midpoint or center of mass) and the corresponding angle, sometimes referred to as scattering coordinates. Depending on the problem to be solved, the entire three-dimensional surface (very seldom) or a representative section thereof is calculated in *ab initio* work. The most interesting questions in an HAB study are: (1) is the system bent or linear in the various states, (2) what is the energy difference for the two isomers HAB and ABH and what is the barrier to possible unimolecular conversion, and (3) how is the stability situation in the excited states?

In order to set the stage for a study of HAB systems, one can play the following game with numbers. A combination of the atomic species C, N, O, Si, P and S leads to 15 heteroatomic diatomics and six of type A_2. If each is combined with an H atom, this results in 30 HAB–ABH pairs and six of the HA_2 class. If positive and negative ions are also included in this list, a total of 108 such species results. If one wants to include the three equivalent atoms from the third row (Ge, As and Se), one quickly arrives at 243 radicals and ions. Even though not all of these species are expected to be stable entities, in particular not the negative ions, one quickly arrives at a number of stable HAB radicals and ions close to 200.

Of the first family of 108 radicals, about 20 compounds have been observed experimentally although some (HNO^+, NOH^+, C_2H^+ for example) have only been characterized indirectly by ionization processes. Hence the number of HAB systems for which spectroscopic information is available from measurements is reduced further, and good measured data for ground and electronically excited states are probably only available for the radicals C_2H, HCN, HCO, HCP, HNO, HPO, HSO, HO_2 and HF_2. Actual knowledge of HAB systems stems therefore primarily from *ab initio* calculations and even though various groups have recently worked on this project, because of the large number of systems much more work in this direction is required.

1. The System HCN–HNC

Just as in the diatomics with 10 valence electrons (VE), the 10 VE system HCN is one of the most stable species, occupying all A–B bonding MOs, and it has therefore received considerable attention in a number of experimental studies. It is linear in its ground state according to Walsh's rules.[54-56] The isomer HNC was first observed by Milligan and Jacox[82] about 20 years ago and later an emission line observed at 90 665 MHz in interstellar space[83] was attributed to the $J = 1 \leftarrow 0$ transition in this isomer. Extensive *ab initio* calculations have also been performed on the isomerization surface HCN–HNC.[84,85] The energy difference between the two conformers was calculated to be of the order of $15 \, \text{kcal} \, \text{mol}^{-1}$, while the barrier height seen by the HNC molecule was predicted to be around $35 \, \text{kcal} \, \text{mol}^{-1}$.

The potential surface can be employed as the basis for a theoretical study of the kinetics of HCN–HNC interconversion. In the standard manner, such unimolecular processes are treated theoretically with the help of the RRKM (Rice–Ramsperger–Kassel–Marcus) theory; the hypersurface is partitioned into three distinct regions corresponding to the stable reactant and product and the activated complex (AC) which possesses the minimal energy in the internal degree of freedom connecting the two local minima. It is assumed in this model that the density of states of the complex relative to the corresponding density of the reactant determines the microscopic rate for the reaction to occur; the reaction coordinate is considered as a pure translation in the region of the AC and in generally associated with a given vibrational mode, which in the case of HAB–ABH isomerization would be the bending mode involving large-amplitude motion. Knowledge of the geometrical structure and vibrational frequencies (generally assumed in the harmonic approximation) of the AC then allows one to calculate the partition function and its relation with the same quantities in the reactant species. While information on the AC is traditionally gained only indirectly by assuming a reasonable structure that gives (within this RRKM model) overall agreement with the measured reaction rate (or its variation with pressure), the *ab initio* work has the advantage of giving a direct determination including the vibrational modes of the AC. Nevertheless, in this RRKM approach there remains the artificial separation between the two points on the same potential surface (AC of HAB and product ABH) from the standpoint of the calculation of the corresponding partition functions. Hence a modified approach has been suggested[85] which avoids all the distinctions with respect to the labeling of vibrations on the hypersurface. It can be looked upon simply as motion of a hydrogen in the field of a CN radical, but always below the dissociation limit H + CN. The vibrational levels within the energy range of interest are calculated for the energy hypersurface by employing an appropriate form of the vibrational–rotational part of the Hamiltonian. In this model it is found that the density of levels near and above the barrier is somewhat higher than the value calculated by the traditional RRKM approach in the harmonic approximation. This finding implies that the microscopic isomerization rate constant is predicted to be up to one order of magnitude larger relative to the value derived on the basis of the same *ab initio* surface in the simple RRKM approximation. This result, exemplified for HCN, should also be kept in mind in the evaluation of the kinetics of isomerization processes of other HAB systems.

2. The System HSO–HOS

One HAB molecule for which a fairly complete *ab initio* potential surface is available[86] not only in the ground but also in the first excited state as well as for its positive and negative ions is the HSO radical. It is of great interest in atmospheric chemistry, in particular with respect to pollution, and has therefore attracted considerable interest in past years. The most important data are collected in Fig. 7 and Table XIII.

First of all, experimentally only the isomer HSO has so far been seen. Its formation in the laboratory occurs in reactions $H_2S + O \rightarrow SH + OH$ fol-

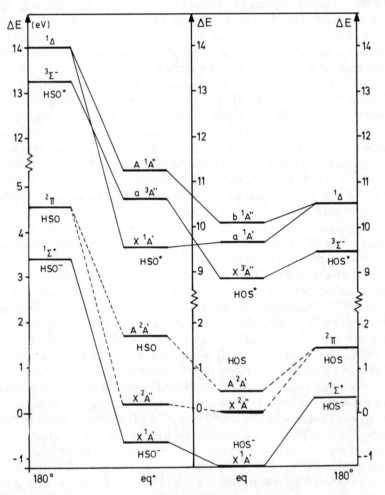

Fig. 7. Calculated energies of various extreme geometries of HOS, HOS⁺, HOS⁻ and their corresponding HSO isomers.

TABLE XIII
Calculated structural data for the two HSO isomers.

Harmonic vibrational frequencies, ν_2, ν_3 (cm^{-1})

State		Ref.	HSO ν_2	HSO ν_3	HOS ν_2	HOS ν_3
Natural	$^2A''$ Expt	86	1133	1010	1223	811
		87	1154	1026	–	–
		88	1063	1013	–	–
		89	>1100	1009	–	–
	Theor	90	1091	805	1308	1035
		91	1187.8	841.2	1326.4	925.2
	$^2A'$ Expt	86	878	822	1019	703
		87	852	672	–	–
		88	828	702	–	–
	Theor	90	959	755	–	–
Pos. ion	$^1A'$		1127	1088	1163	1030
	$^3A''$		1187	792	962	819
	$^1A''$		984	782	1046	890
Neg. ion	$^1A'$		1011	792	783	760

Geometries (X–H, S–O in bohr, θ in degrees)[a]

State	HSO S–H	HSO S–O	HSO θ	HOS O–H	HOS S–O	HOS θ
$^2A''$	2.63	2.87	104.1	1.81	3.11	109.5
$^2A'$	2.53	3.19	93.5	1.81	3.17	110.4
$^1A'$	2.58	2.77	104.8	1.85	2.87	118.3
$^3A''$	2.58	2.95	97.8	1.85	2.89	124.4
$^1A''$	2.58	2.99	96.5	1.85	2.89	124.0
$^1A'$	2.55	3.01	105.9	1.80	3.35	104.0

[a] All data are from Ref. 86.

lowed by $SH + O_3 \rightarrow HSO^* + O_2$ (Ref. 88) or $H_2S + O \rightarrow HSO + H$ (Ref. 87) and hence the HS bond is already formed. Its first excited state has also been observed[87,88] and the theoretical and experimental spectra have been evaluated in close cooperation.[88,90] Calculations[90-92] find HOS to be more stable by $7 \, \text{kcal mol}^{-1}$ in a rather careful study[86] employing a number of different AO basis sets. In spite of various efforts, HOS has not so far been seen, suggesting that the unimolecular transfer HSO–SOH is not likely. Indeed, the

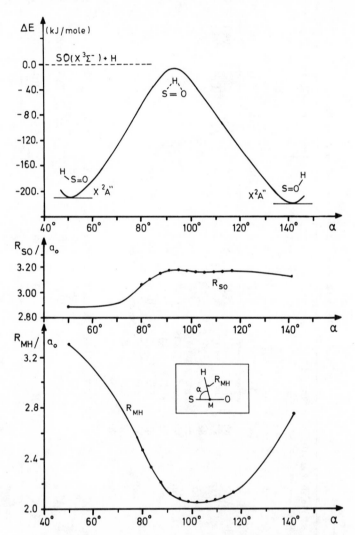

Fig. 8. Minimum energy path in the isomerization of HSO–SOH and the corresponding behavior of the S–O bond length and the angle α.

barrier to interconversion (Fig. 8) is calculated to be roughly $200 \, kJ \, mol^{-1}$, comparable with the $SO(^3\Sigma^-) + H$ dissociation limit, and hence the only chance to obtain HOS seems to be from a reaction in which the OH bond is already formed.

Both HSO and SOH are bent (Fig. 7, Table XIII), in agreement with Walsh's rules;[55,56] the in-plane component of the π_g-type MO is able to mix with the hydrogen AOs upon bending, becomes more stable in this manner and is doubly occupied in both $^2A'$ and $^2A''$ states. The SO separation is larger in the HOS component, as discussed earlier.[90] It is also interesting that the transition energy to the first excited state is quite different in both isomers, around 1.6 eV in HSO and only 0.6 eV in HOS, which makes possible a clear identification of which isomer is present under experimental conditions. It is also seen that the barrier to linearity is much smaller in the energetically preferred system HOS.

The systems HSO^+ and HOS^+ are very interesting in view of their electronic structure. Both possess the diatomic π_g^2 configuration, but while the preferred situation for HSO^+ is double occupancy of the in-plane π_g component with a strong bending trend, the HOS^+ prefers X^3A'' in which both π_g-type components are singly occupied (and since the out-of-plane π_g in HOS is essentially non-bonding with respect to angular character, HOS^+ possesses a larger bond angle). Thus there is the interesting situation that the triplet surface is lower for HOS^+ but that it is the singlet which is the preferred state for the HSO^+ surface. A simple interconversion conserving spin is thus not possible on the HSO^+-SOH^+ surface, even in the case that it would be allowed energetically. The same situation has been found for the isovalent system HNO–ONH,[93] in which HNO is most stable in the X^1A' while NOH possesses a X^3A'' ground state. Finally, there is no choice for variable occupation in the negative ion, which possesses a π_g^4-type configuration, and thus the behavior of its stability (Fig. 7) is roughly parallel to that of the neutral

TABLE XIV

Calculated ionization potentials (IP) and electron affinities (EA) of HSO and HOS (values in eV).

	State	HSO	HOS
Pos. ion	$^1A''$	11.11	10.22
	$^1A'$	9.34	9.67
	$^3A''$	10.44	9.18
Neutral	$^2A''$	0.00	0.00
	$^2A'$	1.42(1.53)[a]	0.57(0.49)
Neg. ion	$^1A'$	−2.59	−1.97

[a] Different AO basis sets.

HSO–SOH system. The calculated energy differences (IP and EA) are summarized in Table XIV.

3. General Trends in HAB Surfaces

A number of HAB systems containing atoms from the first and second rows including their ions have been treated in previous work.[92,94] In most cases only a portion of the potential surface has been investigated, in particular around the minimum of the isomers in order to determine the isomerization energy. In first-row species the favored isomer has the more electronegative atom at the terminal position; in systems possessing constituent atoms from the first and second rows, the second-row atom is generally also in this terminal position. For systems containing only second-row atoms, recent investigations suggest that behavior opposite to that of the first-row atoms seems to be likely, namely that the hydrogen is bound to the more electronegative system, although not enough experience is available to be sure about this fact.

Previous theoretical studies have suggested that the isomerization energy in mixed HAB systems (those with constituent atoms from first and second rows) is surprisingly large for molecules containing eight, nine and ten valence electrons compared to their purely first-row counterparts, while in the bent isomer pairs containing more valence electrons the relative energies are more or less similar for both types. Furthermore, it has been suggested that isomerization energies in HAB systems stemming only from second-row atoms are probably similar in magnitude to those of the first-row analogs, although not enough experience is available yet on this point. All this information was obtained under the assumption of a linear geometry of 8–10 VE HAB systems. This assumption is correct for the first-row species HCO^+ and HCN as well as for COH^+ and HNC, for example, but there seem to be exceptions for higher-row systems: the isovalent compounds $HSiS^+$ and HSiP both possess a linear conformation at equilibrium, but it is found[95] that $SiSH^+$ and SiPH possess their minima for strongly bent structures, and furthermore that these are the absolute minima on the surface. Thus ab $initio$ studies predict the existence of HAB isomers possessing 10 VE with a strongly bent geometry, in contrast to all the rules used so far. The isomerization energies of $SiSH^+$(bent)–$HSiS^+$(linear) and SiPH(bent)–HSiP(linear) are predicted by calculations to be in the 16 and 10 kcal range. Along the same lines the energy difference between the two optimal linear structures of 10 VE HCS^+–CSH^+ and $HOSi^+$–$OSiH^+$ species is 110 and 70 kcal mol^{-1} respectively but early studies[92,96] have already suggested that the linear CSH^+ arrangement is not a minimum and that this structure will simply relax to the absolute HCS^+ minimum. A more detailed study of Berthier et $al.$[95] shows that the highly unstable CSH^+ structure can indeed approach a minimum for a very bent nuclear conformation ($84°$), approximately 40 kcal more stable than the linear counterpart, but still 72 kcal above the preferred HCS^+ conformers. For the $HOSi^+–OSiH^+$ couple Berthier[95] and the earlier *ab initio* calculation[92] agree on a linear conformer in both systems with an isomerization energy between 66 and 70 kcal mol^{-1}. The more recent value for HCS^+ and the earlier values[94] for 8–10 VE systems $HCSi^+, HNSi^+, HCP^+, HCSi, HOSi^+, HNP^+$, HCP and HNSi would then all lie in the 70 kcal mol^{-1} range, i.e. always higher than in the pure first- and second-row analogs. As seen in the previous section, the isomerization energies of bent systems possessing more than 10 VE are comparable to prior experience with first-row analogs. Another example is HOSi, which lies about 15 kcal[63,97] below the less stable HSiO; it has recently been investigated experimentally[98] and the results suggest an energy difference of 20 kcal mol^{-1} between the two isomers.

Finally, it is also clear that the ground-state configurations of second-row isomers might differ from those of the first row for systems with less than 10 electrons, in the manner described for the diatomic skeleton. In accordance with the preferred σ stabilization in second-row species relative to π, it is found, for example,[94] that the $^2\Pi(\sigma^2\pi^3)$ state is lower in second-row HAB molecules while $^2\Sigma^+$ occupying $\sigma\pi^4$ is preferred in first-row HAB systems. A typical example is $HNSi^+(X^2\Pi)$ or $HCP^+(X^2\Pi)$ in comparison to $C_2H(X^2\Sigma^+)$ where it is known[99] that the a $^2\Pi$ state of C_2H is also very low (~ 4000 cm^{-1}) in energy. The flexible ground-state pattern observed for 8 VE diatomics carries also over to the corresponding HAB systems. Calculations predict a $^3\Sigma^-(\sigma^2\pi^2)$ ground state for $HCSi^+$ but a $^3\Pi(\sigma\pi^3)$ ground state for HC_2^+. These examples show the importance of the diatomic study for the understanding of the electronic structure of molecules in the HAB family.

C. Systems of H_nAB Type

The character of the diatomic skeleton is still visible in the H_nAB systems with many hydrogen atoms. In essence the hydrogens modify only the stability of the various diatomic MOs as a result of mixing between hydrogen AOs and the diatomic partners. Depending on the geometrical arrangement, they can preferentially influence one MO over the others, and this not only leads to a predictable shift in the electronic spectrum of the H_nAB system relative to that of the isoelectronic AB but also causes the appearance of various structural isomers.

As discussed in Section III. A, the C_2 radical possesses maximal π and minimal σ occupation while in contrast Si_2 shows the opposite behavior in its ground state by fully occupying σ and only partially occupying the π MO. In other words one can say that C_2 has the possibility of forming σ bonds while Si_2 is already unable (or less capable) to do so and prefers a hydrogen AO combination with its π orbitals. This difference has a direct consequence on the

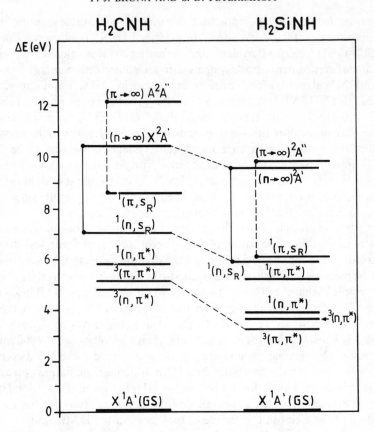

lowest structure of H_2C_2, H_2CSi and H_2Si_2, for example. If one hydrogen atom is formally added to carbon in its $X^1\Sigma^+$ (π_u^4) ground state, it is placed such that it forms a σ bond, or in other words can mix with carbon MOs to stabilize the unpopulated σ MOs. As a result HC_2 is linear in its $X^2\Sigma^+$ ($\pi^4\sigma$) state, as mentioned earlier,[99] and also in the $^2\Pi$ ($\pi^3\sigma^2$) excited state. Formal addition of the second hydrogen also prefers the location for σ bonding or σ stabilizing and leads to the known linear structure of acetylene. In contrast in Si_2 the σ is already doubly occupied and hence it is the π MO that needs stabilization by hydrogen AO admixture and this is achieved by a bridged (although not planar according to *ab initio* work[100,101]) structure. The intermediate compound CSi prefers the $\sigma\pi^3$ occupation in its ground state; the first hydrogen helps the carbon σ bond in the linear HCSi structure and the next is placed such that both σ and π are helped relative to the $\sigma\pi^3$ configuration, which is achieved best by a H_2CSi nuclear arrangement in which mixing with σ and π MOs is possible. (Carbon has a larger tendency for sp^3 hybridization than Si and thus this atom prefers to be tetravalent.) Another

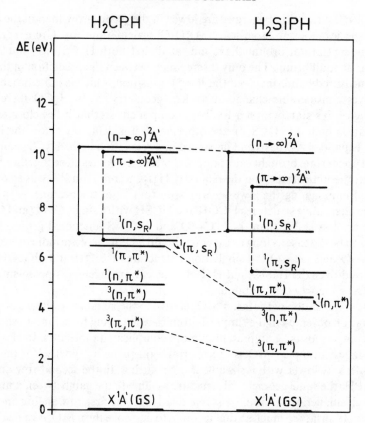

Fig. 9. Calculated vertical energies in various states of H_2CPH, H_2SiPH, H_2CNH and H_2SiNH.

possibility for admixture to the π MOs would be a *trans*-silaacetylene structure which is above the H_2CSi structure by 43 kcal mol^{-1}.[102] In other words the diversity in the C_2, CSi and Si_2 ground-state electronic configurations is fully carried over to the corresponding H_2AB system in which it causes differences in the geometrical structures.

A major structural difference is less likely if the parent diatomic fully occupies the MOs in block B (Fig. 2) as is the case in H_2CS, H_2CO or $(CH_3)_2CS$ and related compounds. Differences appear in these cases in the excitation energies due to the different gaps between the π and π^* energies and the different stabilities of the in-plane (n) and out-of-plane (π^*) MOs. One-dimensional sections of the potential surfaces for CS and H_2CS bending modes are given in the literature for the ($n\pi^*$) and ($\sigma\pi^*$) and ($\pi\pi^*$) states[103] and the location of the ^{3,1}($n\pi^*$) and n → Rydberg states were predicted within an error of less than 0.2 eV long before measured values became available.[104] As expected, all transition energies are lower in the second-row than in the first-row species. Recent large-scale MRD-CI calculations by Grein *et al.*[105] employing natural orbitals have indicated that both $H_2CS(n\pi^*)$ states are planar at equilibrium. The only discrepancy between the prediction of theory and measured bands involved the $^1(\pi\pi^*)$ transition, which is quite broad and involves considerable change in nuclear geometry.[106] In H_2CO there is a mixing of this state within a Rydberg component so that it lies close to the ionization limit. In H_2CS, on the other hand, it is definitely below the IP. A more expanded AO basis and in particular a relaxation in the H_2CS geometry of the upper state brought calculated and measured values close together.[105] A similar theoretical study on the related $(CH_3)_2CS$ compound[107] was also very helpful in assigning the lower-energy spectrum[108] of thioacetone.

The alternative structures HCOH and HCSH with a single C—O and C—S bond are less stable than the $>$C$=$O and $>$C$=$S functional groups. This seems to be the case as long as carbon is involved but there are indications that in Si—O and Ge—O the single-bond structures HSiOH and HGeOH are preferred over the H_2SiO and H_2GeO nuclear arrangements possessing the Si$=$O or Ge$=$O double bond.

The stabilization of the last σ MO relative to the π and the decrease in $\pi-\pi^*$ separation observed in the simple diatomics when going from first- to second-row compounds is again present in more complicated molecules. In H_2CNH the highest occupied MO is the n orbital (in-plane π_g type) while the π is considerably lower with a sizeable $\pi-\pi^*$ splitting. In the second-row analog H_2SiPH (the same geometrical structure assumed), the gap between π and π^* is much smaller, i.e. π^* is more stable relative to the first-row analog and π is less stable, in fact so much so that it comes to lie above the n MO. As a result[109] the first-row compound possesses a low-lying $^3(n\pi^*)$ state followed by $^3(\pi\pi^*)$ while the reverse is true in H_2SiPH, as seen in Fig. 9. In a similar manner one expects that the positive ions possess different ground states, $X^2A'(n \to \infty)$ for H_2CNH and $X^2A''(\pi \to \infty)$ for H_2SiPH. It is of interest in this connection that, while for about one year no compound with an Si$=$P double bond had been isolated in a pure state, very recently the preparation of the first phosphasilaalkene in the laboratory has been reported[110] and within six months three independent *ab initio* studies on the prototype H_2SiPH have been published.[109,111,112] The last two have also studied possible isomeric forms, but the planar H_2SiPH arrangement has been found to be the most stable.

The isomeric compounds H_2SiNH and H_2CPH (Fig. 9) show an intermediate behavior in the spectrum; even though in both the $^3(\pi\pi^*)$ state is the lowest, at least in H_2SiNH the $^3(n\pi^*)$ state is in the close energetic neighborhood and the $^2A'$ is predicted to be the ground state of the ion just as it is in the first-row compound H_2CNH. Even though the functional C$=$P group has received some attention from experimental organic chemists, little is known about Si$=$ N and neither of the two compounds has yet been isolated. In the isovalent compound HP=PH, the lowest vertical transition has also been reported[113] to be $^3(\pi\pi^*)$ followed by the higher $^3(n\pi^*)$ excitations, consistent with the finding in H_2SiPH; the first-row analog HN=NH is known to possess the reverse structure,[114] as expected from the study of H_2CNH and the change in orbital stability in going from N_2^+ or N_2 to the second-row diatomic P_2^+ or P_2. All the second-row trends, discussed in this section, will very likely also be present in compounds with constituent atoms from higher rows in the periodic table.

D. Larger Systems

Among the larger systems for which excited-state potentials have been investigated in detail are triatomic molecules and larger organic systems, predominantly π systems that play a role in organic photochemistry. Generally the need for the study stems from problems in assigning a measured spectrum or from the desire to understand details about a photophysical process. An attractive area for excited-state studies outside of this spectroscopic field is the investigation of radicals with very low-lying electronic states which can monitor the outcome of a thermochemical reaction. Such radicals can often be distinguished simply on the basis of the charge distribution of their odd electron; if this is predominantly localized in an orbital perpendicular to the molecular plane, they are referred to as π radicals, and analogously they are called σ radicals if the charge density is mostly localized in some area of the molecular plane. Generally it is not obvious which of the radicals is the lower and hence it is a challenge for *ab initio* work to settle this question and at the same time determine the geometrical structure of the various radicals and their relative energies.

Among such radicals that have received a relatively large amount of interest[115-117] are the formyl radical HCO_2, the acetyle radical CH_3CO_2 and the succinimidyl radical $N(CO)_2(CH_2)_2$. For the first two it is obvious that they are related to a triatomic molecule for which the orbital and excited-state structure are generally quite well known. For such systems, with 23 electrons, the uppermost MOs available to house the odd electron are $6a_1$, $4b_2$ and $1a_2$ and these differ very much in their angular bonding characteristics. Hence this suggests that the angular potential curves are most important to determine the minima in the potential surface of the various states. Such curves are seen in Fig. 10 for HCO_2 under the assumption of fixed CO and HC bond lengths and a symmetric structure.[118]

It is obvious that the $^2B_2$ (σ) radical is the lowest in energy. Its minimum is at an angle of 113°, which has been confirmed by a later calculation[119] which finds 112.6° in a full geometry optimization. The π radical is found to lie higher by 10 kcal (MRD-CI) or 13.7 kcal (full CI estimate), whereas the calculation allowing for a full geometrical relaxation[119] in a larger AO basis gives

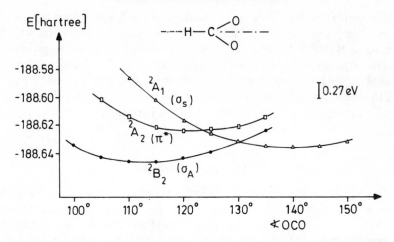

Fig. 10. Calculated angular potential energy curve for the three lowlying states of HCO$_2$. All other geometrical parameters are held constant.

9.16 kcal mol^{-1}. The angles in the two calculations are again practically identical, 122° (Ref. 118) and 121.0° (Ref. 119). The π radical is computed to be symmetric, a finding supported by the extensive MC-SCF and CI calculations of McLean et al.[119] but challenged in another theoretical study.[120] The situation is somewhat different in the σ radical since two such species of different symmetry can exist in the symmetric C$_{2v}$ point group but of the same symmetry in asymmetric distortions. McLean et al. confirm the symmetric structure; nevertheless it is clear that vibronic interaction will occur at a combination of bending and asymmetric stretching vibrations so that a certain probability exists to pass from the $^2B_2$ curve to that for $^2A_1$. For a more quantitative treatment, the non-adiabatic coupling elements would have to be evaluated and a treatment of the dynamics would have to be performed.

Essentially the same angular curves are obtained for CH$_3$CO$_2$ and it is interesting that the large-angle state has a very shallow potential curve for C—C stretch, which means that CO$_2$ can be given off very readily, as has indeed been observed.

Finally, even though the succinimidyl radical is much larger, the triatomic skeleton still plays a role. Essentially analogous curves are found (Fig. 11) for the σ MOs, but due to the presence of the nitrogen the odd electron can also be placed in a π MO localized at nitrogen (b$_1$ symmetry), and this situation is actually found to be preferred.[121] In other words, the succinimidyl radical also possesses at least two low-energy states separated by not more than 10 kcal mol^{-1}, but in this case the lower corresponds to a π radical.

In summary, then, ab initio calculations can also be employed to generate potential energy surfaces for systems with a number of heavy atoms. In this case, however, it is generally too expensive to carry out a straightforward

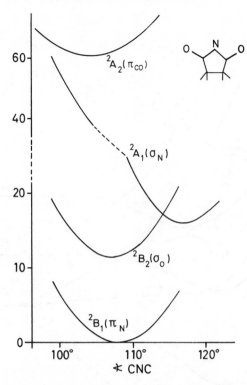

Fig. 11. Calculated angular potential energy curve for the lowest four states in the succinimidyl radical.

variation of all parameters and a screening based on some intuitive or theoretical grounds seems to be in order.

IV. INTERACTION OF EXCITED STATES

One of the great advantages of theoretical approaches is that an interaction of states is readily seen from the solution of the Schrödinger equation. If the interacting states are of different symmetry, they are obtained from different secular equations, and care must only be taken that each individual calculation is up to the given standard, thereby warranting an approximately equal accuracy for all of them. The nature of the interaction matrix elements can vary and the matrix elements themselves can be calculated in a step following the calculation of the potential hypersurface. When the states are of the same symmetry, they must be obtained from the same secular equation. The calculation must be designed in such a way that all interacting states are treated in an equivalent manner, which means in practice that considerable

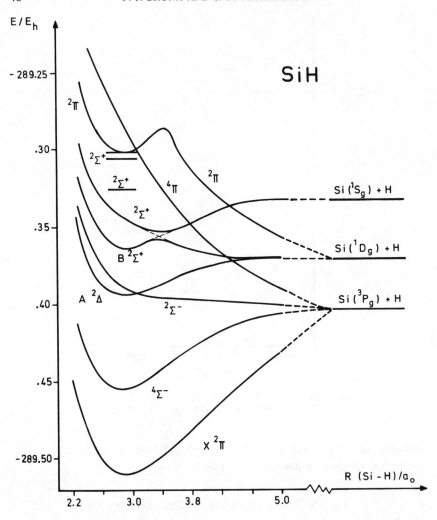

caution must be exercised in determining the reference set, which must be representative for each of the states under consideration. A slight imbalance in this set may lead to a much larger imbalance in the generated CI space and thus to a definite priority in treating one state over the others. Interacting states are often characterized according to their electronic character, i.e. valence-shell, Rydberg or ionic species. In polyatomic molecules, interactions can occur along the various degrees of freedom, and hence interactions among highly excited states are quite likely. On the other hand, the interactions can be correlated best to measured effects in small molecules, for which detailed experimental work often exists; hence most of the examples given here stem

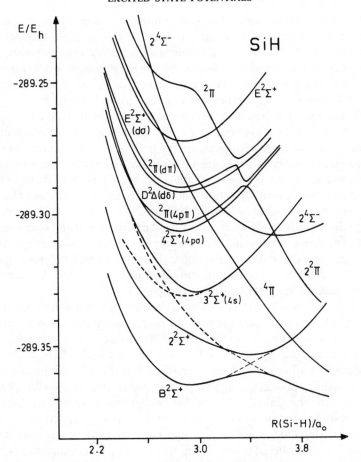

Fig. 12. Calculated potential energy curves of SiH. The unperturbed diabatic $^2\Sigma^+$ states are indicated by broken curves.

from diatomic molecules. The analysis will be equally valid for larger systems possessing a complicated potential energy surface.

A. Interaction of Valence-shell States and Interaction Matrix Elements

A typical example of avoided crossings of states that play a role in spectroscopy[122] is given in Fig. 12 for SiH. There are two $^2\Sigma^+$ states with dominant configurational character $4\sigma^2 5\sigma 2\pi^2$ and $4\sigma^2 5\sigma^2 6\sigma$ (or $4\sigma^2 5\sigma 6\sigma^2$ at large internuclear separations since 5σ becomes hydrogenic at large R and 6σ approaches $3p\sigma$ on silicon). The lower of these possesses a double minimum in the Born–Oppenheimer approximation due to an avoided crossing and changes its character from $5\sigma\pi^2$ to $5\sigma^2 6\sigma$ and finally to $5\sigma 6\sigma^2$ or $H(5\sigma)+$

$Si(^1D_g)$. The second state shows a minimum rather than its original repulsive character and is made up in a complementary manner. The minimum of the lower state at small internuclear separation must correspond to what has been assigned spectroscopically as the B state to which transitions from the $X^2\Pi$ ground state have been observed in the 3.82–3.95 eV range; the MRD-CI values for the vertical energy separation between both states is 3.94 eV and 3.89 eV if the difference is based on the estimated full CI value. The strong predissociation assumed in the B state, which did not allow experimental evaluation of the zero-point energy for this state, strong perturbations and the large number of unassigned lines in the absorption spectrum can be understood easily on the basis of the calculated potential curve and the interaction with a second $^2\Sigma^+$ state. The outer minimum of the lower $^2\Sigma^+$ state can be correlated to what has been called the C state with a shallow minimum at large Si–H separations. In other words, the two states B and C assumed on the basis of measurements must be correlated to one state possessing a double minimum (provided the adiabatic approximation can be employed), and the entire B and C state spectrum in conjunction with the extra lines must be looked upon as one (although quite complex) band.

There are numerous such examples in the literature in which the calculated potential surfaces have been the basis for correct assignment of spectral lines. A very similar situation in which the avoided crossing between a repulsive and a bound state is important can be found in NF and NCl.[123] The $b^1\Sigma^+$ state, whose configuration $\pi^4\pi^{*2}$ would correspond to the third dissociation limit $N(^2P_u) + X(^2P_u)(X = F, Cl)$, shows at large separation an avoided crossing (Fig. 13) with the repulsive $^2\Sigma^+$ $\pi^2\pi^{*2}\sigma^2$ configuration, which correlates with the second dissociation limit $N(^2D_u) + X(^2P_u)$. As a result the lower $b^1\Sigma^+$ state dissociates into the second limit, but only after it overcomes a barrier; this has not been considered (nor is it known) in prior determinations of $b^1\Sigma^+$ dissociation under the incorrect assumption of a simple Morse-type potential curve. Furthermore, this barrier will also have some influence on gas kinetic (recombination) processes of excited states. In a complementary way, the $2^1\Sigma^+$ state possesses a minimum at large separations that was not known prior to the theoretical work. The systems NF and NCl are excellent examples, which show many other avoided crossings in their potential surfaces, as discussed in detail elsewhere;[123] in particular, the reduced symmetry relative to the homonuclear isoelectronic species O_2 formally removes the distinction between *gerade* and *ungerade* states, but the corresponding character is still present in the wavefunctions so that many of the irregularities in the shapes of the Π or Σ^- potential curves can simply be understood as an avoided crossing of what has been Π_u and Π_g or Σ_u^- and Σ_g^- states in the analogous molecule possessing inversion symmetry.

The charge-transfer reaction $N^+ + CO \rightarrow N + CO^+$ also involves an interaction of (at least) two close-lying states, as seen in Fig. 14. In this case the two

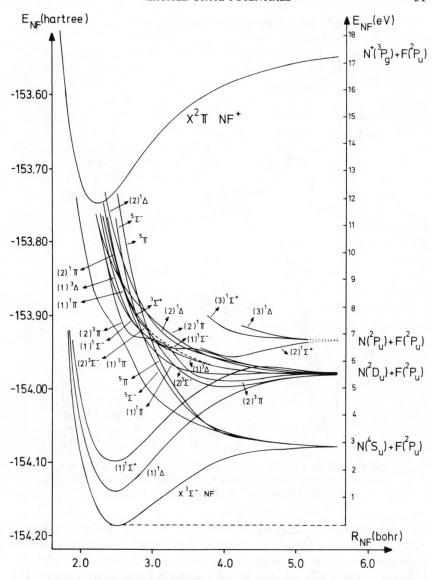

Fig. 13. Calculated potential energy curves of NF.

$^3\Sigma^-$ states involved possess the very different configuration structures $7\sigma^2 1\pi^4 2\pi^2$ and $7\sigma 1\pi^4 2\pi^2 8\sigma$, i.e. the first possesses two open shells $\pi_x\pi_y$ while in the second four shells are only singly occupied. From the technical point of view, a balanced and reliable description of both states is not an easy matter. Natural orbitals (averaged for the two states) are employed[124] to keep the

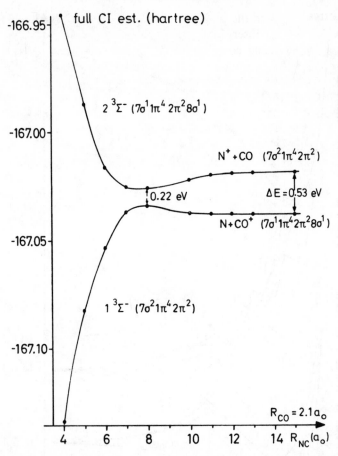

Fig. 14. Calculated potential curves of NCO$^+$ as a function of the N–C distance (constant C–O $= 2.1a_o$), showing an avoided crossing which plays a role in the charge-exchange process.

reference set relatively compact. Nevertheless, the total MRD-CI space (56 AOs) amounts to 1135 991 in this case, from which a selected subset of approximately 15 000 has been diagonalized directly at each point, while the influence of the remaining ones is extrapolated in the standard manner. The minimal distance of 0.22 eV (MRD-CI) and 0.16 eV (estimated full CI) is calculated between the two curves at a CO distance of 2.1 a.u. around an N–CO separation of 8 a.u. The calculated energy separation of 0.53 eV (MRD-CI) and 0.37 eV (estimated full CI) at larger N–CO separation gives an indication of the accuracy of the calculation when compared to the known experimental separation of 0.52 eV between the species. Even though the non-adiabatic (radial) coupling elements have not been calculated, it appears very likely that the small separation in the crossing area favors considerable interaction to explain part of the observed charge-exchange reaction.[125,126] A further process, namely angular coupling with a $^3\Pi$ state not shown in the figure, might also play a role in the total charge-exchange reaction.

A further charge-exchange reaction is that of $Mg^+(^2S) + Na(^2S) \rightarrow Na^+(^1S) + Mg(^3P)$ involving electronically excited states of the $[MgNa]^+$ system. The potential curves are obtained (Fig. 15) by correlating 14 electrons in a carefully chosen AO basis, discussed already in Section II. They show first of all[15] that the energetic separation between entrance and exit channels is very small (2.506 eV and 2.711 eV above the $X^1\Sigma^+$ ground-state asymptote represented as 2.52 eV and 2.63 eV in the present calculations) and that interaction of states appears (neglecting spin–orbit effects) between $^3\Pi$ and $^3\Sigma^+$ in the 4–5 a.u. area, and possibly between the two $^3\Sigma^+$ states at very small and very large

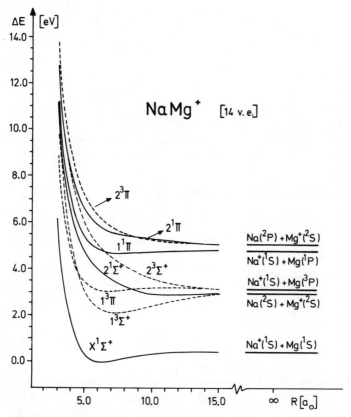

Fig. 15. Calculated potential energy curves for the system $NaMg^+$ by correlating 14 electrons.

separations. The non-adiabatic coupling element (see Eq. (10) in Section II)

$$\langle \Psi_i(\mathbf{r}, \mathbf{R}) | \nabla(\mathbf{R}) | \Psi_j(\mathbf{r}, \mathbf{R}) \rangle$$

reduces in $C_{\infty v}$ cylindrical symmetry to elements in which $\nabla(\mathbf{R})$ becomes $\partial/\partial R$ and $(1/R)\partial/\partial \Psi = (i/R)L_y$ (if y is perpendicular to the A–B axis). The so-called radial coupling element

$$\langle \Psi_i(\mathbf{r}, \mathbf{R}) | \partial/\partial \mathbf{R} | \Psi_j(\mathbf{r}, \mathbf{R}) \rangle$$

couples states of the same symmetry, while the rotational coupling element

$$\langle \Psi_i(\mathbf{r}, \mathbf{R}) | L_y | \Psi_j(\mathbf{r}, \mathbf{R}) \rangle$$

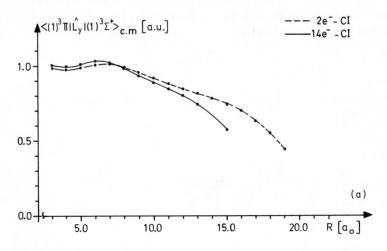

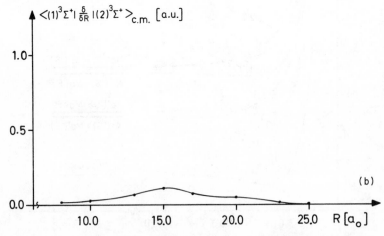

Fig. 16. Calculated rotational and radial coupling matrix elements for various states in MgNa$^+$ as a function of internuclear separation.

couples those whose angular momentum quantum numbers Λ differ by unity. The two elements have been calculated for $1^3\Sigma^+-2^3\Sigma^+$ and $^3\Sigma^+-^3\Pi$ interactions (Fig. 16) and show that the radial coupling element is considerably smaller than that coupling $^3\Sigma^+$ and $^3\Pi$. The simple indication is that charge exchange occurs in the neighborhood of an Na–Mg separation of 4.5 a.u. Computation of the inelastic cross-section on the basis of this matrix element and the calculated potential energy curves[127] gives very good agreement with the corresponding measured cross-sections and hence supports the idea that the charge-exchange mechanism occurs as the calculations predict.

In this connection, it is also interesting to compare the results of these 14-electron calculations with those correlating only two electrons (i.e. the frozen core in the CI consists of 10 MOs rather than four as before). In this case the potential curves are roughly parallel to those of the more extensive calculations (Fig. 17) and interestingly the matrix elements for rotational coupling

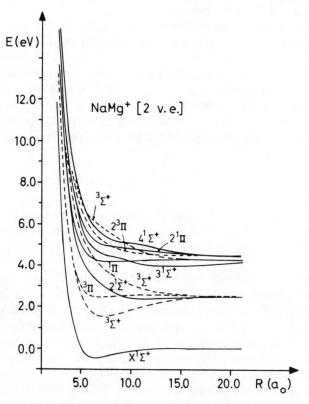

Fig. 17. Calculated potential energy curves for the system NaMg$^+$ by explicitly correlating only two electrons.

are also in quite good agreement with those derived from the larger calculation (Fig. 16a), at least in the area of most concern. This finding indicates that, upon proper evaluation of the electronic structure of states, it may be possible to calculate such interaction matrix elements in a less expensive manner, and this experience might be quite important for pseudopotential calculations. Finally, the internal consistency of the matrix element calculation has also been checked to evaluate the error limits or the significance of the deviations between the 2 VE and 14 VE calculations. In the standard 14 VE treatment, secular equations of order 3000 have been solved directly (corresponding to a selection threshold T of 5 μhartree); increasing the CI expansion lengths by a factor of almost 3 ($T = 1$ μhartree) produces the following changes: $\langle {}^3\Sigma^+ |L_y|{}^3\Pi \rangle$ at 4.5 a.u. is 0.987 a.u. for selected secular equations of 2930 (${}^3\Sigma^+$) and 3158 (${}^3\Pi$); and it is 0.988 a.u. upon increase to 7221 (${}^3\Sigma^+$) and 8742 (${}^3\Pi$). Similar tests for the radial coupling matrix element, for which a variation of the steps in the finite difference–differentiation has also been undertaken, suggest that the error limit for the radial matrix element is approximately 1.5×10^{-2} a.u. at its maximum.

Interaction matrix elements between states of different symmetry involving the spin–orbit operator have also been evaluated in various instances. A typical example is the predissociation of the $b^4\Sigma_g^-$ state in O_2^+ by the ${}^4\Sigma_g^+$ and $f^4\Pi_g$ states.[128] The $\langle b^4\Sigma_g^- |H_{so}|{}^4\Sigma_g^+ \rangle$ interaction is very much dependent on the internuclear separation and varies from zero around 2.4 a.u. to approximately 60 cm^{-1} at 3.5 a.u., where the potential surface crossing occurs in the neighborhood of 3 a.u. Hence in a detailed treatment of the probability γ for non-radiationless transition, the explicit form of this matrix element has to be taken into account as

$$\gamma = \frac{2\pi}{\hbar\varepsilon} \left(\int_0^\infty \Phi_n(R) |\langle H_{so} \rangle| \Phi_E(R) \, dR \right)^2$$

rather than in the standard Franck–Condon type of treatment

$$\bar{\gamma} = \frac{2\pi}{\hbar\varepsilon} |\langle H_{so} \rangle|^2 \left(\int_0^\infty \Phi_n \langle R| \Phi_E(R) \, dR \right)^2$$

where Φ_n are the vibrational functions of the bound $b^4\Sigma_g^-$ state and $\Phi_E(R)$ the energy-normalized functions in the repulsive area of the ${}^4\Sigma_g^+$ state. By contrast the $\langle b^4\Sigma_g^- |H_{so}|f^4\Pi_g \rangle$ matrix element connecting the second state is fairly constant ($- 25$ cm^{-1}) across the region of internuclear separations of interest, so that in this case only its magnitude is really important for the qualitative description of the predissociation—at least as far as the spin–orbit contribution is concerned. Quite often such differences can be traced to changes in configurations with the geometrical variable, and this difference in character directly affects the magnitude of such spin–orbit contributions. Details of this investigation can be found in the literature.[128]

Another example is the predissociation of OH^+ in its $b^1\Sigma^+$ state due to

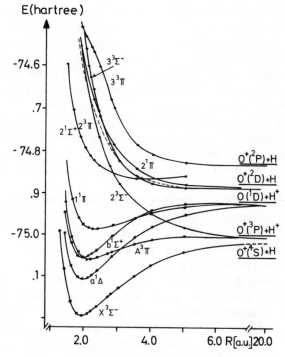

Fig. 18. Calculated potential energy curves for OH$^+$.

spin–orbit interaction with the A$^3\Pi$ state (Fig. 18). The matrix element $\langle b^1\Sigma^+ | H_{so} | A^3\Pi \rangle$ is fairly constant (around 52 cm^{-1}) between 1.4 a.u. and 2.7 a.u. Calculations involving these data and the calculated potential energy curves[129] give high predissociation probabilities for $v > 5$, many orders of magnitude larger than for the radiative b$^1\Sigma^+$–X$^3\Sigma^-$ process.[129,130]

Finally, an example in which all three types of coupling (radial, rotational and spin–orbit) are operative is the charge-transfer reaction Ar$^+$(^{2}P) + H → Ar(^{1}S) + H$^+$. All three have been calculated as a function of internuclear separation.[15] The entrance channel correlates with B$^1\Sigma^+$, a$^3\Pi$, A$^1\Pi$ and b$^3\Sigma^+$ states, while the outgoing channel possesses only $^1\Sigma^+$ symmetry.(The angular coupling involves the term $\langle A^1\Pi | L_y | X^1\Sigma^+ \rangle$ with a maximum around 0.4 a.u.; the radial coupling $\langle X^1\Sigma^+ | \partial/\partial R | B^1\Sigma^+ \rangle$ is only about half of this; and finally the spin–orbit operator couples the output channel X$^1\Sigma^+$ with the upper $^3\Pi$ state of the ingoing partners.

B. Interaction of Valence-shell States and Rydberg States

In many saturated molecules—or molecules in which the valence shell is close to being filled—the excited valence-shell states populate antibonding MOs in lieu of bonding (or less antibonding) MOs and therefore show a repulsive character. At the same time, there is a manifold of Rydberg states available (resulting from a transition of generally an antibonding or a bonding orbital into one of non-bonding character) whose potential energy surfaces are parallel to that of the respective ionic state (as the Rydberg series limit) and generally exhibit potential wells in the region not far away from the equilibrium of the ground state. In such cases, one finds numerous interactions between the valence-shell states and the Rydberg manifold. Again, from a technical point of view, a balanced description of the Rydberg and the valence-shell states is necessary, and this is possible in careful work. If no Rydberg functions are present in the AO basis, it is clear that the interaction is not found and erroneous potential surfaces may result, as can still be found in the present-day literature.

Typical examples of such Rydberg–valence mixing have been discussed in detail in the literature[131] whereby the interaction (if states of the same symmetry are involved) can take place on the CI level, i.e. by a change in the importance of configurations which make up the total wavefunction, or on the MO level, i.e. by a change in a given molecular orbital generally being favored in lower-symmetry situations; the latter has often been called 'Rydbergization' of an orbital.[132] The various double-minima states of H_2 (E–F and G–H) as well as the double minimum in the $B^3\Sigma_u^-$ state of O_2, which explains the structure at the high-energy end of the Schumann–Runge continuum, fall[133] into this category of Rydberg–valence mixing as well as some interesting photochemical states in butadiene.

Representative examples for the interaction of valence-shell states and the manifold of Rydberg states are also found in simple hydrides. The ground states of the molecules from CH and SiH to FH and ClH can be characterized by $\sigma^2\pi^n$ configurations with $1 < n < 4$. Excited states resulting from $\sigma \leftarrow \pi$ excitations in these systems are generally bound and often show interactions with other (valence-shell) states at larger internuclear separations. The $\sigma \rightarrow \sigma^*$ and $\pi \rightarrow \sigma^*$ transitions lead to states that are generally repulsive. The high-multiplicity states $^4\Pi$ $(n = 1, 3)$ and $^5\Sigma^-$ $(n = 2)$ are strongly repulsive and correlate—at least in the second-row hydrides—with the lowest dissociation limit; as such, they are responsible for predissociation of a number of electronic states, as discussed elsewhere.[134] The $\sigma \rightarrow \sigma^*$ states of lower multiplicity and the simple Rydberg $\pi \rightarrow p\pi$ states, on the one hand, and the $\pi \rightarrow \sigma^*$ and $\pi \rightarrow s$ Rydberg states, on the other, lead to common spin and spatial symmetry, so that various avoided crossings (at configurational or orbital level) between the Rydberg species and those of the valence category occur at smaller internuclear separations.

The calculated potential curves for HF,[135] for example, given in Fig. 19, show very convincingly the interaction of the strongly repulsive $^3\Sigma^+$ valence state with the Rydberg manifold; the interaction with the lowest Rydberg state of the same symmetry $(\pi \rightarrow 3p\pi)$ occurs almost at the minimum of the latter, that with the $\pi \rightarrow 3d\pi$ state $(^3\Sigma_{R2}^+)$ and higher $\pi \rightarrow np\pi$ members (not calculated explicitly) always in the repulsive parts. The $^3\Delta$ Rydberg state is the only one unaffected. The $^3\Pi$ Rydberg states show small but distinct interaction with the $\pi \rightarrow \sigma^*$ $^3\Pi$ valence species. The situation is more pronounced in the singlet manifold and, furthermore, can be tested more easily by spectroscopic

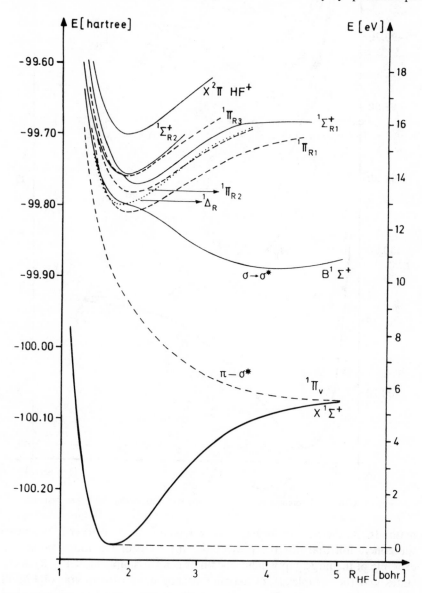

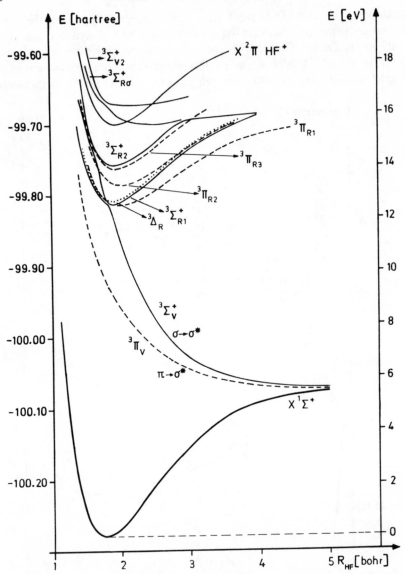

Fig. 19. Calculated potential energy curves for HF in its singlet and triplet systems.

methods. On the basis of measurements, Douglas and coworkers[136] came to the conclusion that 'for the present the spectrum of HF must remain an example of the large gap which exists between the simple model of Rydberg states and the complexity of actual Rydberg states'. Inspection of Fig. 19 makes it clear that all $^1\Pi$ and $^1\Sigma^+$ Rydberg states are heavily perturbed

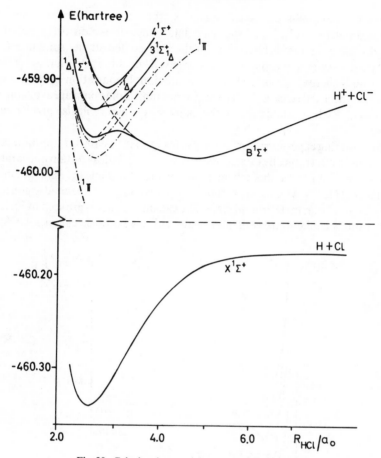

Fig. 20. Calculated potential energy curves for HCl.

compared to the unperturbed $^1\Delta$ species. The interaction of the $\sigma \to \sigma^*$ valence configuration leads to saddle points and double minima in the $^1\Sigma^+$ Rydberg family and the interaction of the $\pi \to \sigma^*$ valence state leads to a distinct shift toward a more repulsive nature in the inner branches of the $^1\Pi$ Rydberg states.

Interactions in the SiH molecule as shown in Fig. 12 are self-explanatory in this context. Figure 20 shows the double-minima excited-state potential in HCl, which is now under high-resolution experimental investigation. In the present context, we want to stress that the correct relative stability of the two minima is quite difficult to obtain; standard AO basis sets are not able to describe the outer well as reliably as the inner Rydberg part and only an f polarization and a fairly contracted d polarization function in conjunction with a semi-diffuse p and d function for the description of the charge distribution in the negative ion Cl$^-$ reduce the errors in excitation energies beyond 0.1 eV. Without such functions, the outer well is too high by approximately $0.5 \, \text{eV}$[137] and such calculation then gives a distorted picture of the double-minimum arrangement of the first excited singlet state in HCl.

Rydberg states of positive ions have received very little attention so far; they lie at high energy, i.e. somewhat below the double ionization of the system, which in the usual sense is the Rydberg series limit. Some data are available for the $C_2H_6^+$ Rydberg states[138] and the following example will also involve such states.

In scattering experiments on argon gas using a He^+ beam, the appearance of Ar^+ ions in electronically excited states can be observed, at certain appearance potentials.[20] From a theoretical point of view, this means that potential surfaces of $He^+ + Ar$ must cross those which dissociate into excited argon ions plus helium atom in its ground state. The calculated curves are seen in Fig. 21. The various configurations that correlate with the entrance channel $Ar(^1S) +$

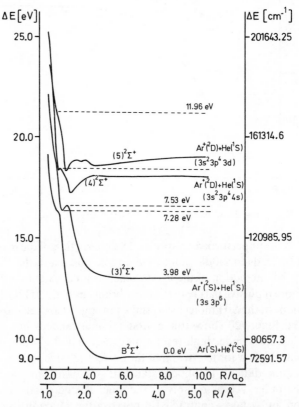

Fig. 21. Calculated potential energy curves for the system $(ArHe)^+$ showing the various avoided crossings of valence shell with Rydberg states.

He$^+$(^{2}S) and the other dissociation limit Ar$^+$(^{2}S, 3s3p^6) + He(^{1}S) lead (with the exception of a van der Waals-type minimum) to strongly repulsive curves. They interact with the curves corresponding to the first members of the p → 4s and p → 3d Rydberg series of Ar$^+$, 3s^{2}3p^{4}4s and 3s^{2}3p^{4}3d, and in this manner produce various avoided crossings as seen in Fig. 21. A transition between B$^2\Sigma^+$ and 3$^2\Sigma^+$ is most likely in the energy range between 7.28 and 7.53 eV according to the calculations, and indeed the appearance potential for Ar$^+$(^{2}S$_g$, 3s3p^6) ions is measured at 7.35 eV.[20] The second appearance potential in the scattering experiments is found at 9.3 eV under the assumption of a release of Ar$^+$ in a 4s or 3d excited state. The potential curves find the strongest interaction between B$^2\Sigma^+$ and 4$^2\Sigma^+$ around 9.38 eV, which would

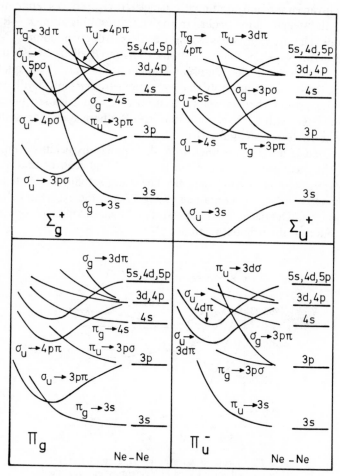

Fig. 22. Schematic behavior of electronic states in Ne$_2$.

entirely fit in with the 9.3 eV measured and would predominantly favor the $Ar^+(^2D, 3s^23p^44s)$ dissociation channel. There is also a certain probability for crossing to the $5^2\Sigma^+$ state at 9.38 eV, but dissociation is only likely into $Ar^+(^2D, 3s^23p^43d)$ at 9.9 eV; the strongest diabatic interaction between the $5^2\Sigma^+$ and $B^2\Sigma^+$ states is around 12 eV according to the calculations, but no further analysis has been undertaken.

The excited states of rare-gas dimers also show mixing of states whereby all of them could be characterized as Rydberg in the sense that they involve only orbitals outside the normal valence shell. The ground-state configuration is $\sigma_g^2\pi_u^4\pi_g^4\sigma_u^2$ (or $\sigma^2\pi^4\pi^{*4}\sigma^{*2}$ if the antibonding MOs are denoted by asterisks). Removal of an electron from the strongly antibonding σ_u and placing it into a mostly non-bonding Rydberg MO is expected to lead to a bound state, while conversely removal of the bonding π and σ electrons produces repulsive states whereby the steeper slope occurs in the states resulting from $\sigma \rightarrow$ Rydberg excitations. The situation is less clear for the less strongly bonding π_g MO but calculations on Ne_2 show that excitations from π_g to Rydberg MOs reduce the character of the original repulsive configuration only. For qualitative purposes, the rule is then simply that excitations from σ_u into Rydberg MOs lead to bound states whereas excitations from π_g, π_u and σ_g lead to increasingly repulsive states, with the exception that $\pi_g \rightarrow$ Rydberg states can sometimes give rise to shallow potential wells.

If this rule is applied together with the known dissociation limits, one obtains the qualitative picture given in Fig. 22 and realizes that standard Morse-type potentials are not realistic. The states actually calculated for the $^3\Pi_u$ symmetry in Ne_2 are contained in Fig. 23 and the expected behavior is fully borne out. Details of the other states and their mixing with one another can be found in the original reference.[139]

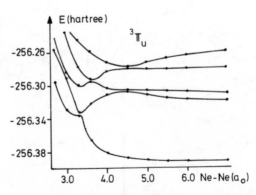

Fig. 23. Calculated states for the $^3\Pi_u$ symmetry in Ne_2. They should be compared with the corresponding states in Fig. 22.

C. Interaction Involving Ionic States

In HCl the $^1\Sigma^+$ $\sigma \rightarrow \sigma^*$ state dissociates at large internuclear separations into $H^+ + Cl^-$ and as such can be characterized at large H–Cl separations as an ionic state. Its potential curve shows a strict $1/R$ dependence in the asymptotic region at large R, and deviates in the bonding area from this behavior due to mixing with other states. In an analogous manner, all the states in Cl_2 which populate the σ^* MO twice as well as the $\sigma \rightarrow \sigma^*$ $^1\Sigma_u^+$ state correlate with the ionic limits $Cl^+(^3P_g) + Cl^-(^1S_g)$ and $Cl^+(^1D_g) + Cl^-(^1S_g)$, and therefore show the $1/R$ dependence for large internuclear separations and can be called ionic states (see figures in Ref. 140). In Cl_2 and HCl the ionic limits are the lowest dissociation limits above that of the ground state and hence these states are normally spectroscopically important and can be interpreted easily in the adiabatic approximation. If the ionic limit is above the asymptote of various other states, the ionic state may cross at large internuclear separations through a manifold of other states and may manifest itself to various degrees by an interaction with the lower states.

A typical example for this behavior is NaH.[141] The $^1\Sigma^+$ ionic $Na^+ + H^-$ asymptote lies above a large number of excited (Na*H) Rydberg-like states and its course through this manifold is clearly seen from the computed

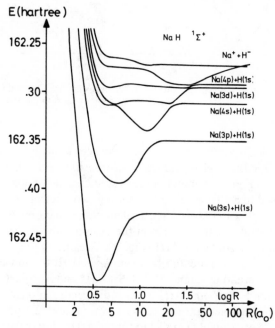

Fig. 24. Calculated potential energy curves of NaH showing the interaction of the ionic component with the manifold of other states.

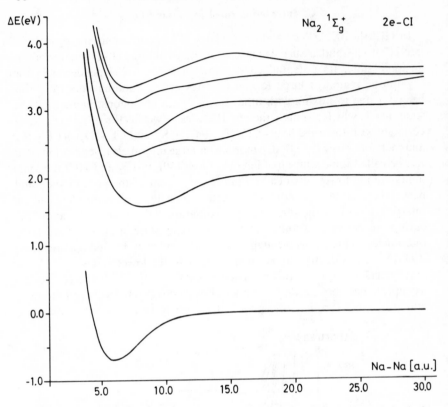

Fig. 25. Calculated potential energy curves of Na_2 showing the interaction of the ionic component with the other states. Only two electrons are correlated in the MRD-CI treatment.

potentials in Fig. 24. Similar behavior has also been found by Konowalow and coworkers[142] for Li_2, and its influence on transition probabilities has been studied in detail. Figure 25 shows the corresponding curves for the $^1\Sigma_g^+$ states in Na_2 obtained in a CI treatment correlating two electrons only; these data compare quite well with the results of a calculation that correlates all electrons outside the 1s shell, as already seen in the example for the $NaMg^+$ states.

Another example studied is ArH (Fig. 26). Because of the relatively small difference in ionization potential between Ar and H, the low-energy states correspond in alternating manner to $Ar(^1S_g) + H^*$ and $Ar^* + H(^2S_g)$ up to the first Rydberg limit, this being ArH^+ which dissociates into $Ar(^1S_g) + H^+$. In the area of small internuclear separations, all the $^2\Pi$ and $^2\Sigma^+$ potential energy curves in Fig. 26 are roughly parallel to that of ArH^+. Large deviations occur for distances greater than 4.0 a.u. due to the interaction with the Ar^+H^- state, which passes through the entire manifold of states twice, as seen in the figure.

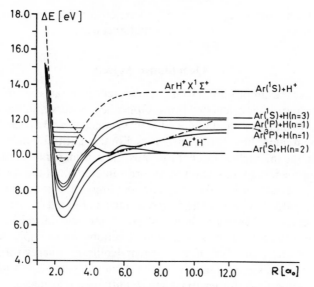

Fig. 26. Calculated excited-state potential energy curves for ArH. The interaction with the Ar^+H^- component is indicated.

All these interactions are important for the understanding of the dynamics of the $Ar + H^-$ system, but this will not be pursued any further in the present context. It should be mentioned, however, that there is a large radial coupling between the first excited $A^2\Sigma^+$ state and the $X^2\Sigma^+$ state (not shown), which leads to considerable predissociation. Technically, this interaction is seen solely at the MO but not at the CI level of treatment.

From a technical point of view, it is worth pointing out that such calculations, which describe the interaction of ionic states with a large manifold of other states, are close to the limit of what can be achieved with a standard program package. In ArH the AO basis set must include all hydrogen functions up to $n = 3$ and argon functions including 4s, 4p and 3d. Furthermore, for large separations, additional functions representing a H^- system must also be present. Furthermore, there are many states of the same symmetry which at each point must be treated in a balanced manner. At small Ar–H values, for example, this is relatively easy to achieve but as soon as the Ar^-H^+ state comes into play, the selection of the reference set is not self-evident any more (sometimes the Ar^+H^- configuration cannot even be easily recognized) and the requirement that up to 6–8 roots are treated simultaneously leads to quite large secular equations. In the calculations for the ArH system, the argon basis consists of (14s10p3d) contracted to (8s6p3d), while the hydrogen functions (12s8p2d) are grouped as (9s6p1d). This amounts to a total basis of 77 functions and total configuration spaces for the Σ and Π states of over a million employing 46 and 37 reference configurations respectively whereby always six roots are treated simultaneously.

D. Quantitative Aspects

In most of the examples given so far, an interaction of states could be seen by direct inspection of the calculated Born–Oppenheimer curves; either there was a genuine crossing between states of different symmetry or an avoided crossing became apparent in the shape of the curves. For a quantitative measure of the interaction, the interaction mechanism must be considered as in the various examples in the foregoing section: spin–orbit interaction if the states couple via the H_{so} operator or radial or angular coupling in the linear system if the coupling operator is ∇ or ∇^2. The matrix elements required can be calculated in simple cases as has been shown above; there are difficulties in more complicated situations such as conical intersections, for example. A detailed treatment of the dynamics of the system (collision, energy-transfer and dissociation processes or simple vibrations) should follow the study of the stationary properties in order to obtain quantitative information about the physical or chemical behavior of the systems, and it can in principle be undertaken on the basis of the known potential surfaces and interaction matrix elements. In this connection, it is important to calculate in an avoided crossing not only the lower potential surface but also those complementary to it, so that in addition to the calculated adiabatic surfaces the diabatic states can be constructed at least approximately, which is often done by simple inspection. If a rigorous dynamical study is not undertaken or is not feasible, it is sometimes possible to estimate whether the diabatic or adiabatic potentials are dominant for the process: one expects that for high-energy collisions the diabatic surfaces are important while the opposite should hold for slow collisions; the vibrational pattern is generally dominated by the Born–Oppenheimer surface if the ratio between zero-point vibration ω_0 in the upper curve and the energy splitting ΔE between the two curves is $\omega_0/\Delta E > 1$.

There are certain interactions due to nuclear motion which are not readily apparent from the appearance of the potential curves. A typical example is the Renner–Teller effect in triatomic molecules, in which the two components of an orbitally degenerate state split into two states of different symmetry upon bending motion, as seen in Fig. 27. A strong coupling between the two components is observed, however, and spectral patterns can be understood only by coupling the electronic and nuclear motions, as has been pointed out already in numerous textbooks. From an *ab initio* point of view, one can treat this problem in analogy to the previous examples by calculating the Born–Oppenheimer potential surfaces and the interaction matrix element, in this case $\langle \Psi_A | \partial/\partial\Psi | \Psi_B \rangle$ and $\langle \Psi_A | \partial^2/\partial\Psi^2 | \Psi_A \rangle$ (and the same for Ψ_B) as exemplified for the NH_2 molecule.[143] The angle Ψ is thereby the variable which rotates the x into the y component of the degenerate states. The

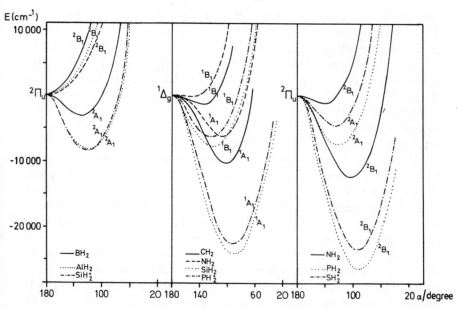

Fig. 27. Calculated angular potential energy curves of the lowest $^2\Pi$ and $^2\Delta$ states in various AH$_2$ molecules showing Renner–Teller splitting.

interaction element involving $\partial/\partial\Psi$ does not vary much with bending (its absolute value is 1.0 at linear geometry, and in NH$_2$ it changes to 0·8 at a molecular angle of 60°) and hence its limiting value at 180° is often assumed in treatments of the Renner–Teller effect. In NH$_2$ the matrix elements for the second derivatives are found to vary considerably with bending angle, but, since they furnish only a diagonal contribution and no coupling of states, they are normally incorporated in the potential energy term (adiabatic approximation). In this manner the behavior of vibrational levels due to vibronic interactions has been studied in a considerable number of AH$_2$ molecules.[144]

In summary, then, the interaction of electronically excited states can appear in many variations. *Ab initio* calculations are an excellent means of studying this phenomenon; the interaction can normally be seen very readily from the form of the potential surfaces, from an analysis of the wavefunction (combined MO and CI expansion) or from direct calculation of the respective interaction matrix element. All this information can be employed as the basis for a solid treatment of the dynamics of the system.

V. POTENTIAL ENERGY SURFACES FOR SHORT-LIVED NEGATIVE-ION (RESONANCE) STATES

Negative-ion states can be treated by standard SCF and CI calculations as long as their potential energy surface is energetically more stable than that of the corresponding neutral system. If this is not the case, the SCF solution will invariably place the extra electron into the most diffuse orbital available in a given basis, thereby simulating the neutral system plus a free electron; in this manner the potential surface of the negative system approaches that of the neutral system from above. Ample evidence for this behavior can be found in the literature.[33,145-147] The CI treatment bases on such SCF sometimes converges on a stable root (specified by a compact upper MO) above the neutral system, i.e. on one that is not affected by any further AO basis set expansion, in particular not by addition of further diffuse functions. This is especially true for a relatively localized negative-ion charge distribution, i.e. for one in which the extra electron can be considered to be in a high-lying well defined MO with little coupling to the continuum and therefore a small probability for decay. Nevertheless, such calculations are generally expensive since the proof of stability requires a number of SCF-CI calculations employing variable sets of diffuse basis functions.

In many of the systems of interest, the negative-ion curve is below that of the neutral system in certain ranges of the nuclear conformation and above it in others. A typical example is the CO_2 molecule, in which the neutral system is linear and has a rising potential energy curve with bending (Fig. 28), while the negative system possesses a minimum in the bent conformation with an energy

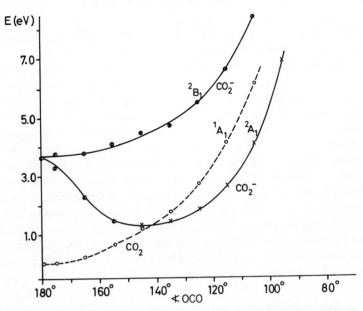

Fig. 28. Calculated potential energy curves for the ground state of CO_2 and two negative-ion states as a function of internuclear angle. The C–O distance is kept constant. The modified CI treatment is employed for the calculation of the CO_2 points above the energy of CO_2.

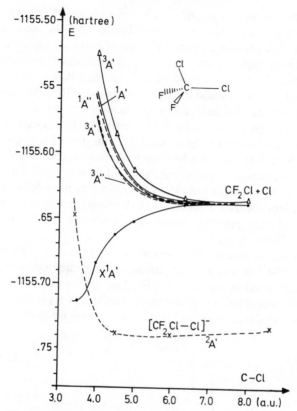

Fig. 29. Calculated potential energy curves for a removal of Cl and Cl⁻ from the neutral or negative-ion system CF_2Cl_2.

below that of CO_2; the energy of the negative system is above that of neutral CO_2 in the linear conformation. Other representative examples are freons (Fig. 29) such as CF_2Cl_2, $CFCl_3$, $CHCl_3$, etc., in which the negative-ion curve also crosses that of the neutral ground state along the coordinate of Cl-atom removal and as such allows the important dissociation (Cl⁻ removal) as a result of electron attachment. These examples show that it would be desirable to employ the same methods to describe negative-ion states below (stable) and above ('resonance') the neutral system rather than employing different procedures that are available from scattering theory for the area of the short-lived resonance states.

This goal can be accomplished by a modification of the standard CI-type programs as shown elsewhere.[33,148] In such an approach, the nuclear charge $Z(1 + \lambda)$ of all atoms in a system is simply perturbed by a positive value of λ and this small increase in nuclear charge is able to transform the discrete component of a resonance state into a true bound state in the sense used above.

In the $^2\Pi_g$ state of N_2^-, for example, values of λ above 0.01 already give SCF solutions below that of $N_2 + e$, i.e. are able to bind the electron. Such a procedure can be carried out for several values of λ, for example (note that only the one-electron part in the AO integral evaluation is affected), and extrapolation to $\lambda = 0$ gives a good estimate for the location of the resonance state. Alternatively, a standard CI at $\lambda = 0$ employing the CSCF MOs of the $\lambda > 0$ calculation can be undertaken for the states in which the stabilized resonance decomposes; or more refined extrapolation methods to $\lambda = 0$ accounting for a relaxation of the molecular charge distribution upon removal of the nuclear perturbation can be employed which lead to small correction terms for the negative-ion energy, as discussed elsewhere.[148] The resonance width can also be obtained from the CI calculations provided the AO basis contains appropriate diffuse functions as the coupling between the compact component of the resonance and the scattering states in which it decomposes.[148]

Results of such treatments are given in Fig. 30 for the $^2\Pi_g$ state of N_2^-. The standard MRD-CI treatment is employed for large internuclear separations, while for smaller values of R the modified procedure is used. Without the correction for orbital relaxation, the energy values are (as expected) somewhat too high; the curve including this correction is very close to that derived in a

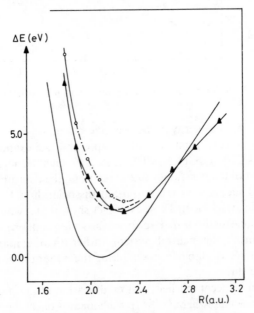

Fig. 30. Calculated potential energy curve for N_2^-: $\bigcirc$, uncorrected values; $\blacktriangle$, values obtained from an extrapolation;------, values taken from the work of Hazi[149];———, RKR curve of N_2.

completely different manner (Stieltjes imaging with core polarization[149]). The calculation of the response width can also be undertaken and comparison with other methods shows very good accord with all those treatments accounting for core polarization, as discussed elsewhere.[148] From a technical point of view, important data are as follows: the standard Dunning (9s5p) set for nitrogen in the contraction [5s3p] is employed with two d and bond s and p polarization functions; the diffuse part is described by three p functions on each nitrogen with exponents 0.08, 0.02 and 0.01 and three s (0.09, 0.025, 0.015) and one d (0.004) functions placed in the molecular center. The latter functions have been primarily included to describe the $^2\Pi_u$ $(\pi_u^3 3s^2)$ resonance state of N_2.

A further example is the potential energy curve of the $^2\Pi_g$ state of isoelectronic C_2H_2 displayed in Fig. 31 for a change in the symmetric stretching coordinate v_2.[150] In this instance the negative state is above the $X^1\Sigma_g^+$ ground-state curve in the entire range considered. The minimum in the negative system is as expected (due to occupation of the antibonding π_g MO) at a somewhat larger (0.065 Å) C—C distance than in the $X^1\Sigma_g^+$ ground state and is calculated to lie 2.9 eV (measured as $E(F_0^-) - E(R_0)$) above the ground-state minimum. This value can be compared to results of the most recent measurements of vibrational excitation of acetylene by electron collision in the 0–3.6 eV range[151] which place the $^2\Pi_g$ resonance at 2.6 eV. The calculated resonance width also seen in the figure is considerably larger than in N_2^- ($^2\Pi_g$), a finding that is also consistent with the experiments since the measured cross-

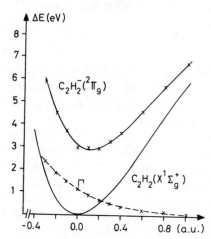

Fig. 31. Calculated potential energy curve for the C–C stretching coordinate for the neutral C_2H_2 molecule and its negative ion. The calculated width of the resonance state is also given.

sections do not show the vibrational structure known for the corresponding collision processes in N_2.

In summary, then (note that the curves in CO_2 and CF_2Cl_2 are also obtained in this manner), there is enough evidence that potential energy surfaces for negative-ion states of reasonably large molecules can be handled by essentially the same, although somewhat modified methods as bound ground and excited states of molecules, and this achievement is expected to open important new areas of applications for *ab initio* treatments, in particular in connection with studies of reaction processes.

VI. FINE-STRUCTURE EFFECTS

In the standard calculation of potential energy surfaces, only the multiplicity of a given state is considered but not the splitting between various spin multiplets of the same state. This is sufficient for the study of a large number of problems. In certain instances, however, the location and character of the individual spin components are required because (a) the splitting is relatively large, as in heavy atoms, or (b) the reactive behavior is different for the different components, or (c) the radiative lifetime depends on the particular spin component, or (d) the splitting changes considerably with nuclear conformation so that it affects the spectroscopic levels, for example. Present-day state-of-the-art calculations are also able to cope with these problems. The major part of the splitting is a consequence of the spin–orbit interaction. In all studies to be discussed, the spin–orbit Hamiltonian is taken in the Breit–Pauli form

$$H_{so} = \frac{e^2\hbar^2}{2m^2c^2}\left(\sum_i \sum_\alpha \frac{Z_\alpha}{r_{i\alpha}^3}(\mathbf{r}_i \times \mathbf{p}_i)\mathbf{s}_i - \sum_i \sum_{j \neq i} \frac{(\mathbf{r}_{ij} \times \mathbf{p}_i)(\mathbf{s}_i + 2\mathbf{s}_j)}{r_{ij}^3}\right)$$

In principle this contribution must be added to the Hamiltonian of Eq. 4. If the spin–orbit effect is small compared to the electronic contribution, it can be treated by perturbation theory, and evaluation of the various matrix elements $\langle \Psi_A | H_{so} | \Psi_B \rangle$ is required. This procedure has been followed in most applications, in which the electronic wavefunctions Ψ_A and Ψ_B are MRD-CI expansions of the order of 1000 to 10 000 depending on the problem at hand. In a few cases, the energies have actually been obtained by diagonalizing a secular equation corresponding to $H_e + H_{so}$ whereby the solutions of the unperturbed problem have been chosen as functional basis.[152,153] Such a treatment can also give an indication of the convergence of the perturbation sum, which is the main uncertainty in the first (perturbation) approach. Initial results have been obtained by employing a spin–orbit integral program over Gaussian lobe functions,[153] and for more recent studies the integral package employing Cartesian Gaussians as described by Chandra and Buenker[154] is used. All spin–orbit integrals have been evaluated explicitly.

TABLE XV

Calculated spin–orbit splittings (in cm^{-1}) in various molecules at their equilibrium conformation and comparison with measured data.[a]

Molecule	State	Calc.	Remarks[b]	Expt
HF^+	$X^2\Pi$	-280	Core $= 0$	-292.6
		-281	Core $= 1$	
HCL^+	$X^2\Pi$	-617	Core $= 0$	-648.1
		-617	Core $= 2$	
		-574	Core $= 5$	
C_2^-	$A^2\Pi$	-22	Core $= 0$	-24^c
CSi^-	$A^2\Pi$	-68	Core $= 0$	
		-68	Core $= 2$	$-$
		-64	Core $= 6$	
Si_2^-	$A^2\Pi$	-100	Core $= 2$	$-$
		-98	Core $= 10$	$-$
H_2S^+	$^2\Pi_u$	-418.4	Perturb. theory	-432.5 for HS^+
		-418.5	H_{so}-CI	
BO	$A^2\Pi_r$	-120.5	Perturb. theory	$-122.3, -116.7$
		-120.6	H_{so}-CI	$-123.13, -122.2$
OH	$X^2\Pi_i$	-138.0	First-order perturb	$-139.2, 139, 7$
		-138.1	Higher-order perturb.	
SH	$X^2\Pi_i$	-378.4	Perturb. theory	-378.5
S_2	$X^3\Sigma_q^-$	23.4	H_{so}-CI	23.6
O_1	$X^3\Sigma_q^-$	3.93	$H_{so} + H_{ss}$, perturb.	3.97
SO	$X^3\Sigma^-$	10.86	$H_{so} + H_{ss}, H_{so}$-CI	10.56

[a] Unless mentioned otherwise, the experimental data are from Ref. 53.
[b] Core refers to the number of orbitals held doubly occupied in the CI treatment. H_{so}-CI refers to a variational–perturbation treatment in which a large number of states are considered and the eigenvalues stem from solving a secular equation.[155]
[c] From Ref. 173.

Typical calculated spin–orbit splittings are contained in Table XV, and comparison with measured data, as far as available, shows that the theoretical values are reliable. The table also shows a few variants in the technical treatment of the CI, which are self-explanatory. As mentioned in Section II, the contribution of the doubly occupied 1s or 2s shell is negligible while the 2p shell of second-row atoms shows enough influence to change the spin–orbit results by a few per cent if this shell is kept doubly occupied in the CI procedure.

The spin–orbit splitting and structural data for C_2^- had been predicted by *ab initio* computations ($X^2\Sigma_g^- - A^2\Pi_u = 0.44$ eV, $R_e = 1.32$ Å and $\omega_e = 1653$ cm^{-1}) and have since been confirmed by experimental techniques ($T_e = 0.50 \pm 0.02$ eV, $R_e = 1.313$ Å and $\omega_e = 1656 \pm 20$ cm^{-1}). So far no experimental information is available for the negative ions of CSi and Si$_2$. Calculations predict an EA for both in the region of 1.8–2 eV, i.e. 1.4 eV smaller than in C$_2$. The first excited state $A^2\Pi$ in CSi is predicted to lie 0.38 eV above the $X^2\Sigma^+$ ground state, but unlike in C$_2^-$ no third stable state is found. In Si$_2^-$ both low-energy states $^2\Sigma_g$ and $^2\Pi_u$ are found at almost equal energy.

One of the advantages of the theoretical treatment is that the splittings can be obtained not only at the potential minima but also with the same accuracy in the other regions of geometrical variables; this can be achieved by spectroscopic measurements only if higher vibrational levels are populated. An example for possible comparison is the $^3\Sigma_g^-$ state of S$_2$. The behavior of the splitting is derived experimentally as

$$\lambda_v(\text{expt}) = 11.82 + 0.05(v + \tfrac{1}{2}) + 0.000\,24(v + \tfrac{1}{2})^2$$

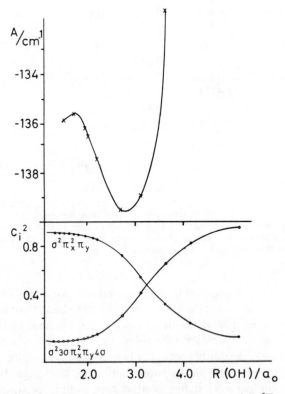

Fig. 32. Calculated spin–orbit splitting parameter of the $^2\Pi$ state of OH (upper part) and weight of the dominant contributions in the CI expansion.

and is reproduced very well by the corresponding formula for the theoretical work[155] which is

$$\lambda_v(H_{so}\text{ calc}) = 11.70 + 0.055(v + \tfrac{1}{2}) + 0.0011(v + \tfrac{1}{2})^2$$

where the contribution of the spin–spin splitting to λ_v is probably negligible. A similar example is the ground state $X^2\Pi_g$ of O_2^+, for which the splitting has been calculated over a large range of internuclear separations.[128] A distinct change in the matrix element $\langle H_{so} \rangle$ is usually observed if a change in the CI wavefunction occurs. This is seen in Fig. 32 for the $X^2\Pi_i$ state of OH[129] where, around an internuclear O–H separation of 3.0, the configuration $3\sigma^2\pi^3$, which has been dominant for small separations, is exchanged for $3\sigma\pi^34\sigma$ occupation, which allows dissociation into $O(^3P) + H(^2S)$. It is furthermore obvious that in such cases an SCF evaluation of the multiplet splitting is not sufficient but that a multi-reference CI wavefunction is required for a proper description. This can be seen quite instructively from Fig. 33 in which the value $A = 2\langle ^2\Pi_x|H_{so}|^2\Pi_y \rangle$ for the $X^2\Pi$ state of SiH is plotted as obtained from a simple CI(SD) and from a proper treatment with various reference configurations.[122] While the potential energy surface itself only rises too steeply at larger Si–H bond separations in the single-reference treatment, the decrease in

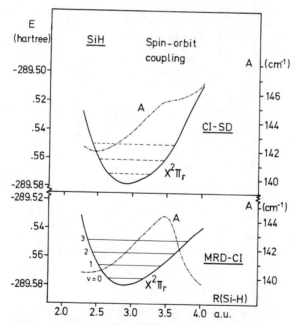

Fig. 33. Comparison of calculated spin–orbit splitting in the $X^2\Pi_r$ state of SiH by employing a single- and a multi-reference CI treatment.

the splitting is not given at all by such a simple CI procedure based on single-reference configuration alone.

Very drastic examples for large changes in the spin–orbit matrix element with change in configurations around avoided crossings have been found in BeF^{156} and in $O_2{}^{157,158}$ as shown in Fig. 34. The wavefunction for the $X^3\Sigma_g^-$ is dominated by the same π_g^2 configuration in the range of 0–0 bond lengths considered but the first two $^1\Pi_g$ and $^3\Pi_g$ states exchange their dominant character as seen from Fig. 35, and this behavior is entirely reflected in the $\langle\, ^{1,3}\Pi_g | H_{so} | ^3\Sigma_g^- \rangle$ matrix elements, which are of importance in evaluating the radiative lifetime (in the dipole-forbidden process) of the $c^1\Sigma_u^-$, $A^3\Delta_u$ and $A^3\Sigma_u^+$ states. The lowest $^3\Pi_g$ state, for example, undergoes two avoided crossings around 2.2 a.u. and 3.1 a.u., and this pattern is enhanced in the respective matrix element, which has values around $-10\,cm^{-1}$ for distances smaller than 2.2 a.u., around $+80\,cm^{-1}$ until around 3.2 a.u. and then falls off

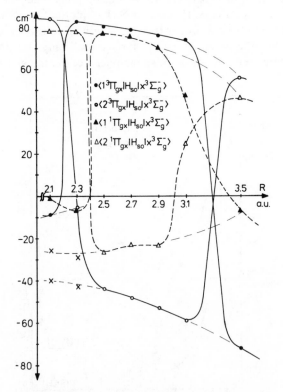

Fig. 34. Calculated spin–orbit splitting matrix elements for various states in O_2. The strong variation is due to various avoided crossings (Fig. 35). The matrix element resulting from the corresponding configurations is also indicated and varies relatively smoothly.

to $-70\,\mathrm{cm}^{-1}$ for larger bond lengths. It should be noted, however, that the change in the matrix element for a given configuration varies only gradually, as also indicated in Fig. 35.

In all the examples presented so far, the multiplet splitting has been followed in the area around the minimum of a given state. It is also of interest to look at

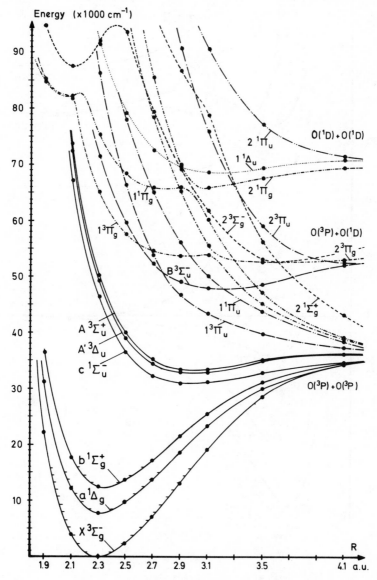

Fig. 35. Calculated potential energy curves for O_2.

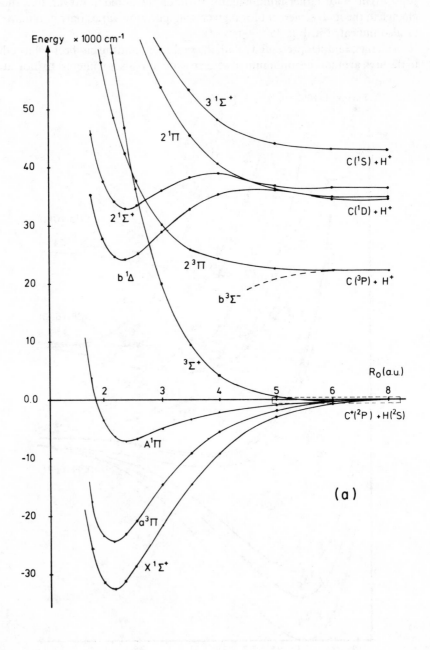

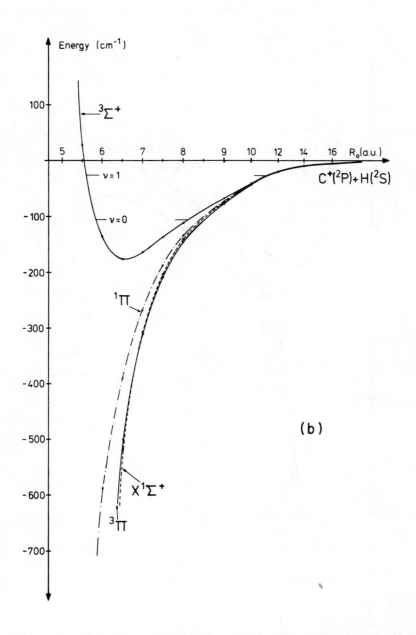

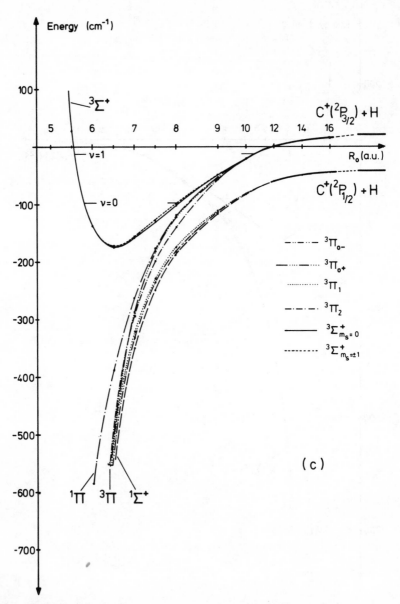

Fig. 36. (a) Calculated potential energy curves for CH^+. The calculated dissociation behavior (c) with and (b) without spin–orbit splitting is enlarged.

this effect in the area of dissociation, in order to find out the way in which the various states correlate with the multiplet structure of the corresponding atoms. This might be of considerable importance for gas-phase reactions, as for example found in rare-gas dimers,[159] in which various atomic multiplets show different reactive behavior. It should also be pointed out that this dissociative behavior cannot always be determined from symmetry rules alone.

Results of such work are seen in Fig. 36 for CH^+. Figure 36a shows the global picture of the potential energy curves of the system; Figs. 36b and 36c show an enlargement of the curves in the first dissociation channel between 5.0 and 16.0 a.u. In the standard treatment without spin–orbit interaction all curves go to the same limit $C^+(^2P) + H(^2S)$. The spin–orbit matrix elements between the various states have been calculated for a series of large internuclear separations (at very large distances they approach the same value) and, together with the corresponding calculated energy differences (which become zero at the dissociation limit), the perturbation matrix has been constructed by transforming to Ω eigenfunctions and has then been diagonalized. These results are plotted in Fig. 36c. The curves in the dissociation channel are split by the spin–orbit contribution into two levels corresponding to $C^+(^2P_{1/2})$ and $C^+(^2P_{3/2})$ and the calculated energy difference is 60.5 cm^{-1} compared to the experimentally known ^2P splitting of the C^+ ground state of 64 cm^{-1}. The molecular states $^1\Sigma_{0+}^+$, $^3\Pi_1$ and $^3\Pi_{0-}$ correlate with $^2P_{1/2}$ while the states $^3\Sigma_{0-}$, $^3\Sigma_1$, $^1\Pi_1$, $^3\Pi_2$ and $^3\Pi_{0+}$ correlate with the higher $^2P_{3/2}$ component. The very shallow minimum of the $^3\Sigma^+$ state is influenced most by the treatment allowing for spin–orbit interaction.

Similar studies have been made for HF^+ and SH. The (inverted) splitting in the $H^+ + F(^2P)$ channel (correlating with $^2\Sigma^+$ and $^2\Pi$) is calculated[160] as -393 cm^{-1} compared to the measured splitting of 404 cm^{-1}. In SH the molecules dissociates into $S(^3P) + H(^2S)$ and the calculated and measured data for the $^3P_0$, $^3P_1$ and $^3P_2$ splittings are also in similar good accord as the values for the Se atom already mentioned in Section II. The latter belongs to the 'heavier' atoms for which the spin–orbit contribution becomes quite sizeable. As indicated in Fig. 37 the first-order contribution is not enough in this case. It splits $^3P_1$ and $^3P_2$ symmetrically, and the calculated value of $2A$ is 1816 cm^{-1}. The second-order effect lowers $^3P_2$ due to interaction with $^1D_2$ by 140 cm^{-1}, while the $^3P_0$ component is affected by the $^1S_0$ state and lowered by 311 cm^{-1}. The combined results are 1955 cm^{-1} (1989 cm^{-1} expt) for $^3P_2 - ^3P_1$ and 597 cm^{-1} (545 cm^{-1} expt) for $^3P_0 - ^3P_1$ splitting.[161]

This last example shows that larger splittings as they occur in heavy atoms can also be treated by the theoretical methods. In such cases, first-order corrections, which have often been found to be sufficient to describe spin–orbit effects in the area of potential minima in light molecules, have to be

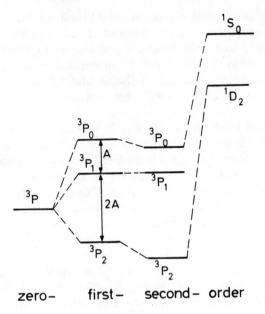

Fig. 37. Schematic picture showing the contribution of
states to the ^3P splitting in the Se atom.

supplemented by higher-order corrections because of the close neighborhood
of other states. This can be done most efficiently by including the various
components in a matrix blocked according to Ω (or J) eigenfunctions. In
bromine, for example, ^4P_g and ^2P_g states are relatively close in energy so that
$^4P_{3/2}$ and $^2P_{3/2}$ components are influenced quite heavily by one another just as
are $^4P_{1/2}$ and $^2P_{1/2}$.[162] A first-order treatment alone would find an ordering of
$^4P_{5/2}$, $^4P_{3/2}$, $^2P_{3/2}$, $^4P_{1/2}$, $^2P_{1/2}$, with a large splitting between the first two,
while an H_{so}-CI between the states influences the $^4P_{3/2}$ and $^2P_{3/2}$ most and
places the $^2P_{3/2}$ correctly above $^4P_{1/2}$ and reduces the energy gap between
$^4P_{5/2}$ and $^4P_{3/2}$ considerably, as shown in Ref. 162.

In summary, then, experience of the past few years has shown that the fine
structure of potential energy surfaces due to spin–orbit splitting can also be
treated in multi-reference CI procedures. The examples chosen are again
primarily from diatomics, since the most reliable experimental data are
available for those, and in the first stages of a theoretical development
comparison with measured data is essential. In the meantime, enough
representative material has been accumulated so that the method can be
applied in the future to a large area of applications and the same reliability of
the calculated results as shown in the various examples discussed in this
section is to be expected.

VII. RELATION BETWEEN CALCULATED AND MEASURED DATA

In comparing calculated results with measured data, one has to be certain that there is a strict one-to-one correspondence between the two types of results. Spectroscopic measurements give energy differences directly, but further data such as equilibrium distances and angles, R_e, θ_e, zero-point frequencies ω_e, anharmonicities $\omega_e x_e$ and rotational constants B_e (commonly referred to as 'spectroscopic constants') are all secondary data for the condensed characterization of a potential surface and are *derived* from the measured results on the basis of a given model. The same holds for so-called experimental potential energy surfaces, which are generally derived from the vibrational pattern of a spectrum or from features in the differential or total cross-section of atom–molecule scattering experiments. The data reduction is usually achieved by fitting the experimental results on the basis of an assumed potential form containing several free parameters. Such a procedure is especially problematic for surfaces in polyatomic systems in which not only the potential term of the model hamiltonian $H = T + V$ but also the kinetic energy operator often contains free parameters or at least is only approximate, so that the resultant V is strictly speaking only the potential surface under the assumption of the particular model Hamiltonian employed, and could easily vary if the form of T is modified somewhat. For diatomics, the conventional RKR (Rydberg–Klein–Rees) method is also favored, which constructs the classical turning points from the vibrational spectrum and assumes them to be equivalent to the potential data.

In light of this, the best comparison is made with the original data, which means that in addition to the solution of the electronic Schrödinger equation it is necessary also to solve the Schrödinger equation for the nuclear motion in order to obtain the vibrational (rotational) energy levels and wavefunctions. This is an easy matter for diatomics but causes problems in polyatomics, in particular if large-amplitude motion (as in many excited states) and coupling of various modes occur. The difficulty is not so much in the potential energy term, which must be fitted for further usage in some analytical form from the calculated potential energy points, but rather the form of the kinetic energy operator, which is available only for certain categories of nuclear arrangements.

A. Potential Energy Surface Characterization

The computed potential energy surfaces are often expanded in polynomials of nth ($n > 3$) degree and the minimum R_e, the zero-point frequency ω_e, the anharmonicity $\omega_e x_e$ (and corresponding terms in the other dimensions) and other terms are determined from the various derivatives of the polynomial:

$$V(R) = V(R_e) + (1/2!)(\partial^2 V/\partial R^2)_{R_e}(R - R_e)^2 + (1/3!)(\partial^3 V/\partial R^3)_{R_e}(R - R_e)^3$$
$$+ (1/4!)(\partial^4 V/\partial R^4)_{R_e}(R - R_e)^4 + \cdots \tag{14}$$

or

$$V(\rho) = V(0) + (1/2)f_2\rho^2 + (1/6)f_3\rho^3 + (1/24)f_4\rho^4 + \cdots \tag{15}$$

where $\rho = R - R_e$ is the displacement from equilibrium.

From measurements, it has been found that the energy of a rotational and vibrational level with quantum numbers J and v can be best represented as

$$E(v, J)/hc = T(v, J)$$
$$= \omega_e(v + \tfrac{1}{2}) - \omega_e x_e(v + \tfrac{1}{2})^2 + \cdots + [B_e - \alpha_e(v + \tfrac{1}{2}) + \cdots]J(J + 1)$$
$$+ [D_e + \beta_e(v + \tfrac{1}{2}) + \cdots]J^2(J + 1)^2 + \cdots \tag{16}$$

Even though there are formulas[163] to relate the quantities f in Eq. (15) with those of Eq. (16), as for example

$$f_2 = 4\pi^2 m\omega_e^2 c^2$$
$$f_3 = -3f_2/R_e(1 + \alpha_e\omega_e/6B_e^2) \tag{17}$$
$$f_4 = f_2/R_e^2[15(1 + \alpha_e\omega_e/6B_e^2)^2 - 8\omega_e x_e/B_e]$$

it is clear that the coefficients themselves in both expansions depend on the order of the polynomial, since the basis functions (x^n) in the expansion are not orthogonal. Furthermore, extrapolation to $v = 0$ from high vibrational quantum numbers (in case the lower are not known) will not always result in the same values ω_e and $\omega_e x_e$ because of uncertainties in vibrational numbering, as if lower quantum numbers are employed. Considering all this, it should be clear that one cannot expect that quantities from an nth-order polynomial via Eqs. (15) and (17) coincide with those derived from a series of term values $T(v, J)$ via Eq. (16).

If one is interested in a strict comparison, the more parallel procedure is to generate the theoretical vibrational and rotational energy levels $E(v, J)$ on the basis of the computed potential V and then use these data (if possible for the same values v, J as employed in the experimental extrapolation) to derive the quantities according to Eq. (16). This has been done for PH and corresponding results are given in Table XVI and show what one would call (according to experience in such comparisons) an overall 'very good agreement'. The best documented data are naturally available for the ground state. Since the $A^3\Pi$ state's potential curve is relatively flat, only a few vibrational levels could be obtained, and hence the uncertainty in the extrapolation is greater than for the other three states. The table shows also another commonly occurring feature, namely that constants for excited states are often not known at all from experiment.

TABLE XVI

Comparison of spectroscopic data for the low-lying states of PH obtained from theoretical and experimental work. For more detail see the original reference.[168]

State		R_e(Å)	B_e(cm^{-1})	B_0(cm^{-1})	α_e(cm^{-1})	ω_e(cm^{-1})	$\omega_e x_e$ (cm^{-1})
$X^3\Sigma^-$	Theor	1.4226	8.533	8.406	0.254	2380	43.9
	Expt	1.4223	8.537	8.411	0.251	2365.2	44.5
$a^1\Delta$	Theor	1.4178	8.591	8.457	0.269	2415	48.0 ± 3
	Expt	1.4303(R_0)	–	8.440	–	–	—
$b^1\Sigma^+$	Theor	1.4141	8.637	8.507	0.260	2438	48 ± 2
	Expt	–	–	—	–	–	–
$A^3\Pi_i$	Theor	1.4420	8.306	8.081	0.45	2071	80, 87.8
	Expt	1.4458	8.259	8.023	0.473	2030.6	90.5
	Expt	1.4673(R_0)	–	8.019	–	–	–

A supposed disagreement between theoretical and measured data has also occurred in the literature for some time for the BH_2 molecule, but a better analysis of what must actually be compared plus a renumbering of vibrational levels removes the discrepancy entirely. Transitions occur between the bent X^2A_1 and the upper (linear) $^2B_1$ ($^2\Pi$) state and are seen in the 6400–8700 Å region. On the basis of an assumed numbering of upper bending levels $v_2' = 7, 9, 11$ and the representation

$$G(v_2') = \omega_2'(v_2' + 1) + x_{22}'(v_2' + 1)^2 + T_{00}$$

Herzberg and Johns[164] found $\omega_2' = 954.65 \, \text{cm}^{-1}$, $x_{22}' = -1.0 \, \text{cm}^{-1}$ and $T_{00} = 4194.1 \, \text{cm}^{-1}$. The electronic energy difference between $^2A_1$ and $^2B_1$ would then amount to $T_e = T_{00} + \frac{1}{2}\omega_2'' = 4709 \, \text{cm}^{-1}$, whereby the ground-state bending frequency is taken to be $\omega_2'' = 1030 \, \text{cm}^{-1}$. This T_e value does not fit with data obtained from ab initio calculations, i.e. a CI value of $2260 \, \text{cm}^{-1}$,[165] an IEPA value of $2740 \, \text{cm}^{-1}$,[166] and an MRD-CI and estimated full CI value of 2712 and 2675 cm^{-1} respectively.[167] The independent electron-pair approach (IEPA) study challenged the T_{00} value of 419.1 cm^{-1} and put it at 3035 cm^{-1} by taking all calculated zero-point energies into account but applying the incorrect formula $E_0 = T_e + \frac{1}{2}\omega_2'$ for the upper-state bending, overlooking the fact that in the two-dimensional oscillator the lowest quantum number is $v_2' = 1$ and not $v_2' = 0$ and hence $G(v_2' = 1) = 2\omega_2'$. All discrepancies disappear if the alternative numbering $v_2' = 9, 11, 13$ is chosen for the observed lines, which leads to a T_{00} value of 2285 cm^{-1} and a T_e value of 2800 cm^{-1} (under the assumption of harmonic approximation and the same stretching frequencies for upper and lower states), in reasonable accord with the theoretical energy difference between the potential minima. An actual calculation of all bending levels represents the measured lines in the BH_2 spectrum quite well. This example again shows that values derived from measured data should not always be taken at their face value for comparison with *ab initio* results but should be further analyzed if it seems necessary.

B. Averaging of Properties

Measured properties refer to a special vibrational (rotational) level. This has to be taken into account if the calculated data are required with high accuracy. For this reason, the vibrational wavefunctions and levels have to be determined from the potential energy surface, and the respective properties have to be averaged over the corresponding vibrational functions.

The fact that there is a difference between R_0 and R_e is well known. In ground-state potentials with deep minima and almost harmonic behavior, this difference is relatively small, however. On the other hand, excited-state potentials are often very shallow and anharmonic, in particular if mixing of

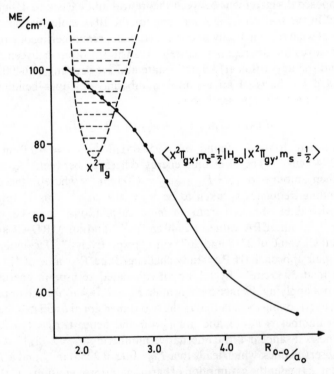

Fig. 38. Variation of the $^2\prod_g$ spin–orbit matrix element in O_2^+ with internuclear separation. The corresponding potential energy curve is also indicated.

states has taken place. In such cases, large-amplitude vibrations occur, and one has to expect considerable differences between properties taken at the potential curve minimum and those for the first vibrational level. Because of the difficulty in calculating vibrational wavefunctions for general polyatomic molecules, it may happen that a disagreement between a measured and a properly averaged computed quantity occurs, which does not necessarily reflect an error in the calculated potential but rather in the vibrational treatment.

The dependence of properties such as transition moments of spin–orbit splittings on an internuclear coordinate is determined experimentally from the corresponding quantities measured as a function of vibrational quantum number v. A small dependence on v is normally interpreted as a small variation of the property with the interatomic coordinate. This may not necessarily be the case, as is seen from Figs. 38 and 39. The first shows the calculated spin–orbit matrix element for the $X^2\Pi_g$ ground state of O_2^+ (Ref. 128) as a function of the internuclear separation, and a strong dependence on this variable is obvious. Nevertheless, an averaging of this quantity over the vibrational wavefunctions in the $X^2\Pi_g$ potential also displayed in Fig. 38 results in a splitting constant A_v that is only slowly varying with v, as obvious from Fig. 39. The calculated curve is thereby roughly parallel to the measured data. The absolute value is somewhat smaller but this is not surprising since only a relatively limited AO basis has been taken for the computation of the spin–orbit matrix element. A somewhat larger AO basis improves the value

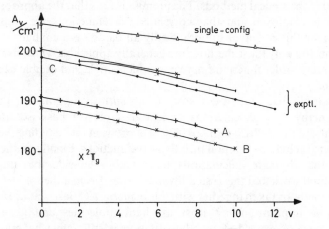

Fig. 39. Calculated spin–orbit splitting in the $^2\Pi_g$ state of O_2^+ as a function of the vibrational quantum numbers by employing the matrix element displayed in Fig. 38. Curve A refers to the calculation that employs the *ab initio* potential curve for vibrational averaging; B employs the experimental RKR curve for vibrational averaging, and C shows the experimental points from various measurements.

markedly, as also seen in the figure. Furthermore, it is seen that the form of the potential curve is also critical, since the values obtained by employing the computed potential, on the one hand, and those of an RKR curve, on the other, do show differences for A_v of the order of $2\,\text{cm}^{-1}$ even though the same $\langle H_{\text{so}} \rangle$ matrix element is taken in both cases.

Such has also been the experience in the calculation of the Λ-doubling for OH and SH.[129] The p_0 and q_0 values can vary between $0.3104\,\text{cm}^{-1}$ and $0.3374\,\text{cm}^{-1}$ (-0.0094 and -0.01107 for q_0) for example, depending on which potential curve is taken for SH; experimental data are $0.3004\,\text{cm}^{-1}$ and $-0.0094\,\text{cm}^{-1}$. Since quite often spin–orbit properties are derived from a smaller AO basis (simply because of computer time expenditure) than potential energy curves, it is important to realize that distortions of this magnitude as given in the examples can occur if potentials of lower accuracy are chosen.

In summary, then, the few examples given in this chapter should show that a careful use of potential surfaces is necessary in order to derive the physical data; alternatively, it should also be kept in mind that special procedures might have been employed in deducing data from a large body of measurements.

VIII. CONCLUSIONS

In the preceding sections it has been shown that potential energy surfaces of small polyatomic molecules, positive and negative ions and radicals can be obtained quite reliably over the entire range of geometrical variables by present-day theoretical methods. The premise is thus that the appropriate AO basis set is employed and that correlated wavefunctions are constructed. Methods for this endeavor are practically all based on a configuration-type expansion and only those methods are generally applicable which use a multi-reference zero-order function relative to which single and double excitations are taken.

If one looks back into the history of *ab initio* computations, the 1960s showed many SCF calculations and studies of AO basis sets for such treatments. In the 1970s, it looks as if most effort went into devising theoretical methods to include correlation into the wavefunctions. Finally, in the present decade, the *ab initio* calculations have had a considerable impact on experimental work and this is steadily increasing. From a theoretical point of view, the methods taken together with the available AO basis sets are adequate for routine applications, at least for light molecules and ground-state phenomena. For excited states, the situation is still somewhat more complicated; the calculations need a more careful design, and good AO basis sets for excited atoms are not available in standard reference data sets. Hence the bottleneck, which still does not allow routine excited-state calculations lies primarily in the unavailability of appropriate excited-state AO basis sets, but as time goes on it is hoped that this situation will be remedied so that not every calculation requires first a basis set optimization. For heavier systems beyond the second and third rows, however, the usual electrostatic Hamiltonian is not sufficient any more, and relativistic effects—at the level of the kinetic, potential energy and spin–orbit operator—have to be accounted for. There seems to be good progress in this area so that in the not-too-distant future similar calculations as are now possible for lighter atoms will become frequent.

Finally, there is a wide area in molecular physics and chemistry in which excited-state potential surfaces can be employed for the study of phenomena. Many spectroscopic investigations to date are already coupled with *ab initio* computations. A typical examples is the study of the Cl_2 molecule,[40,140,169] for which the computed potential curves were the main guideline to decipher the spectrum. Others are C_2H, whose spectrum has been analyzed a number of years after the first appearance of the theoretical data, or HCN, which has recently been reinvestigated[170] with the main conclusion being that the calculations which about 10 years ago had denied the existence of a β state were correct. A similar situation has recently been reported for the SiN radical[171] in which calculations showed that the so-called $K\,^2\Sigma$ state in reality involves only a vibrational progression of the low-lying $B^2\Sigma$ state. The assignment of photoelectron spectra (PES) is also helped very much by *ab initio* predictions, as exemplified a long time ago by the complex PES of ethane[172] and more recently by the PES of NCO. Reaction mechanisms, such as ion–molecule or atom–molecule recombinations or charge-exchange processes, rely these days heavily on the availability of *ab initio* potential energy surfaces, and finally, even work on thermochemical processes involving radicals in low-lying excited states has been made in direct cooperation with *ab initio* computations. The list of pertinent examples could be extended quite far to include a large number of systems. The important observation is that, as time goes on, the theoretically predicted potential surfaces and their interaction become more and more detailed, in particular with the possibility of including non-adiabatic and spin–orbit effects. We assume that in the foreseeable future the calculation of excited-state phenomena will develop more and more as just another instrument to be added to experimental equipment in the laboratory for the study of phenomena in chemistry and molecular physics.

Acknowledgements

The authors wish to thank R. J. Buenker for his collaboration over many years, which has been essential to many of the results presented in this work. They thank their coworkers and students, in particular C. M. Marin, R. Klotz and R. de Vivie who are responsible for a large portion of the spin–orbit work, B. Nestmann and V. Krumbach for their contribution to the electron–

molecule scattering problem and H. Dohmann and D. Stärk for many calculations involving aspects of the charge-exchange reactions. M. Bettendorff, S. Sczech, U. Meier and many others have also contributed to the present results. Thanks are due to those who installed EARN and BITNET, which was the major communication line between Canada and Germany in the writing of this article.

References

1. Löwdin, P. O., and Pullmann, B. (Eds.), *New Horizons of Quantum Chemistry*, Reidel, Dordrecht, 1983.
2. *Faraday Symp. Chem. Soc.*, **19** (1983).
3. Murrel, J. N., Carter, S., Farantos, S. C., Huxley, P., and Varandas, A. J. C., *Molecular Potential Energy Functions*, John Wiley and Sons, Chichester, 1984.
4. Hirst, D. M., Potential Energy Surfaces, *Molecular Structure and Reaction Dynamics*, Taylor and Francis, London, 1985.
5. Berkowitz, J., and Groeneveld, K. O. (Eds.), *Molecular Ions*, Plenum, New York, 1983.
6. Buenker, R. J., *Gazz. Chim. Ital.*, **108**, 245 (1978).
7. Hirsch, G., Bruna, P. J., Buenker, R. J., and Peyerimhoff, S. D., *Chem. Phys.*, **45**, 335 (1980).
8. Buenker, R. J., Hirsch, G., Peyerimhoff, S. D., Bruna, P. J., Römelt J., and Bettendorff, M., in *Current Aspects of Quantum Chemistry* (Ed. R. Carbo), Studies in Physical and Theoretical Chemistry, Vol. 21, pp. 81, Elsevier Science, Amsterdam, 1982.
9. (a) Poirier, P., Kari, R., and Csizmadia, I. G., *Handbook of Gaussian Basis Sets*, Physical Sciences Data 24, Elsevier, Amsterdam, 1985.
   (b) Huzinaga, S., *Gaussian Basis Sets for Molecular Calculations*, Physical Sciences Data 16, Elsevier, Amsterdam, 1984.
   (c) Roos, B., and Siegbahn, P. E. M., *Theor. Chim. Acta*, **17**, 199 (1970).
   (d) Frisch, M. J., Pople, J. A., and Brinkley, J. S., *J. Chem. Phys.*, **80**, 3265 (1984); recommend f function exponents, see also their Ref. 13 for negative ions.
10. Dunning, T. H., and Hay, P. J., in *Modern Theoretical Chemistry* (Ed. H. F. Schaefer III), Vol. 3, p. 1, Plenum, New York.
11. Strömberg, A., Wahlgren, U., Pettersson, L., and Siegbahn, P. E. M., *Chem. Phys.*, **89**, 323 (1984).
12. Jankowski, K., Becherer, R., Scharf, P., Schiffer, H., and Ahlrichs, R., *J. Chem. Phys.*, **82**, 1413 (1985).
13. Weiss, A. W., *Correlation in Excited State Atoms*, Adv. At. Mol. Phys., Vol. 9, p. 1, Academic Press, New York, 1973.
14. Fischbach, U., Buenker, R. J., and Peyerimhoff, S. D., *Chem. Phys.*, **5**, 265 (1974).
15. Dohmann, H., this laboratory.
16. Veillard, A., *Theor. Chim. Acta*, **12**, 405 (1968).
17. McLean, A. D., and Chandler, G. S., *J. Chem. Phys.*, **72**, 5639 (1980).
18. Spiegelmann, F., and Malrieu, J. P., *Chem. Phys. Letters*, **57**, 214 (1978).
19. (a) Peyerimhof, S. D., Buenker, R. J., and Allen, L. C., *J. Chem. Phys.*, **45**, 734 (1978).
   (b) Huzinaga, S., *J. Chem. Phys.*, **42**, 1293 (1965).
   (c) Dunning, T. H., *J. Chem. Phys.*, **55**, 3958 (1971).

20. Schindler, M., Schlier, Ch., and Teloy, E., *Symp. on Atomic and Surface Physics*, p. 122, Maria Alm, Salzburg, 1984.
21. Stärk, D., *Diplomarbeit*, University of Bonn, 1985.
22. Banichevich, A., this laboratory.
23. Bruna, P. J., and Anglada, J., *Quantum Chemistry: The Challenge of Transition Metals and Coordination Chemistry* (Ed. A. Veillard), NATO Adv. Res. Workshop, Strasbourg, 1985 in press.
    See also Anglada, J., Bruna, P. J., Peyerimhoff, S. D., and Buenker, R. J., *J. Mol. Struct.* (*Theochem*), **93**, 299 (1983); *ibid.*, **107**, 163 (1984).
24. Klotz, R., Marian, C. M., Peyerimhoff, S. D., Hess, B. A., and Buenker, R. J., *Chem. Phys.*, **76**, 367 (1983).
25. Matsushita, T., Marian, C. M., Klotz, R., Hess, B. A., and Peyerimhoff, S. D., *Chem. Phys.*, **96**, 371 (1985).
26. Dunning, T. H., Jr., *Chem. Phys.*, **66**, 1382 (1972); Appendices I and II.
27. Buenker, R. J., Chandra, P., and Hess, B. A., *Chem. Phys.*, **84**, 1177 (1984).
28. Shih, S. K., Peyerimhoff, S. D., Buenker, R. J., and Perić, M., *Chem. Phys. Lett.*, **55**, 206 (1978).
29. Anglada, J., Bruna, R. J., Peyerimhoff, S. D., and Buenker, R. J., *J. Phys. B: At. Mol. Phys.*, **16**, 2469 (1983).
30. Raghavachari, K., *J. Chem. Phys.*, **82**, 4142 (1985).
31. Feller, D., and Davidson, E. R., *J. Chem. Phys.*, **82**, 4135 (1985).
32. Lewerenz, M., Bruna, P. J., Peyerimhoff, S. D., and Buenker, R. J., *J. Phys. B: At. Mol. Phys.*, **16**, 4511 (1983).
33. Nestmann, B., and Peyerimhoff, S. D., *J. Phys. B: At. Mol. Phys.*, **18**, 615 (1985).
34. Harrison, R. J., and Handy, N. C., *Chem. Phys. Lett.*, **95**, 386 (1983).
35. Handy, N. C., Knowles, P. J. and Somasundram, K., *Theor. Chim. Acta*, **68**, 87 (1985).
36. Handy, N. C., *Faraday Symp. Chem. Soc.*, **19**, 17 (1984).
37. Bartlett, R. J., Sekino, H., and Purvis, G. D., *Chem. Phys. Lett.*, **98**, 66 (1983).
38. Hirao, K., and Hatano, Y., *Chem. Phys. Lett.*, **100**, 519 (1983).
39. Pulay, P., and Saebo, S., *Chem. Phys. Lett.*, **117**, 37 (1985).
40. Brown, F. B., Shavitt, I., and Shepard, R., *Chem. Phys. Lett.*, **105**, 363 (1984).
41. Phillips, R. A., Buenker, R. J., Bruna, P. J., and Peyerimhoff, S. D., *Chem. Phys.*, **84**, 11 (1984).
42. Buenker, R. J., and Peyerimhoff, S. D., *Theor. Chim. Acta*, **35**, 33 (1974); *ibid.*, **39**, 217 (1975).
43. Buenker, R. J., and Peyerimhoff, S. D., in *New Horizons of Quantum Chemistry* (Eds. P. O. Löwdin and B. Pullmann), p. 183, Reidel, Dordrecht, 1983.
44. Buenker, R. J., Peyerimhoff, S. D., and Butscher, W., *Mol. Phys.*, **35**, 771 (1978).
45. Burton, P. G., Buenker, R. J., Bruna, P. J. and Peyerimhoff, S. D., *Chem. Phys. Lett.*, **95**, 379 (1983).
46. Bruna, P. J., Peyerimhoff, S. D., and Buenker, R. J., *Chem. Phys. Lett.*, **72**, 278 (1980).
47. Buenker, R. J., Peyerimhoff, S. D., and Bruna, P. J., in *Computational Organic Chemistry* (Eds. I. G. Csizmadia and R. Daudel), p. 55, Reidel, Dordrecht, 1981.
48. (a) Davidson, E. R., in *The World of Quantum Chemistry* (Eds. R. Daudel and B. Pullmann), p. 17, Reidel, Dordrecht, 1974.
    (b) Langhoff, S. R., and Davidson, E. R., *Int. J. Quantum Chem.*, **81**, 61 (1974).
49. Pople, J.-A., Seeger, R., and Krishnan, R., *Int. J. Quantum Chem.*, **S11**, 149 (1977).
50. Siegbahn, P. E. M., *Chem. Phys. Lett.*, **55**, 386 (1978).

51. Jankowski, K., Meissner, L., and Wasilewski, J., *Int. J. Quantum Chem.*, **28**, 931 (1985).
52. Prime, S., Rees, C., and Robb, M. A. *Mol. Phys.*, **44**, 173 (1981).
53. Huber, K. P., and Herzberg, G., *Molecular Specta and Molecular Structure*, Vol. IV, *Constants of Diatomic Molecules*, Van Nostrand, Princeton, 1979.
54. Mulliken, R. S., *Rev. Mod. Phys.*, **14**, 204 (1942).
55. Walsh, A. D., *J. Chem. Soc.*, 2260 (1953).
56. Buenker, R. J., and Peyerimhoff, S. D., *Chem. Rev.*, **74**, 127 (1974).
57. Petrongolo, C., Bruna, P. J., Peyerimhoff, S. D., and Buenker, R. J., *J. Chem. Phys.*, **74**, 4594 (1981).
58. O'Keefe, A., Derai, R., and Bowers, M. T., *Chem. Phys.*, **91**, 161 (1984).
59. Karma, S. P., and Grein, F., *Mol. Phys.*, **56**, 641 (1985).
60. Bruna, P. J., Petrongolo, C., Buenker, R. J., and Peyerimhoff, S. D., *J. Chem. Phys.*, **74**, 4611 (1981).
61. Hirsch, G., Wuppertal, private communication to one of us (P.J.B.).
62. Kouba, J. E., and Öhrn, Y., *J. Chem. Phys.*, **53**, 3923 (1970).
63. Bruna, P. J., this laboratory.
64. Kirby, K., and Liu, B. L., *J. Chem. Phys.*, **70**, 893 (1979).
65. Zeitz, M., Peyerimhoff, S. D., and Buenker, R. J., *Chem. Phys. Lett.*, **58**, 487 (1978).
66. Chabalowski, C. F., Peyerimhoff, S. D., and Buenker, R. J., *Chem. Phys. Lett.*, **81**, 57 (1983).
67. Karma, S. P., and Grein, F., *Chem. Phys.*, **89**, 207 (1985) and private communication.
68. Dohmann, H., Bruna, P. J., Peyerimhoff, S. D., and Buenker, R. J., *Mol. Phys.*, **51**, 1109 (1984).
69. Bruna, P. J., Peyerimhoff, S. D., and Buenker, R. J., *J. Chem. Phys.*, **72**, 5437 (1980).
70. Peyerimhoff, S. D., and Buenker, R. J., *Chem. Phys.*, **72**, 111 (1982).
71. Pacchioni, G., *Mol. Phys.*, **49**, 727 (1983).
72. Pacchioni, G., *Mol. Phys.*, **55**, 211 (1985).
73. Krauss, M., and Neumann, D. B., *Mol. Phys.*, **32**, 101 (1976).
74. Verma, R. D., *Can. J. Phys.*, **62**, 569 (1984).
75. Bredohl, H., Dubois, I., Houbrechts, V., and Vzohabonayo, P., *J. Mol. Spectrosc.*, **112**, 430 (1985).
76. Bruna, P. J., Dohmann, H., and Peyerimhoff, S. D., *Can. J. Phys.*, **62**, 1508 (1984).
77. Bruna, P. J., Dohmann, H., Anglada, J., Krumbach, V., Peyerimhoff, S. D., and Buenker, R. J., *J. Mol. Struct.* (*Theochem*), **93**, 309 (1983).
78. Dohmann, H., and Krumbach, V., this laboratory.
79. Bruna, P. J., Dohmann, H., and Krumbach, V., unpublished results, this laboratory.
80. Partridge, H., Bauschlicher, C. W., and Stallcop, J. R., *J. Quant. Spectrosc. Radiat. Transfer*, **33**, 653 (1985).
81. Itoh, H., Qzahki, Y., Nagata, T., Kondow, T., and Kuchitsu, K., *Can. J. Phys.*, **62**, 1586 (1984).
82. Milligan, D. E., and Jacox, M. E., *J. Chem. Phys.*, **39**, 712 (1963).
83. Snyder, L. E., and Buhl, D., *Bull. Am. Astron. Soc.*, **3**, 388 (1971).
84. Pearson, P. K., Schaefer, H. F., III, and Wahlgren, U., *J. Chem. Phys.*, **62**, 350 (1975).
85. Perić, M., Mladenović, M., Peyerimhoff, S. D., and Buenker, R. J., *Chem. Phys.*, **82**, 317 (1983); *ibid.*, **86**, 85 (1984).
86. Meier, U., *Staatsexamensarbeit*, Bonn, 1984, and additional work.

87. Ohashi, N., Kakimoto, M., Saito, S., and Hirota, H., *J. Mol. Spectrosc.*, **84**, 204 (1981).
88. Schurath, U., Weber, M., and Becker, K. H., *J. Chem. Phys.*, **67**, 110 (1977).
89. Sears, T. J., and McKellar, A. R. W., *Mol. Phys.*, **49**, 25 (1983).
90. (a) Sannigrahi, A. B., Thunemann, K. H., Peyerimhoff, S. D., and Buenker, R. J., *Chem. Phys.*, **20**, 25 (1977).
    (b) Sannigrahi, A. B., Peyerimhoff, S. D., and Buenker, R. J., *Chem. Phys.*, **20**, 381 (1977).
91. Luke, B. T., and McLean, A. D., *J. Phys. Chem.*, **89**, 4592 (1986).
92. Buenker, R. J., Bruna, P. J., and Peyerimhoff, S. D., *Isr. J. Chem.*, **19**, 309 (1980).
93. Bruna, P. J., *Chem. Phys.*, **49**, 39 (1980).
94. Bruna, P. J., Hirsch, G., Buenker, R. J., and Peyerimhoff, S. D., in *Molecular Ions* (Eds. J. Berkowitz and K. O. Groenefeld), p. 309, Plenum, 1983.
95. Berthier, G., Pauzat, F., and Yuanqi, T., *J. Mol. Struct. (Theochem)*, **107**, 39 (1984).
96. Bruna, P. J., Peyerimhoff, S. D., and Buenker, R. J., *Chem. Phys.*, **27**, 33 (1978).
97. Frenking, G., and Schaefer, H. F., *J. Chem. Phys.*, **82**, 4585 (1985).
98. van Zee, R. J., Ferrante, R. F., and Weltner, W., *J. Chem. Phys.*, **83**, 6181 (1985).
99. Shih, S. K., Peyerimhoff, S. D., and Buenker, R. J., *J. Mol. Spectrosc.*, **74**, 124 (1979); *ibid*, **64**, 167 (1977).
100. Lischka, H., and Koehler, H. J., *J. Am. Chem. Soc.*, **105**, 6646 (1983).
101. Luke, B. T., Pople, J. A., Krogh-Jespersen, M. B., Apeloig, Y., Karni, M., Chandrasekhar, J., and von Rague Schleyer, P., *J. Am. Chem. Soc.*, **108**, 270 (1986).
102. Colvin, M. E., and Schaefer, H. F., III, *Faraday Symp. Chem. Soc.*, **19**, 39 (1984).
103. Bruna, P. J., Peyerimhoff, S. D., Buenker, R. J., and Rosmus, P., *Chem. Phys.*, **3**, 35 (1974).
104. Steer, R. P., Knight, A. R., Clouthier, D. J., and Moule, D. C., *J. Photochem.*, **9**, 157 (1978).
105. Karma, S. P., and Grein, F., *Mol. Phys.*, **57**, 939 (1986).
106. Burton, P. G., Peyerimhoff, S. D., and Buenker, R. J., *Chem. Phys.*, **73**, 83 (1982).
107. Bruna, P. J., Buenker, R. J., and Peyerimhoff, S. D., *Chem. Phys.*, **22**, 375 (1977).
108. Judge, R. H., Moule, D. C., Bruno, A. E., and Steer, R. P., *Chem. Phys. Lett.*, **102**, 385 (1983).
109. Bruna, P. J., Krumbach, V., and Peyerimhoff, S. D., *Can. J. Chem.*, **63**, 1594 (1985).
110. Smit, C. N., Lock, F. M., and Bickelhaupt, F., *Tetrahedron Lett.*, **25**, 3011 (1984).
111. Dykema, K. J., Truony, T. N., and Gordon, M. S., *J. Am. Chem. Soc.*, **107**, 4535 (1985).
112. Lee, J., Goggs, J. E., and Cowley, A. H., *J. Chem. Soc. Chem. Commun.*, 773 (1985).
113. Allen, T. L., Schreiner, A. C., Yamaguchi, Y., and Schaefer, H. F., III, *Chem. Phys. Lett.*, **121**, 154 (1985).
114. (a) Perić, M., Buenker, R. J., and Peyerimhoff, S. D., *Can. J. Chem.*, **55**, 1533 (1977).
    (b) Vasudevan, K., Peyerimhoff, S. D., Buenker, R. J., Kammer, W. E., and Hsu, H., *Chem. Phys.*, **7**, 187 (1975).
115. Skell, P. S., and Day, J. C., *Acc. Chem. Res.*, **11**, 381 (1978).
116. Skell, P. S., Teumark, R. L., and Seshadri, S., *J. Am. Chem. Soc.*, **105**, 5125 (1983).
117. Walling, C., El-Taliawi, G. M., and Zhao, C., *J. Am. Chem. Soc.*, **105**, 5119 (1983).
118. Peyerimhoff, S. D., Skell, P. S., May, D. D., and Buenker, R. J., *J. Am. Chem. Soc.*, **104**, 4515 (1982).
119. McLean, A. D., Lengsfield, B. H., III, Pacansky, J., and Ellinger, Y., *J. Chem. Phys.*, **83**, 3567 (1985).

120. Feller, D., Huyser, E., Borden, W. T., and Davidson, E. R., *J. Am. Chem. Soc.*, **105**, 1459 (1983).
121. Petrongolo, C., and Peyerimhoff, S. D., *J. Mol. Struct.* (*Theochem*), in press.
122. Lewerenz, M., Bruna, P. J., Peyerimhoff, S. D., and Buenker, R. J., *Mol. Phys.*, **49**, 1 (1983).
123. Bettendorff, M., and Peyerimhoff, S. D., *Chem. Phys.*, **99**, 55 (1985); *Chem. Phys.*, **104**, 29 (1985).
124. Sczech, S., this laboratory.
125. Schlier, Ch., *J. Chem. Phys.*, **67**, 5505 (1977).
126. Leone, S., *J. Chem. Phys.*, **83**, 601 (1985).
127. von Busch, F., Universität Bonn, private communication.
128. Marian, C. M., Marian, R., Peyerimhoff, S. D., Hess, B. A., Buenker, R. J., and Seger, G., *Mol. Phys.*, **46**, 779 (1982).
129. de Vivie, R., *Diplomarbeit*, Bonn, 1984.
130. For example, Peyerimhoff, S. D., *Gazz. Chim. Ital.*, **108**, 411 (1978).
131. Wolnieiwiez, L., and Dressler, K., *J. Mol. Spectrom.*, **67**, 416 (1977).
132. Mulliken, R. S., *Chem. Phys. Lett.*, **46**, 197 (1977).
133. (a) Yoshimine, M., Tanaka, K., Tatawaki, H., Obara, S., Sasaki, F., and Ohno, K., *J. Chem. Phys.*, **64**, 2254 (1976).
    (b) Buenker, R. J., Peyerimhoff, S. D., and Perić, *Chem. Phys. Lett.*, **42**, 383 (1976).
134. Bruna, P. J., and Peyerimhoff, S. D., *Bull. Soc. Chim. Belg.*, **92**, 525 (1983).
135. Bettendorff, M., Buenker, R. J., Peyerimhoff, S. D., and Römelt, J., *Z. Phys. A*, **304**, 125 (1982).
136. Douglas, A. E., and Greening, F. R., *Can. J. Phys.*, **57**, 1650 (1979).
137. Bettendorff, M., Peyerimhoff, S. D., and Buenker, R. J., *Chem. Phys.*, **66**, 261 (1982).
138. Richartz, A., Buenker, R. J., and Peyerimhoff, S. D., *Chem. Phys.*, **28**, 305 (1978).
139. Grein, F., Peyerimhoff, S. D., and Buenker, R. J., *J. Chem. Phys.*, **82**, 353 (1985).
140. Peyerimhoff, S. D., and Buenker, R. J., *Chem. Phys.*, **57**, 279 (1981).
141. Connor, D. C., and Römelt, J., to be published.
142. (a) Konowalow, D. D., and Julienne, P. S., *J. Chem. Phys.*, **72**, 5815 (1980).
    (b) Konowalow, D. D., Rosenkrantz, M. E., and Olson, M. L., *J. Chem. Phys.*, **72**, 2612 (1980).
    (c) Konowalow, D. D., Rosenkrantz, M. E., and Hochhauser, D. S., *J. Mol. Spectrosc.*, **99**, 321 (1983).
143. Buenker, R. J., Perić, M., Peyerimhoff, S. D., and Marian, R., *Mol. Phys.*, **43**, 987 (1981).
144. Perić, M., Peyerimhoff, S. D., and Buenker, R. J., *Int. Rev. Phys. Chem.*, **4**, 85 (1985).
145. Bruna, P. J., Peyerimhoff, S. D., and Buenker, R. J., *Chem. Phys.*, **39**, 211 (1976).
146. Bettendorff, M., Buenker, R. J., and Peyerimhoff, S. D., *Mol. Phys.*, **50**, 1363 (1983).
147. Taylor, H. S., Goldstein, E., and Segal, G. A., *J. Phys. B: At. Mol. Phys.*, **10**, 2253 (1977).
148. Nestmann, B., and Peyerimhoff, S. D., *J. Phys. B: At. Mol. Phys.*, **18**, 4309 (1985).
149. Hazi, A. U., Rescigno, T. N., and Kurilla, M., *Phys. Rev. A*, **23**, 1089 (1981).
150. Krumbach, V., this laboratory.
151. Kochem, K.-H., Sohn, W., Jung, K., Ehrhard, H., and Chang, E. S., *J. Phys. B: At. Mol. Phys.*, **18**, 1253 (1985).
152. Hess, B. A., *PhD Thesis*, Bonn, 1981.
153. Marian, C. M., *PhD Thesis*, Bonn, 1981.

154. Chandra, P., and Buenker, R. J., *J. Chem. Phys.*, **79**, 358–66 (1983).
155. Hess, B. A., Marian, C. M., Buenker, R. J., and Peyerimhoff, S. D. *Chem. Phys.*, **71**, 79 (1982).
156. Marian, C. M., *Chem. Phys.*, **100**, 13 (1985).
157. Klotz, R., *PhD Thesis*, Bonn, 1984.
158. Klotz, R., Peyerimhoff, S. D., *Mol. Phys.*, **57**, 573 (1986).
159. Hotop, H., private communication.
160. Fey, S., this laboratory.
161. Matsushita, T., Marian, C. M., Klotz, R., Hess, B. A., and Peyerimhoff, S. D., *Chem. Phys.*, **96**, 371 (1985).
162. Hess, B. A., Chandra, P., and Buenker, R. J., *Mol. Phys.*, **52**, 1177 (1984).
163. Herzberg, G., in *Molecular Spectra and Molecular Structure*, Vols. 1, 3, and 4. Van Nostrand, Princeton, 1978.
164. Herzberg, G., and Johns, J. W. C., *Proc. R. Soc. A*, **298**, 142 (1967).
165. Bender, C. F., and Schaefer, H. F., III, *J. Mol. Spectrosc.*, **37**, 423 (1971).
166. Staemmler, V., and Jungen, M., *Chem. Phys. Lett.*, **16**, 187 (1972).
167. Perić, M., Peyerimhoff, S. D., and Buenker, R. J., *Can. J. Chem.*, **59**, 1318 (1981).
168. Bruna, P. J., Hirsch, G., Peyerimhoff, S. D., and Buenker, R. J., *Mol. Phys.*, **42**, 875 (1981).
169. Castex, M. C., Le Calvé, J., Haaks, D., Jordan, B., and Zimmerer, G., *Chem. Phys. Lett.*, **70**, 106 (1980).
170. Bickel, G. A., and Innes, K. K., *Can. J. Phys.*, **62**, 1763 (1984).
171. Forster, S. C., *J. Mol. Spectrosc.*, **106**, 369 (1984).
172. Richartz, A., Buenker, R. J., Bruna, P. J., and Peyerimhoff, S. D., *Mol. Phys.*, **33**, 1345 (1977).
173. Mead, R. D., Hefter, U., Schulz, P. A., Lineberger, W. C., *J. Chem. Phys.*, **82**, 1723 (1985).

Ab Initio Methods in Quantum Chemistry—I
Edited by K. P. Lawley
© 1987 John Wiley & Sons Ltd.

MOLECULAR PROPERTY DERIVATIVES

ROGER D. AMOS

University Chemical Laboratory, Lensfield Road, Cambridge CB2 1EW, UK

CONTENTS

I. INTRODUCTION

This chapter is concerned primarily with *property surfaces*, that is the variation of molecular properties with geometry. The properties that will be considered in greatest detail are the dipole moment and the polarizability. Comparatively little attention will be given to the total energy of a system, though obviously potential energy surfaces are very important, and are implicit in discussions of vibrational effects or in relating dipole moment or polarizability derivatives to infrared or Raman intensities. The viewpoint of the chapter is that of computational chemistry. In particular, the intention is to review the development of analytic gradient methods as applied to property calculations. These are approaches designed to give all the derivatives of a given property, including the energy, in the course of a single calculation, without recourse to finite-difference methods. These analytic gradient methods have considerable advantages in terms of numerical stability and efficiency, particularly for larger molecules. The chapter commences with a brief review of the experimental sources and uses of property surfaces. The last part is an equally brief review of some recent calculations, including the results of some analytic dipole moment and polarizability derivative computations.

II. EXPERIMENTAL USES AND SOURCES OF PROPERTY DERIVATIVES

Although the emphasis of this chapter will be on *ab initio* methods and calculations, it will begin with a survey of experimental sources. This survey is by no means exhaustive and is intended to remind the *ab initio* quantum chemist of the standards by which his calculations must be judged, and the uses to which calculated values may be put. The range of precision in experimental techniques is probably much wider than that of theoretical methods. The most accurate experimental values, such as the very precise measurements of dipole moments using microwave spectroscopy and molecular-beam techniques, exceed the current capabilities of *ab initio* calculations. At the other extreme, some of the results that emerge from liquid-state measurements and collision-induced spectroscopy are so dependent upon the models used to interpret these experiments that nearly any *ab initio* calculation is capable of producing values that can be beneficially used to improve these models. Somewhere in between these extremes lies the large and important area of infrared and Raman intensities, where the interaction of experiment and theory can be highly productive.

A. Vibrational and Isotopic Dependence

A basic use for property derivatives is to determine the value of a property in a particular vibrational or rotational state of a molecule. Starting with the simplest case, diatomic molecules, the internuclear potential energy is written[1] as

$$V(\xi) = a_0 \xi^2 (1 + a_1 \xi + a_2 \xi^2 + \cdots) \tag{1}$$

where

$$\xi = (R - R_e)/R$$

If a given property is expanded similarly as

$$P = P_e + P_1 \xi + \tfrac{1}{2} P_2 \xi^2 + \cdots \tag{2}$$

then[2]

$$\langle vJ|P|vJ \rangle = P_e + (v + \tfrac{1}{2})(B_e/\omega_e)(P_2 - 3a_1 P_1)$$
$$+ 4J(J + 1)(B_e/\omega_e)^2 P_1 + \cdots \tag{3}$$

where B_e and ω_e are the equilibrium rotational and vibrational spectroscopic constants. The higher-order terms in this series are given by Herman and Short.[3]

The formula may be extended to polyatomic molecules. If the potential energy surface is expressed in terms of the dimensionless normal coordinates:

$$V/hc = \sum_i \tfrac{1}{2}\omega_i q_i^2 + \sum_{ijk} \tfrac{1}{6}k_{ijk}q_i q_j q_k + \cdots \tag{4}$$

and the property is

$$P = P_e + \sum_i P_i q_i + \sum_{ij} \tfrac{1}{2}P_{ij}q_i q_j + \cdots \tag{5}$$

then the expectation value of P in a given vibrational state is[4]

$$\langle v|P|v\rangle = P_e + \sum_i \tfrac{1}{2}(v_i + \tfrac{1}{2})\left(P_{ii} - \sum_j (k_{iij}/\omega_i)P_j\right) \tag{6}$$

Again, this is just the leading term in the expansion. More complete formulae can be found in Toyama et al.,[5] Krohn et al.[6] and Riley et al.[7] Particular expressions for the vibrational and rotational effects in symmetric and spherical tops have been given by Fowler.[8] These expressions are all based upon perturbation theory, and if extended to very high vibrational levels will eventually break down. Under these circumstances, it will be better to solve explicitly for the vibrational wavefunctions and to evaluate the expectation values of the properties by direct integration.

The same basic expressions also provide explanations for the temperature dependence of a property through the variations in the population of particular vibrational or rotational states, and for the variations occurring upon isotopic substitution as the normal coordinates depend upon the atomic masses.

One source of high-precision experimental information is molecular-beam spectroscopy, which is capable of measuring properties, particularly dipole moments, in specific vibrational or rotational states. Detailed results are available for a variety of small molecules including SO_2,[9] HCN,[10,11] OCS,[12-14] CO,[15] H_2O,[16] H_2S,[17] H_2CO,[18-20] NH_3,[21] HF and HCl,[22] among others. These techniques can even make measurements on systems such as van der Waals dimers which are particularly interesting as the weakness of the intermolecular potentials means that comparatively large changes in the molecular property can occur upon isotopic substitution. Recent examples of this type of study may be found elsewhere.[23-26]

Another property for which there is a large amount of experimental information concerning isotopic effects is the nuclear magnetic resonance (NMR) chemical shift. This topic has been reviewed at yearly intervals by Raynes[27] for 1978–79 and Jameson[28] for 1980–85, and so will not be considered here except to mention some recent work.[29-32]

B. Infrared Intensities

One of the most familiar uses of molecular property derivatives is the calculation of the infrared (IR) intensity from the derivatives of the dipole moment. A widely used expression is that for the integrated absorption coefficient

$$A = (1/cl) \int \ln(I_0/I) \, dv \tag{7}$$

where I_0 is the intensity of the incident radiation, I that of the transmitted radiation, c is the concentration and l is the path length. The integration over the frequency range can correspond either to a specific single absorption line, or to a whole vibration–rotation band. To relate the intensity to molecular properties, it is necessary to evaluate the transition dipole moment $\langle v|\mu|v'\rangle$. To do this, a Taylor series expansion of the dipole moment in the normal coordinates may be used

$$\mu = \mu_e + \sum_i (\partial\mu/\partial q_i)q_i + \sum_{ij} \tfrac{1}{2}(\partial^2\mu/\partial q_i q_j)q_i q_j + \cdots \tag{8}$$

If this series is truncated after the first derivative, the approximation is known as 'electrical harmonicity', and gives the simplified expression

$$\langle v|\mu|v'\rangle = (\partial\mu/\partial q_i)\langle v|q_i|v'\rangle \tag{9}$$

If the further approximation is made that the vibrational states are described by harmonic-oscillator wavefunctions ('mechanical harmonicity'), then $v = v' \pm 1$ and the resulting expression for the integrated intensity of a band is

$$A_i = (4\pi\varepsilon_0)^{-1}(N_A\pi/3c^2)d_i(\partial\mu/\partial q_i)^2 \tag{10}$$

where N_A is Avogadro's number, d_i is the degeneracy of the mode and the other symbols have their usual meanings. Some authors, e.g. Overend,[33] prefer the expression

$$A_i = (4\pi\varepsilon_0)^{-1}(N_A\pi/3c^2)(v_i/\omega_i)d_i(\partial\mu/\partial q_i)^2 \tag{11}$$

where v_i is the fundamental and ω_i the harmonic frequency of the mode.

Measurement of infrared intensities yields the dipole derivatives with respect to the normal coordinates. However, it is more useful if these can be expressed in some other coordinate system, such as the symmetry or internal coordinates of the molecule, or Cartesian coordinates. One form of analysis that is particularly suitable for comparison with *ab initio* results uses the 'atomic polar tensors' (APT).[34,35] This particular quantity is just the set of derivatives of the dipole, usually in Cartesian coordinates (despite its name), corresponding to the displacements of one particular atom, i.e.

$$P^1 = \begin{bmatrix} (\partial\mu_x/\partial R_{1x}) & (\partial\mu_y/\partial R_{1x}) & (\partial\mu_z/\partial R_{1x}) \\ (\partial\mu_x/\partial R_{1y}) & (\partial\mu_y/\partial R_{1y}) & (\partial\mu_z/\partial R_{1y}) \\ (\partial\mu_x/\partial R_{1z}) & (\partial\mu_y/\partial R_{1z}) & (\partial\mu_z/\partial R_{1z}) \end{bmatrix} \tag{12}$$

This is obtained directly in *ab initio* calculations, particularly those based upon the analytic derivative methods discussed later. Another advantage of working with the polar tensors is that the relationship between the Cartesian and the normal coordinate derivatives is straightforward

$$(\partial\mu/\partial q_i) = \sum_j (\partial\mu/\partial x_j)(\partial x_j/\partial q_i) \tag{13}$$

where $\{x\}$ is the set of all $3N$ Cartesian coordinates, or

$$x = Lq$$

with

$$L_{ji} = (\partial x_j/\partial q_i)$$

The matrix L is also straightforward to obtain from *ab initio* calculations.

One difficulty in obtaining values of the dipole moment derivative from experimental measurements is that, as the intensity depends upon the square of the derivative, the sign cannot be determined directly. A variety of ways to get round this problem have been developed, including the use of *ab initio* or semi-empirical calculations, and the use of isotopic substitution. The latter technique can be used in two ways. First, though the intensities vary, there are certain sum rules that must be obeyed,[36,37] and these can be used to check for consistency in the choice of signs. Secondly, information can be obtained from the variation of the dipole moment in the various isotopically substituted species, or in particular vibrational states, as this determines the sign as well as the magnitude of the dipole derivative. Recent examples of this last technique are the determination of the dipole surfaces in HCN[10] and OCS.[14] The volume edited by Person and Zerbi[38] contains several useful articles on the relationships between infrared intensities and various molecular parameters, and discusses in detail the difficulties inherent in extracting these parameters from measured intensities. It also contains a wide selection of experimental results. Another recent compendium of IR intensities may be found in the article by Bishop and Cheung[39] on vibrational polarizabilities.

In the harmonic approximation, overtones and combination bands have zero intensity. To derive expressions for the overtone, it is necessary to consider both higher derivatives of the dipole moment function and anharmonic terms in the potential. Detailed equations have been given by Overend.[40-42] It is also possible to find expressions for the intensities of particular lines within a band; for example, Harman and Wallis[43] consider the differences in the rotational lines in the separate branches of a rotation–

vibration band. A recent example of the precision with which individual lines can be studied is measurement of transition dipole moments in HF by Pine *et al.*[44] However, in order to make full use of measurements on overtones and combination bands, a detailed knowledge of the potential energy surface of the molecule is required, including accurate anharmonic constants. This requirement has limited knowledge of higher derivatives of the dipole surface to a small number of molecules. Accurate information is available only for some diatomics, such as HF,[45] HCl[46,47] and CO,[50,51] and a few triatomic molecules, particularly CO_2,[51-54] SO_2,[9] HCN[10,48,49] and OCS.[14]

C. Raman Intensities

The Raman intensity, in the Placzek theory, is proportional to the derivatives of the polarizability, and is often described in terms of the coefficient

$$S = d[45(\alpha')^2 + 7(\gamma')^2] \tag{14}$$

where α' is the derivative of the isotropic polarizability and γ' is that of the anisotropic polarizability:

$$\alpha' = \tfrac{1}{3}\sum_i \alpha'_{ii} \tag{15}$$

$$(\gamma')^2 = \sum_{i>j} (\alpha'_{ii} - \alpha'_{jj})^2 + 6(\alpha'_{ij})^2 \tag{16}$$

Another observable is the depolarization ratio

$$\rho = \frac{3(\gamma')^2}{45(\alpha')^2 + 4(\gamma')^2} \tag{17}$$

The constants in the above expressions are in fact dependent upon the experimental conditions—plane-polarized incident light viewed at a scattering angle of 90°—and expressions appropriate to alternative conditions are tabulated by Long.[55] The same approximations of electrical and mechanical harmonicity as were mentioned for infrared intensities have been assumed. An additional assumption which is often made is that the intensities in the Raman effect can be calculated using the *static* polarizabilities rather than the polarizabilities at the frequency of the incident light. As with the IR intensities it is more convenient for comparison with *ab initio* calculations if the polarizability derivatives are expressed in Cartesian coordinates.[56] There are few reliable values for polarizability derivatives, either experimental or from *ab initio* calculations. There are data for a small number of diatomic molecules such as H_2 and N_2,[57] and a few polyatomic molecules such as CH_4, C_2H_2 and C_2H_6,[58-61] but for most other molecules the degree of uncertainty is large. For diatomic molecules, Buckingham and Szabo[62] have developed a theory for rotation–vibration interaction, akin to that of Herman and Wallis[43] for the infrared case, and some information on the derivatives of the anisotropic polarizability has been obtained from this.[63,64] In principle, information about the derivatives of higher-order polarizabilities are obtainable from the hyper-Raman effect and from Raman optical activity.[55]

D. Induced Spectra and Intermolecular Effects

Infrared and Raman spectroscopy are not the only phenomena which involve the derivatives of the dipole and the polarizability. As the energy of a molecule in the presence of an electric field is

$$E = E_0 - \sum_i \mu_i F_i - \tfrac{1}{2}\sum_{ij} \alpha_{ij} F_i F_j \qquad (18)$$

then it follows that the derivative of the energy involves the derivatives of the multipole moments and polarizabilities. Thus the potential

$$V_0 = \sum_{ij} \tfrac{1}{2}(\partial^2 E/\partial q_i \partial q_j) q_i q_j + \cdots \qquad (19)$$

becomes

$$\begin{aligned} V = V_0 &+ \sum_{ij}(\partial^2 E/\partial q_i F_j)q_i F_j + \sum_{ijk} \tfrac{1}{2}(\partial^3 E/q_i F_j F_k)q_i F_j F_k \\ &+ \sum_{ijk} \tfrac{1}{2}(\partial^3 E/q_i q_j F_k)q_i q_j F_k + \cdots \\ = V_0 &- \sum_{ij}(\partial\mu_j/\partial q_i)F_j q_i - \sum_{ijk}\tfrac{1}{2}(\partial\alpha_{jk}/q_i)F_j F_k q_i \\ &- \sum_{ijk}\tfrac{1}{2}(\partial^2\mu_k/\partial q_i \partial q_j)F_k q_i q_j \qquad (20) \end{aligned}$$

The changes in the potential result in changes in the molecule's equilibrium geometry and vibrational frequencies, which can be related to the dipole and polarizability derivatives. The infrared intensities can also be modified by external fields as the dipole moment derivatives become

$$(\partial\mu_j/\partial q_i) = (\partial\mu_j/\partial q_i)_0 + \sum_k (\partial\alpha_{jk}/\partial q_i)F_k \qquad (21)$$

If the molecule was originally infrared-inactive, then it follows that an applied electric field can result in an induced infrared spectrum proportional in strength to the polarizability derivative. All of the above effects are observable. The theory is comparatively straightforward and may be found in the literature.[65-69] A recent example is a study of the effects of electric fields on CH_4.[70,71]

The perturbing electric fields can also be produced by molecular interactions. This is obviously a very general effect, and, just as the multipole moments have widespread applications in the descriptions of molecular interactions,[72] so may their derivatives be applied to describing interaction-induced spectra. The dipole moment of a molecule in the presence of a non-uniform electric field is

$$\mu_i = \mu_i^0 + \sum_j \alpha_{ij} F_j + \sum_{jk} \tfrac{1}{2} \beta_{ijk} F_j F_k + \sum_{jk} \tfrac{1}{3} A_{i,jk} F'_{jk} + \cdots \qquad (22)$$

where β is the first hyperpolarizability, A is the dipole–quadrupole tensor, and F' is the electric field gradient.[72] The electric fields and field gradients can be produced by the multipole moments of neighbouring molecules, and since the molecules are vibrating it follows that induced dipole moment derivatives will arise, having the frequencies of the normal molecular vibrations. For example, isolated centrosymmetric molecules such as H_2 or N_2 are not infrared-active; however, in a dense medium these molecules do absorb in the infrared at the fundamental vibration frequency. The mechanism is that the quadrupole moment of these molecules produces a field which results in an induced dipole moment derivative in neighbouring molecules, proportional to the polarizability derivative. This was the first type of collision-induced spectrum to be observed.[73] It is also possible to have interaction-induced Raman spectra. For example, the bending mode in molecules such as CO_2 or CS_2 can be Raman-active in liquids and high-pressure gases. This has been observed experimentally[74] and investigated theoretically.[75] The mechanism in this case involves the derivatives of β and A. Collision-induced absorption can also occur in systems such as rare gases where the isolated atoms have no multipole moments. This results from a pair of rare-gas atoms having a polarizability and a polarizability anisotropy that varies with internuclear distance and consequently gives rise to a Raman spectrum. Various models have been proposed for this type of induced polarizability, the simplest of which, the 'dipole-induced dipole' model,[76] gives the polarizability anisotropy as a function of distance as

$$\Delta\alpha = 6\alpha_0^2/R^3 + 6\alpha_0^3/R^6 \qquad (23)$$

where α_0 is the polarizability of the isolated atom. Some more sophisticated models have also been proposed.[77–80] This area has been reviewed by Tabisz.[81]

III. METHODS

It might seem that the methods needed to obtain a property derivative are trivial. Quantum-chemistry programs capable of evaluating virtually any property at a nominated geometry are widely available, and therefore the derivative of a property with respect to a geometrical parameter may be obtained by calculating it at several different geometries. Finite-difference methods, i.e. numerical differentiation, do have some advantages. In particular, they are simple to implement, both for geometric dependences and for the effect of external static electric fields (magnetic properties are more difficult, but not impossible). One disadvantage is the loss of precision that occurs in a numerical differentiation. Though purely technical, this is not a negligible problem. Finite-field methods are familiar in polarizability calculations. Their success is due in part to the fact that the required magnitude for the perturbing field is well understood, which follows from the fact that the magnitude of a molecular polarizability is fairly easy to estimate. Dipole moment and polarizability derivatives, and force constants, on the other hand, can vary considerably from molecule to molecule, especially if calculated in terms of internal or symmetry coordinates. It is therefore more difficult to choose the magnitudes of the displacements in order to achieve the desired precision, particularly if calculating second derivatives. Nevertheless, it is usually possible to obtain the correct values by Finite-difference techniques, given sufficient care, and sufficient time. It is the last point—the time required—that is the real disadvantage of finite-difference methods.

The alternatives to numerical differentiation are the analytic gradient methods. These methods calculate quantities such as the gradient (first derivative) or the second derivative of the energy, and are very widely used to search potential energy surfaces to locate minima and transition states, and to calculate harmonic frequencies. Though this chapter is primarily concerned with molecular property derivatives, it will be necessary to make some mention of the techniques used in analytic energy derivative calculations, both because a property such as a dipole derivative may be regarded as a special case of the general expression for the derivative of the energy, and because it is often desirable to express properties as derivatives with respect to the normal coordinates, for which a force-constant calculation may be required.

A. Analytic Gradients

Analytic gradient methods became widely used as a result of their implementation for closed-shell self-consistent field (SCF) wavefunctions by Pulay,[82] who has reviewed the development of this topic.[83] Since then, these methods have been extended to deal with all types of SCF wavefunctions,[84,85] as well as multi-configuration SCF (MC-SCF),[83,86-88] configuration-interaction (CI) wavefunctions,[89-91] and various non-variational methods such as Moller–Plesset (MP) perturbation theory[85,92,93] and coupled-cluster (CC) techniques.[94,95] In short, it is possible to obtain analytic energy derivatives for virtually all the standard *ab initio* approaches. The main use of analytic gradient methods is, and will remain, the location of stationary points on a potential energy surface, to obtain equilibrium and transition-state geometries. However, there is a specialized use in the calculation of quantities such as dipole derivatives.

The dipole moment may be regarded as the derivative of the energy with respect to an external electric field. One standard method of calculating dipole moments, and also polarizabilities, is to use a finite-field perturbation method[96] and obtain the energy

$$E(F) = \langle \Psi | H - \mu \cdot F | \Psi \rangle \qquad (24)$$

for several specific values of the electric field F. As was pointed out by Komornicki and McIver,[97] since

$$(\partial \mu / \partial R) = -(\partial^2 E / \partial R \partial F) = -(\partial / \partial F)(\partial E / \partial R) \qquad (25)$$

it follows that one can obtain dipole moment derivatives by calculating the gradient of the energy in the presence of an electric field

$$(\partial E(F) / \partial R) = (\partial / \partial R) \langle \Psi | H - \mu \cdot F | \Psi \rangle \qquad (26)$$

If one has available an analytic derivative program, this approach has a distinct advantage over calculating the dipole moment at many geometries, especially for a polyatomic molecule, as instead it is only necessary to calculate the gradient of the energy for a relatively small number of field strengths. As a minor variation upon this idea, instead of using $\mu \cdot F$ as the perturbation, the molecule can be perturbed by placing one or more point charges at a large distance. When Komornicki and McIver first suggested this idea, the only analytic gradient methods in use were for SCF wavefunctions. However, as indicated above, these techniques are now possible for most *ab initio* methods, and consequently it should be possible to obtain dipole moment derivatives relatively easily for nearly any type of wavefunction, greatly extending the range of systems that can be studied accurately. Obviously the derivatives of other properties such as polarizabilities or quadrupole moments may be obtained by similar methods. However, though better than evaluating the dipole moment at many geometries, this approach is not ideal, as it is still necessary to perform at least one, and probably several, calculations for each component of the applied electric field. There are also more subtle disadvantages such as the fact that it is difficult to make full use of molecular symmetry in such a scheme. These problems are addressed in the next section.

B. Analytic Dipole Moment Derivatives

The methods mentioned above for the calculation of the first derivative of the energy have been extended to provide analytic evaluations of the second derivative. The necessary theory was first considered some time ago,[98] but practical implementations of these schemes can be regarded as starting with those of Pople et al.[85] in 1979. This important paper demonstrated how the complete second derivative of the molecular energy with respect to geometric parameters could be obtained in a single calculation. This paper used the unrestricted Hartree–Fock (UHF) method, but the methodology has since been extended to deal with most other types of wavefunction. The importance of this kind of calculation is that the whole force-constant matrix, and hence the vibrational frequencies and normal modes, is obtained in a manner that is both considerably faster and also numerically more accurate than the earlier finite-difference approaches. Obtaining the normal coordinates is of course important in the context of molecular property calculations. However, of even greater relevance is the fact that the development of these analytic derivative methods opened up the possibility of obtaining *property surfaces*, or at least the leading terms in these surfaces, in an efficient manner. The simplest, and probably the most important, of these property surfaces is that for the dipole moment. Although this section will refer constantly to the dipole moment and the dipole moment derivative, it should be borne in mind that the theory is in fact identical for the derivative of any first-order property, and applies equally well to the derivative of any multipole moment, or indeed the derivatives of quantities such as the diamagnetic part of the nuclear magnetic shielding or the magnetizability.

The theory for the analytic evaluation of dipole moment derivatives will be considered first for the simplest case, that of closed-shell SCF wavefunctions. The electronic contribution to the dipole moment is

$$\sum_{pq} D_{pq} \langle p | \mu_b | q \rangle \tag{27}$$

where D is the density matrix in the atomic-orbital (AO) basis, given in terms of the molecular-orbital (MO) coefficients as

$$D_{pq} = \sum_i f_i C_{pi} C_{qi} \tag{28}$$

where f_i is the occupation number of the ith orbital (for closed-shell SCF $f_i = 2$ or 0). The derivative with respect to some parameter is

$$\sum_{pq} D_{pq} \langle p | \mu_b | q \rangle^a + \sum_{pq} D_{pq}^a \langle p | \mu_b | q \rangle \tag{29}$$

The derivative integrals

$$\langle p | \mu_b | q \rangle^a = (\partial / \partial a) \langle p | \mu_b | q \rangle \tag{30}$$

are straightforward to evaluate. The derivative of the density matrix D^a requires the use of coupled Hartree–Fock (CHF) theory. One of the first formulations of CHF theory as applied to nuclear displacements, by Gerratt and Mills,[98] considered dipole moment derivatives. However, there was no practical implementation at the time. CHF theory for closed-shell SCF wavefunctions is particularly well known, so only a brief summary will be given, to introduce the ideas used in later, more complicated approaches. The first-order changes in the molecular orbitals due to some external perturbation are expressed by writing the perturbed orbitals as linear combinations of the original orbitals:

$$\phi_i^a = \sum_j U_{ji}^a \phi_j \tag{31}$$

The coefficients U^a are then determined by a set of simultaneous equations

$$\sum_{kl} A_{ij,kl} U_{kl}^a = B_{ij}^a \tag{32}$$

where the matrix A is

$$A_{ij,kl} = (\varepsilon_j - \varepsilon_i)\delta_{ik}\delta_{jl} - (4\langle ij|kl \rangle - \langle ik|jl \rangle - \langle il|jk \rangle) \tag{33}$$

and B^a is defined as

$$B_{ij}^a = \varepsilon_{ij}^a - \varepsilon_j S_{ij}^a - \sum_{kl}^{\text{occ}} S_{kl}^a(2\langle ij|kl \rangle - \langle ik|jl \rangle) \tag{34}$$

Here S^a is the derivative of the overlap matrix, and ε^a is

$$\varepsilon_{pq}^a = h_{pq}^a + \sum_{rs}(2\langle pq|rs \rangle^a - \langle pr|qs \rangle^a)D_{rs} \tag{35}$$

$$\varepsilon_{ij}^a = \sum_{pq} C_{pi}C_{qj}\varepsilon_{pq}^a$$

In these and subsequent expressions i,j,k,l are used to denote molecular orbitals whereas p,q,r,s are atomic orbitals (basis functions). Thus the quantity ε^a is formed in the AO basis and transformed into the MO basis. This quantity has the same structure as a Fock operator but is constructed using the derivatives h^a and $\langle pq|rs \rangle^a$ of the one-electron and two-electron integrals. A compact notation[99] for this operation is

$$\varepsilon^a = h^a + g^a * D \tag{36}$$

where the term $g^a * D$ symbolizes the contraction of the two-electron integrals with the density matrix. The solution of the simultaneous equations (32) is routine even with large basis sets using the iterative method suggested by Pople et al.[85] Having solved these equations, the perturbed density matrix, in the MO basis, is

$$D_{ij}^a = U_{ij}^a f_j + U_{ji}^a f_i \tag{37}$$

Given the perturbed density matrices, the complete expression for the dipole moment derivative may be assembled. The derivatives will generally be with respect to all possible Cartesian displacements of the nuclei, so the result of such a calculation is that all the atomic polar tensors for the molecule will be obtained simultaneously.

Before considering more complicated types of wavefunction, it is worth discussing some of the practical considerations involved in the implementation of such a scheme. The most important stage, because it is the most time-consuming, is the evaluation of the derivatives of the integrals, particularly the two-electron integrals. It is essential that these integrals be evaluated efficiently. A significant step towards achieving this was the introduction by Dupuis and King[100,101] of the Rys polynomial method for integrals involving Gaussian basis sets. The essential step in this approach is the use of a quadrature scheme to express two-electron integrals as a sum over products of much simpler subsidiary integrals

$$\langle pq|rs \rangle = \sum_i I_x(u_i)I_y(u_i)I_z(u_i)w_i \tag{38}$$

The efficiency in this approach comes from exploiting the shell structure of basis sets, that is the basis will contain shells of functions, for example a set of p functions, $P = \{p_x, p_y, p_z\}$, in which all members of the set have the same exponents and contraction coefficients. This allows a relatively large number of two-electron integrals to be constructed from a small set of subsidiary integrals I_x, I_y and I_z. It was quickly realized that this method also made it simple to construct the derivatives of the two-electron integrals. In the original implementation by Dupuis and King[102] this was done by noting that all the derivatives of a shell of p functions on one centre yield an s function and a set of d functions at the same centre, and with the same exponents and contraction coefficients. Therefore, to obtain the derivatives of the integrals involving, to take a particular example, four p functions $\langle P_A P_B | P_C P_D \rangle$, one would evaluate all integrals of the form $\langle S_A P_B | P_C P_D \rangle$ and $\langle D_A P_B | P_C P_D \rangle$, and by taking suitable linear combinations of these one would obtain $\langle P'_A P_B | P_C P_D \rangle$. This approach was incorporated into the successful HONDO program. Later authors, however, have preferred to evaluate the derivatives of the subsidiary integrals, so that

$$(\partial/\partial x)\langle pq|rs \rangle = \sum_i I'_x(u_i)I_y(u_i)I_z(u_i)w_i \tag{39}$$

The advantage of this is that once the sets of integrals I'_x, I'_y and I'_z have been obtained, then all derivatives of $\langle pq|rs \rangle$ may be assembled simultaneously using methods that are structurally identical to the assembly of the original integral. This simplifies the logic of the program and leads to gains in efficiency. Various methods have been proposed for making the derivative subsidiary integrals I'. Saxe et al.[103] and Schlegel et al.[104] have adopted methods based on differentiating the recursion relationships used to construct the subsidiary integrals. The present author has suggested[105] evaluating a slightly larger set of I_x, I_y and I_z and getting the I' as linear combinations of these original integrals. This has the advantage that it is easily extended to higher derivatives. This method has also been used in the latest edition of the HONDO program.[106] It is also possible to obtain special formulae for certain classes of integrals, such as those involving only s and p functions.[107] Whatever the details of the method used, all these approaches have one feature in common, which is efficiency. Generally, it is possible to obtain the gradient of he energy of a molecule in about the same time, or less, as was required for the initial energy evaluation. Clearly this is much faster than the energy derivative could be obtained by numerical methods. In the context of CHF methods, the derivative integrals are used to form the matrices ε^a defined in Eq. (35). These can be constructed as the derivative integrals are being evaluated so that it is not necessary to store the latter. Given the ε^a the remainder of the right-hand side (34) of the simultaneous equations (32) is straightforward.

The solution of the simultaneous equations (32) requires some care because of their size. The number of independent elements of the matrix U^a which describes the perturbed orbitals is, for closed-shell SCF wavefunctions, the product of the number of occupied and the number of virtual molecular orbitals. This can be a large number—several thousand for a calculation on a big molecule. This makes it difficult to solve (32) by the conventional techniques used for small sets of simultaneous equations. However, an iterative approach introduced by Pople et al.[85] provides the answer. This method constructs the solution U as a linear combination of trial vectors

$$U = \alpha_1 U^{(1)} + \alpha_2 U^{(2)} + \cdots + \alpha_n U^{(n)} \tag{40}$$

The noteworthy point is that to obtain successive vectors $U^{(n)}$ it is not necessary to store the matrix A, but only the products

$$(AU^{(1)}), \quad (AU^{(2)}), \quad \ldots, \quad (AU^{(n)}) \tag{41}$$

In this it resembles the techniques used in the large CI calculations to determine an eigenvalue without having to store the entire Hamiltonian matrix. This approach is rapidly convergent, at least in SCF calculations where the matrix A is diagonally dominant. For a molecule with N nuclei it is necessary to solve $3N$ CHF equations for the perturbations due to the nuclear displacements (ignoring symmetry for the moment). All of these equations can be solved at the same time.[108]

In the form given, the CHF equations require a four-index transformation to obtain the two-electron integrals in the MO basis. Actually this is not a strict necessity, as it is possible to solve these equations in the AO basis basis.[108] However, in practice the transformation is not the major step in the calculation. The reason for this is that the transformed integrals $\langle ij|kl \rangle$ required do not include any involving either three or four virtual orbitals. This greatly reduces the computational effort necessary for this step. Optimum algorithms for partial four-index transformations have been discussed by Saunders and van Lenthe.[109] Another factor that has reduced the difficulties associated with four-index transformations is the availability of vector-processing computers. Since a four-index transformation is essentially just a series of matrix multiplications, this stage of the calculation is ideally suited to the capabilities of vector machines, as is the iterative scheme for solving CHF equations mentioned above. Other sections of the process can also be adapted to make use of vector-processing capabilities. For example, these techniques can be used beneficially in the evaluation of integrals and their derivatives. The strategy adopted by the author exploits the tendency to use basis sets consisting of contracted Gaussian functions, i.e. functions that are linear combinations of primitive Gaussians. The vectorization scheme essentially consists of evaluating the integrals for many different primitive functions in a single batch. This concept is also used in a recent version of the HONDO program.[106]

Another factor that benefits all stages of a major *ab initio* calculation is the use of molecular symmetry. Dupuis and King[110-113] have written a series of papers which show how most stages in SCF, gradient, coupled Hartree–Fock and force-constant calculations can be achieved using only the *unique*, i.e. symmetry-distinct, integrals and derivative integrals. For example, the matrices ε^a can be constructed from a reduced list of integral derivatives and symmetrized afterwards.[113] This type of approach is very useful in gradient calculations as it can be applied effectively *in the atomic-orbital basis*. Symmetry can also be used in the solution of the CHF equations by the obvious tactic of solving only for the unique nuclear displacements — translational and rotational invariance can also be used at this stage to reduce the number of equations. Symmetry can also be of assistance in the four-index transformation. This step consists of producing the two-electron integrals in the MO basis from those in the AO basis according to

$$\langle \phi_i \phi_j | \phi_k \phi_l \rangle = \sum_{pqrs} C_{pi} C_{qj} C_{rk} C_{sl} \langle \chi_p \chi_q | \chi_r \chi_s \rangle \tag{42}$$

For Abelian point groups, the use of symmetry is particularly simple. Under each operation R in the group, each basis function transforms entirely into another, apart from a possible change of sign, i.e.

$$R\chi_p = \pm \chi_{p'} \tag{43}$$

and the molecular orbitals can do no more than change sign

$$R\phi_i = \pm \phi_i \tag{44}$$

Similarly the coefficients of two basis functions related by symmetry must have the relationship

$$C_{pi} = \pm C_{p'i} \tag{45}$$

Now consider two terms in the above summation

$$C_{pi} C_{qj} C_{rk} C_{sl} \langle pq | rs \rangle \tag{46}$$

and

$$C_{p'i} C_{q'j} C_{r'k} C_{s'l} \langle p'q' | r's' \rangle \tag{47}$$

where p', q', r' and s' are related to p, q, r and s by a particular symmetry operation. For those integrals $\langle \phi_i \phi_j | \phi_k \phi_l \rangle$ that *do not vanish* according to symmetry criteria, all the factors of ± 1 must multiply together to yield $+1$, with the result that the above two terms make precisely equal contributions to the integral in the MO basis. This means that one could omit one term, and include only that from the 'unique' integral, multiplied by a factor consisting of the number of terms related by symmetry. This is precisely the condition required to make use of what Dupuis and King call a 'reduced' list of integrals. The only snag is that the integrals $\langle ij|kl \rangle$ that are *exactly zero* by symmetry criteria are so because of the exact *cancellation* of terms, which cannot be achieved if only one term in the summation is included. However, it is a simple matter to identify these integrals and reject them explicitly. Also, transformations are usually done in stages, for example converting the list of AO integrals into a 'half-transformed' list $\langle pq|kl \rangle$ and then into the fully transformed $\langle ij|kl \rangle$, and for efficiency it is essential to make use of symmetry when constructing, and rejecting, elements of the 'half-transformed' list. For example, it is only necessary to produce the $\langle pq|kl \rangle$ for the unique pairs (pq) of AO labels. Point groups containing degenerate representations pose much greater problems in a four-index transformation; however, some progress has been made in this area.[114]

Although the points discussed above are purely technical, they are worth mentioning for two reasons. The first is that efficient methods of integral evaluation, the use of symmetry and the consequences of the existence of the large vector-processing supercomputers will apply to all analytic derivative calculations, not just to dipole moment derivatives for closed-shell SCF wavefunctions. The second point is that it is only because of these technical improvements that is has become possible to carry out calculations using large basis sets on reasonably large molecules (see Refs. 115 and 116 and the results discussed later in this Chapter)—the theory needed has existed for many years.[98]

Now consider the formula for the analytical second derivative of a molecular system using closed-shell SCF wavefunctions

$$\partial^2 E / \partial a \partial b = \sum h_{ii}^{ab} + \varepsilon_{ii}^{ab} - 2\varepsilon_i S_{ii}^{ab} + 4 \sum \sum U_{ki}^b (\varepsilon_{ki}^a - \varepsilon_i S_{ki}^a) - 2 \sum \varepsilon_{ij}^{(b)} S_{ij}^a \quad (48)$$

with

$$\varepsilon^{(b)} = h^{(b)} + g^b * D + g * D^b \quad (49)$$

and ε^{ab} is a Fock operator constructed with the second derivatives of the one-electron and two-electron integrals

$$\varepsilon^{ab} = h^{ab} + g^{ab} * D \quad (50)$$

In addition to the second derivatives of the integrals, the results of the CHF equations already discussed are needed to evaluate this expression. It is not proposed to go into details of analytic second derivative calculations, except to point out that they are now possible for many types of wavefunction.[85,92,98,99,108,113,117-122] Their importance when dealing with property surfaces is that these methods provide an efficient method of obtaining the *normal coordinates*, which are virtually essential for the analysis of properties such as infrared or Raman intensities. It is also worth examining this formula to see what occurs in the special case where one perturbation is an electric field and the other is a nuclear displacement, as the formula is not symmetric in the two perturbations (it is possible to use a symmetric formula but there is no advantage in doing so).

If in Eq. (48) the perturbation 'a' is the electric field and 'b' is the nuclear displacement, then

$$\partial^2 E/\partial a \partial b = \sum_i^{occ} 2h_{ii}^{ab} + \sum_i^{occ} \sum_k^{all} U_{ki}^b h_{ki}^a \tag{51}$$

This is just the formula (29) for the dipole derivative expressed in the MO basis. In this context the h^a are the dipole moment integrals and the h^{ab} are the derivatives of the dipole integrals. (It has been assumed here and at most points in this chapter that the basis set used does not consist of functions with an explicit dependence upon the external electric field—if this is not the case then the full formula for the second derivative of the energy (48) must be used for the dipole derivative as well.) Taking the perturbations in the other order,

$$\partial^2 E/\partial b \partial a = \sum_i^{occ} 2h_{ii}^{ab} + 4\sum_i^{occ} \sum_k^{all} U_{ki}^a(\varepsilon_{ki}^b - \varepsilon_i S_{ki}^b) - 2\sum_{ij}^{occ} \varepsilon_{ij}^{(a)} S_{ij}^{(b)} \tag{52}$$

This is a formula for the electric field dependence of the gradient, which is of course equivalent to the dipole moment derivative. This is the analytic counterpart of the finite-field method proposed by Komornicki and McIver[97] and mentioned earlier. In some circumstances, there can be an advantage is using Eq. (52) rather than Eq. (29). It should be noted that the most time-consuming steps are the same in both approaches, that is the evaluation of the integral derivatives and the construction of the ε^a matrices. However, in (52) the CHF equations that have to be solved are those with the electric field as the perturbation rather than the nuclear displacements, and there are only three of these equations rather than $3N$. This could be significant in very large molecules. If it is necessary to calculate the harmonic force constants and normal coordinates by an *ab initio* analytic technique, then all $3N$ CHF equations have to be solved anyway, so there is little difference between the formulae (29) and (52). However, force constants can be obtained from other sources, or it could be reasonably argued that the dipole moment derivatives should be calculated with a larger basis set than the force constants, in which case the alternatives (29) and (52) should be kept in mind.

The expression (52) is also a useful way to introduce a simplification suggested by Lazzeretti,[123] which is that the dipole derivative can be obtained as the electric field dependence of the Hellmann–Feynman force. Buckingham and Fowler have pointed out that this approach is extendable to any multipole moment or polarizability derivative.[124] In this case, all basis function derivatives drop out of the equations and the result is just

$$\partial \mu_b / \partial a = \sum_{ki} 4 U_{ki}^b h_{ki}^a \tag{53}$$

where the h^a are just the integrals for the Hellmann–Feynman force. This formula is true for the exact wavefunction, and consequently is a valid way of interpreting dipole moment derivatives. However, in the artificial world of quantum-chemical calculations and approximate wavefunctions, there are difficulties with this approach, as the Hellmann–Feynman force is, in a calculation with a small basis set, a poor estimate of the gradient of the energy, and dipole moment derivatives calculated in this manner are unreliable. On the other hand, some preliminary investigations by Lazzeretti[123] indicate that, with a large enough basis, reasonable results for $\partial \mu / \partial R$ may be obtained, so the topic needs further study. It may well turn out, however, that 'large enough' in this context will be too large to allow calculations on big molecules. One possible approach is to take a small basis and expand it by including all the derivatives of the basis functions with respect to the nuclear coordinates.[125] This will result in a basis that will nearly satisfy the Hellmann–Feynman theorem. However, the basis will be quite large; for example, starting with a double-zeta (DZ) basis, i.e. 4s2p on a first-row atom, one would obtain 6s6p2d. If the intention is simply to avoid calculating the derivatives of the integrals in the initial DZ basis, then this approach is futile, as the undifferentiated integrals in the expanded basis will take considerably longer than the derivatives of the integrals in the original basis. It is for this reason that choosing a basis in this manner is of no help in geometry optimization.[126] On the other hand, dipole moment derivatives need larger basis sets anyway—*at least* two shells of polarization functions per atom are advisable. Consequently, it may turn out that choosing a basis set by expanding a small number of functions through the addition of derivative functions will produce a basis that is competitive *with one of similar size* but obtained by more 'conventional' means. It should be noted, however, that to obtain intensities the normal coordinates are required. Calculating force constants by use of the Hellmann–Feynman theorem is unlikely to be competitive as this would need the inclusion of the *second derivatives* of the generating basis set. Unless one always uses experimental force constants, this implies that solving sets of CHF equations for the nuclear displacements cannot be avoided at some stage.

Most of the general features discussed above for closed-shell SCF will also apply to calculations using other types of wavefunction. For other varieties of

SCF, the only significant change will be that the CHF equations (33) and (34) will be different. The relevant equations for UHF have been given by Pople *et al.*,[85] for high-spin open-shell wavefunctions the theory has been examined by Saxe *et al.*;[103] and Osamura *et al.* have considered general open-shell wavefunctions.[108,117] This last case is worth giving in detail as it includes a wide variety of SCF structures. It deals with all SCF methods for which the energy expression can be placed in the form

$$E = \sum_i f_i h_{ii} + \sum_i \sum_j \alpha_{ij} \langle ii|jj \rangle + \beta_{ij} \langle ij|ij \rangle \tag{54}$$

This includes as special cases both closed-shell and high-spin open-shell, and also cases such as open-shell singlets, symmetry-equivalenced SCF and even some very simple types of MC-SCF wavefunction, such as two-configuration (TC-SCF) wavefunctions.[127] The CHF equations are derived by differentiating the convergence criterion, which is

$$\varepsilon_{ij} = \varepsilon_{ji} \tag{55}$$

where

$$\varepsilon_{ij} = f_i h_{ij} + \sum_k \alpha_{ik} \langle ij|kk \rangle + \beta_{ik} \langle ik|jk \rangle \tag{56}$$

As with the closed-shell case the result is a set of simultaneous equations

$$AU^a = B^a$$

where the A matrix is

$$A_{ij,kl} = 2(\alpha_{il} - \alpha_{jl} - \alpha_{ik} + \alpha_{jk}) \langle ij|kl \rangle + (\beta_{il} - \beta_{jl} - \beta_{ik} + \beta_{jk})(\langle ik|jl \rangle + \langle il|jk \rangle)$$
$$+ (\varepsilon_{ik} - \zeta_{ik}^j)\delta_{jl} - (\varepsilon_{jk} - \zeta_{jk}^i)\delta_{il} - (\varepsilon_{il} - \zeta_{il}^j)\delta_{jk} + (\varepsilon_{jl} - \zeta_{jl}^i)\delta_{ik} \tag{57}$$

with

$$\zeta_{jk}^i = f_i h_{jk} + \sum_l \alpha_{il} \langle jk|ll \rangle + \beta_{il} \langle jl|kl \rangle \tag{58}$$

The right-hand side of the simultaneous equations is

$$B_{ij}^a = \varepsilon_{ji}^a - \varepsilon_{ij}^a + \sum_{k>l} \sum S_{kl}^a [2(\alpha_{ik} - \alpha_{jk}) \langle ij|kl \rangle + (\beta_{ik} - \beta_{jk})(\langle ik|jl \rangle + \langle il|jk \rangle)$$
$$+ (\varepsilon_{il} - \zeta_{il}^j)\delta_{jk} - (\varepsilon_{jl} - \zeta_{jl}^i)\delta_{ik}]$$
$$+ \tfrac{1}{2} \sum_k S_{kk}^a [2(\alpha_{ik} - \alpha_{jk}) \langle ij|kk \rangle + 2(\beta_{ik} - \beta_{jk}) \langle ik|jk \rangle] \tag{59}$$

Despite the complications of the coupling coefficients α and β, it can be seen that the basic principles of these equations are the same as the closed-shell equations discussed earlier. The left-hand side of the equations is independent of the perturbation, and is a function of the one- and two-electron integrals for the system. The right-hand side depends upon the derivatives of the integrals and, as with the closed-shell case, most of the work goes into constructing the term ε^a, which is

$$\varepsilon_{ij}^a = f_i h_{ij}^a + \sum_k \alpha_{ik} \langle ij|kk \rangle^a + \beta_{ik} \langle ik|jk \rangle^a \tag{60}$$

As with the closed-shell case, this matrix should be constructed from the derivative integrals in the atomic-orbital basis. Indeed, it is possible to solve the entire set of equations in the AO basis if desired.[108] From these equations, it can be seen that properties such as dipole moment derivatives can be obtained at the SCF level as easily for open-shell systems as is the case for closed-shell systems. Analytic second derivatives are also quite straightforward for all types of SCF wavefunction,[85,103,108,117] and consequently force constants, vibrational frequencies and normal coordinates can be obtained as well. It is also possible to use the full formulae for the second derivative of the energy to construct alternative expressions for the dipole derivative.

Though the SCF methods outlined above will be capable of providing a great deal of useful information, there will be cases where the SCF method is inadequate, and an approach that takes electron correlation into account will become necessary. One such case is the well known example of HCN[128,129] where SCF calculations give dipole moment derivatives with the wrong sign, and there must be other similar failures. As already mentioned, analytic second derivatie methods are possible for several types of energy expression, including CI[122] and second-order Moller–Plesset perturbation theory,[92] though most effort has concentrated on MC-SCF wavefunctions.[99,118–121] Obviously a formula for the dipole moment derivative can be obtained simply by using the full expression for the analytic second derivative of the energy for each of these approaches. However, some caution is needed in the way this is done, as there are two steps in the evaluation of the full second derivative of a correlated wavefunction which are particularly complex and time-consuming, and which it is desirable to avoid if at all possible. One is the transformation of the *derivative* integrals into the MO basis; the other is the solution of the CI or MC-SCF equivalents of the coupled Hartree–Fock equations with the nuclear displacements as the perturbations. As an illustration, the MC-SCF case will be examined. There are several formulations of this problem. The version given here will follow that of Hoffmann et al.[121] as their notation most closely resembles the SCF cases already discussed.

If the MC-SCF energy is written in terms of one- and two-particle density matrices

$$E = \sum_{ij} \gamma_{ij} h_{ij} + \sum_{ijkl} \Gamma_{ijkl} \langle ij|kl \rangle \tag{61}$$

where the density matrices are defined

$$\gamma_{ij} = \sum_{IJ} C_I C_J a_{ij}^{IJ} \tag{62}$$

and

$$\Gamma_{ijkl} = \sum_{IJ} C_I C_J b_{ijkl}^{IJ} \tag{63}$$

with the C being the MC-SCF configuration coefficients and a^{IJ} and b^{IJ} being the spin-coupling coefficients, then the gradient is

$$(\partial E/\partial a) = \sum_{ij} \gamma_{ij} h_{ij}^a + \sum_{ijkl} \Gamma_{ijkl} \langle ij|kl \rangle^a - \sum_{ij} \varepsilon_{ij} S_{ij}^a \tag{64}$$

Here ε is the MC-SCF Lagrangian

$$\varepsilon_{ij} = \sum_k \gamma_{ik} h_{jk} + \sum_{klm} 2\Gamma_{iklm} \langle jk|lm \rangle \tag{65}$$

Thus, as with SCF wavefunctions, the first derivative of an MC-SCF energy does not need the derivative of the wavefunction. However, to proceed to second derivatives, it is necessary to solve the MC-CHF equations (multi-configuration coupled Hartree–Fock). Unlike SCF, however, not only the orbital changes are required but also the changes in the configuration weights. Thus the set of equations has the structure

$$\begin{bmatrix} A^{(11)} & A^{(21)} \\ A^{(12)} & A^{(22)} \end{bmatrix} \begin{bmatrix} U^a \\ (\partial C/\partial a) \end{bmatrix} = \begin{bmatrix} B^{(1)} \\ B^{(2)} \end{bmatrix} \tag{66}$$

with the left-hand side of these equations being

$$A_{ij,kl}^{(11)} = Y_{ijkl} - Y_{jikl} - Y_{ijlk} + Y_{jilk} + \delta_{il}\varepsilon_{jk} - \delta_{ik}\varepsilon_{jl} - \delta_{jl}\varepsilon_{ik} + \delta_{jk}\varepsilon_{il}$$

$$A_{I,ij}^{(21)} = \sum_J C_J (X_{ij}^{IJ} - X_{ji}^{IJ}) \tag{67}$$

$$A_{I,J}^{(22)} = H_{IJ} - \delta_{IJ} E^0 + C_I C_J$$

In these equations

$$X_{ij}^{IJ} = \sum_k 2a_{ik}^{IJ} h_{jk} + \sum_{klm} 4b_{jklm}^{IJ} \langle ik|lm \rangle \tag{68}$$

and

$$Y_{ijmn} = \gamma_{jn} h_{im} + \sum_{kl} 2\Gamma_{jnkl} \langle in|kl \rangle + 4\Gamma_{jknl} \langle ik|ml \rangle \tag{69}$$

Though more complicated than the equivalent matrices for SCF wavefunctions (Eqs. (33) or (57)), the left-hand side of the MC-CHF equations is not a major problem as it consists of combinations of quantities that must be formed in a normal MC-SCF calculation of the energy. The right-hand side, however, causes more difficulty. This is

$$B_{ij}^{(1)} = \varepsilon_{ji}^a - \varepsilon_{ij}^a + \sum_{m>n} S_{mn}^a (Y_{ijnm} - Y_{jinm} + \delta_{im}\varepsilon_{jn} - \delta_{jm}\varepsilon_{in})$$

$$+ \tfrac{1}{2} \sum_m S_{mm}^a (Y_{ijmm} - Y_{jimm} + \delta_{im}\varepsilon_{jm} - \delta_{jm}\varepsilon_{im}) \tag{70}$$

and

$$B_I^{(2)} = \sum_J C_J \left(\sum_{i>j} S_{ij}^a X_{ij}^{IJ} + \tfrac{1}{2} \sum_i S_{ii}^a X_{ii}^{IJ} \right) + C_I E_0^a - \sum_J C_J H_{IJ}^a \qquad (71)$$

There are two quantities in these expressions that cause considerable technical difficulty, namely the derivative MC-SCF Lagrangian

$$\varepsilon_{ji}^a = \sum_k \gamma_{ik} h_{jk}^a + \sum_{klm} 2\Gamma_{iklm} \langle jk|lm \rangle^a \qquad (72)$$

and the derivative Hamiltonian elements

$$H_{IJ}^a = \sum_{ij} a_{ij}^{IJ} h_{ij}^a + \sum_{ijkl} b_{ijkl}^{IJ} \langle ij|kl \rangle^a \qquad (73)$$

The complexity of these equations means that the 'obvious' expression or the dipole derivative obtained by a literal-minded differentiation of the formula

$$\langle \mu_b \rangle = \sum_{ij} \gamma_{ij} h_{ij}^b$$

for the dipole moment, giving

$$(\partial \mu_b / \partial a) = \sum_{ij} \gamma_{ij} h_{ij}^{ab} + \sum_{ijk} 2\gamma_{ij} U_{ki}^a h_{kj}^b + \sum_{ij} (\partial \gamma_{ij} / \partial a) h_{ij}^a \qquad (74)$$

will result in a very considerable amount of computation in solving the MC-CHF equations for the nuclear perturbations. However, if the perturbation is an electric field, then most of the difficulties in the construction of the MC-CHF equations are removed, as then the right-hand side consists of just dipole moment integrals. The resulting equations are solved regularly in MC-SCF polarizability calculations (e.g. Refs. 130–133). Consequently, one should return to the expression for the MC-SCF gradient and differentiate this with respect to an external electric field, solving the MC-CHF equations with the field as the perturbation to obtain the changes in the orbitals and configuration coefficients. Note that, since the gradient is usually evaluated with the derivative integrals in the AO basis, i.e.

$$E^a = \sum_{pq} \gamma_{pq} h_{pq}^a + \sum_{pqrs} \Gamma_{pqrs} \langle pq|rs \rangle^a - \sum_{pq} \varepsilon_{pq} S_{pq}^a \qquad (75)$$

where

$$\gamma_{pq} = \sum_{ij} C_{pi} C_{qj} \gamma_{ij} \qquad (76)$$

etc., and since the electric field does not change the basis functions (usually), then it must be possible to collect all the field-dependent terms together to produce a modified set of density matrices, resulting in a dipole derivative expression with the general form

$$(\partial \mu_b / \partial a) = \sum_{pq} P_{pq}^b h_{pq}^a + \sum_{pqrs} Q_{pqrs}^b \langle pq|rs \rangle^a - \sum_{pq} R_{pq}^b S_{pq}^a + \sum_{pq} \gamma_{pq} h_{pq}^{ab} \qquad (77)$$

with

$$P_{pq}^b = (\partial/\partial b)\left(\sum_{ij} C_{pi} C_{qj} \gamma_{ij}\right) \tag{78}$$

and so on. Thus it must be possible to obtain MC-SCF dipole moment derivatives analytically without transforming the derivative integrals to the MO basis and without solving the MC-CHF equations for the $3N$ nuclear displacements. This means that a calculation of this type cannot be much more difficult than separate MC-SCF polarizability and gradient calculations, both of which are standard procedures.

Similar considerations will hold when using the CI method in this case, it seems to be necessary to solve for the *orbital* changes with respect to both nuclear displacements and electric fields, but the changes of the configuration coefficients are required only with respect to the field—the CI-CHF equations required are given by Daborn and Handy.[134] Moreover, it is not necessary to solve any second-order CHF equations—this is avoided by using the interchange theorem put forward by Handy and Schaefer.[135] The MP2 case is very similar, in that is it necessary to solve for both the field and nuclear displacement perturbations. In both CI and MP2 cases, it is possible to manipulate the expression for the dipole derivative such that it is not essential to transform the derivative integrals into the MO basis.

The use of the Hellmann–Feynman theorem has already been mentioned in connection with the proposal by Lazzeretti.[123,125] With CI wavefunctions, and also with methods like MBPT (many-body perturbation theory), the Hellmann–Feynman theorem is not satisfied when the perturbation is an electric field, even with field-independent basis sets. Consequently

$$\langle \Psi|\mu|\Psi \neq -(\partial E/\partial F)\rangle \tag{79}$$

This raises an interesting question; namely, in circumstances where there is a choice, is the dipole moment better calculated as an expectation value or as a derivative of the energy. Most authors favour the latter, and I would agree with this, though the situation is not as simple as it is sometimes made out to be. It is sometimes argued that the derivative of the energy is preferable as it reflects more closely what occurs in an experimental measurement. This argument is untenable. Experimental measurements are made in the real world, with real molecules, for which the Hellmann–Feynman theorem in all its forms, for all perturbations, is obeyed exactly. One cannot appeal to experiment to distinguish between the two definitions when as far as experiment is concerned the two definitions are identical. On the other hand, there are good arguments for favouring the derivative of the energy. One is consistency. Consider, first, a geometrical distortion. It is reasonable to require that the derivative of the energy, or of the dipole moment, when calculated analytically, should correspond to the value that would be obtained by pointwise calculation of the energy or dipole at several geometries. This is primarily an argument against replacing the gradient, i.e. the exact derivative, by the Hellmann–Feynman force, since the latter is often a poor approximation to the exact derivative, and does not even have the correct translational and rotational invariance.[82] If this argument is used for geometrical distortions, then it would be illogical not to use it for electric field distortions also. Admittedly there is an assumption in this reasoning, namely that the concept of a potential energy surface remains valid under all circumstances, and that other properties of a molecule should be derived from consideration of the properties of the energy surface. Then there are considerations based on pragmatic grounds. Consider the dipole moment derivative again. The discussions given earlier indicate that, if this property is to be obtained by an analytic technique, then the simplest way is through a specialized version of the formula for the second derivative of the energy, expressed as the field dependence of the gradient. Contradicting this, it would seem that, if a pointwise calculation is used, then the expectation value of the dipole operator would be simplest. This is true, but by borrowing a technique from analytic gradient methods the difference between the two forms is not as great as generally believed. The gradient of a CI wavefunction may be written[89,90]

$$(\partial E/\partial a) = \sum_{ij} \gamma_{ij} h^a_{ij} + \sum_{ijkl} \Gamma_{ijkl} \langle ij|kl \rangle^a + \sum_{ij} 2X_{ij} U^a_{ij} \qquad (80)$$

where γ and Γ are the one- and two-particle density matrices, h^a and $\langle ij|kl \rangle^a$ are the derivatives of the one- and two-electron integrals, X is the CI Lagrangian and the U^a represent the changes in the orbitals caused by the perturbation. At first glance, it looks as if this requires the solution of a set of CHF equations for each perturbation to obtain the U^a matrices. However, as pointed out by Handy and Schaefer,[135] there is a way of avoiding this. The CHF equations for U^a are of the form

$$AU^a = B^a$$

where details were given earlier. If instead the set of equations

$$A^T Z = X \qquad (81)$$

is solved, then

$$\sum_{ij} X_{ij} U^a_{ij} = \sum_{ij} Z_{ij} B^a_{ij} \qquad (82)$$

Note that Z is independent of the perturbation. If this is applied to the derivative of the energy with respect to an electric field then[136]

$$(\partial E/\partial a) = \sum_{ij} \gamma_{ij} h^a_{ij} + \sum_{ij} 2X_{ij} U^a_{ij} = \sum_{ij} \gamma_{ij} h^a_{ij} + \sum_{ij} 2Z_{ij} h^a_{ij} \qquad (83)$$

where in this case the h^a are just the dipole integrals. The first term ($\gamma \cdot h^a$) in this expression is just the expectation value of the dipole operator. The matrix Z will not be particularly time-consuming to construct (about the equivalent of an extra iteration of the CI). This particular discussion does not favour one possible definition of the dipole moment over the other, but simply points out that, as the two can be calculated with similar effort, there is no point in calculating just one. Obviously the approach described works for any property. It has been suggested that the derivative of the energy is a preferable way to calculate a molecular property as this allows for the relaxation of the orbitals. The expression for the CI dipole moment (Eq. 83) shows that the term added to the expectation value can either be interpreted as a relaxation of the orbitals due to the field, or to a partial re-optimization of the orbitals. This latter interpretation of the matrix Z arises because A is essentially the Hessian for orbital rotations, and the Lagrangian X is the gradient. There are also comments in the literature to the effect that expectation values are linear in the error in the wavefunction, while the energy and the energy gradient are of second-order accuracy.[137] This statement is not clear, but if correct is a further reason for preferring a definition based on the energy derivative.

Some of the above reasons for preferring the energy derivative over the expectation value will only hold for variational wavefunctions. However, Diercksen et al.[139] have argued that this technique is more suitable with MBPT methods as well. The techniques developed in analytic derivative methods can also be applied to the calculation of MBPT properties. In Moller–Plesset theory (the simplest form of MBPT), the zeroth-order wavefunction is SCF and the Hamiltonian is partitioned so that

$$H_0 = \sum_i E_i$$

is the sum of the Fock operators, and

$$H_1 = H - H_0$$

It is possible to define a property such as the dipole either in terms of the derivatives of the energy series

$$(\partial E/\partial \lambda) = (\partial E_{SCF}/\partial \lambda) + (\partial E^{(2)}/\partial \lambda) + (\partial E^{(3)}/\partial \lambda) + \cdots \tag{84}$$

or by collecting terms of equal order in the expansion of the wavefunction

$$\Psi = \Psi^{(0)} + \Psi^{(1)} + \Psi^{(2)} + \cdots \tag{85}$$

For example, the second-order correction to the expectation value is[138]

$$\langle \Psi^{(1)} | \mu | \Psi^{(1)} \rangle + 2 \langle \Psi^{(0)} | \mu | \Psi^{(2)} \rangle \tag{86}$$

and similarly with higher-order terms. There is no term-by-term equivalence between these two series, though both start at the same point, i.e. the SCF value of the dipole, and presumably both converge to the same point, which implies that any differences must eventually cancel out. The expression (86) is easily evaluated. In practice, it is used to define a second-order correction to the SCF one-particle density matrix, i.e. it becomes

$$\sum_{pq} D_{pq}^{(2)} \langle p | \mu | q \rangle \tag{87}$$

Assuming that the basis functions do not depend upon the electric field, then the derivative of $E^{(2)}$ can be manipulated into a form very similar to (86), i.e.

$$(\partial E^{(2)}/\partial \lambda) = \langle \Psi^{(1)} | (\partial H_0/\partial \lambda) | \Psi^{(1)} \rangle + 2 \langle \Psi^{(0)} | (\partial H_0/\partial \lambda) | \Psi^{(2)} \rangle \tag{88}$$

i.e. it is the expectation value of the derivative of the Fock operator. The interesting thing about this formula is that it does *not* contain the derivatives of $\Psi^{(1)}$ or $\Psi^{(2)}$. Instead, the dependence on the external field is through the perturbed Fock operator, which is easily evaluated. In this respect, it is the equivalent of the *first* of the two forms for the CI dipole given in Eq. (83). However, as with the CI case, the field-dependent term can be removed by use of the 'Z-matrix' idea of Handy and Schaefer.[135] The essential point of this idea is that Z is *independent* of the perturbation (it is determined by the orbital optimization conditions in the SCF wavefunction) and once determined can be used to obtain *any* first derivative of $E^{(2)}$, including that with respect to a nuclear displacement,[92] without having to solve any further CHF equations. Similar ideas can be used to simplify the higher-order terms. Thus the derivatives of $E^{(3)}$, $E^{(4)}$, etc., can either be expressed as *expectation values* of $(\partial H_0/\partial \lambda)$ and $(\partial H_1/\partial \lambda)$ or, even better, the 'Z-matrix' idea can be used.[93] This should make it much easier to obtain the MBPT corrections to any first-order property. When it comes to dipole derivatives and similar properties, it has already been mentioned that the simplest formula is that for the field dependence of the gradient. It is worth pointing out that one can also calculate Moller–Plesset corrections to the *expectation value* of the Hellmann–Feynman force, and that this approach could presumably be used to estimate dipole derivatives also. Whether there is any advantage in doing this would require extensive numerical tests. Further comments on this interesting area, not all of which the author agrees with, can be found in Refs. 139–141.

The conclusion of this section is that analytic calculation of dipole moment derivatives is practicable and efficient for the most widely used *ab initio* methods, i.e. SCF, MC-SCF, CI and Moller–Plesset perturbation theory. The author is not aware of any calculations of this type to date using correlated wavefunction methods, but it cannot be long before examples of these appear.

C. Polarizability Derivatives

There are several alternative formulae for the polarizability derivatives. One can be obtained from the following symmetric formula for the complete third derivative of the energy for closed-shell SCF wavefunctions[142]

$$(\partial^3 E/\partial a \partial b \partial c) = \sum h_{ii}^{abc} + \varepsilon_{ii}^{abc} - 2\varepsilon_i S_{ii}^{abc}$$
$$+ 4 \sum \sum (\varepsilon_{ji}^{(ab)} - \varepsilon_i S_{ji}^{ab}) U_{ji}^c + (cab) + (bca)$$
$$+ 4 \sum \sum (\varepsilon_{jk}^{(a)} - \varepsilon_i S_{jk}^a) U_{ji}^b U_{ki}^c + (cab) + (bca)$$
$$- 2 \sum (\varepsilon_{ij}^{(a)} - \varepsilon_i S_{ij}^a) \zeta_{ij}^{bc} + (cab) + (bca) \tag{89}$$

where (cab), (bca) denote permutations of the superscripts. This expression requires the first, second and third derivatives of the integrals. Thus

$$\varepsilon^{abc} = h^{abc} + g^{abc} * D \tag{90}$$

is a Fock operator constructed with third derivative integrals. The term $\varepsilon^{(ab)}$ is part of the second derivative of the Fock operator, but lacking the terms involving the second derivative of the wavefunction

$$\varepsilon^{(ab)} = h^{ab} + g^{ab} * D + g^a * D^b + g^b * D^a \tag{91}$$

and the factor ζ^{ab} is

$$\zeta_{ij}^{ab} = S_{ij}^{ab} + \sum_k^{\text{all}} U_{ik}^a U_{jk}^b + U_{ik}^b U_{jk}^a - S_{ik}^a S_{jk}^b - S_{ik}^b S_{jk}^a \tag{92}$$

The other factors, ε^a and $\varepsilon^{(a)}$, have already been encountered in the discussion on dipole moments. This expression is essentially the same as that given by Moccia[143] and Pulay.[99] Another expression, which after some manipulation can be shown to be equivalent to the above, has been given by Gaw *et al.*[144,145] It should be noted that the above expression (89) does not require the second derivative of the wavefunction. It does, however, require the solution of all first-order CHF equations.

If the above expression is used for polarizability derivatives, then no third derivative integrals appear, and the only new terms not present in dipole derivative calculations are $\varepsilon^{(ab)}$ and ζ^{ab}. The latter is simply constructed from the U^a and U^b matrices, and the former reduces to

$$\varepsilon^{(ab)} = h^{ab} + G^a * D^b \tag{93}$$

where a is a nuclear displacement and b an electric field, as there are no second derivatives of the two-electron integrals (assuming the basis functions have no explicit dependence upon the field). Terms such as ε^a, $\varepsilon^{(ab)}$, etc., can always be evaluated in the AO basis and consequently derivative integrals need never be transformed or stored. As $\varepsilon^{(ab)}$ can be constructed simultaneously with ε^a during the calculation of the first derivative integrals, polarizability derivative calculations require only a little more time than dipole derivatives.

An alternative approach would be to differentiate directly the expression

(29) for the dipole moment derivative

$$\partial^2 \mu_b / \partial a \partial c = \sum_{pq} D_{pq}^c \langle p | \mu_b | q \rangle^a + D_{pq}^{ac} \langle p | \mu_b | q \rangle \tag{94}$$

The term that would cause most difficulty here would be the second derivative D^{ac} of the density matrix. In the MO basis, this is

$$D_{ij}^{ac} = U_{ij}^{ac} f_j + U_{ji}^{ac} f_i + \sum_k (U_{ik}^a U_{jk}^c + U_{ik}^c U_{jk}^a) f_k \tag{95}$$

The second-order coefficients U^{ac} would be obtained by solving a set of equations of the form

$$AU^{ac} = B^{ac} \tag{96}$$

However, it is not necessary to solve these equations, as use may be made of an exchange theorem, which can be expressed most simply using the technique of Handy and Schaefer.[35] The coefficients U^{ac} would multiply elements of the dipole moment integrals $\langle p | \mu_b | q \rangle$, but these integrals are the right-hand side of a set of simultaneous equations which have already been solved, i.e.

$$AU^b = B^b$$

and from this equation and the previous one it follows that

$$U^{ac} B^b = U^b B^{ac} \tag{97}$$

The term B^{ac} is[92]

$$B_{ij}^{ac} = \varepsilon_{ij}^{(ac)} - \zeta_{ij}^{ac} \varepsilon_j$$

$$+ \sum_k^{all} (U_{ki}^a \varepsilon_{kj}^{(c)} + U_{kj}^a \varepsilon_{ik}^{(c)} + U_{ki}^c \varepsilon_{kj}^{(a)} + U_{kj}^c \varepsilon_{ik}^{(a)})$$

$$- \sum_k^{all} (U_{ki}^a U_{jk}^b \varepsilon_j + U_{ki}^b U_{jk}^a \varepsilon_j + U_{kj}^a U_{ik}^b \varepsilon_i + U_{kj}^b U_{ik}^a \varepsilon_i + U_{ki}^a U_{kj}^b \varepsilon_k$$

$$+ U_{ki}^b U_{kj}^a \varepsilon_k)$$

$$+ \sum_{kl}^{all} \sum_m^{occ} U_{km}^a U_{lm}^c (4 \langle ij | kl \rangle - \langle ik | jl \rangle - \langle il | jk \rangle)$$

$$- \tfrac{1}{2} \sum_{kl}^{occ} \zeta_{kl}^{ac} (4 \langle ij | kl \rangle - \langle ik | jl \rangle - \langle il | jk \rangle) \tag{98}$$

Thus instead of solving (96) explicitly, it is only necessary to construct the term (98). The resulting expression for the polarizability derivative obtained from (94) will be very similar to (89) except that it will not be symmetrical in the perturbations a, b and c.

A third possibility is to differentiate the expression for the electric field dependence of the gradient with respect to an additional field perturbation. This will also produce an expression containing second-order terms U^{bc} but in this case b and c will both be electric fields. These U^{bc} terms could be manipulated away in the same manner as described above, but as the solution of the second-order CHF equations (96) is comparatively simple for the case where both perturbations are electric fields, explicit calculation of U^{bc} is a practical possibility. The author's own preference, and the approach used in the calculations that will be described later, is the symmetrical formula (89), as this can be applied to any combination of perturbations. For example, it can equally well describe the hyperpolarizability (when (89)) reduces to the expression given by Lazzeretti[146] or the second derivative of a first-order property such as the dipole or quadrupole moment.

No detailed consideration has yet been made of fully analytic expressions for polarizability derivatives with correlated wavefunctions. Some general expressions exist,[140,141] though it should be noted that some of these were derived before the significance of the exchange theorem proposed by Handy and Schaefer became apparent. Of course, it is possible to obtain polarizability derivatives with the aid of an analytic gradient program through the technique of Komornicki and McIver,[97] or alternatively an analytic dipole moment derivative program could be used in the same manner. However, the history of analytic derivative techniques suggests that if something is useful, and polarizability derivatives with correlated wavefunctions certainly come into this category, then an efficient implementation will eventually be achieved. MC-SCF wavefunctions are the most probable candidate for the first such implementation, as all the necessary components are available at the end of an analytic second derivatives calculation.

Though this section has concentrated on analytic derivative methods, it should always be remembered that there are occasions when these are either not possible or not appropriate. For example, the complexity of these approaches means that there is always a time lag between the development of a particular method that can evaluate a property at one geometry and the development of the corresponding method for the derivative of the same property. To take a specific example, it is possible to use fully analytic methods to obtain derivatives of the *static* polarizability for SCF wavefunctions. However, there is as yet no equivalent for the *dynamic* polarizability. However, it is possible to calculate the dynamic polarizability with either SCF or MC-SCF wavefunctions (e.g. Refs. 130–133, 147–149) at specific geometries and determine the derivative of this quantity numerically. The frequency dependence of the polarizability may be of importance when determining the Raman intensities of some molecules, especially those with low-lying excited states.[149] A similar situation exists with paramagnetic quantities. As it currently stands, the only way to obtain the derivatives of important properties such as the magnetizability or the NMR shielding constants is by numerical differentiation. No doubt analytic approaches will be developed for these properties, but until they are, good calculations of the geometry dependence of $\alpha(\omega)$, χ^p or σ^p are limited to comparatively small systems, as a numerical approach would be very time-consuming for a large polyatomic. Having said that, it should not be assumed that an analytic approach will always be the best. Even if it is possible to use such an approach to obtain the first and second derivatives of a property, this actually describes only a small region of the property surface, and this would remain true even if third and fourth derivatives were available. If one wished to obtain the entire dipole moment function for a diatomic molecule, nobody would use a derivative-based method, and indeed one would not express such a function as an expansion about a single point. Similarly a truly definitive study of the properties of any molecule would necessitate a calculation of the complete potential energy and property surfaces over a wide range of geometries. For a large molecule, this would require an enormous amount of computing time, but there is probably no avoiding this. Meanwhile, the aim of the analytic derivative methods will be to provide geometries, harmonic, cubic and perhaps quartic force constants, first and second derivatives of dipole moments and polarizabilities, and similar properties, for polyatomic molecules, initially at the SCF level, eventually with MC-SCF of CI wavefunctions. This is not a minor goal but is one that will provide a great deal of information of use to both theoretician and experimentalist.

IV. RESULTS

A. A Brief Survey

This section is not intended to provide a complete bibliography of *ab initio* property calculations. Rather the intention is to give a guide to the most recent and the most accurate work in this field.

As might be expected, the molecule for which the most accurate results are available is H_2. Some highly detailed calculations have been made on the quadrupole moment,[150,151] the polarizability, both static[151–153,160] and dynamic,[154,155] and the magnetizability.[156,157] These results are in excellent agreement with the experimental values.[161–164] There are also calculations on some excited states for the quadrupole moment[158] and the polarizability.[80,159] Calculations on the first triplet state of H_2 are of interest as it is a model of the 'smallest van der Waals molecule', and a prototype for various collision-induced phenomena. Hunt and Buckingham[80] and Rychlewski[159] have made accurate calculations of the polarizability of this system, and compared the results with various models based on perturbation theory.

Other than H_2, the most accurate calculations are on the dipole moment derivatives and transition probabilities in small molecules, mostly diatomics. There are two groups of calculations of particular note. First, there is the work based on the Table CI program of Buenker and Peyerimhoff.[165–174] These are large-scale multi-reference CI calculations. Buenker and Peyerimhoff and coworkers have specialized in detailed calculations of the potential energy curves of small molecules, including many excited states. Among these calculations are some transition probability studies, including some work on spin-forbidden transitions, which has not been extensively investigated elsewhere.

Another group of calculations has used the CEPA (coupled electron-pair approximation) or SCEP (self-consistent electron pair) methods.[175-178] This is a large body of work to which many authors have contributed. It includes calculations on CO,[179] HF, HCl and HBr,[180] HI,[181] OH, OH^+ and OH^-,[182,183] C_2^-,[184] N_2^+,[185] SiO and SiO^+,[186,187] CN^-,[188] HF^+ and HCl^+,[182] NO^+,[189] CO^+,[190] AlF and AlF^+,[191] BeH, BH, CH, MgH and SiH,[192] and LiF.[193] Obviously these are not the only calculations in this area (though they are probably the most consistent), and some other recent work on small molecules may be found elsewhere.[194-205] A detailed review of this area has been given by Werner and Rosmus.[206] For some molecules, there have been a variety of calculations using different techniques, e.g. for OH.[177,182,183,197] The most recent of these calculations are in good agreement with each other, and also with experiment. Another much studied molecule is CO,[179,196,201-204] which is also one of the few for which the experimental dipole curve is well known.[50] The most detailed calculations are those by Diercksen et al.[201] using fourth-order MBPT. However, none of these calculations are in satisfactory agreement with experiment, though the dipole moment of CO is a notoriously tricky problem. This system probably needs a very large multi-reference CI calculation like those for OH.[183,197]

For triatomic molecules, there are inevitably fewer high-accuracy calculations, though obviously some of the standard 'theoretical' molecules such as H_2O have been studied (e.g. Ref. 207). An interesting group of papers has been initiated recently by Botschwina, who has been extending the CEPA/SCEP methods that have been successful on diatomics to larger systems. Calculations so far include HCN, HCP and C_2N_2,[208,209] CS and HCS^+,[210] CH_3,[211] NH_2^+[212] and diacetylene.[213] One interesting feature of some of these calculations is that the dipole moments are obtained as both expectation values and derivatives of the energy, and Botschwina concluded that the energy derivative method was preferable, as the result converged more rapidly to a stable value as the number of configurations was increased. The HCN molecule has received some study, as this is known to be a case where the SCF method fails, producing dipole derivatives with the wrong sign. Calculations have been made with various kinds of CI wavefunctions,[128,129,214] CEPA,[208] and most recently a large CA-SCF and CI calculation of a considerable part of the dipole surface, which was used to evaluate both fundamental and overtone intensities.[215] One odd feature of these calculations is that none correctly predict the intensity of the v_2 bending vibration. Another recent CAS-SCF calculation is on the ozone molecule, where again the dipole surface was obtained for a wide range of surfaces, and the fundamental and overtone intensities were evaluated.[217] In this case agreement with the experimental data is much better. Both of the previous examples have used a dipole surface and avoided use of the harmonic approximation. For molecules with low-frequency modes and large-amplitude motions, anharmonic effects will be particularly pronounced. Two examples of this are studies of KCN and LiCN[218] and of the $(HF)_2$ dimer.[216]

For molecules with more than three nuclei, there are few calculations using correlated wavefunctions. Apart from the calculations of Botschwina already mentioned, and $(HF)_2$,[216] the only results known to the author are some GVB (generalized valence bond) calculations on CH_4, C_2H_2, C_2H_4 and C_2H_3 by Dupuis and Wendoloski.[219] Doubtless more such calculations will be reported shortly. There are of course many SCF calculations on larger molecules. A representative sample of recent results can be found in Refs. 220–237. Many of these calculations use comparatively small basis sets—DZP (double-zeta plus polarization) seems typical—however, this does not necessarily invalidate such studies, as in many cases the purpose is to obtain *qualitative* information, for example the signs of the dipole derivatives, in order to assist the interpretation of experimental results. As yet, there are few calculations that have used fully analytic methods.[115,116,238] However, as several such programs now exist, it can be anticipated that the number of such calculations will increase rapidly. At the SCF level, the primary use of such programs will be in applications to polyatomic molecules, where the improved efficiency will either allow larger molecules to be studied or allow larger basis sets, thereby improving the qualitative results to give more plausible values. Genuinely quantitative calculations will, however, require something beyond the SCF level.

As will be gathered from the above, there is a fairly large literature on dipole moment derivatives and infrared intensities. Unfortunately, the situation with all other properties is much less healthy. Most calculations of the derivatives of quadrupole, or higher-order, moments are on small molecules,[75,115,200,201,203,239–242] and most of these could easily be improved with current techniques, including the author's own work on CO[203] and N_2.[241] Two recent studies that have made more detailed calculations are on LiH and CO,[200,201] the former using CAS-SCF and the latter MBPT techniques. These are probably the best of the geometry studies of quadrupole moments that have been completed. It is just as easy to obtain a quadrupole derivative using an analytic approach as it is to calculate a dipole derivative, and the theory is identical. The main reason why more calculations have not been done is probably not that it is difficult, but that there is a lack of interest in the results. Derivatives of the higher-order multipole moments have occasionally been given (e.g. Refs. 203, 240 and 241).

In the case of polarizability derivatives, however, the sparsity of results is not due to lack of interest, as this is a property that is just as important as the dipole moment derivative. Here the problem is that the calculations are more difficult, though not so much more difficult as to justify the comparatively small number of calculations in this area. There was a brief period of activity some five or six years ago in which various MC-SCF and CI methods were tried on small molecules.[68,240-245] Some earlier calculations are listed elsewhere.[243] As with the quadrupole moment results, most of these could easily be improved upon with the aid of a large-scale multi-reference CI calculation, which would be well within current capabilities. Some more recent polarizability derivative calculations, mostly SCF, may be found in Refs. 220 and 246-257. The most detailed of these is an MBPT calculation by Diercksen and Sadlej[256] on CO. Another interesting group of calculations has considered the derivatives of the *frequency-dependent* polarizability. This shows some expected effects, for example that the frequency dependence in Cl_2 is noticeable,[149] and some unexpected results, for example that the intensity of the v_4 Raman-active mode of CH_4 has a very marked frequency dependence.[257] Dacre[251-255] has provided some calculations on the polarizability of rare-gas dimers, which is of interest to the collision-induced Raman spectrum of such systems. Calculations of hyperpolarizabilities are confined to small systems. A recent example is for LiH.[258] An example of the use of hyperpolarizability derivatives can be found[75] where some fairly crude calculations were nevertheless useful in distinguishing two possible mechanisms in the collision-induced Raman spectrum of CO_2.

There has been very little study of magnetic properties recently, and one of the few to consider the geometry dependence is the calculation on H_2O by Fowler and Raynes.[259]

B. Some Analytic Calculations

First, some large basis set SCF calculations of the infrared and Raman intensities in H_2O, NH_3 and CH_4 will be given. The choice of basis set is one of computational chemistry's perennial problems. For some particularly simple systems, it is possible to draw some conclusions from symmetry. For example, in an atom with occupied s and p orbitals, the dipole polarizability cannot be obtained reliably without d functions, and the quadrupole polarizability would require f functions. Similarly in a linear molecule with occupied σ and π orbitals, a polarizability calculation would require δ orbitals. The same argument implies that the geometric derivative of the wavefunction in the same linear molecule would also require δ orbitals. However, these arguments cannot be absolute. For a start, the language used is based upon the SCF model. More particularly, most molecules have no symmetry—it is possible in principle to reach the SCF limit for a molecule with no symmetry by using only s-type basis functions, provided their positions are not confined to be those of the nuclei, though obviously an exorbitant number of such functions would be required. However, most calculations will use basis functions centred on the nuclei, in which case the empirical rules derived from high-symmetry systems provide a good practical guide to the requirements elsewhere, i.e. if s and p functions are important in the zeroth-order wavefunction, then d functions (or higher angular momentum functions) will be useful to describe perturbed systems. The weakness of arguments based on symmetry or perturbation theory is that, though they can tell one that it is impossible to reach the basis set limit unless certain types of functions are present, they cannot predict *how many* functions are required, nor what their exponents should be. When it comes to deciding how many polarization functions are required, and what their exponents should be, some other criteria are needed. The most basic requirement is to minimize the total energy of the system. This must take precedence. However, other rules can be used, particularly in property calculations. For example, various sum rules or closure relationships exist which check the completeness of a basis set.[260] These can point to weaknesses in a basis and indicate where an additional function should be added. When calculating polarizabilities, it is advisable to have equally good descriptions of the zeroth-order and perturbed wavefunctions. One possibility is to choose some basis function exponents to minimize the energy of the molecule in the presence of a small perturbing electric field. This approach was used by the author in calculations on CO and N_2.[203,243] Similar ideas have been suggested for quadrupole perturbations.[261] Another idea is the 'geometric series' (or even-tempered) approach, i.e. that one should use a basis set containing sets of functions whose exponents are related by a constant ratio. This idea was used by Werner and Meyer[262] in one of the first adequate studies of molecular polarizabilities, and has since been used by many other authors. However, though all the above ideas have been used to assist in the design of basis sets, the actual choice is usually determined on much cruder grounds. Basis sets are adopted not according to what is desirable, but by what is computationally achievable. That is, the real rule by which basis sets tend to be chosen is that the basis set used is the largest possible given the available computational resources. Obviously this depends upon the molecule being studied. It is also a function of time—a decade ago calculations with a DZP basis were described as 'near Hartree–Fock'; five years ago calculations tended to be about the 'triple-zeta' level. i.e. something like 6s4p2d on a first-row atom; currently a calculation cannot be safely described as 'near Hartree–Fock' unless one uses a large s and p basis, plus several sets of d functions and one or two sets of f functions. No doubt in a few years g functions will become a regular feature of small-molecule calculations. These developments do not represent any increased understanding—they are purely the result of the increase in computational resources over the last decade. The following calculations are designed to investigate basis set effects in H_2O, NH_3 and CH_4. They start with small basis sets and are increased in size till the basis sets are approaching the limit of what is currently reasonable. The main purpose of doing this is to see if the infrared and Raman intensities stabilize, in the hope that the lessons learned from this can be transferred to larger molecules.

The smallest basis sets used on H_2O are simple 'double-zeta' sets, which can be expected to be inadequate for all but qualitative purposes. To be more specific, these are the standard 4-31G and 6-31G sets,[263] and the Dunning[264] double-zeta basis. The next stage is to add a set of polarization functions on each atom, giving 4-31G**, 6-31G** and DZP basis sets. Following this, some basis sets with two shells of polarization functions were used. One of these used the Dunning 5s4p contraction,[265] with d functions of exponents 1.2 and 0.4 on the oxygen, and the Dunning 3 s basis set[265] on hydrogen, with p functions of exponent 0.75 and 0.25. At a similar level is the basis denoted '6-31G extended'. This is generated by adding low-exponent s and p functions to the 6-31G basis—the exponents were taken to be one-third of those of the most diffuse functions in that basis—and using two sets of polarization functions, with exponents the same as those added to the 5s4p set. The largest basis sets are all built around an 8s6p contraction of the van Duijneveldt[266] 13s8p set of primitive Gaussians. This was gradually extended to 8s6p4d2f—the oxygen d functions had exponents 2.7, 0.9, 0.3 and 0.1; the f function exponents were 1.5 and 0.5. The final hydrogen basis was 6s3p1d, being a 6s contraction of the 10s van Duijneveldt set, plus three p functions with exponents 1.8, 0.6 and 0.2, and a d function with exponent 1.0. The basis sets for NH_3 and CH_4 were generated in a similar manner. In the 5s4p2d and 6-31G ext. sets, the nitrogen d function exponents were 1.0 and 0.33; the equivalent carbon basis sets had d functions with exponents 0.8 and 0.27. The largest nitrogen and carbon basis sets were 8s6p4d1f. The nitrogen d functions were 2.4, 0.8, 0.27 and 0.09, with an f function 1.0; the carbon d function exponents were 1.8, 0.6, 0.2 and 0.07, and the f function exponent was again 1.0. The energies obtained are -76.06749, -56.22489 and -40.21678 for H_2O, NH_3 and CH_4 respectively.

With each of these basis sets, the geometry of each molecule was fully optimized and the dipole derivatives, polarizability derivatives and normal modes were obtained using the analytic methods described in Section III. It was probably not necessary to calculate the force constants with the largest basis sets, as these were changing less rapidly than the dipole or polarizability derivatives; however, for the sake of consistency, every calculation was performed in the same manner, so that the differences are only due to the changes in the basis sets. The infrared intensities are given in Tables I–III. First of all, consider just the variation with basis set without any regard to the final accuracy. It is clear that the smallest basis sets are inadequate except for qualitative purposes (if the intensities are ordered according to strength, all basis sets give the same order). It is also clear that with the larger basis sets the results are approaching some kind of limit. The principal difficulty is in deciding what is meant by converged, and what is an acceptable degree of deviation from the final result, particularly with low-intensity modes. The overall conclusions are that all results with three or more sets of d functions are essentially converged, and that f functions do not make a great deal of difference; that the results with two sets of polarization functions are nearly correct (although the 5s4p2d basis for CH_4 seems out of step), and that the 6-

TABLE I

Infrared intensities (km mol^{-1}) for H_2O for various basis sets.[a]

	v_2	v_1	v_3
4–31G	120	2.8	50
6–31G	123	2.9	54
DZ	136	3.4	65
4–31G**	89	19	51
6–31G**	90	20	56
DZP	96	24	74
6–31G ext.	90	14	81
5s4p2d/3s2p	91	14	80
8s6p3d/6s2p	101	17	88
8s6p4d/6s3p	98	15	92
8s6p4d1f/6s3p	97	15	92
8s6p4d2f/6s3p1d	97	15	92
Experimental[b]	53.6	2.2	44.6
D_2O			
8s6p4d2f/6s3p1d	51	10	54

[a] See text for details of basis sets used in all tables.
[b] From Zilles and Person.[222]

TABLE II

Infrared intensities (km mol^{-1}) for NH_3 for various basis sets.

	v_2	v_4	v_1	v_3
6–31G**	192	34	0.29	1.1
6–31G ext.	187	38	0.6	10
5s4p2d/3s2p	181	37	0.6	11
8s6p3d/6s2p	188	46	0.7	10
8s6p4d/6s3p	176	37	1.2	11
8s6p4d1f/6s3p1d	179	38	1.2	13
ND_3				
8s6p4d1f/6s3p1d	104	22	0.04	11

TABLE III
Infrared intensities (km mol^{-1}) for CH_4 for various basis sets.

	v_4	v_2	v_1	v_3
4–31G	61	0	0	114
6–31G	64	0	0	113
4–31G**	23	0	0	128
6–31G**	25	0	0	120
6–31G ext.	26	0	0	115
5s4p2d/3s2p	30	0	0	98
8s6p3d/6s2p	28	0	0	115
8s6p4d/6s3p	28	0	0	114
8s6p4d1f/6s3p1d	28	0	0	114
Experimental[a]	33	0	0	67
CD_4				
8s6p4d1f/6s3p1d	18	0	0	54

[a] From Hiller and Straley.[270]

31G ext. basis in particular is quite successful; and that basis sets with only one or no polarization functions are unlikely to be reliable.

Before comparing the SCF results with available experimental data, Tables IV and V give two more examples of the variation of infrared intensities with basis set size, for C_2H_2 and C_2H_4. The basis sets used are not quite as large as in Tables I–III, and in particular no f functions have been used, and no d functions on the hydrogens. Accordingly, the convergence of the results to a limit cannot be demonstrated so convincingly. However, the basic features are

TABLE IV
Infrared intensities (km mol^{-1}) for C_2H_2 for various basis sets.

	$v_3(\sigma)$	$v_5(\pi)$
4–31G	94	265
6–31G	98	282
4–31G**	86	180
6–31G**	92	193
6–31G ext.	100	246
5s4p2d/3s1p	106	238
5s4p2d/3s2p	97	225
7s5p4d/5s3p	104	228
Experimental[a]	76 ± 1	194 ± 5
C_2D_2		
5s4p2d/3s2p	52	121

[a] From Smit et al.[271]

TABLE V
Infrared intensities (km mol^{-1}) for C_2H_4 for various basis sets.

	$v_7(b_{1u})$	$v_9(b_{2u})$	$v_{10}(b_{2u})$	$v_{11}(b_{3u})$	$v_{12}(b_{3u})$
4–31G	131	44	1.0	22	7.9
6–31G	140	43	1.1	21	8.6
4–31G**	88	42	0.09	26	7.0
6–31G**	96	38	0.1	23	7.5
6–31G ext.	122	28	0.007	20	12
5s4p2d/3s1p	111	22	0.02	15	11
5s4p2d/3s2p	106	24	0.0	16	12
7s5p4d/5s3p	121	19	0.05	18	12
Experimental^a	84	26	0.03	14	10
C_2D_4 7s5p4d/5s3p	60	13	0.0	7.3	7.1

^{a}From Nakanaga et al.[272]

the same as with H_2O, NH_3 and CH_4, and in particular the largest changes come when going from one to two sets of polarization functions. Ideally, still larger calculations should be done on these molecules, and this would be feasible with current programs.

When it comes to comparison with the experimental values, the results for C_2H_4 are fair—agreement is not perfect, but the results are plausible. For C_2H_2 the results are less good, both intensities being high by 15–20%. The real disaster, however, is H_2O. Here the calculated SCF values are seriously in error for all modes. Indeed, qualitatively the results are worse than for HCN, which is usually cited as the standard case where the SCF method fails in dipole moment derivative calculations. The CH_4 values are not convincing either, with one mode being in error by about 100%. It is disconcerting that the errors are greatest for the two smallest molecules for which the most detailed calculations have been made.

For completeness, the Cartesian dipole derivatives are also given. These are quoted in their atomic polar tensor (APT) form (Tables VI–X). For H_2O, NH_3, CH_4, C_2H_2 and C_2H_4, the APT need only be given for one of the hydrogen atoms in each molecule, as all other derivatives can be generated from this using a combination of symmetry and the translational sum rules, which are obeyed exactly in an analytic calculation, provided that the geometry is known. The results are only given for the largest basis sets.

Two further sets of calculated IR intensities are given. Table XI gives results for H_2CO with a 5s4p2d basis set (3s2p on H). Table XII gives some results for benzene, with 6–31G** basis, and with a larger basis obtained by adding an extra set of d functions. This last example, with 156 basis functions, shows that the approach described for intensity calculations is practicable for fairly large

TABLE VI

APT for H_2O with a 8s6p4d2f/6s3p1d basis. The oxygen atom is at the origin, and the hydrogens are at ($\pm$1.421 623, 0, 1.065 461) a_0. The derivatives are quoted for the H at (1.421 623, 0, 1.065 461). In atomic units.

	$\partial/\partial x$	$\partial/\partial y$	$\partial/\partial z$
μ_x	0.2870	0	-0.0580
μ_y	0	0.3578	0
μ_z	-0.0531	0	0.2198

TABLE VII

APT for NH_3, using 8s6p4d1f/6s3p1d basis. The origin is at the centre of mass, with the nitrogen at $(0, 0, -0.119 033)$, and the hydrogens at $(1.762 927, 0, 0.551 302)$ and $(-0.881 463, \pm 1.526 740, 0.551 302)$ a_0. The derivatives quoted are for the hydrogen in the xz plane. In atomic units.

	$\partial/\partial x$	$\partial/\partial y$	$\partial/\partial z$
μ_x	0.0988	0	-0.0986
μ_y	0	0.1609	0
μ_z	-0.1312	0	0.2081

TABLE VIII

APT for CH_4 with 8s6pd1f/6s3p1d basis. The carbon atom is at the origin and the hydrogens are in the xz and yz planes. The derivatives are given for the hydrogen at (1.668 767, 0, 1.179 996) a_0. In atomic units.

	$\partial/\partial x$	$\partial/\partial y$	$\partial/\partial z$
μ_x	-0.0951	0	-0.1084
μ_y	0	0.0582	0
μ_z	-0.1084	0	-0.0184

TABLE IX

Dipole derivatives of C_2H_2 in atomic units. The positions of the nuclei are: carbon $(0, 0, \pm 1.1368)$, hydrogen $(0, 0, \pm 3.1411)$. The derivatives given are those with respect to the hydrogen at $(0, 0, 3.1411)$. For the C:7s5p4d, H:5s3p basis.

	$\partial/\partial x$	$\partial/\partial y$	$\partial/\partial z$
μ_x	0.2337	0	0
μ_y	0	0.2337	0
μ_z	0	0	0.2234

TABLE X

Dipole derivatives for C_2H_4 in atomic units. The positions of the nuclei are: carbon $(0, 0, \pm 1.2652)$, hydrogen $(0, \pm 1.7556, \pm 2.3142)$. The derivatives given are those with respect to the hydrogen at $(0, 1.7556, 2.3142)$. For the C:7s5p4d, H:5s3p basis.

	$\partial/\partial x$	$\partial/\partial y$	$\partial/\partial z$
μ_x	0.1634	0	0
μ_y	0	−0.0579	−0.0309
μ_z	0	−0.0873	0.0104

systems. As with the results given earlier, the values for H_2CO and C_6H_6 have mixed success—for some modes there is good agreement with the experimental values, but with others the calculated results can be in error by a factor of 2 or more. This seems to be typical of SCF results, and indicates that, although they will be of value in obtaining estimates of IR intensities,

TABLE XI

Calculated infrared intensities $(km\,mol^{-1})$, Raman intensities $(A^4\ amu^{-1})$ and depolarization ratios for H_2CO, with a 5s4p2d/3s2p basis.

	IR calc	IR expt[a]	Raman	ρ
$v_1(a_1)$	58	75	124	0.12
$v_2(a_1)$	148	74	15	0.21
$v_3(a_1)$	20	11	7	0.46
$v_4(b_2)$	91	87	52	0.75
$v_5(b_2)$	20	9.9	1.5	0.75
$v_6(b_1)$	3.5	6.5	0.03	0.75

[a]From Nakanaga et al.[273]

TABLE XII

Infrared intensities $(km\,mol^{-1})$ in benzene with 6–31G** and 6–31G(2d, p) basis sets.

	6–31G**	6–31G(2d, p)	Expt[a]
v_{11}	101	93	88
v_{18}	6.2	5.0	8.8
v_{19}	24	24	13.0
v_{20}	116	116	59.8

[a]From Spedding and Whiffen.[274]

particularly in larger molecules where there are few reliable experimental data, the use of correlated wavefunctions will be essential for accurate predictions. The basis set investigations indicate that, to ensure the values are *fully* converged, then large basis sets are needed, including several sets of polarization functions. This implies that basis sets of this size will be needed when correlated wavefunctions are used. However, for SCF, if it is accepted that the results may be significantly in error no matter what basis is used, then

TABLE XIII

Raman intensities ($A^4 amu^{-1}$) of the H_2O molecule with various basis sets.

	v_2	v_1	v_3
4–31G	10	100	39
6–31G	10	71	25
DZ	12	91	42
4–31G**	15	58	20
6–31G**	6.4	73	35
DZP	7.3	73	36
6–31G ext.	2.0	81	27
5s4p2d/3s2p	2.2	73	25
8s6p3d/6s2p	1.2	73	28
8s6p4d/6s3p	0.67	85	24
8s6p4d1f/6s3p	0.67	85	25
8s6p4d2f/6s3p1d	0.68	85	24
Experimental[a]	0.9 ± 0.2	108 ± 14	19.2 ± 2.1
D_2O			
8s6p4d2f/6s3p1d	0.38	44	13

[a] From Murphy.[275]

TABLE XIV

Raman intensities ($A^4 amu^{-1}$) for NH_3 with various basis sets.

	v_2	v_4	v_1	v_3
6–31G**	11	20	119	110
6–31G ext.	1.1	3.8	139	86
5s4p2d/3s2p	2.5	5.6	130	79
8s6p3d/6s2p	1.6	4.7	122	82
8s6p4d/6s3p	0.47	1.9	148	79
8s6p4d1f/6s3p1d	0.42	1.8	147	78
ND_3				
8s6p4d1f/6s3p1d	0.25	1.0	75	42

there is no point in using an extremely large basis just to approach an SCF limit. A useful compromise appears to be the TZ + 2P type of basis, i.e. triple-zeta using two sets of polarization functions.

Polarizability derivatives and Raman intensities have also been calculated for the same set of molecules used in the IR investigations. Tables XIII–XV

TABLE XV
Raman intensities (A^4 amu^{-1}) for CH_4 with various basis sets.

	v_4	v_2	v_1	v_3
4–31G	8.4	77	138	203
6–31G	9.0	79	146	200
4–31G**	5.7	53	137	189
6–31G**	5.8	55	143	186
6–31G ext.	0.03	13	211	168
5s4p2d/3s2p	0.36	15	179	133
8s6p3d/6s2p	0.009	14	193	173
8s6p4d/6s3p	0.06	6.7	227	151
Experimental[a]	0.24	7.0	230	128

[a]From Bermejo et al.[59]

TABLE XVI
Raman depolarization ratios of the H_2O molecule with various basis sets.

	v_2	v_1	v_3
4–31G	0.40	0.72	0.75
6–31G	0.61	0.07	0.75
DZ	0.41	0.21	0.75
4–31G**	0.19	0.22	0.75
6–31G**	0.50	0.17	0.75
DZP	0.51	0.16	0.75
6–31G ext.	0.56	0.08	0.75
5s4p2d/3s2p	0.57	0.10	0.75
8s6p3d/6s2p	0.74	0.07	0.75
8s6p4d/6s3p	0.75	0.07	0.75
8s6p4d1f/6s3p	0.75	0.07	0.75
8s6p4d2f/6d3p1d	0.75	0.07	0.75
Experimental[a]	0.74	0.03	0.75
D_2O 8s6p4d2f/6s3p1d	0.73	0.07	0.75

[a]From Murphy[275]

TABLE XVII
Raman depolarization ratios for NH_3 calculated with various basis sets.

	v_2	v_4	v_1	v_3
6–31G**	0.145	0.75	0.066	0.75
6–31G ext.	0.726	0.75	0.030	0.75
5s4p2d/3s2p	0.091	0.75	0.035	0.75
8s6p3d/6s2p	0.749	0.75	0.040	0.75
8s6p4d/6s3p	0.621	0.75	0.026	0.75
8s6p4d1f/6s3p1d	0.650	0.75	0.027	0.75
ND_3				
8s6p4d1f/6s3p1d	0.583	0.75	0.028	0.75

TABLE XVIII
Raman intensities $(A^4 \, amu^{-1})$ for C_2H_2 with various basis sets.

	v_4	v_2	v_1
4–31G	10	55	56
6–31G	12	60	55
4–31G**	5.0	60	56
6–31G**	6.5	64	53
6–31G ext.	14	122	41
5s4p2d/3s2p	15	98	41
Experimental[a]	4.1 ± 1.1	125 ± 11	75 ± 15
C_2D_2			
5s4p2d/3s2p	20	87	0.2
Experimental[a]	8.4 ± 1.5	119 ± 16	6.8 ± 1.5

[a]From Orduna et al.[61]

TABLE XIX
Raman intensities $(A^4 \, amu^{-1})$ in C_2H_4 with various basis sets.

	$v_8(b_{2g})$	$v_6(b_{1g})$	$v_3(a_g)$	$v_2(a_g)$	$v_1(a_g)$	$v_5(b_{1g})$
4–31G	5.2	1.0	49	15	184	136
6–31G	5.7	1.2	50	16	189	129
4–31G**	2.9	1.0	46	18	185	133
6–31G**	3.5	1.2	46	19	186	126
6–31G ext.	7.4	0.5	58	55	195	119
5s4p2d/3s2p	9.1	0.9	49	38	168	97
C_2D_4						
5s4p2d/3s2p	11.2	2.4	16	75	56	47

TABLE XX

Calculated frequencies (cm^{-1}), infrared intensities (km mol^{-1}), Raman intensities (A^4 amu^{-1}) and depolarization ratios for C_2H_6 with a 6–31G ext. basis.

Mode	ω	IR	Raman calc	Raman expt[a]	ρ
a_{1u}	330	0	0	0	0
e_u	874	4.2	0	0	0
a_{1g}	1051	0	16	13	0.21
e_g	1317	0	0.3	0.6	0.75
a_{2u}	1512	0.1	0	0	0
a_{1g}	1543	0	0.2	0.2	0.01
e_g	1613	0	23	17.8	0.75
e_u	1615	15	0	0	0
a_{2u}	3158	74	0	0	0
a_{1g}	3164	0	332	305	0.1
e_g	3207	0	270	290	0.75
e_u	3234	198	0	0	0.

[a] From Snyder.[276]

TABLE XXI

Cartesian polarizability derivatives in H_2O, in atomic units. Geometry as in Table VI.

	$\partial/\partial x$	$\partial/\partial y$	$\partial/\partial z$
α_{xx}	4.4602	0	2.5717
α_{yy}	1.1149	0	0.5729
α_{zz}	2.2988	0	2.1355
α_{xy}	0	0.4006	0
α_{xz}	2.0106	0	1.7238
α_{yz}	0	0.2443	0

TABLE XXII

Cartesian polarizability derivatives in NH_3, in atomic units. Geometry as in Table VII.

	$\partial/\partial x$	$\partial/\partial y$	$\partial/\partial z$
α_{xx}	7.075	0	2.2757
α_{yy}	1.2255	0	0.4671
α_{zz}	2.0705	0	0.2683
α_{xy}	0	0.5468	0
α_{xz}	2.1906	0	0.6976
α_{yz}	0	0.2953	0

contain Raman intensities for H_2O, NH_3 and CH_4, and Tables XVI and XVII the depolarization ratios for H_2O and NH_3 (depolarization ratios in CH_4 are all symmetry-determined). Tables XVIII and XIX contain results for C_2H_2 and C_2H_4 but with a smaller range of basis sets, and Table XX contains the results of a calculation on C_2H_6. Looking at all these results, two features emerge. The first, which is to be expected, is that the polarizability derivatives converge more slowly than the IR results. Somewhat less expected is the fact that the answers are in quite good agreement with the available experimental data. Overall the degree of agreement, though not complete, is much better than was found for the dipole moment derivatives. It is not clear why this should be the case, as the effects of electron correlation on polarizabilities is

TABLE XXIII

Cartesian polarizability derivatives in CH_4, in atomic units. Geometry as in Table VIII.

	$\partial/\partial x$	$\partial/\partial y$	$\partial/\partial z$
α_{xx}	5.6978	0	3.4850
α_{yy}	1.3547	0	0.4866
α_{zz}	2.8084	0	3.0011
α_{xy}	0	0.7179	0
α_{xz}	3.0347	0	2.1715
α_{yz}	0	0.0363	0

TABLE XXIV

Frequencies (cm^{-1}) of the vibrational modes of the water dimer, with the changes from the monomers, with a 6–31G ext. basis. The intensities $(km\, mol^{-1})$ of the intra- and intermolecular modes are given, with the monomer intensities.

		Intensity	
Mode[a]	Frequency	Dimer	Monomer
v_1A	4123(−5)	22	14
v_2A	1751(+6)	110	91
v_3A	4226(−9)	103	81
v_1D	4076(−52)	191	14
v_2D	1767(+21)	65	91
v_3D	4213(−22)	127	81
a″ shear	589	154	
a′ shear	356	88	
a′ stretch	174	257	
a″ torsion	153	17	
a′ bend	142	64	
a″ bend	129	180	

[a]A = acceptor; D = proton donor molecule.

TABLE XXV

The Raman intensity factors (A^4 amu^{-1}) and the depolarization ratios for the $(H_2O)_2$ dimer.

	Intensity		Depolarization ratio	
Mode[a]	Dimer	Monomer	Dimer	Monomer
$v_1 A$	62	81	0.09	0.08
$v_2 A$	1.5	2.0	0.70	0.56
$v_3 A$	25	27	0.75	0.75
$v_1 D$	111	81	0.11	0.08
$v_2 D$	1.0	2.0	0.62	0.56
$v_3 D$	37	27	0.46	0.75
a'' shear	0.003		0.75	
a' shear	0.41		0.71	
a' stretch	0.03		0.51	
a'' torsion	0.86		0.75	
a' bend	0.01		0.63	
a'' bend	0.66		0.75	

[a] A = acceptor; D = proton doner molecule.

not negligible, and there are other factors that have been neglected, such as anharmonic effects and the frequency dependence of the polarizability tensor. Unfortunately, it will be difficult to test whether this is a general feature, or something peculiar to the molecules considered here, as there are very few reliable experimental results for Raman intensities. Conclusions regarding depolarization ratios are similar to those for intensities, i.e. convergence and accuracy are about the same. One thing to note is the large fluctuations with small basis sets, i.e. there are large changes in the details of the calculation with considerable basis set effects on the ratio of the anisotropic-to-isotropic derivatives. The Cartesian polarizability derivatives for H_2O, NH_3 and CH_4 are given in Tables XXI–XXIII.

As a final qualitative example, Tables XXIV and XXV contain calculations on the water dimer.[142,267] One interesting feature of hydrogen-bonded dimers is that the infrared intensity of the proton donor molecule is significantly enhanced, often by an order of magnitude.[267–269] This is clearly seen in the $(H_2O)_2$ calculation. The behaviour of the Raman intensities is less well known, though there have been calculations on $(H_2O)_2$.[142,249] It seems that the effect of the hydrogen-bond formation are much less marked than in the IR case. The intermolecular modes have very little Raman activity.

V. CONCLUSIONS

This chapter has considered the impact on molecular property derivative calculations of the rapid development in recent years of analytic gradient methods. This field continues to advance rapidly, and future developments will be most significant. All approximate wavefunctions have advantages and disadvantages, and whether these are serious depends upon the type of calculation being attempted. Property surfaces require large basis sets, which in turn necessitates efficiency. If only the first few derivatives are required—and in many circumstances this will be all that is needed—then the analytic methods have a clear advantage. This still leaves unanswered the question of which type of wavefunction is best.

SCF approaches can yield much useful information, and will be widely used, especially for larger molecules. The available evidence suggests that these methods are sufficiently accurate to answer many qualitative questions, providing information on the signs of dipole moment derivatives, showing which modes are weak and which strong, and giving approximate, and often plausible, values for the intensities. The polarizability derivative calculations reported here, which are among a very small group that have used large basis sets, have proved surprisingly effective. Ultimately, however, SCF methods are insufficiently accurate, and a method that considers some degree of electron correlation will be essential.

MC-SCF methods look very attractive. Several rapidly convergent large-scale MC-SCF programs are now available. As these methods dissociate correctly, and consequently describe large areas of the potential energy surface, they are also suitable for the calculation of property surfaces. Some impressive results have already been achieved with CAS-SCF wavefunctions (e.g. Ref. 217). The development of analytic second derivative programs for MC-SCF wavefunctions (e.g. Refs. 99 and 118–121) means that similar approaches must be possible for property surfaces as well. Dipole moment derivatives are certainly possible; polarizability derivatives should also be achievable. There are likely to be a considerable number of calculations of this type in the next few years. However, like SCF, MC-SCF ultimately does not have sufficient accuracy—the percentage of the correlation energy recovered is not high enough.

Both CI and MBPT methods are capable of obtaining much larger fractions of the correlation energy than MC-SCF methods, but these methods too are not without difficulties. Gradients and analytic second derivatives have been implemented for second-order Moller–Plesset calculations.[85,92] Dipole moment derivative calculations at the same level would be fairly straightforward, and some preliminary investigations reveal that these correct many of the deficiencies of SCF calculations. Consequently, this approach is likely to be useful as MP2 is sufficiently simple and cheap to be applied to large molecules, which is where dipole derivative calculations are most needed. MP2 will account for a significant fraction of the correlation effects, perhaps 60–70%. However, it cannot account for it all and it is here that the main defect of MBPT becomes apparent, for once past second order the series is slowly convergent, particularly for properties. This can be seen from some of the calculations of Diercksen and Sadlej, e.g. the very detailed study of CO^{256} which concluded that to achieve definitive values one would have to go to fifth order or beyond. Obviously CI wavefunctions do not have this particular problem, but a difficulty with restricted CI wavefunctions that is potentially just as serious is the lack of size consistency/extensivity.

Size consistency refers to the fact that as a system is pulled apart the energy should tend to the sum of the energies of the isolated subsystems. Size extensivity requires that the energy has the correct functional dependence on the number of electrons. Neither of these conditions is exactly satisfied by a restricted CI wavefunction, the problems being a result of the truncation of the CI series at a given excitation level. There is also the question of the Hellmann–Feynman theorem mentioned earlier. The practical, if not the conceptual, problem with the Hellmann–Feynman theorem has been largely solved through the use of analytic methods as there should be no difficulty in calculating and comparing both definitions. Size consistency and extensivity are more fundamental long-term problems, which will also affect certain categories of MC-SCF wavefunction, as these arise from an actual error in the *physics* of these wavefunctions, whereas the Hellmann–Feynman theorem is essentially a choice of approximation. Another difficulty, which will be shared by MBPT, is that, if the CI calculation uses SCF orbitals, then the incorrect dissociation of the underlying SCF calculation will degrade the CI results when away from the equilibrium geometry. In the context of gradient calculations, this presumably means that the higher-order derivatives will become inaccurate. This last feature can be largely eliminated by using a multi-reference CI calculation based on MC-SCF orbitals. The size consistency problem can be reduced by including some estimate of the effects of higher-order excitations. It is probably no accident that the most consistently successful dipole moment derivative calculations in small molecules has used an approach, CEPA/SCEP, which has these last two features. Such wavefunctions are not ideal—the size consistency is not exact, and most forms of CEPA are not completely variational; however, they seem to give very good results in practice. The current main difficulty is that applications have been limited to comparatively small systems. One reason for this is that fully analytic gradient and second derivative methods for these types of wavefunction have not yet been established in a computationally efficient form. However, the knowledge to do this probably now exists, and it is only a matter of time before these approaches become established techniques. When this occurs, the effects upon high-accuracy calculations on polyatomic molecules will be very significant.

REFERENCES

1. Dunham, J. L., *Phys. Rev.*, **41**, 713 (1932).
2. Buckingham, A. D., *J. Chem. Phys.*, **36**, 3096 (1962).

3. Herman, R. M., and Short, S., *J. Chem. Phys.*, **48**, 1266 (1968); *J. Chem. Phys.*, **50**, 572 (1969).
4. Buckingham, A. D., and Urland, W., *Chem. Rev.*, **75**, 113 (1975).
5. Toyama, S. M., Oka, T., and Morino, Y., *J. Mol. Spectrosc.*, **13**, 193 (1964).
6. Krohn, B. J., Ermler, W. C., and Kern, C. W., *J. Chem. Phys.*, **60**, 22 (1974).
7. Riley, G., Raynes, W. T., and Fowler, P. W., *Mol. Phys.*, **38**, 877 (1979).
8. Fowler, P. W., *Mol. Phys.*, **43**, 591 (1981); *Mol. Phys.*, **51**, 1423 (1984).
9. Patel, D., Margolese, D. M., and Dyke, T. R., *J. Chem. Phys.*, **70**, 2740 (1979).
10. DeLeon, R. L., and Muenter, J. S., *J. Chem. Phys.*, **80**, 3992 (1984).
11. Ebenstein, W. L., and Muenter, J. S., *J. Chem. Phys.*, **80**, 3989 (1984).
12. Reinartz, J. M. L. J., and Dymanus, A., *Chem. Phys. Lett.*, **24**, 346 (1974).
13. Tanaka, K., Ito, H., Harada, K., and Tanaka, T., *J. Chem. Phys.*, **80**, 5893 (1984).
14. Tanaka, K., Tanaka, T., and Suzuki, I., *J. Chem. Phys.*, **82**, 2835 (1985).
15. Meertz, W. L., de Leeuw, F. H., and Dymanus, A., *Chem. Phys.*, **22**, 319 (1977).
16. Dyke, T. R., and Muenter, J. S., *J. Chem. Phys.*, **59**, 3125 (1973).
17. Viswanathan, R., and Dyke, T. R., *J. Mol. Spectrosc.*, **103**, 231 (1984).
18. Johns, J. W. C., and McKellar, A. R. W., *J. Mol. Spectrosc.*, **64**, 327 (1977).
19. Allegrini, M., Johns, J. W. C., and McKellar, A. R. W., *J. Mol. Spectrosc.*, **66**, 69 (1977).
20. Fabricant, B., Krieger, D., and Muenter, J. S., *J. Chem. Phys.*, **67**, 1576 (1977).
21. Weber, W. H., *J. Mol. Spectrosc.*, **107**, 405 (1984).
22. de Leeuw, F., and Dymanus, A., *J. Mol. Spectrosc.*, **48**, 427 (1973).
23. Joyner, C. H., Dixon, T. A., Baiocchi, F. A., and Klemperer, W., *J. Chem. Phys.*, **74**, 6550 (1981).
24. Baiocchi, F. A., Dixon, T. A., Joyner, C. H., and Klemperer, W., *J. Chem. Phys.*, **74**, 6544 (1981).
25. Fraser, G. T., Nelson, D. D., Charo, A., and Klemperer, W., *J. Chem. Phys.*, **82**, 2535 (1985).
26. Maroncelli, M., Hopkins, G. A., Nibler, J. W., and Dyke, T. R., *J. Chem. Phys.*, **83**, 2129 (1985).
27. Raynes, W. T., in *Nuclear Magnetic Resonance*, Vols. 8 and 9, Chemical Society Specialist Periodical Reports, Chemical Society, London, 1978 and 1979.
28. Jameson, C. J., in *Nuclear Magnetic Resonance*, Vols. 10 to 14, Chemical Society Specialist Periodical Reports, Chemical Society, London, 1980 to 1985.
29. Osten, H. J., and Jameson, C. J., *J. Chem. Phys.*, **82**, 4595 (1985).
30. Osten, H. J., and Jameson, C. J., *J. Chem. Phys.*, **81**, 4288 (1984).
31. Jameson, C. J., and Osten, H. J., *J. Chem. Phys.*, **81**, 4293 (1984).
32. Jameson, C. J., and Osten, H. J., *J. Chem. Phys.*, **81**, 4300 (1984).
33. Overend, J., in *Infrared Spectroscopy and Molecular Structure* (Ed. M. M. Davies), Elsevier, Amsterdam, 1963.
34. Person, W. B., and Newton, J. H., *J. Chem. Phys.*, **61**, 1040 (1974).
35. Newton, J. H., and Person, W. B., *J. Chem. Phys.*, **64**, 3036 (1976).
36. Crawford, B. L., *J. Chem. Phys.*, **20**, 977 (1952).
37. de Barros Neto, B., and Bruns, R. E., *J. Chem. Phys.*, **71**, 5042 (1979).
38. Person, W. B., and Zerbi, G. (Eds.), *Vibrational Intensities in Infrared and Raman Spectroscopy*, Elsevier, Amsterdam, 1982.
39. Bishop, D. M., and Chueng, L. M., *J. Phys. Chem. Ref. Data*, **11**, 119 (1982).
40. Yao, S. J., and Overend, J., *Spectrochim. Acta*, **A32**, 1059 (1976).
41. Overend, J., *J. Chem. Phys.*, **64**, 2878 (1976).
42. Person, W. B., and Overend, J., *J. Chem. Phys.*, **66**, 1442 (1977).
43. Herman, R., and Wallis, R. F., *J. Chem. Phys.*, **23**, 637 (1955).
44. Pine, A. S., Fried, A., and Elkins, J. W., *J. Mol. Spectrosc.*, **109**, 30 (1985).

45. Sileo, R. N., and Cool, T. A., *J. Chem. Phys.*, **65**, 117 (1976).
46. Kaiser, E. W., *J. Chem. Phys.*, **53**, 1686 (1970).
47. Kaiser, E. W., *J. Mol. Spectrosc.*, **77**, 143 (1977).
48. Kim, K., and King, W. T., *J. Chem. Phys.*, **71**, 1967 (1979).
49. Smith, I. W. M., *J. Chem. Soc. Faraday Trans. 2*, **77**, 2357 (1981).
50. Chackerian, C., Farrenq, R., Guelachvili, G., Rossetti, C., and Urban, W., *Can. J. Phys.*, **62**, 1579 (1984).
51. Chackerian, C., and Tipping, R. H., *J. Mol. Spectrosc.*, **99**, 431 (1983).
52. Millie, P., *Can. J. Chem.*, **63**, 278 (1985).
53. Brown, K. G., and Person, W. B., *J. Chem. Phys.*, **65**, 2367 (1976).
54. Suzuki, I., *J. Mol. Spectrosc.*, **80**, 12 (1980).
55. Long, D. A., *Raman Spectroscopy*, McGraw-Hill, New York, 1977.
56. Bogaard, M. P., and Haines, R., *Mol. Phys.*, **41**, 1281 (1980).
57. Schrotter, H. W., and Klockner, H. W., *Raman Spectroscopy of Gases and Liquids*, Topics in Current Physics, Vol. 11, Springer-Verlag, Berlin, 1979.
58. Montero, S., and Bermejo, D., *Mol. Phys.*, **32**, 1229 (1976).
59. Bermejo, D., Escribana, R., and Orza, J. M., *J. Mol. Spectrosc.*, **65**, 345 (1977).
60. Martin, J., and Montero, S., *J. Chem. Phys.*, **80**, 4610 (1984).
61. Orduna, F., Domingo, C., Montero, S., and Murphy, W. F., *Mol. Phys.*, **45**, 65 (1982).
62. Buckingham, A. D., and Szabo, A., *J. Raman Spectrosc.*, **7**, 46 (1978).
63. Hamaguchi, H., Suzuki, I., and Buckingham, A. D., *Mol. Phys.*, **43**, 963 (1981).
64. Hamaguchi, H., Buckingham, A. D., and Jones, W. J., *Mol. Phys.*, **43**, 1311 (1981).
65. Dows, D. A., and Buckingham, A. D., *J. Mol. Spectrosc.*, **12**, 189 (1964).
66. Hush, N. S., and Williams, M. L., *J. Mol. Spectrosc.*, **50**, 349 (1974).
67. Courtois, D., and Jouvre, P., *J. Mol. Spectrosc.*, **55**, 18 (1975).
68. Gready, J. E., Bacskay, G. B., and Hush, N. S., *Chem. Phys.*, **24**, 333 (1977).
69. Gready, J. E., Bacskay, G. B., and Hush, N. S., *Chem. Phys.*, 31, 467 (1978).
70. Cohen de Lara, E., Kahn, R., and Seloudoux, R., *J. Chem. Phys.*, **83**, 2646 (1985).
71. Kahn, R., Cohen de Lara, E., and Moller, K., *J. Chem. Phys.*, **83**, 2653 (1985).
72. Buckingham, A. D., *Adv. Chem. Phys.*, **12**, 107 (1967).
73. Crawford, M. F., Welsh, H. L., and Locke, J. L., *Phys. Rev.*, **75**, 1607 (1949).
74. Cox, T. I., and Madden, P. A., *Mol. Phys.*, **39**, 1487 (1980).
75. Amos, R. D., Buckingham, A. D., and Williams, J. W., *Mol. Phys.*, **39**, 1519 (1980).
76. Silberstein, L., *Phil. Mag.*, **33**, 92 (1917).
77. Buckingham, A. D., and Clarke, K. L., *Chem. Phys. Lett.*, **57**, 321 (1978).
78. Clarke, K. L., Madden, P. A., and Buckingham, A. D., *Mol. Phys.*, **36**, 301 (1978).
79. Oxtoby, D. W., and Gelbart, W. M., *Mol. Phys.*, **30**, 535 (1975).
80. Hunt, K. L. C., and Buckingham, A. D., *J. Chem. Phys.*, **72**, 2832 (1980).
81. Tabisz, G. C., *Molecular Spectroscopy*, Vol. 6, Chemical Society Specialist Periodical Reports, Chemical Society, London, 1979.
82. Pulay, P., *Mol. Phys.*, **17**, 197 (1969).
83. Pulay, P., in *Modern Theoretical Chemistry* (Ed. H. F. Schaefer), Vol. 4, Plenum Press, New York, 1977.
84. Goddard, J. D., Handy, N. C., and Schaefer, H. F., *J. Chem. Phys.*, **71**, 1525 (1979).
85. Pople, J. A., Krishnan, R., Schlegel, H. B., and Binkley, J. S., *Int. J. Quantum Chem. Symp.*, **13**, 225 (1979).
86. Kato, S., and Morokuma, K., *Chem. Phys. Lett.*, **65**, 19 (1979).
87. Schlegel, H. B., and Robb, M. A., *Chem. Phys. Lett.*, **93**, 43 (1982).
88. Knowles, P. J., Sexton, G. J., and Handy, N. C., *Chem. Phys.*, **72**, 337 (1982).
89. Brooks, B. R., Laidig, W. D., Saxe, P., Handy, N. C., and Schaefer, H. F., *Phys. Scr.*, **21**, 312 (1980).

90. Brooks, B. R., Laidig, W. D., Saxe, P., Goddard, J. D., Yamaguchi, Y., and Schaefer, H. F., *J. Chem. Phys.*, **72**, 4652 (1980).
91. Krishnan, R., Schlegel, H. B., and Pople, J. A., *J. Chem. Phys.*, **72**, 4654 (1980).
92. Handy, N. C., Amos, R. D., Gaw, J. F., Rice, J. E., and Simandiras, E. D., *Chem. Phys. Lett.*, **120**, 151 (1985).
93. Fitzgerald, G., Harrison, R. J., Laidig, W. D., and Bartlett, R. J., *J. Chem. Phys.*, **82**, 4379 (1985).
94. Adamowicz, L., Laidig, W. D., and Bartlett, R. J., *Int. J. Quantum Chem. Symp.*, **18**, 245 (1984).
95. Fitzgerald, G., Harrison, R. J., Laidig, W. D., and Bartlett, R. J., *Chem. Phys. Lett.*, **117**, 433 (1985).
96. Cohen, H. D., and Roothaan, C. C. J., *J. Chem. Phys.*, **43**, S34 (1965).
97. Komornicki, A., and McIver, J. W., *J. Chem. Phys.*, **70**, 2014 (1979).
98. Gerratt, J., and Mills, I. M., *J. Chem. Phys.*, **49**, 1719 (1968).
99. Pulay, P., *J. Chem. Phys.*, **78**, 5043 (1983).
100. Dupuis, M., Rys, J., and King, H. F., *J. Chem. Phys.*, **65**, 111 (1976).
101. Rys, J., Dupuis, M., and King, H. F., *J. Comput. Chem.*, **4**, 154 (1983).
102. Dupuis, M., and King, H. F., *J. Chem. Phys.*, **68**, 3998 (1978).
103. Saxe, P., Yamaguchi, Y., and Schaefer, H. F., *J. Chem. Phys.*, **77**, 5647 (1982).
104. Schlegel, H. B., Binkley, J. S., and Pople, J. A., *J. Chem. Phys.*, **80**, 1976 (1984).
105. Amos, R. D., *Chem. Phys. Lett.*, **108**, 347 (1984).
106. Dupuis, M., in *Geometrical Derivatives of Energy Surfaces and Molecular Properties* (Eds. J. Simons and P. Jorgensen), Reidel, Dordrecht, 1986.
107. Schlegel, H. B., *J. Chem. Phys.*, **77**, 3676 (1982).
108. Osamura, Y., Yamaguchi, Y., Saxe, P., Fox, D. J., Vincent, M. A., and Schaefer, H. F., *J. Mol. Struct.* (*Theochem*), **103**, 234 (1983).
109. Daunders, V. R., and van Lenthe, J. H., *Mol. Phys.*, **48**, 923 (1983).
110. Mupuis, M., and King, H. F., *Int. J. Quantum Chem.*, **11**, 613 (1977).
111. Dupuis, M., and King, H. F., *J. Chem. Phys.*, **68**, 3998 (1978).
112. Takada, T., Dupuis, M., and King, H. F., *J. Chem. Phys.*, **75**, 332 (1981).
113. Takada, T., Dupuis, M., and King, H. F., *J. Comput. Chem.*, **4**, 234 (1983).
114. Carsky, P., Hess, B. H., and Schaad, L. S., *J. Comput. Chem.*, **5**, 280 (1984).
115. Amos, R. D., *Chem. Phys. Lett.*, **108**, 185 (1984).
116. Amos, R. D., *Chem. Phys. Lett.*, **114**, 10 (1985); *Chem. Phys. Lett.*, **122**, 180 (1985).
117. Osamura, Y., Yamaguchi, Y., Saxe, P., Vincent, M. A., Gaw, J. F., and Schaefer, H. F., *Chem. Phys.*, **72**, 131 (1982).
118. Camp, R. N., King, H. F., McIver, J. W., and Mullally, D., *J. Chem. Phys.*, **79**, 1088 (1983).
119. Page, M., Saxe, P., Adams, G. F., and Lengsfield, B. H., *J. Chem. Phys.*, **81**, 434 (1984).
120. Jorgensen, P., and Simons, J., *J. Chem. Phys.*, **79**, 334 (1983).
121. Hoffmann, M. R., Fox, D. J., Gaw, J. F., Osamura, Y., Yamaguchi, Y., Grev, R. S., Fitzgerald, G., Schaefer, H. F., Knowles, P. J., and Handy, N. C., *J. Chem. Phys.*, **80**, 2660 (1984).
122. Fox, D. J., Osamura, Y., Hoffmann, M. R., Gaw, J. F., Fitzgerald, G., Yamaguchi, Y., and Schaefer, H. F., *Chem. Phys. Lett.*, **102**, 17 (1983).
123. Lazzeretti, P., and Zanasi, R., *Chem. Phys. Lett.*, **112**, 103 (1984); *J. Chem. Phys.*, **83**, 1218 (1985).
124. Fowler, P. W., and Buckingham, A. D., *Chem. Phys.*, **98**, 167 (1985).
125. Lazzeretti, P., and Zanasi, R., *J. Chem. Phys.*, **84**, 3916 (1986).
126. Nakatsuji, H., Kanda, K., and Yonezawa, T., *Chem. Phys. Lett.*, **75**, 340 (1980).

127. Yamaguchi, Y., Osamura, Y., and Schaefer, H. F., *J. Am. Chem. Soc.*, **105**, 7506 (1983).
128. Gready, J. E., Bacskay, G. B., and Hush, N. S., *J. Chem. Phys.*, **70**, 1071 (1979).
129. Lie, G. C., Peyerimhoff, S. D., and Buenker, R. J., *J. Mol. Spectrosc.*, **93**, 74 (1982).
130. Dalgaard, E., *J. Chem. Phys.*, **72**, 816 (1980).
131. McWeeny, R., *Int. J. Quantum Chem.*, **23**, 405 (1983).
132. Yeager, D. L., Olsen, J., and Jorgensen, P., *Int. J. Quantum Chem. Symp.*, **15**, 151 (1981).
133. Reinsch, E. A., *J. Chem. Phys.*, **83**, 5784 (1985).
134. Daborn, G. T., Ferguson, W. I., and Handy, N. C., *Chem. Phys.*, **50**, 255 (1980).
135. Handy, N. C., and Schaefer, H. F., *J. Chem. Phys.*, **81**, 5031 (1984).
136. Rice, J. E., Amos, R. D., Handy, N. C., Lee, T. J., and Schaefer, H. F., *J. Chem. Phys.*, **85**, 963 (1986)
137. Rosmus, P., and Werner, H. J., in *Geometrical Derivatives of Energy Surfaces and Molecular Properties* (Eds. J. Simons and P. Jorgensen), Reidel, Dordrecht, 1986.
138. Amos, R. D., *Chem. Phys. Lett.*, **73**, 602 (1980); *Chem. Phys. Lett.*, **88**, 89 (1982).
139. Diercksen, G. H. F., Roos, B. O., and Sadlej, A. J., *Chem. Phys.*, **59**, 29 (1981).
140. Almlof, J., and Taylor, P. R., *Int. J. Quantum Chem.*, **27**, 743 (1985).
141. Simons, J., and Jorgensen, P., *Int. J. Quantum Chem.*, **25**, 1135 (1984).
142. Amos, R. D., *Chem. Phys. Lett.*, **124**, 376 (1986).
143. Moccia, R., *Chem. Phys. Lett.*, **5**, 260 (1970).
144. Gaw, J. F., Yamaguchi, Y., and Schaefer, H. F., *J. Chem. Phys.*, **81**, 6395 (1984).
145. Gaw, J. F., and Handy, N. C., *Chem. Phys. Lett.*, **121**, 321 (1985).
146. Lazzeretti, P., and Zanasi, R., *J. Chem. Phys.*, **74**, 5216 (1981).
147. Jorgensen, P., *Annu. Rev. Phys. Chem.*, **26**, 359 (1975).
148. Amos, R. D., Handy, N. C., Knowles, P. J., Rice, J. E., and Stone, A. J., *J. Phys. Chem.*, **89**, 2186 (1985).
149. Oddershede, J., and Svendsen, E. N., *Chem. Phys.*, **64**, 359 (1982).
150. Kolos, W., and Wolniewicz, L., *J. Chem. Phys.*, **43**, 2429 (1965).
151. Poll, J. D., and Wolniewicz, L., *J. Chem. Phys.*, 3053 (1978).
152. Hunt, J. L., Poll, J. D., and Wolniewicz, L., *Can. J. Phys.*, **62**, 1719 (1984).
153. Kolos, W., and Wolniewicz, L., *J. Chem. Phys.*, **46**, 1426 (1967).
154. Cheung, L. M., Bishop, D. M., Drapcho, D. L., and Rosenblatt, G. M., *Chem. Phys. Lett.*, **80**, 445 (1981).
155. Rychlewski, J., *J. Chem. Phys.*, **78**, 7252 (1983).
156. Rychlewski, J., and Raynes, W. T., *Mol. Phys.*, **41**, 843 (1980).
157. Raynes, W. T., Riley, J. P., Davies, A. M., and Cook, D. B., *Chem. Phys. Lett.*, **24**, 139 (1974).
158. Rychlewski, J., *J. Chem. Phys.*, **80**, 2643 (1984).
159. Rychlewski, J., *J. Chem. Phys.*, **81**, 6007 (1984).
160. Hunt K. L. C., and Buckingham, A. D., *J. Chem. Phys.*, **72**, 2832 (1982).
161. McAdam, K. B., and Ramsey, N. F., *Phys. Rev. A*, **6**, 898 (1972).
162. Buckingham, A. D., and Cordle, J. E., *Mol. Phys.*, **28**, 1037 (1974).
163. Barnes, R. G., Bray, P. J., and Ramsey, N. F., *Phys. Rev.*, **94**, 893 (1954).
164. Quinn, W. E., Baker, J. M., la Tourrette, J. T., and Ramsey, N. F., *Phys. Rev.*, **112**, 1929 (1958).
165. Buenker, R. J., and Peyerimhoff, S. D., *Theor. Chim. Acta*, **35**, 33 (1974).
166. Buenker, R. J., and Peyerimhoff, S. D., *Theor. Chim. Acta*, **39**, 217 (1975).
167. Buenker, R. J., and Peyerimhoff, S. D., in *New Horizons of Quantum Chemistry* (Eds. P. O. Löwdin and B. Pullman), Reidel, Dordrecht, 1983.
168. Peyerimhoff, S. D., *Faraday Symp.*, **19**, 63 (1983).

169. Chabalowski, C. F., Peyerimhoff, S. D., and Buenker, R. J., *Chem. Phys. Lett.*, **83**, 441 (1981).
170. Dohmann, H., Bruna, P. J., Peyerimhoff, S. D., and Buenker, R. J., *Mol. Phys.*, **51**, 1109 (1984).
171. Chabalowski, C. F., Peyerimhoff, S. D., and Buenker, R. J., *Chem. Phys.*, **81**, 57 (1983).
172. Klotz, R., Marian, C. M., Peyerimhoff, S. D., Hess, B. A., and Buenker, R. J., *Chem. Phys.*, **89**, 223 (1984).
173. Marian, C. M., and Klotz, R., *Chem. Phys.*, **95**, 213 (1985).
174. Zeitz, M., Peyerimhoff, S. D., and Buenker, P. R., *Chem. Phys. Lett.*, **64**, 243 (1979).
175. Meyer, W., *J. Chem. Phys.*, **58**, 1017 (1973).
176. Meyer, W., *J. Chem. Phys.*, **64**, 2901 (1976).
177. Werner, H. J., *J. Chem. Phys.*, **80**, 5080 (1984).
178. Werner, H. J., and Reinsch, E. A., *J. Chem. Phys.*, **76**, 3144 (1982).
179. Werner, H. J., *Mol. Phys.*, **44**, 111 (1981).
180. Werner, H. J., and Rosmus, P., *J. Chem. Phys.*, **73**, 2319 (1980).
181. Werner, H. J., Reinsch, E. A., and Rosmus, P., *Chem. Phys. Lett.*, **78**, 311 (1981).
182. Werner, H. J., Rosmus, P., and Reinsch, E. A., *J. Chem. Phys.*, **79**, 905 (1983).
183. Werner, H. J., Rosmus, P., Schatzl, W., and Meyer, W., *J. Chem. Phys.*, **80**, 831 (1984).
184. Rosmus, P., and Werner, H. J., *J. Chem. Phys.*, **80**, 5085 (1984).
185. Werner, H. J., Kalcher, J., and Reinsch, E. A., *J. Chem. Phys.*, **81**, 2420 (1984).
186. Botschwina, P., and Rosmus, P., *J. Chem. Phys.*, **82**, 1420 (1985).
187. Werner, H. J., Rosmus, P., and Grimm, M., *Chem. Phys.*, **73**, 169 (1982).
188. Botschwina, P., *Chem. Phys. Lett.*, **114**, 58 (1985).
189. Werner, H. J., and Rosmus, P., *J. Mol. Spectrosc.*, **96**, 362 (1982).
190. Rosmus, P., and Werner, H. J., *Mol. Phys.*, **47**, 661 (1982).
191. Klein, R., and Rosmus, P., *Theor. Chim. Acta*, **66**, 21 (1984).
192. Meyer, W., and Rosmus, P., *J. Chem. Phys.*, **63**, 2356 (1975).
193. Werner, H. J., and Meyer, W., *J. Chem. Phys.*, **74**, 5802 (1981).
194. Langhoff, S. R., Bauschlicher, C. W., and Partridge, *Chem. Phys. Lett.*, **102**, 292 (1983).
195. Langhoff, S. R., and Arnold, J. O., *J. Chem. Phys.*, **70**, 852 (1979).
196. Cooper, D. M., and Langhoff, S. R., *J. Chem. Phys.*, **74**, 1200 (1981).
197. Langhoff, S. R., van Dischoek, E. F., Wetmore, R., and Dalgarno, A., *J. Chem. Phys.*, **77**, 1379 (1982).
198. Cooper, D. L., Gerratt, J., and Raimondi, M., *Chem. Phys. Lett.*, **118**, 580 (1985).
199. Partridge, H., and Langhoff, S. R., *J. Chem. Phys.*, **74**, 2361 (1981).
200. Roos, B. O., and Sadlej, A. J., *Chem. Phys.*, **94**, 43 (1985).
201. Diercksen, G. H. F., and Sadlej, A. J., *Chem. Phys.*, **96**, 17 (1985).
202. Kirby-Docken, K., and Lui, B., *J. Chem. Phys.*, **66**, 4309 (1977).
203. Amos, R. D., *Chem. Phys. Lett.*, **68**, 536 (1979).
204. Bauschlicher, C. W., *Chem. Phys. Lett.*, **118**, 307 (1985).
205. Larsson, M., *Chem. Phys. Lett.*, **117**, 331 (1985).
206. Werner, H. J., and Rosmus, P., in *Comparison of Ab Initio Calculations With Experiment—The State of the Art* (Ed. R. Bartlett), Reidel, Dordrecht, in press.
207. Rosenberg, B. J., Ermler, W. C., and Shavitt, I., *J. Chem. Phys.*, **65**, 4072 (1976).
208. Botschwina, P., *Chem. Phys.*, **81**, 73 (1983).
209. Botschwina, P., *J. Mol. Struct.*, **88**, 371 (1982).
210. Botschwina, P., and Sebald, P., *J. Mol. Spectrosc.*, **110**, 1 (1985).
211. Botschwina, P., Flesch, J., and Meyer, W., *Chem. Phys.*, **74**, 321 (1983).

212. Botschwina, P., *Chem. Phys. Lett.*, **107**, 535 (1984).
213. Botschwina, *Mol. Phys.*, **47**, 241 (1982).
214. Lie, G. C., Peyerimhoff, S. D., and Buenker, R. J., *J. Chem. Phys.*, **75**, 2892 (1981).
215. Jorgensen, U. G., Almhof, J., Gustafsson, B., Larsson, M., and Siegbahn, P., *J. Chem. Phys.*, **83**, 3034 (1985).
216. Michael, D. W., Dykstra, C. E., and Lisy, J. M., *J. Chem. Phys.*, **81**, 5998 (1984).
217. Adler-Golden, S. M., Langhoff, S. R., Bauschlicher, C. W., and Carney, G. D., *J. Chem. Phys.*, **83**, 255 (1985).
218. Brocks, G., Tennyson, J., and van der Avoird, A., *J. Chem. Phys.*, **80**, 3223 (1984).
219. Dupuis, M., and Wendoloski, J. J., *J. Chem. Phys.*, **80**, 5696 (1984).
220. Bacskay, G. B., Saebo, S., and Taylor, P. R., *Chem. Phys.*, **90**, 215 (1984).
221. Swanton, D. J., Bacskay, G. B., and Hush, N. S., *Chem. Phys.*, **82**, 303 (1983).
222. Zilles, B. A., and Person, W. B., *J. Chem. Phys.*, **79**, 65 (1983).
223. Fogarasi, G., and Pulay, P., *Acta Chim. Acad. Sci. Hung.*, **108**, 55 (1981).
224. Hess, B. A., Schaad, L. J., and Polavarapu, P. L., *J. Am. Chem. Soc.*, **106**, 4348 (1984).
225. Polavarapu, P. L., Hess, B. A., and Schaad, L. J., *J. Mol. Spectrosc.*, **109**, 22 (1985).
226. Hess, B. A., Carsky, P., and Schaad, L. J., *J. Am. Chem. Soc.*, **105**, 695 (1983).
227. Carsky, P., Hess, B. A., and Schaad, L. J., *J. Am. Chem. Soc.*, **105**, 396 (1983).
228. Wiberg, K. B., and Wendoloski, J. J., *J. Am. Chem. Soc.*, **100**, 723 (1978).
229. Pulay, P., Fogarasi, G., Pang, F., and Boggs, J. E., *J. Am. Chem. Soc.*, **101**, 2550 (1979).
230. Pulay, P., Fogarasi, G., Porigar, G., Boggs, J. E., and Varga, A., *J. Am. Chem. Soc.*, **105**, 7037 (1983).
231. Colvin, M., Raine, G. P., Schaefer, H. F., and Dupuis, M., *J. Chem. Phys.*, **79**, 1551 (1983).
232. von Carlowitz, S., Zeil, W., Pulay, P., and Boggs, J. E., *J. Mol. Struct.*, **87**, 113 (1982).
233. Komornicki, A., and Jaffe, R. L., *J. Chem. Phys.*, **71**, 2150 (1979).
234. Fredkin, D. R., Komornicki, A., White, S. R., and Wilson, K. B., *J. Chem. Phys.*, **78**, 7077 (1983).
235. Pulay, P., Fogarasi, G., and Boggs, J. E., *J. Chem. Phys.*, **74**, 3999 (1981).
236. Jalsovszky, G., and Pulay, P., *J. Mol. Struct.*, **26**, 277 (1975).
237. Almhof, J., and Faegri, K., *J. Chem. Phys.*, **79**, 2284 (1983).
238. Lee, T. J., and Schaefer, H. F., *J. Chem. Phys.*, **83**, 1784 (1985).
239. Williams, J. H., and Amos, R. D., *Chem. Phys. Lett.*, **66**, 370 (1979).
240. Amos, R. D., *Mol. Phys.*, **35**, 1765 (1978).
241. Amos, R. D., *Mol. Phys.*, **39**, 1 (1980).
242. Morrison, M. A., and Hay, P. J., *J. Chem. Phys.*, **70**, 4034 (1979).
243. Amos, R. D., *Chem. Phys. Lett.*, **70**, 613 (1980).
244. Kendrick, J., *J. Phys. B.; At. Mol. Phys.*, **11**, L601 (1978).
245. Martin, R. L., Davidson, E. R., and Eggers, D. F., *Chem. Phys.*, **38**, 341 (1979).
246. John, I. G., Bacskay, G. B., and Hush, N. S., *Chem. Phys.*, **51**, 49 (1980).
247. Sadlej, J., and Sadlej, A. J., *Faraday Disc. Chem. Soc.*, **64**, 112 (1977).
248. John, I. G., Bacskay, G. B., and Hush, H. S., *Chem. Phys.*, **38**, 319 (1979).
249. Swanton, D. J., Bacskay, G. B., and Hush, N. S., *Chem. Phys.*, **83**, 69 (1984).
250. van Hemert, M. C., and Blom, C. E., *Mol. Phys.*, **43**, 229 (1981).
251. Dacre, P. D., *Mol. Phys.*, **36**, 541 (1978).
252. Dacre, P. D., *Mol. Phys.*, **45**, 17 (1982).
253. Dacre, P. D., *Can. J. Phys.*, **59**, 1439 (1981).
254. Dacre, P. D., *Mol. Phys.*, **45**, 1 (1982).

255. Dacre, P. D., *Mol. Phys.*, **47**, 193 (1982).
256. Diercksen, G. H. F., and Sadlej, A. J., *Chem. Phys.*, **96**, 43 (1985).
257. Stroyer, T., and Svendsen, E. N., *Int. J. Quantum Chem. Symp.*, **18**, 519 (1984).
258. Bishop, D. M., and Lam, B., *Chem. Phys. Lett.*, **120**, 69 (1985).
259. Fowler, P. W., and Raynes, W. T., *Mol. Phys.*, **43**, 65 (1981).
260. Arrighini, G. P., Maestro, M., and Moccia, R., *J. Chem. Phys.*, **49**, 892 (1968).
261. Diercksen, G. H. F., and Sadlej, A. J., *Theor. Chim. Acta* **63**, 69 (1983).
262. Werner, H. J., and Meyer, W., *Mol. Phys.*, **31**, 855 (1976).
263. Ditchfield, R., Hehre, W. J., and Pople, J. A., *J. Chem. Phys.*, **54**, 724 (1971); Hehre, W. J., Ditchfield, R., and Pople, J. A., *J. Chem. Phys.*, **56**, 2257 (1972); Hariharan, P. C., and Pople, J. A., *Theor. Chim. Acta*, **28**, 213 (1973).
264. Dunning, T. H., *J. Chem. Phys.*, **53**, 2823 (1970).
265. Dunning, T. H., *J. Chem. Phys.*, **55**, 716 (1971).
266. van Duijneveldt, F. B., IBM Research Report RJ945 (1971).
267. Amos, R. D., *Chem. Phys.*, **104**, 145 (1986).
268. Kollman, P., in *Modern Theoretical Chemistry* (Ed. H. F. Schaefer), Vol. 4, Plenum, New York, 1977.
269. Beyer, A., Karpfen, A., and Schuster, P., in *Hydrogen Bonds* (Ed. P. Schuster), Topics in Current Chemistry, Vol. 120, Springer-Verlag, Berlin, 1984.
270. Hiller, R. E., and Straley, J. W., *J. Mol. Spectrosc.*, **5**, 24 (1960).
271. Smit, W. M. A., van Straten, A. J., and Visser, T., *J. Mol. Struct.*, **48**, 177 (1978).
272. Nakanaga, T., Kondo, S., and Saeki, S., *J. Chem. Phys.*, **70**, 2471 (1979).
273. Nakanaga, T., Kondo, S., and Saeki, S., *J. Chem. Phys.*, **76**, 3860 (1982).
274. Spedding, H., and Whiffen, D. H., *Proc. R. Soc. A*, **238**, 245 (1956).
275. Murphy, W. F., *Mol. Phys.*, **33**, 1701 (1977); *Mol. Phys.*, **36**, 727 (1978).
276. Snyder, R. G., *J. Mol. Spectrosc.*, **36**, 204 (1970).

Ab Initio Methods in Quantum Chemistry—I
Edited by K. P. Lawley
© 1987 John Wiley & Sons Ltd.

TRANSITION STRUCTURE COMPUTATIONS AND THEIR ANALYSIS

FERNANDO BERNARDI

Instituto Chimico G. Ciamician, Universita di Bologna, Via Selmi 2, 40136 Bologna, Italy

and

MICHAEL A. ROBB

Department of Chemistry, Kings College London, Strand, London WC2R 2LS, UK

CONTENTS

I. INTRODUCTION

The ability to calculate the energy gradients and second derivatives analytically (see the chapter by Amos in this volume for a review) for self-consistent field (SCF)[1] and multi-configuration SCF[2] and correlated wavefunctions[3] has begun to revolutionize the way in which the quantum chemist probes a potential surface. For systems of more than three atoms, there are too many degrees of freedom to allow the full surface to be mapped out. Rather, one attempts to locate the critical points (minima and saddle points) on the surface and views the surface in terms of molecular structures (minima) and transition structures (saddle points) and a quadratic representation of the surface in the region of these critical points (vibrational frequencies computed from the second derivatives of the energy). The concept of a reaction path can be defined as a path of steepest descent[4] (in mass-weighted coordinates) connecting transition structures and reactants or products. Further, analysis of the surface in the region of this reaction path (using the reaction path Hamiltonian of Miller, Handy and Adams[5]) makes a study of the dynamics of complicated reactions possible. Thus the concept of an experimental mechanism translates into the heights and positions of the cols (transition structures) on the potential surface and the dynamical behaviour of the reactants in relation to the surface in these regions.

Most calculations on chemical systems within the scheme just mentioned are *model calculations* in the sense that they are performed on very simple chemical systems and are intended to serve as a model for a large class of similar systems. Consequently, it is also important to develop qualitative models of the topology and energetics of potential surfaces so that one may interpolate between experimental data that may be available, since only a small number of model situations may be explored with detailed numerical computations. Such qualitative theoretical models need to be formulated so that a numerical calculation of the characteristics of a molecular potential surface can be analysed *a posteriori* within a quantitative model and the results of this analysis carried forward to other systems in a qualitative way. The familiar Woodward–Hoffmann[6a] approach (based on Hückel theory) and the frontier orbital method[6b] have the difficulty that they cannot be used to analyse the results of a molecular structure computation based upon an extensive configuration-interaction (CI) expansion. Thus, one must begin to reformulate[7,8] such qualitative models within the sophisticated models that are now routinely used to perform molecular structure computations.

Our objective in this chapter is to survey the theory and practice of the computation of transition state structure. We take the word 'structure' in this context to encompass the geometry of the saddle point on the surface, a local quadratic approximation to the surface at that point (the second derivative matrix), and the nature of the reaction path in the region of the critical point. Broadly, our contribution is divided into two parts. In Section II we shall give a discussion of the practical computation of wavefunctions that can be relied upon to give an accurate representation of the potential surface in the transition structure region as well as the associated problem of geometry optimization. In Section III we are concerned with chemical applications and the analysis of the electronic structure in the transition structure region. From a theoretical point of view, we shall concentrate on those aspects of transition structure theory that can be distinguished from the considerations that apply to all molecular structure computations (which will be dealt with in other chapters in this volume). In our survey of applications, we shall attempt to pick several interesting chemical problems and illustrate the type of chemical information that can be obtained from molecular transition structure computations. This case-study approach is necessary because the very large number of applications that have been performed in recent years makes a comprehensive survey impossible.

By way of introduction, we begin with a formal definition of transition structure and some brief discussion of the origin of the associated reaction barriers. The definition of the transition structure in relation to the potential surface has been carefully considered by Murell et al.[9] We take the transition state to be the maximum on a minimal energy path which connects reactants and products. Thus the gradient of the potential energy function must be zero in all directions and has only one principal direction of negative curvature. The latter condition arises because if there were two directions of negative curvature then there would be a lower-energy path between reactants and products. When the nuclear kinetic energy is taken into account, then the transition structure has at most one imaginary frequency. As we shall discuss later, this definition is not necessarily coincident with the definition that has had to be used in dynamics studies. At the transition structure, the *transition vector* is the normal coordinate corresponding to the imaginary frequency and is parallel to the reaction path. Various symmetry conditions on the transition vector have been discussed by Stanton and McIver.[10]

From the outset, it is convenient to have some model which gives the origin of the energy barriers separating energy minima that lie on potential surfaces. For our purposes, it will be convenient to divide the barriers into two broad classes—conformational barriers that arise from steric/electrostatic effects, and electronic barriers that arise from the fact that the bonding situation in the reactants is different from that in the products. In this study, we shall be concerned exclusively with transition structures where the barrier arises predominantly from the electronic rearrangement. It is very useful to model these electronic barriers with a two-valued potential surface[8]—a surface describing the bonding situation of the reactants in all regions of configuration space, and an analogous surface for the bonding situation of the products—and an interaction matrix element. The non-interacting sur-

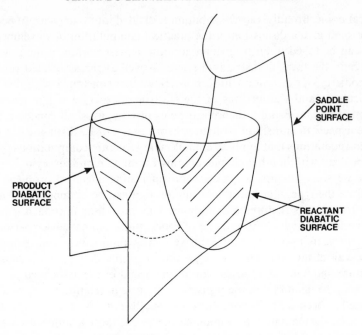

Fig. 1. Interaction of two diabatic surfaces to produce a saddle point surface.

faces are referred to as diabatic surfaces and were first introduced by Evans and Warhurst.[11] In our own work[8] we have been able to formulate this model in the context of the multi-configuration SCF (MC-SCF) method. This concept has also been applied in a qualitative manner by Dewar.[12] In the diabatic surface model, when the interaction matrix element is ignored, the transition structure lies on the minimum of the *seam of intersection* of the two diabatic surfaces (one product-like and one reactant-like). This idea is illustrated in Fig. 1. As we shall discuss subsequently, this simple concept provides us with (a) a criterion for the choice of an appropriate zeroth-order wavefunction (both diabatic states must be represented equally well), (b) a method for the approximate location of transition structures (seam following) and (c) a method for the *a posteriori* analysis of the electronic rearrangement effects that give rise to the activation barrier. Since we hope to use this concept to unify much of our latter discussion, we now give a brief presentation of this model (the mathematical details can be found in Ref. 8c).

We assume that for a reaction

$$R \to P \tag{1}$$

The adiabatic (interacting) surface can be descried in terms of two sets of variables, R_r and R_p. One then models this surface with two diabatic surfaces—

a reactant-like surface $E_r(R_r, R_p)$ and a product-like surface $E_p(R_r, R_p)$. The surface E_r must correctly describe the energy minimum of the species R and must represent the *bonding situation* of R at all regions of configuration space (including the highly distorted reactant geometry that corresponds to the minimum of the surface E_p). The surface E_p must describe the corresponding *bonding situation* of the products. In practice, the diabatic surfaces are computed from wavefunctions that have been built from configurations constructed from the orbitals of non-interacting fragments. Thus, it is the language of valence-bond theory that is most appropriate. At infinite inter-fragment separation, the diabatic surfaces correlate with Heitler–London configurations for open-shell states or no-bond configurations for closed-shell states (we refer to these configurations collectively as interacting fragment configurations (IFC)), while at finite interfragment separations these basic configurations are augmented by charge-transfer configurations (CTC) (electron transfer between fragments). Of course, in the calculation of the adiabatic surface itself, the fragment orbitals distort and mix to form mole-cular orbitals (MO) associated with the supermolecule. In order to represent the transition structure region of the surface in a balanced way, the wave-function constructed from these delocalized MO must be capable of representing both diabatic surfaces equally well. This probem is discussed in Section II.c.

II. THEORETICAL METHODS

A. Physical Significance of Transition Structures: Comparison of Theory and Experiment

As suggested by Salem,[13] it is convenient to divide a chemical reaction into a dynamic part (vibrational excitation of the reactants, exchange of vibrational energy among the various modes of the molecule and the deforma-tion of the molecule along the reaction coordinate) and a static part (the transition structure on the surface). The ultimate objectives of transition structure computations must be the computation of the kinetic parameters of the reaction within the framework of transition state theory and a study of dynamic effects in the region of the transition structure itself. A full study of dynamic effects for general molecular systems using dynamical trajectories is not possible since a point-by-point representation of the surface is unfeasible because of the large number of internal degrees of freedom. Rather, one is limited to the unique curve referred to as the intrinsic reaction coordinate[14] which passes through reactants, transition structure and products. It is not our intention to give an extensive discussion of reaction dynamics or transi-tion state theory in this chapter. However, for completeness we should indicate the ways in which the results of transition structure computations can be used to rationalize experimental kinetic data.

Transition state theory can be formulated in many ways (see, for example, Refs. 15), which differ mainly in the interpretation of the Arrhenius-like parameters in the usual rate-constant expression. Many of these are rooted in the theory of chemical dynamics (see, for example, the variational transition state theory developed by Truhlar and coworkers[15f] in which the transition state is optimized and may not correspond to the transition structure itself). However, for general molecular systems, we are limited to a quadratic representation of the potential surface in the region of the transition structure. Thus we are limited to the use of the barrier height and the vibrational frequencies of the transition structure itself. The thermodynamic properties (including thermal effects and zero-point vibrational corrections) of the transition structure itself are easily computed from the transition structure vibrational frequencies and the transition structure geometry. A comprehensive discussion of the calculation of thermodynamic properties from molecular structure results is to be found in the recent book by Daudel et al.[16] Since SCF computations give good vibrational frequencies at the reactant equilibrium geometry, one can expect that reasonable vibrational frequencies at the saddle point will be obtained with a reasonable zeroth-order wavefunction (see, for example, Ref. 17). Often, a simplified estimate of the Arrhenius pre-exponential factor as the ratio of the product of the vibrational frequencies in the reactants to the product of the vibrational frequencies in the transition structure (omitting the imaginary frequency of the transition vector) can provide enough information to discriminate between one mechanism and another.[18]

The intrinsic reaction coordinate[14] and the associated reaction path Hamiltonian[5] provide computationally feasible methods of obtaining dynamical information about chemical reactions from transition structure computations (see the review of Morokuma and Kato[19]). The intrinsic reaction coordinate is defined[5,14] as the steepest-descent path in mass-weighted coordinates that connects transition state and reactants/products. It is invariant to coordinate transformations and its direction corresponds to the normal coordinate having an imaginary frequency at the transition structure. The dynamical features of chemical reactions are analysed in terms of the curvature of this intrinsic reaction path. This is carried out[5] by diagonalizing the force-constant matrix in a subspace where the intrinsic reaction coordinate has been projected out. By following the remaining $3n - 7$ vibrational frequencies as the intrinsic reaction coordinate changes, one observes the manner in which the curvature of this coordinate causes the transfer of energy between the intrinsic reaction coordinate and the other $3n - 7$ normal modes. This information in turn can be used to rationalize the vibrational and rotational distribution of the products.

B. Special Features of Geometry Optimization Algorithms as Applied to Transition Structure Optimization

Since the subject of geometry optimization is discussed in a separate chapter by H. B. Schlegel in this volume, we shall limit our discussion to those particular features of geometry optimization algorithms that are particularly relevant to the optimization of transition structures.

The use of quantum-chemistry computer codes[20] for the determination of the equilibrium geometries of molecules is now almost routine owing to the availability of analytical gradients at SCF,[1] MC-SCF[2] and CI[3] levels of theory and to the robust methods available from the field of numerical analysis for the unconstrained optimization of multi-variable functions (see, for example, Ref. 21). In general, one assumes a quadratic Taylor series expansion of the energy about the current position X_0

$$E(X_0 + \delta) = E(X_0) + \mathbf{q}^\dagger \boldsymbol{\delta} + \tfrac{1}{2} \boldsymbol{\delta}^\dagger \mathbf{H} \boldsymbol{\delta} \tag{2}$$

where $\mathbf{q}$ is the gradient, $\mathbf{H}$ is the Hessian and $\boldsymbol{\delta}$ is the step to the stationary point. If both $\mathbf{q}$ and $\mathbf{H}$ are known, the step is computed as

$$\boldsymbol{\delta} = -\mathbf{H}\mathbf{q} \tag{3}$$

Of course, the surface is not quadratic and the Hessian is not constant from step to step. However, near a critical point, the Newton–Raphson method (Eq. (2)) will converge rapidly. The main difficulty is that the convergence of the Newton–Raphson method is local. Thus the method will converge to the *nearest* critical point to the starting point. Consequently, one must start the optimization with a Hessian that contains no or one negative eigenvalue according to whether a minimum or a saddle point structure is required. For a minimum, the steepest descent direction

$$\boldsymbol{\delta} = -\alpha \mathbf{g} \tag{4}$$

will always lower the energy (for a sufficiently small scalar α). However, for a saddle point, the gradient information provides no information about whether the saddle point is higher or lower than the current position. Thus it is clear that one must start the geometry optimization for a transition structure in a region where the Hessian has the correct number of negative eigenvalues and that this starting point may not be very easy to locate. Further, the accuracy of the Hessian is of vital importance in locating transition structures, since it supplies a search direction that is not available from the gradient itself. Methods for the location of critical points that do not use the Hessian are not likely to be successful. Thus transition structure optimizations differ from equilibrium structure optimizations in that (a) a very good initial starting geometry where the Hessian has one negative eigenvalue is required and (b) the accuracy of the Hessian during the geometry optimization itself is very important. The availability of analytical methods for the computation of the Hessian at the SCF and MC-SCF levels and sophisticated update methods (see the chapter by Schlegel in this volume) for improving the Hessian from one iteration to the next will undoubtedly remove the need for the often unreliable and inefficient computation of Hessian matrices by finite difference, with the result that, given a reasonable starting geometry, transition structure determinations will soon be as routine as the determination of equilibrium geometries. Further, the Hessian matrix can often be computed quite cheaply at a low level of theory/small basis set and used reliably in an extended basis set computation with updating. Thus the major problem to be faced in transition structure determination is that of locating an approximate starting structure. Thus we shall be concerned with this problem for the remainder of this section.

It is obvious that between two minima at least one saddle point must exist. Thus one should be able to search the line joining the two minima for a maximum. This idea is the basis of the *synchronous transit* methods.[22] An even simpler method is to associate the negative eigenvalue of the Hessian with a single variable (distinguished coordinate) and to look for a maximum in this coordinate while minimizing others. However, the reaction path is often highly curved so that one must connect the two minima with a quadratic function and define the minimum as the $n-1$ directions conjugate to this quadratic synchronous transit path.[21b,c] Unfortunately, there is no guarantee that this synchronous transit path has a maximum at the saddle point, since the synchronous transit path may not be parallel to the negative direction of curvature at the saddle point (i.e. the transition vector can be almost orthogonal to the synchronous transit direction at the saddle point, which would lead to a minimum along the synchronous transit search direction rather than a maximum). Thus one would be faced with the task of the computation of the Hessian at many points along a synchronous transit path, which could be very expensive. In spite of their inherent difficulties, synchronous transit methods in various approximations can often lead to good starting geometries where there is not much information available about the transition structure.

There is, however, an alternative method using diabatic surfaces which can be used to locate the transition structure approximately. We have used this method in our own work quite successfully. The model of interacting diabatic surfaces is represented in three diamensions in Fig. 1 in a schematic manner. If we have one product-like diabatic surface E_p that represents the product in the region of the product equilibrium structure P and a second reactant-like diabatic surface E_r that represents the reactant in the region of the reactant equilibrium structure R, then the transition structure TS lies at the minimum of the surface of intersection of the reactant-like and product-like diabatic surfaces provided the *derivative* (with respect to nuclear displacement) of the interaction matrix element between the two diabatic surfaces is *zero*. Thus one can find the transition structure by minimizing the energy on one surface (say E_p) with the constraint that the energy lies on the intersection of the two surfaces (i.e. $E_p - E_r = 0$). Of course, one must perform a constrained energy minimization (rather than an unconstrained transition structure optimization). Morokuma[23] has recently given an algorithm for such an optimization and the convergence is quite good. In spite of the obvious disadvantage of the fact that one must deal with two surfaces rather than one, there is no requirement for an initial direction of negative curvature, and so the method may exhibit convergence from quite poor initial starting geometries. In one of our first applications[24] of this technique (in a rather crude fashion), we were able to locate a starting geometry for a sigmatropic hydrogen shift with a very simple grid search.

C. The Choice of Zeroth-order Wavefunction

As we have just discussed, the efficient determination of transition structure is most readily accomplished if the gradient and second derivatives of the energy can be computed analytically. However, the calculation of these quantities is very expensive for highly correlated wavefunctions. The resolution of this accuracy–expense dilemma is usually accomplished with a compromise. The stationary points of the potential surface are usually characterized at some lower level of theory/basis set and then the relative energetics are improved by performing single-point calculations at the geometry so obtained using highly correlated wavefunctions with large basis sets. Obviously, the choice of the wavefunction (which we shall refer to as the zeroth-order wavefunction) used for characterizing the critical points is vital for the success of this procedure. For reasons of cost, the zeroth-order wavefunction is usually constructed within the single- or multi-configuration SCF (MC-SCF) theory. Thus a consideration of when a single-configuration SCF approach is expected to be adequate and/or the selection of configurations in an MC-SCF approach is the starting point for most theoretical computations of molecular structure. In this section we shall give some guidelines as to how such decisions can be made for transition structure computations.

As a prerequisite to our discussions we must decompose the correlation energy into two parts as first suggested by Sinanoglu[25] in the 1960s— a dynamical (largely structure-independent) part and a non-dynamical (strongly structure-dependent) part. The non-dynamical part of the electron correlation is accounted for using an MC-SCF wavefunction and one assumes that at this level of theory the topology of the potential surface will be correct (i.e. the number of minima and saddle points will be correct). Thus the non-dynamical correlation is associated with a zeroth-order wavefunction that is capable of describing appropriate bond dissociations, diradicaloid intermediates, etc., in a balanced way. The dynamical part of the electron correlation is recovered using a multi-reference configuration-interaction (MR-CI) calculation where the MC-SCF wavefunction is taken as the reference. The CI calculation in this case is correcting for the incorrect treatment of the instantaneous electron repulsions in the SCF method (either single- or multi-configuration). The magnitude of changes in the dynamical part of the electron correlation is essentially a reflection of the small changes in the electron density of the inactive orbitals (i.e. those that are not involved in bond making/breaking or the formation of diradical centres). Thus the magnitude of these changes is very important in the determination of barrier heights, but the presence or absence of the topological feature that gives rise to the barrier (i.e. the transition structure) is assumed to be represented at the SCF level (or MC-SCF level).

A criterion for the choice of zeroth-order wavefunction for transition structure computation is most easily described in the context of the diabatic surface model that we have discussed previously. The zeroth-order wavefunction should be capable of describing the reactant-like diabatic surface both in the region of the reactants and in the region of the products. Similarly, the product-like diabatic surface should be properly represented in the region of both reactants and products. If this criterion is met, then, in the model of interacting diabatic surfaces, the products, the transition structure and the reactants will have been treated in a balanced way. The accuracy of the SCF method for the equilibrium geometries of species that are well represented by a single configuration is well documented (see, for example, the compilation in Ref. 26). One's aim is the construction of a zeroth-order wavefunction of a similar accuracy in the calculation of transition structures.

The nature of the problem can be easily illustrated using an example of a two-bond cycloaddition. Let us consider a process that involves the breaking of two bonds and the formation of two new ones, or one new one and a diradical centre, as illustrated schematically in Fig. 2. We shall consider the wave function for this system from two complementary points of view—a fragment-in-molecule approach using configurations built from the orbitals of the isolated fragments (diabatic model), and a supermolecule approach where the orbitals of the fragments are allowed to mix in the formation of the supermolecule using a simple 2×2 interaction diagram approach.

Let us assume that we can treat the system in terms of four electrons distributed in four orbitals. In the diabatic model we take as a basis the highest occupied MO (HOMO), and lowest unoccupied MO (LUMO) of each of the isolated fragments AB or CD in Fig. 2. The reactants can be represented by a configuration in which both HOMO have double occupancy, configuration Ia in Fig. 3. As one stretches one of the bonds AB or CD, then the configuration Ib mixes with Ia and, in the limit of dissociation, the two configurations must have equal weight. Thus configurations Ia–Id plus charge-transfer configurations are

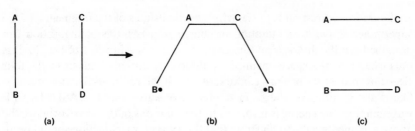

Fig. 2. A schematic representation of a $2 + 2$ cycloaddition reaction.

capable of representing the bonding situation given in Fig. 2a at all regions of configuration space on including the product geometry. The product can be represented by configuration II in Fig. 3. Here each of the fragments is singly excited with the spins coupled to a triplet (corresponding to no bond between ab and cd) and coupled to a singlet overall. At infinite interfragment separation, this

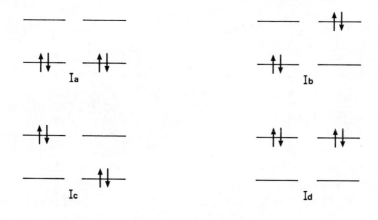

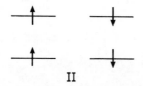

Fig. 3. Interacting fragment configurations (IFC) for reactant and product diabatic wavefunctions for a $2 + 2$ addition. Ia–Id are no-bond reactant-like configurations while II is a triplet–triplet Heitler–London configuration.

configuration corresponds to the triplet excited states of the fragments. In the supermolecule (with the addition of charge transfer), this configuration can describe both the diradical of Fig. 2a or the two-bond product of Fig. 2b. It is thus clear that the space spanned by the configurations built from the four electrons in the four orbitals can describe both the reactant-like and product-like diabatic surfaces. Since, in a complete active space (CAS) MC-SCF wavefunction, the energy is invariant to the rotations of the active orbitals, the above arguments serve to illustrate that this type of wavefunction should be an adequate zeroth-order wavefunction for describing the transition structure region of the surface.

Now let us examine the adequacy of the SCF model for this same problem. We shall use a very simple approach in which we form the molecular orbitals of the supermolecule by allowing 2×2 interfragment orbital mixing only, occupying the lowest configuration. We can distinguish a number of limiting situations. In Fig. 4a we allow only interfragment HOMO–HOMO and LUMO–LUMO orbital mixing (e.g. when HOMO and LUMO have different

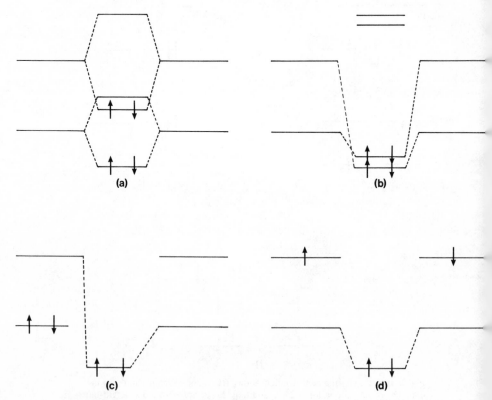

Fig. 4. Relative energies of various isovalent diatomics in their $^3\Pi$, $^1\Sigma^+$ and $^3\Sigma^-$ states.

symmetry). Clearly SCF will not be adequate here since the HOMO and LUMO of the supermolecule can become quasi-degenerate and at least a two-configuration wavefunction may be required. The case where only interfragment HOMO–LUMO interaction is allowed is shown in Fig. 4b. Here the SCF will be acceptable as a zeroth-order wavefunction. Finally, in Figs. 4c and 4d we give the case where only one HOMO–LUMO or HOMO–HOMO interaction is allowed, corresponding to zwitterionic and diradical structures. In this case, an SCF calculation would bias the structure of the zwitterion (Fig. 4c). Clearly, an SCF wavefunction will only be adequate if the orbital mixing is dominated by the HOMO–LUMO interactions of the type shown in Figs. 4b or 4c. Finally, it should be apparent that, while one pathway may be adequately described in an SCF approximation, a competing pathway may not. For example, in the case just discussed, the path leading to the product shown in Fig. 2c may be well described in an SCF approximation. However, the diradical structure will involve the configuration d of Fig. 4, which is not represented at the SCF level. If this diradical lies considerably below the transition structure for the formation of the structure shown in Fig. 2c, then the MC-SCF critical point (for Fig. 2c) could be a saddle point with two directions of negative curvature, while the SCF would show only one because the diradical cannot be represented. Thus, while it is easy to argue that a CAS MC-SCF with four orbitals should be an adequate zeroth-order approximation, SCF must be used with some considerable caution.

The unrestricted Hartree–Fock (UHF) method appears at first sight to be an attractive alternative to the use of the much more expensive MC-SCF method as a zeroth-order approximation in transition structure computations. It readily permits the evaluation of analytic gradients and second derivatives and it is possible to perform subsequent CI and perturbation calculations. In the UHF method, each electron occupies a different orbital so that diradical character can be represented. However, while the wavefunction has a well defined S_z eigenvalue, it will in general be a mixture of spin multiplicities. Thus a singlet spin wavefunction is contaminated with triplet components. Because of this the diradical character will be overemphasized when comparing to closed-shell species, leading to a very unbalanced representation of the surface. It is of course possible to project the wavefunction after optimization. However, the resulting wavefunction is no longer optimal and the analytic determination of gradients and second derivatives is not possible. In practice, a UHF calculation of a singlet closed-shell species can be carried out starting from the orbitals of an SCF calculation by taking the alpha MO to be those of the SCF and exchanging certain occupied and unoccupied levels (e.g. HOMO–LUMO) in the beta MO. If the energy is significantly lower than the restricted SCF wavefunction, then one has a strong indication that the restricted SCF solution needs diradical character added to it. Thus SCF wavefunctions should be tested in this way before being accepted as a good zeroth-order wavefunction for the transition structure. In spite of the problems just discussed, it should be apparent that, while a UHF wavefunction cannot represent the quasi-degenerate situation such as that which arises in the configuration of Fig. 4a it should be quite accurate in the representation of the configuration in Fig. 4d if the diradical centres are well separated (so that singlet and triplet have approximately the same energy). However, this has not been tested extensively.

In view of the preceding discussion, it seems likely that the MC-SCF method (particularily in its CAS form) will evolve to become the preferred zeroth-order wavefunction in transition structure studies. However, this method is often criticized because one must make a selection of the configurations (or in the case of CAS MC-SCF a choice of the active orbitals) from the outset. In this sense it is less objective than ordinary SCF theory. Thus the choice of the active orbital subspace is critical to the success of the method in practical applications. For this reason we now address ourselves to this point.

We begin by observing that the SCF and MC-SCF methods (when formulated in terms of the Newton–Raphson scheme) normally exhibit local convergence and will converge to the solution nearest to the starting orbital set. In fact, this may not be a global minimum and could be a saddle point even for the ground state (if the Hessian in the orbital space has been constructed, this can be tested). One may think of these local minima as corresponding to exchanges of active and inactive orbitals. Clearly one must recognize this problem from the outset. As the size of the active space increases, the number of possible local minima will diminish. However, the cost of the calculation increases very rapidly with the size of the active space, so that it is desirable to have criteria for the choice of small active spaces.

As we have discussed previously, the diabatic surface model provides a simple model for choosing an active space. Let us first describe how this approach can be implemented from an operational point of view and then consider possible alternatives. We assume that we have the SCF/MC-SCF orbitals of the isolated fragments that are to be used as the starting orbitals for an MC-SCF computation at some finite interfragment separation. We distinguish three types of orbital—inactive (core) orbitals which are doubly occupied in all configurations in the MC-SCF wavefunction, active (valence) orbitals that have variable (including open-shell) occupancy, and virtual (unoccupied) orbitals. These orbitals must be orthogonalized in such a way that the orthogonalization has as small an energetic effect as possible. (For example, an admixture of core orbitals with virtual orbitals would raise the energy dramatically.) Thus the orbitals can be orthogonalized as follows: (1) the core orbital set, the valence orbital set and the virtual orbital set are each symmetrically orthogonalized within each set, and (2) the valence orbitals are then Schmidt orthogonalized to the core set and the virtual orbitals subsequently orthogonalized to the core and then to the valence. This procedure unambiguously defines the active orbital set and consequently the configurations for a CAS MC-SCF given that the possible optimized MC-SCF wavefunctions correspond to various different permutations of valence/core–virtual orbitals. Further, if the diabatic surface model has been used to make the initial partition of core/valence/virtual orbitals, then the zeroth-order wavefunction constructed from the starting orbitals just described will be a faithful representation of this assumed model.

If we now consider using the SCF orbitals to define the initial partition into core/valence/virtual orbitals, then we have some possible problems. The SCF MO are normally ordered on the basis of orbital energies. However, there is no guarantee that highest-energy occupied orbitals and lowest-energy empty orbitals of the SCF approximation will lead ultimately to the same local energy minimum as the procedure outlined in the previous paragraph (diabatic surface model). The active (valence) fragment orbitals may combine to yield active MO of lower orbital energy than the SCF core orbitals; thus the two procedures would yield different optimized MC-SCF wavefunctions. While the MC-SCF wavefunction that has been constructed starting from the SCF canonical orbitals may have a lower energy than that constructed starting from the orbitals of the isolated fragments, it will not correctly describe both diabatic surfaces over a large region of configuration space. Rather, the MC-SCF wavefunction constructed from the canonical SCF orbitals may include a mixture of non-dynamical correlation (valence–valence excitation) and dynamical correlation (core–virtual excitation), leading to a worse (more unbalanced) description of the transition structure than the SCF itself. (Of course, with sufficiently large active spaces this dilemma disappears.) The problem is magnified with the use of extended basis sets. The low-energy SCF virtual orbitals are usually very diffuse and the weakly occupied active orbitals of the isolated fragments correlate with SCF virtual orbitals that may have very high energies. Choosing these diffuse SCF virtual orbitals as active orbitals can lead to divergence or convergence to energy saddle points.

The ideas discussed in the preceding paragraphs are best illustrated with a few case studies. We shall use the extensively studied transition structure[2d,27] for the addition of H_2 to CO as our first example. The reaction is Woodward–Hoffmann forbidden and the highly asymmetric transition structure is shown in Fig. 5. The zeroth-order MC-SCF wavefunction can be constructed from four orbitals/four electrons and the choice of active (valence) orbitals in this case is indicated in Fig. 5. The numbering of the orbitals is taken to be in one-to-one correspondence with the labelling given in Figs. 3 and 4. For this choice of active space, the HOMO of one fragment has the same symmetry as the LUMO of the other (in C_s symmetry). Thus the SCF should give an acceptable zeroth-order wavefunction. In rows 1 and 2 of Table I we give the geometrical parameters obtained at the SCF and four-orbital CAS MC-SCF level for a 6-31G* basis[2d,27a] and in rows 4 and 5 the corresponding results for a double-

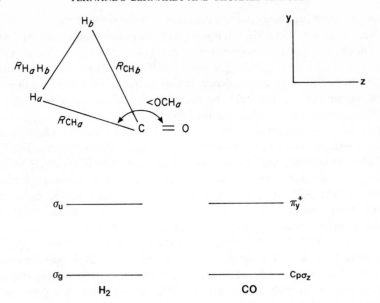

Fig. 5. Parameters and active orbitals used in a transition structure optimization for H₂ + CO.

zeta plus polarization (DZP) basis.[27c] The shifts in the parameters on moving from SCF to MC-SCF can be retionalized using the diabatic surface model. The SCF wavefunction cannot represent the configurations Ib or Ic shown in Fig. 3. Since the H_a–H_b bond is stretched to almost double its value in the H_2 molecule, the reactant diabatic surface is not well represented at the SCF level with respect to the H_a–H_b stretch, leading to the fact that the MC-SCF value for this parameter is significantly larger than the SCF value. Thus the transition state occurs at an earlier stage in the reaction path at the SCF level (i.e. the H_a–H_b bond distance is less for the SCF than for the MC-SCF since the reactant-like diabatic surface is not well represented). In Table I we have also

TABLE I

Transition structure parameters for $H_2 + CO \rightarrow H_2CO$ (distances in Å, angles in deg).

Method/basis	Ref.	R_{CO}	R_{CH_a}	$R_{H_aH_b}$	$\langle OCH_a \rangle$	R_{CH_b}
SCF/6–31G*	27a	1.134	1.094	1.328	164.5	1.739
MC-SCF/6–31G*	2d	1.152	1.092	1.449	163.9	1.726
MP2/6–31G*	27a	1.180	1.092	1.356	164.0	1.786
SCF/SZP	27c	1.151	1.104	1.208	161.9	86.58
MC-SCF/DZP	27c	1.160	1.098	1.350	162.25	86.97
MR-CI(SD)/DZP	27c	1.183	1.096	1.301	162.65	85.70

given the MP2 (Moller–Plesset perturbation theory to second order) geometrical parameters obtained in the 6-31G* basis (row 3) as well as the single- and double-excitation MR-CI(SD) results in a DZP basis (row 6). While the MP2 results are very similar to the SCF, the MR-CI(SD) results show a similar shift from the SCF as the MC-SCF. The good agreement between MR-CI (SD) and MC-SCF particularly indicates the adequacy of the MC-SCF zeroth-order wavefunction. In contrast, the MP2 geometry (which is based on the correction of the SCF wavefunction for dynamic correlation effects) has the same type of error as the SCF geometry. Dupuis et al.[27c] show similar effects for transition structure harmonic vibrational frequencies, where an MC-SCF geometry is required to obtain acceptable results. Thus, while the SCF wavefunction gives results that are qualitatively correct for this example, an MC-SCF zeroth-order wavefunction is required to obtain the same level of accuracy for transition structure computations as for SCF equilibrium geometries.

Now let us examine a second example where the SCF does not provide an adequate zeroth-order approximation—the propene/trimethylene diradical surface[24,28] (to be discussed in detail in Section III). The [1, 3] migration of a hydrogen from the methylene group of CH_3CHCH_2 to form CH_2CHCH_3 may be regarded as a one-step process or may involve a trimethylene diradical $\cdot CH_2CH_2CH_2\cdot$ intermediate/transition structure and two [1, 2] H shifts. In a diabatic surface model, we must have a diabatic surface to describe the closed-shell CH_3CHCH_2 species and a second diabatic that correlates with $\cdot CH_2CH_2CH_2\cdot$ diradical. Here the SCF calculation[28] shows a transition structure for the [1, 3] shift since the diradical surface (see Fig. 4d) is not represented at all. In contrast, an MC-SCF geometry optimization[24] finds only a diradical intermediate and a transition structure for a [1, 2] hydrogen shift. One must emphasize that this result has nothing to do with electron correlation in the usual way in which it is used (i.e. dynamical electron correlation). Rather, the failure of SCF theory in this case is due to the fact that only the relatively unimportant zwitterionic component of the trimethylene diradical diabatic surface can be represented at the SCF level, giving rise to the incorrect surface topology.

D. Basis Set and Correlation Effects

As we have discussed in the preceding subsection, in transition structure computations one should aim to use a zeroth-order wavefunction that is expected to be of a similar accuracy as the SCF approximation for closed-shell species in the region of their equilibrium geometries. Thus the starting point for a discussion of dynamical correlation in the transition structure will often be an MC-SCF reference. However, given an acceptable zeroth-order wavefunction, the effects of electron correlation and the effects of basis set extension should not be significantly different from those of closed-shell species. However, the methodology required to treat dynamical correlation with an MC-SCF reference is much more complicated than with an SCF reference. Thus we begin with a brief summary of basis set and correlation effects on the energies and geometries of closed-shell species at their equilibrium geometries.

Recently, Iwata[26] has attempted to document the reliability of various basis sets in common use as well as electron correlation effects. His global conclusions can be summarized as follows:

1. Minimum basis sets (Slater-type orbitals, three Gaussian (STO-3G)) tend to underestimate bond lengths by 0.05 Å at SCF and overestimate by 0.05 Å with electron correlation, with an uncertainty of $\pm 10°$ in the bond angle.
2. Extended basis sets underestimate bond lengths by about 0.02 Å and give rise to an uncertainty of $\pm 5°$ in bond angles at the SCF level. When electron correlation is included, the agreement improves as the electron correlation method improves and as the basis is made more flexible by the inclusion of polarization effects.
3. Energy differences between stable closed-shell species in a double-zeta plus polarization (DZP) basis are accurate to an uncertainty of 20% at the SCF level (30% at DZ) and this reduces to 10% when electron correlation CI (SD) is introduced. A minimum basis can be very unreliable in computing energy differences.

Given an adequate zeroth-order wavefunction, there is no reason to expect the reliability of transition structure computations to be any different with respect to 1. or 2. above. However, reliable energetic predictions (3. above) present a more subtle problem. While the reactants and products may be stable closed-shell species, the transition structure has partly formed bonds and/or diradical character. Thus correlation effects cannot be expected to cancel in the same way as occurs in closed-shell systems.

The magnitude of this effect is illustrated in the dissociation energy of simple molecules into open-shell fragments. This has been studied using a 6-31G* basis by Krishnan et al.[29] For example, the binding energy of N_2 at the SCF level is 4.7 eV and this is increased to 8.70 eV at the level of MP4 (dynamic plus non-dynamic electron correlation), in reasonable agreement with the experimental value of 9.91 eV (12% error). However, an MC-SCF calculation in the same basis,[30] which included the necessary configurations to allow proper dissociation (six orbitals/six electrons), gives a binding energy of 8.70 eV. Thus the true dynamic correlation contribution to the binding energy is remarkably small. There is no reason to expect that the situation will be very different in computations of transition structure energetics.

Let us now consider another example from an actual transition structure computation on the barrier for the reaction that we have considered

TABLE II

Non-dynamic (MC-SCF) and dynamic (CI(SD)) correlation effects on the barrier height for the reaction $H_2CO \rightarrow H_2 + CO$.

Method/basis	Barrier height (kacl mol^{-1})	Ref.
SCF/6–31G*	105.0	27a
MC-SCF/6–31G*	96.3	2d
MP2/6–31G*	96.1	27a
MP3/6–31G*	98.3	27a
MP4/6–31G*	95.5	27a
SCF/DZP	105.2	27c
CI(SD)/DZP	98.1	27b
CI(SDQ)/DZP	94.2	27b
MC-SCF/DZP	94.1	27c
MR-CI(SD)	91.4	27c

previously—CH_2O decomposition into H_2 and CO. The barrier heights corresponding to the transition structure[27] geometries given previously in Fig. 5 and Table I are summarized in Table II. For this example, as we have discussed, the SCF is a quite reliable zeroth-order wavefunction. Thus we have a much more favourable situation than in the example just discussed where three bonds were dissociated into open-shell fragments. If we regard the MR-CI(SD) (multi-reference CI with single and double replacements) results at the DZP basis level as the most accurate treatment, it is apparent that the MC-SCF recovers 80% of the correction to the barrier height due to electron correlation. Thus the true dynamic correlation effect is quite small. In these calculations, the difference MP4–MP3 for the reactant CH_2O is very small and thus most of this contribution to the barrier comes from the transition structure itself. This fact is consistent with the large size-extensivity correction (CI(SDQ)–CI with single, double and quadruple excitations) seen in the CI calculations reported at the DZP level. (The MC-SCF wavefunction is size-extensive like the SCF.) Thus, even when the SCF is a reasonable zeroth-order wavefunction, CI and MPn wavefunctions may be more poorly converged for transition structures with distorted geometries than for SCF wavefunctions in the region of equilibrium geometries.

Recently Handy et al.[31a] have made a thorough examination of the convergence properties of MPn by comparision with a full CI calculation on H_2O at $R = R_e$, $R = 1.5R_e$ and $R = 2R_e$. At $2R_e$, the MP series converged very erratically and would require 200 terms to converge (in spite of the fact that the correlation energy is smaller than at R_e). In Table III we have collected results for H_2O (from the tabulation of Handy[31b]) for R_e and $2R_e$. This table is indicative of the error in a barrier height that involved a bond length stretch of $2R_e$. The error resulting from the SCF result is 64%, which drops to 20% with CI(SD) and to 2% with CI(SDTQ). However, at the MC-SCF level, the error

TABLE III
Benchmark calculations on H_2O in a DZ basis using various correlated wavefunctions.

Method	$E(R_e) - E(2R_e)\,(\text{kacl mol}^{-1})$
SCF	260.2
CI(SD)	191.5
CI(SDTQ)	161.1
MP4	179.5
MC-SCF	219.0
MR-CI(SD)	177.1
Full CI	158.6

From Handy.[316]

is reduced from the SCF value to 38% and to a more pleasing 11% with MR-CI(SD). It should be noted that the coefficient of the SCF configuration in the full CI at R_e is 0.979, which drops to 0.764 at $2R_e$.

From the preceding discussion, it is apparent that the computation of the contribution of dynamic electron correlation to the barrier heights corresponding to transition structures is a problem that requires special attention. It is apparent that methods using a single-reference function for the computation of the dynamic correlation are poorly converged (in the case of MPn) or require the inclusion of more than double excitation in the case of single-reference CI based upon an SCF reference. Thus, in general, multi-reference methods are usually indicated. Unfortunately, these methods are considerably more expensive than single-reference MPn or CI(SD) because the number of configurations required rapidly becomes very large as the number of references increases. In the case of multi-reference CI(SD) (MR-CI(SD)), the use of unitary group methods (see, for example, the supercomputer implementation of Saunders and van Lenthe[32]) has enabled very large CI expansions to be used. However, in most practical calculations one must make some effort to reduce the number of variational parameters considered. This is usually accomplished by using perturbation theory to select an important subset of configurations (viz. the multi-reference double-excitation CI methods of Buenker and Peyerimhoff[33] or the CIPSI method[34]) or by limiting the variational degrees of freedom by assuming certain relationships among the variables (viz. the contracted CI method of Siegbahn[35]). While the theory of multi-reference Moller–Plesset perturbation theory has been known for a long time (see, for example, the review article of Brandow[36]), these techniques have not yet been applied in molecular structure studies in a routine manner. The main stumbling block appears to be the choice of a suitable zeroth-order Hamiltonian. Clearly a reliable multi-reference formulation of MP3 would be desirable.

Since MC-SCF methods should yield reliable transition structures and the CI(SD) problem rapidly becomes intractable for a multi-reference zeroth-order problem, it is desirable to investigate semi-empirical computation of the dynamic correlation. Ruedenberg and coworkers have made some considerable progress in this direction (see Refs. 37 and the papers cited therein). Their approach is related to the diabatic surface model, which we have already developed, so we shall discuss it in this context. Clearly, our objective is the computation of the dynamic correlation energy difference between the reactants and the transition structure/products. The hypothesis of Ruedenberg et al.[37c] is that the bulk of this difference results from the changes arising from intrafragment correlation (intra-atomic in the case of the diatomic molecules studied by Ruedenberg). The MC-SCF CI Hamiltonian, expressed in terms of distorted atomic orbitals (termed molecule-adapted atomic valence orbitals by Ruedenberg), can be corrected for intrafragment correlation by adding the correlation energies of the various neutral and ionic states of the fragments to the diagonal matrix element corresponding to the Heitler–London (HL), no-bond or charge-transfer configurations. The MC-SCF CI Hamiltonian is then rediagonalized. The effect of the dynamic correlation on the binding energy then results from the difference in the intrafragment correlation between neutral and ionic states of the fragments. For example, in the case of the binding of H_2, the Heitler–London configuration must be augmented at R_e by charge-transfer configurations. Thus, in this case, the effect of the dynamic electron correlation on the binding energy arises from the effect of electron correlation on the H negative ion. The magnitude of the effect varies with the magnitude of the matrix element between the charge-transfer configuration and the HL configuration, falling to zero at infinite interfragment separation. In the case of the transition structure, there will be a separate effect on each constituent diabatic surface. Clearly, the implicit assumption in this procedure is that the intrafragment correlation is independent of interfragment separation and that the magnitude of the off-diagonal matrix elements of the MC-SCF CI Hamiltonian is independent of dynamic correlation. In their calculations on diatomic molecules, Ruedenberg et al. were able to compute very accurate binding energies by correcting MC-SCF results for intra-atomic correlation in this manner. There is no reason not to expect similar accuracy in molecular transition structure computations. The potential of this method arises from the fact that the intrafragment correlation contribution can be computed from single-reference computations on isolated fragments.

In addition to the *a priori* estimation of dynamic electron correlation effects, the model proposed by Ruedenberg, when applied within a diabatic surface model, enables the *a posteriori* rationalization of the magnitude of dynamic electron correlation effects on barrier heights and reaction energies. If we use a 2 + 2 cycloaddition as a model, we can rationalize the correlation energy

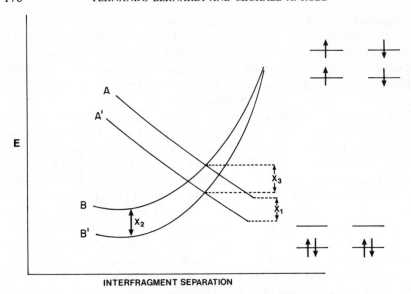

Fig. 6. Dynamic electron correlation effects in a diabatic surface representation.

effects on the barrier and reaction energy. In Fig. 6 we show a schematic representation of the diabatic surfaces for such a cycloaddition (the reaction of a to produce c shown in Fig. 2). The curve A represents the energy profile corresponding to the reactants (configuration Ia to Id plus charge transfer shown in Fig. 3) and the curve B represents the energy profile for the product-like diabatic (configuration II plus charge transfer shown in Fig. 3). The curves A' and B' are intended to show the effects of the addition of intrafragment correlation. Since we are concerned only with energy differences, we have taken the intrafragment correlation energy of the state represented by the curve B to be zero at infinite interfragment separation. Thus the shift of curve A to A' at infinite interfragment separation represents the dynamic correlation energy difference between reactants in the product-like diabatic state (curve B) and reactants in the reactant-like diabatic state (curve A) at infinite interfragment separation. Since there are two more paired electrons in the reactant-like diabatic state at infinite interfragment separation than in the product-like state at infinite interfragment separation, we expect the dynamic electron correlation of the reactant-like diabatic state to be the larger in magnitude. The behaviour of the product-like diabatic curve under the influence of dynamic electron correlation will be similar to that of a diatomic molecule (curve B to curve B'), and the correlation energy will be expected to increase as charge-transfer configurations are added to the wavefunction as the interfragment distance decreases (i.e. the charge-transfer configurations have more electron pairs). On the other hand, the diabatic curve for the reactant-like

TABLE IV
Reaction energetics for cycloaddition reactions—the effect of dynamic electron correlation.

Addition of fulminic acid to acetylene	SCF[38]	CI(SD)[38]
HCCH + HCNO	0.0	0.0
Transition structure	30.8	19.5
Isoxazole	−80.3	−70.3

Addition of H_2 to CO to form H_2CO	SCF[27a]	MC-SCF[2d]	MP4[27a]
H_2 + CO	0.0	0.0	0.0
Transition structure	104.7	96.3	92.7
H_2CO	−0.3	3.11	−2.8

diabatic (curve A) will be almost parallel to curve A' when the effects of intra-atomic correlation are included. This situation arises because, in a charge-transfer configuration derived from the configuration Ia of Fig. 3 by one-electron transfer, the loss in dynamic correlation energy of one fragment is almost equal to the gain in the electron correlation of the other fragment, with the result that the correlation energies of the no-bond configuration and the charge-transfer configuration should be similar. Thus using qualitative reasoning we expect that the effect of intrafragment correlation will stabilize the reactant-like diabatic curve by an amount X_1 relative to the product-like diabatic at infinite interfragment separation, that the product-like diabatic will be stabilized by an amount X_2 in the region of the product interfragment separation, and that the overall reaction barrier will be lowered by an amount X_3. Depending on the relative magnitudes of X_1 and X_2, the intrafragment correlation energy may increase or decrease the overall reaction energy. However, one expects that X_2 will be small relative to X_1 and the effects of dynamic correlation should decrease the overall reaction energy and the barrier. This observation is consistent with many computations where the barrier to reaction is normally overestimated at the SCF and MC-SCF level and overall reaction energies are overestimated. In Table IV we have collected the data for two addition reactions of the type just discussed—the 1, 3 dipolar cycloaddition of fulminic acid to acetylene[38] and the addition of H_2 to CO.[27a] It can be seen that in both cases the reaction barrier is reduced by the inclusion of electron correlation. However, in the case of fulminic acid plus acetylene, the reaction energy is decreased by the effect of electron correlation, whereas in the case of H_2 plus CO it is increased. However, in H_2 plus CO, X_1 is expected to be dominated by the correlation energy of H_2 and the pair correlation energy of the carbon lone pair in CO, and thus is quite small; whereas in fulminic acid plus acetylene, X_1 arises from the pair correlation energies of π orbitals, which will be very large.

E. Qualitative and Quantitative Theoretical Models for the Analysis of Transition Structure Energetics

The philosophy that the whole may be understood from its parts has been extremely successful in structural chemistry where fragment orbitals are used as the building blocks to construct molecular orbitals and thereby understand why molecules have the shapes that they do. In chemical reactivity the building blocks are necessarily diabatic surfaces (or curves) whose combinations and avoided crossings generate the barrier of the reaction. Clearly, understanding the origins of barriers is essential to the understanding of reactivity patterns. It is our intention in this subsection to discuss the way in which transition structure can be represented in a diabatic surface model. In our subsequent discussions of applications, we shall illustrate the utility of this model in the *a posteriori* rationalization of transition structure energetics.

While chemical reactivity is normally discussed in terms of transition state theory, corresponding to motion on an adiabatic potential energy surface, a complementary theory has been formulated (see the review by Metiu *et al.*[39]) in which the passage from reactants to products is formulated in terms of a transition between two diabatic surfaces. We shall refer to this as diabatic transition structure theory. This approach is a very natural one to use in semiclassical trajectory calculations (see, for example, Refs. 40) and, as we shall presently discuss, enables a simple interpretation of transition structure.

The utility of the diabatic surface model in rationalizing the transition structure can be illustrated using a simple two-valued representation given in Fig. 1. Surfaces R and P represent the diabatic surfaces for reactant-like and product-like wavefunctions. Under the influence of the interaction matrix element, a maximum is created on the ground-state adiabatic surface and a minimum is created on the excited adiabatic surface. In many dimensions, the transition structure (in the diabatic picture) is associated with the minimum of the seam of intersection of the diabatic surfaces. As we mentioned previously, the diabatic transition structure corresponds to the optimization of the geometry on either the reactant-like or product-like diabatic surfaces subject to the constraint that the geometry lies on the intersection of the two diabatic curves (an algorithm has recently been given by Morokuma[23]). The transition structure in the adiabatic picture has a different geometry because of the influence of the interaction matrix element between the diabatic surfaces. In recent work[8] we have been able to correlate diabatic surface crossings with the transition structure and we expect that the two definitions of transition structure will be quite similar. However, the adiabatic model has the disadvantage that, while a wavefunction for a diabatic surface changes only very slowly as one passes through the transition structure and has a very simple interpretation, the adiabatic wavefunctions change rapidly and are not easily interpreted.

In a diabatic surface formalism, Metiu et al.[39] have shown that the probability of a transition from one surface to another and hence the reaction probability is given as

$$T = T_1 + T_2 \tag{5}$$

$$T_1 = \langle \psi_e^P | H | \psi_e^R \rangle \langle \Phi_N^P | \Phi_N^R \rangle \tag{6}$$

$$T_2 = \left\langle \psi_e^P \left| \frac{\partial}{\partial R} \right| \psi_e^R \right\rangle \langle \Phi_N^P | \Phi_N^R \rangle \tag{7}$$

where ψ_e^R is the electronic wavefunction of the reactant-like diabatic state, ψ_e^P that of the product-like diabatic state and Φ_N^R and Φ_N^P are nuclear motion wavefunctions. This probability must be weighted by Boltzmann factors in order to obtain reaction rates. Of course, this formula is similar to the non-adiabatic coupling matrix element, except that the term T_1 would be zero in this case. Thus, while the barrier height has its usual interpretation in this formalism, the reaction probability is related to the interaction matrix element between diabatic surfaces. Alternatively, one may use the Landau–Zener formula for the probability of a transition from one diabatic surface to another

$$P = \exp(\pi g^2 / hv\delta s) \tag{8}$$

where v is the nuclear velocity, δs is the difference in slopes between reactant-like and product-like diabatic surfaces and g is the matrix element of H in Eq. (6) above. Thus transition structure dynamics are controlled by the following features:

1. the difference in slopes δs (Landau–Zener) between the intersecting diabatic states (when the difference in slopes is small, the reaction probability is largest),
2. the magnitude of the interaction matrix element g (a large interaction matrix element increases the reaction probability),
3. for non-adiabatic effects, the derivative of the interaction matrix element with respect to a nuclear coordinate T_2.

Most of the theoretical methods used to correlate chemical reactivity trends implicitly (or in some cases explicitly) use a model of interacting diabatic surfaces in which the transition structure is associated with an avoided crossing of a reactant-like or a product-like diabatic surface. In the molecular-orbital correlation diagram approach of Woodward and Hoffmann[6a] or the frontier orbital method of Fukui,[6b] the molecular orbitals of the fragments are first mixed to form the MO of the supermolecule and then the electrons are assigned to various configurations of these supermolecule MO. We shall refer to this approach as the MO diabatic method. In contrast, in the original diabatic surface model proposed by Evans *et al.*[11] and applied in recent work by Epiotis and Shaik,[41] by Dewar[12] and by the present authors,[8] the configurations are constructed from the orbitals of the isolated fragments in a valence-bond-like approach. This latter approach has already been used extensively in this chapter to rationalize the choice of zeroth-order wavefunctions and correlation effects.

The MO diabatic surface method has inherent difficulties which severely limit its applicability. In the Woodward–Hoffmann approach, the orbitals are classified with respect to their symmetry elements that bisect the bonds being broken or made. The orbitals are allowed to mix as a function of distance along the reaction coordinate. Avoided crossings between configurations that differ in the occupancy of doubly occupied MO (Fig. 4a) occur on the state correlation diagram when the orbital energies of orbitals with different symmetry cross on the orbital correlation diagram. The difficulty arises in explaining the nature of avoided crossings when MO theory does not give rise to an orbital crossing and yet a large barrier is still observed (e.g. the *supra–antara* approach of two ethylene molecules). In this case the diabatic surfaces must be constructed within a valence-bond type of approach. Thus the study of transition structure in the context of diabatic surfaces is most generally formulated using orbitals that are localized on the fragments (i.e. valence-bond-type approaches). Accordingly, we now give some discussion of the way in which diabatic surfaces can be computed *ab initio*.

The most general method for the computation of diabatic surfaces involves the use of effective Hamiltonian theory (for a general treatment, the reader is referred to the review article of Brandow[36] or the recent treatment of Spiegelmann and Malrieu[42]). In effective Hamiltonian theory, one seeks a model Hamiltonian in which the diabatic state energies are obtained as the diagonal matrix elements, the interaction matrix element is obtained from the off-diagonal elements, and the adiabatic state energies are obtained as the eigenvalues. In order to illustrate the general approach, we shall consider the case where we are interested only in two diabatic states; however, the theoretical treatment is completely general.

We begin by considering a similarity transformation of the full CI Hamiltonian (Eq. (9))

$$U^{-1}\left(\begin{array}{c|c} H^0 & Z^\dagger \\ \hline Z & W \end{array}\right)U = \left(\begin{array}{c|c} H_{\text{eff}} & 0 \\ \hline 0 & \bar{W} \end{array}\right) \qquad (9)$$

We have partitioned the full CI space into a block H^0 spanned by the reference configurations $|R\rangle, |P\rangle$ and a secondary block spanned by the remainder of the configurations. The matrix Z represents the matrix elements between reference and secondary configurations. We now seek the transformation matrix U which reduces the full CI Hamiltonian (left-hand side of Eq. (9)) to the blocked form (right-hand side of Eq. (9)), where the interaction matrix elements Z_{Ri} between $|R\rangle/|P\rangle$ and the remainder of the configurations are zero. The eigenvalues of the submatrix H_{eff} (the effective Hamiltonian matrix) corresponding to the right-hand side of Eq. (9) will now reproduce two eigenvalues (and thus two adiabatic surfaces) of the original Hamiltonian matrix exactly. Thus the diagonal elements of H_{eff} are the diabatic surface energies, and the off-diagonal element is the *resonance interaction*. In general, the coefficient vector (diabatic wavefunction) corresponding to the Pth or Rth columns of the transformation matrix U will be dominated by the corresponding Pth or Rth components corresponding to the configurations of Fig. 3, with much smaller contributions from charge transfer, etc.

We must now discuss the way in which the transformation can be determined. We shall limit ourselves to a brief discussion of two methods: the Van Vleck method[43,44] and the Bloch equations[45] as formulated by Durand.[45e] The so-called Van Vleck method (see Ref. 44 for a very readable account) is conceptually very simple and we begin with this technique.

In the Van Vleck method, we take the transformation U to be a unitary transformation. The transformation U can be written in terms of a parameter matrix G as

$$U = \exp G \tag{10}$$

where G is skew-symmetric

$$G^\dagger = -G$$

One can then set up equations that determine the parameters from the condition (9) (see Refs. 8c or 43c). The procedure is similar to the familiar Jacobi diagonalization.

While the exponential transformation is appealing by virtue of its computational simplicity, it is desirable to reformulate the problem in a non-orthogonal basis for reasons that will become apparent presently. Following Shavitt et al.,[44] we define a new set of parameters X as

$$X = \tanh G \tag{11a}$$

$$X^\dagger = -X \tag{11b}$$

Using $\operatorname{arctanh} X = \frac{1}{2}\ln[(1 + X)/(1 - X)]$ we obtain

$$G = \tfrac{1}{2}\ln[(1 + X)/(1 - X)] \tag{12}$$

and we can rewrite our unitary transformation as

$$\begin{aligned}\exp G &= [(1 + X)/(1 - X)]^{1/2} \\ &= (1 + X)(1 - X^2)^{-1/2}\end{aligned} \tag{13}$$

Now because of the skew-symmetry of X we can write

$$-X^2 = X^\dagger X \tag{14}$$

and we can identify

$$S = 1 + X^\dagger X \tag{15}$$

as the overlap matrix of the diabatic states. We can now write the transformation U as

$$\exp G = (1 + X)S^{-1/2} \tag{16}$$

where the diabatic wavefunctions are the columns of the matrix and S is the overlap. Equation (9) now takes the equivalent form

$$\exp(-G)H\exp(G) = S^{-1/2}[(1 - X)H(1 + X)] \tag{17}$$

where the $S^{-1/2}$ matrix transformation merely corresponds to symmetric orthogonalization.

If we define[8c] C as

$$C = \left(\begin{array}{c|c} & \\ \hline X & \\ \end{array}\right)$$

then we obtain

$$Z + WC - CH_{eff} = 0 \tag{18}$$

It is useful at this stage to consider some approximations to the scheme just discussed. If we neglect the off-diagonal elements of H_{eff} in Eq. (18), the equations become decoupled and one obtains one equation of the form

$$Z_R + (W - 1 \cdot E_R)C_R = 0 \tag{19}$$

for each reference configuration. (In Eq. (19) Z_R denotes the Rth column of the submatrix Z in Eq. (9)). Equation (19) is in turn equivalent to the augmented eigenvalue problem

$$\left(\begin{array}{c|c} H_{RR}^0 & Z_R^\dagger \\ \hline Z_P & W \end{array}\right)\left(\begin{array}{c} 1 \\ C_R \end{array}\right) = E_R\left(\begin{array}{c} 1 \\ C_R \end{array}\right) \tag{20}$$

In other words if the off-diagonal elements of H_{eff} are neglected, we obtain an independent CI problem for each diabatic state. This is the basis of the method that we have described in Refs. 8a and 8b. In this work one also performed a partition of the full valence CI space (CAS) into packets. Each packet consists of an isolated fragment configuration (either a Heitler–London or no-bond configuration in the language of valence-bond theory) plus all possible one-electron transfer configurations. Thus in terms of Eq. (20) the CI expansion for each diabatic wavefunction was truncated. The truncation of each packet at one-electron charge transfer configurations will clearly begin to break down at small interfragment separations where more than one-electron transfer will become very important and other locally excited configurations will begin to make large contributions. However, the truncation at one-electron transfer has the feature that the packets are mutually exclusive and hence approximately orthogonal.

To summarize, in the preceding discussion we have presented two formalisms for the computation of diabatic wavefunctions using an effective Hamiltonian. In the Van Vleck method the diabatic wavefunctions are orthogonal. In the non-orthogonal transformation method the diabatic wavefunctions are non-orthogonal. In practice, perturbation theory is normally used to solve the equation system (20) (see the review of Spiegelmann and Malrieu[42]) and the method has found its application in this form in the non-adiabatic coupling problem, in particular. However, our objective at this stage is to illustrate how the method can be used in a qualitative and quantitative way to understand the transition structure in terms of diabatic components.

In the effective Hamiltonian formalism just reviewed, the diabatic state energies are obtained as the diagonal matrix elements of the effective Hamiltonian, while the *resonance interaction* between product-like and reactant-like diabatic surfaces is obtained as the off-diagonal matrix elements. The adiabatic states are obtained as the eigenvalues. Thus, the effective Hamiltonian corresponds to the projection of the full CI Hamiltonian onto the subspace of the (product-like and reactant-like) Heitler–London and no-bond configurations. We now wish to comment briefly on the physical interpretation of the effective Hamiltonian computed via Eq. (17).

For the purpose of the present discussion we take the effective Hamiltonian in the reference space to be as in Eq. (17). Clearly, the term

$$H_{\text{eff}}^{\text{NO}} = (1 - X)H(1 + X) \tag{21}$$

represents the effective Hamiltonian in a space of non-orthogonal diabatic states since one recognizes the symmetric orthogonalization matrix $S^{-1/2}$. We are thus led to the following interpretation of the effective Hamiltonian $H_{\text{eff}}^{\text{NO}}$ defined in Eq. (21):

The effective Hamiltonian (of Eq. (21)) represents the matrix elements between the configurations $|P\rangle$ and $|R\rangle$ where the orbitals used in the construction of $|P\rangle$ and $|R\rangle$ have been built from distorted non-orthogonal fragment orbitals.

Similarly the overlap matrix $S^{-1/2}$ represents the overlap matrix between configurations built from non-orthogonal orbitals.

As a consequence of this observation, we are thus led to a very simple interpretation of the diabatic energies. We can now make an empirical association of the matrix elements of our effective Hamiltonian with the formulae for matrix elements for configurations built from non-orthogonal orbitals. This idea is best illustrated with an example. If we take the configuration $|R\rangle$ to be the configuration Ia in Fig. 3, then we have

$$(H_{eff})_{RR} = [(2\varepsilon_1 + 2\varepsilon_3 + 2J_{13} - K_{13})(1 - S_{13}^2)$$
$$+ 2k_{13}(S_{13}^3 - S_{13})](1 - 2S_{13}^2 + S_{13}^4)^{-1} \qquad (22)$$

where S_{13} is the HOMO(1)–HOMO(3) orbital overlap (not to be confused with the configuration overlap in Eqs. (15)–(17)), J_{13} and K_{13} are the usual two-electron Coulomb and exchange integrals, and k_{13} are diagonal and off-diagonal one-electron energy integrals. Since this term should be dominated by $-2k_{13}S_{13}$, this term should be repulsive and proportional to the orbital overlap S_{13}. Thus, as we have discussed previously,[8] one expects the reactant-like diabatic surface to be repulsive with slope proportional to the interfragment MO overlap. From similar arguments we expect the product-like diabatic surface to be attractive with slope proportional to the interfragment overlap. Finally, if we assume that the major component of the resonance interaction arises from the interaction of the two configurations (Ia and II) in Fig. 3, then the interaction is given as

$$H_{RP} = \int \phi_1^{(1)} \phi_4^{(1)} \frac{1}{r_{12}} \phi_3^{(2)} \phi_2^{(2)} \, dr_1 \, dr_2 \qquad (23)$$

where the orbital sequence numbers 2 and 4 correspond to LUMO(2) and LUMO(4) of Fig. 3. Clearly the resonance interaction H_{RP} involves the overlap density of the HOMO of one fragment with the LUMO of the other. However, the dependence of the slope of H_{RP} on interfragment separation should, again, be proportional to the overlap.

III. APPLICATIONS

In the previous section we have provided theoretical arguments which demonstrate that the MC-SCF method is, in general, the more appropriate computational procedure to use for studying the transition structure region of a potential energy surface associated with a chemical reaction. In fact, this procedure, in the CAS form and with a judicious choice of the active orbitals, is the only procedure that can describe concerted and two-step processes with equal accuracy. Further, we have indicated that the electronic rearrangement effects associated with the transition structure can be rationalized at this computational level using a model of interacting diabatic surfaces. It is our intention in this section to illustrate the applicability of these methods in some problems of chemical reactivity.

The problems illustrated here include the dimerization of CH_2 and SiH_2, sigmatropic rearrangements, the cycloaddition of two ethylenes, the 1, 3 dipolar cycloadditions and the Diels–Alder reaction. Therefore, they represent examples of significant importance in organic chemistry. These problems have also been chosen because in all cases the MC-SCF results have been analysed in terms of diabatic surface analysis. This type of analysis provides a clear rationalization *a posteriori* of the origin and nature of the critical points of a potential energy surface. In all cases presented here, the diabatic surfaces have been computed as an independent CI problem for each diabatic state[8a,8b] (see Eq. (20)). This approach is, in fact, much simpler to use than the two general formalisms and for this type of problem provides similar information.

A. Dimerization of CH_2 and SiH_2

The dimerization of methylene (CH_2) to form ethylene is a problem of chemical and theoretical interest. This reaction, in fact, was considered as a textbook example for the non-least-motion path[46] and was also the first polyatomic chemical reaction to be investigated at the MC-SCF level.[47] In recent years, this problem has continued to attract interest and various MC-SCF studies[48,49] have been reported where the behaviour along the various possible reaction paths has been investigated. Because of the small number of atoms involved, this problem has been studied using large basis sets, so that the results obtained in such studies can be considered to be quite reliable. Recently, the dimerization of silylene (SiH_2) has also been studied, as well as the coupling reaction[47] of CH_2 and SiH_2.[49b]

Both methylene and silylene have the same reaction orbitals, denoted by σ and π (see Scheme I):

Scheme I

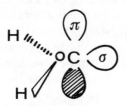

From the various possible occupations of these orbitals, there arise the four configurations[50] shown in Scheme II:

Scheme II

$$\pi \quad \underline{\uparrow} \qquad \underline{\uparrow} \qquad \underline{\quad} \qquad \underline{\uparrow\downarrow}$$

$$\sigma \quad \underline{\uparrow} \qquad \underline{\uparrow} \qquad \underline{\uparrow\downarrow} \qquad \underline{\quad}$$

$$\qquad {}^3B_1 \qquad\quad {}^1B_1 \qquad\quad {}^1A_1,\sigma^2 \qquad {}^1A_1,\pi^2$$

Two of these configurations, namely ${}^3B_1$ and ${}^1B_1$ represent states of methylene and silylene. The other two interact to form the two ${}^1A_1$ states, which are denoted by ${}^1A_1$ and ${}^1A_1^*$.

For methylene it has been found experimentally[51] that the $^3B_1$ state is more stable than the $^1A_1$ state by $9\,\mathrm{kcal\,mol^{-1}}$. The geometrical parameters of methylene in these two states[52] are shown in Scheme III:

Scheme III

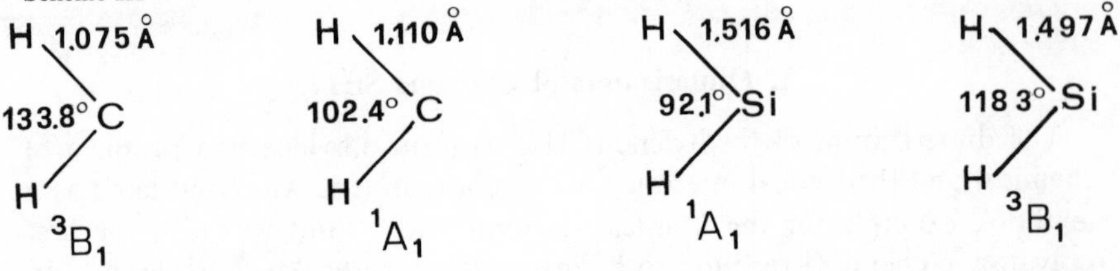

H $\diagdown$ 1.075 Å   H $\diagdown$ 1.110 Å   H $\diagdown$ 1.516 Å   H $\diagdown$ 1.497 Å

133.8° C   102.4° C   92.1° Si   118 3° Si

H $\diagup$ H $\diagup$ H $\diagup$ H $\diagup$

$^3B_1$ $^1A_1$ $^1A_1$ $^3B_1$

The various computational results [53-56] agree well with these data. For silylene the situation is opposite to that for methylene; in fact, a recent experimental nuclear recoil study[57] seems to establish for SiH_2 the existence of a singlet ground state. However, in this case, the energy separation with the $^3B_1$ state is not known, since, in the various spectroscopic studies[58-60] reported for SiH_2, no transitions involving triplet electronic states have been identified.

Therefore, the energy difference is known only at a theoretical level and an accurate value[56] indicates that the $^1A_1$ state is more stable than the $^3B_1$ state by $17.5\,\mathrm{kcal\,mol^{-1}}$. Furthermore, the $^1A_1$ state also seems to be more stable than the $^3B_1$ state even at the optimum structure for $^3B_1$.[49b] Experimentally, only the geometry of the $^1A_1$ state is known[58] and is reported in Scheme III together with the geometry of the $^3B_1$ state determined at a theoretical level.[56]

Two possible reaction paths can be considered (see Scheme IV) for such reactions, namely a least-motion path where both molecules approach each other in a plane and a non-least-motion path where at large distances the methylenes approach each other in perpendicular planes:

Scheme IV

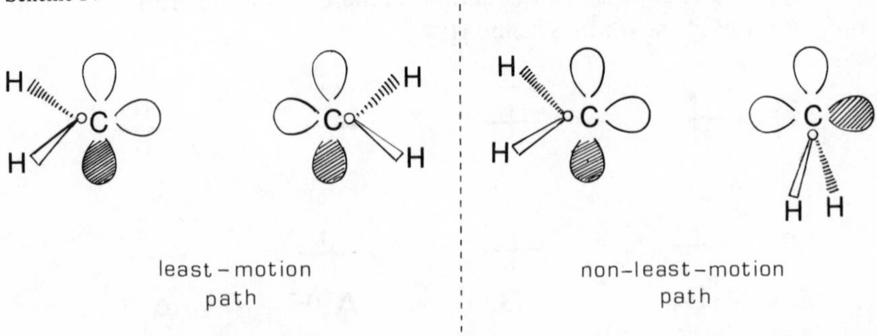

least – motion   non–least–motion
path   path

Approximate non-least-motion paths have been determined by optimization of the geometrical parameters as functions of the C–C, C–Si or Si–Si distance at the extended Hückel[46] or at the *ab initio* SCF level.[49b] The results of the various studies[47-49] can be summarized as follows:

1. Along a least-motion path, two ground-state triplet methylenes dimerize without a barrier to give the ground-state ethylene, while two singlet methylenes dimerize to give a Rydberg excited state of ethylene. Along a non-least-motion path the situation is essentially the same. In fact, two triplet methylenes dimerize to give the ground state of ethylene without a barrier and two singlet methylenes give a Rydberg excited state of ethylene.
2. Two singlet silylenes dimerize to give a ground-state disilene with a substantial barrier via the least-motion path and without a barrier via the non-least-motion path. On the other hand, two triplet silylenes dimerize to give adiabatically an excited state of disilene.
3. A ground-state methylene ($^3B_1$) and an excited-state silylene ($^3B_1$) give a ground-state silaethylene without a barrier in the least-motion path, and an excited-state methylene ($^1A_1$) and a ground-state silylene also give a ground-state silaethylene without a barrier in the non-least-motion path.

These results can be easily rationalized in terms of a diabatic surface analysis.[8b] For illustrative purposes, we report here the diabatic surface analysis for the dimerization of methylenes and for the coupling reaction of methylene and silylene along the least-motion path; the fragments here are in one case two CH_2 and in the other case a CH_2 and a SiH_2. The geometry of the CH_2 fragment in both reactions is assumed to be equal to the geometry of ethylene[61] ($R_{CH} = 1.07$ Å and $\angle HCH = 115.9°$), while that of the SiH_2 fragment has been taken from the STO-3G[62] geometry of disilene[61] ($R_{SiH} = 1.417$ Å and $\angle HSiH = 113.5°$). In our analysis, we consider only two diabatic curves, which are associated with the reactant and product IFC shown in Scheme V:

Scheme V

I   II

Configuration I, the reactant IFC, involves the singlet states of the two fragments, while configuration II, the product IFC, involves the triplet states of the two fragments with the overall spin coupled to a singlet. Thus the two diabatic curves are associated with the wavefunctions containing these two IFC plus the related one-electron CTC. In the case of configuration II, where there are four unpaired electrons, there is the possibility of coupling two pairs in two singlets and then coupling the result to an overall singlet or coupling first to two triplets then to a singlet. The packet II contains both possibilities but is dominated by the triplet–triplet configuration and correlates at infinite

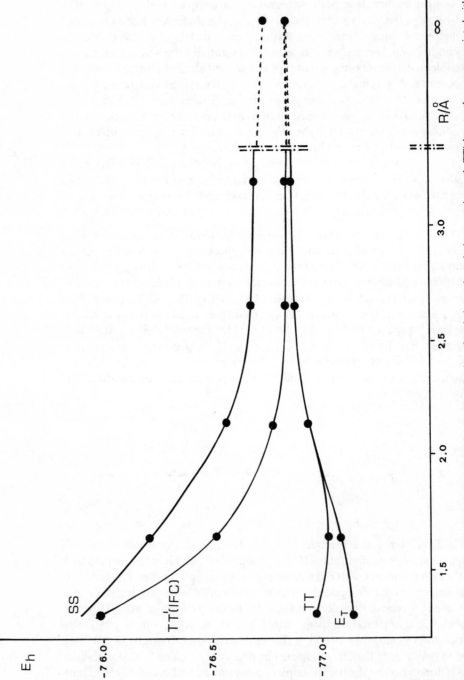

Fig. 7. Diabatic (SS and TT) and adiabatic (E_T) curves for the dimerization of methylenes along the least-motion path. TT′ denotes the curve associated with the triplet–triplet IFC. (*From Bernardi and Robb.*[8a] *Reproduced by permission of Taylor and Francis Ltd, Publishers.*)

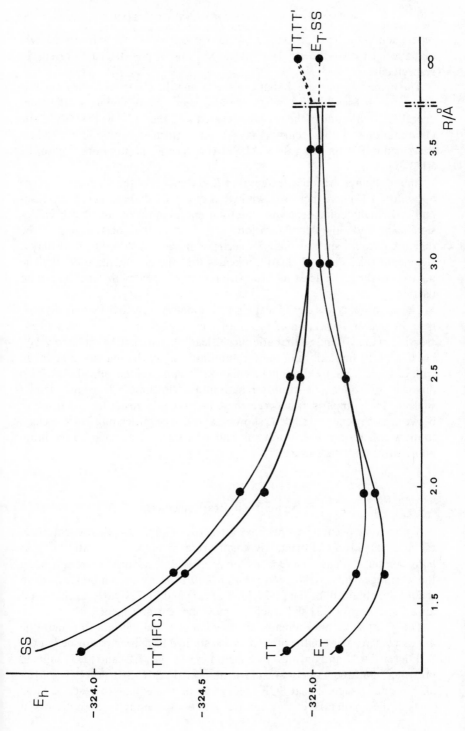

Fig. 8. Diabatic (SS and TT) and adiabatic (E_T) curves for the coupling reaction of CH_2 and SiH_2 along the least-motion path. TT′ denotes the curve associated with the triplet–triplet IFC. *(From Bernardi and Robb.[8a] Reproduced by permission of Taylor and Francis Ltd, Publishers.)*

separation with the triplet states of the two fragments. We refer to the two diabatic curves as the singlet–singlet (SS) and triplet–triplet (TT) curves respectively.

The behaviour of these diabatic curves along the fragment separation R is illustrated in Figs. 7 and 8, where we also show the behaviour of the total energy (E_T) curve and of the curve associated with the ${}^3B_1-{}^3B_1$ IFC (TT'). The computations of these curves have been performed with the procedure described in Ref. 8a using the GAUSSIAN 76 series of programs[63] with the STO-3G basis.

In both figures, with the decreases of R, the energy of the SS diabatic curve and of the TT' curve increases, while that of the TT diabatic curve decreases and is virtually coincident with the total energy until the overlap becomes significant and other configurations neglected in the construction of the diabatic packets, such as the two-electron charge transfers, begin to play a significant role. Figure 7 clearly indicates that the ground state of ethylene dissociates through a least-motion path into two triplet methylenes without a barrier.

The situation is less clear for the reaction between methylene and silylene: here, in fact, an inspection of Fig. 8 for large values of R shows an avoided crossing of the TT and SS diabatic curves, leading to a small barrier. However, because of the limitations of these computations, this finding has to be taken with caution. Actually, a recent calculation[49b] indicates that silaethylene also dissociates through a least-motion path into a triplet methylene and a triplet silylene. The comparison between these two types of result for the reaction between methylene and silylene shows that it can be misleading in some cases to draw conclusions based only on diabatic curves rather than on the more informative diabatic surfaces.[64]

B. Sigmatropic Rearrangements

Sigmatropic rearrangements have attracted considerable theoretical interest since the advent of orbital symmetry rules,[65] but only recently have *ab initio* computationals been reported on processes of chemical interest. Among the rearrangements investigated at the *ab initio* level of particular relevance are the results obtained for the [1, 2], [1, 3], [1, 5] and [1, 7] hydrogen shifts and for an important [3, 3] shift such as the Cope rearrangement.

The [1, 2] shifts have already been reviewed[66] and will not in general be discussed here. Most of the other *ab initio* studied have been performed at the SCF level[28,67,68] and only few recent studies, viz. the [1, 2] and [1, 3] shifts in propene[69] and the Cope rearrangement,[70] at the MC-SCF level. The MC-SCF results[69] suggest that SCF theory can be inadequate to describe large parts of these potential energy surfaces and consequently here we discuss in detail the MC-SCF results and only those parts of the SCF results that appear to be reliable and that to correspond to the 'allowed' pathways only.

1. The [1, 2] and [1, 3] Sigmatropic Shifts in Propene

The reaction discussed in this section is the rearrangement of a propene A to a propene B, i.e. the migration of a hydrogen from a methyl group towards the opposite terminal carbon atom. This model reaction is usually classified as a [1, 3] sigmatropic shift and is said to be allowed in an *antara* fashion (see Fig. 9, path 1) and forbidden in a *supra* one (path 2).

The calculations[69] on this reaction surface have been carried out at the CI and MC-SCF level with minimal (STO-3G[71]) and extended (4-31G[72]) basis

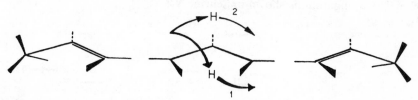

Fig. 9. *Antara* and *supra* [1,3] sigmatropic shifts of hydrogen in propene (1 and 2, respectively). (*Reprinted with premission from Bernardi et al.*[69] *Copyright 1984 American Chemical Society.*)

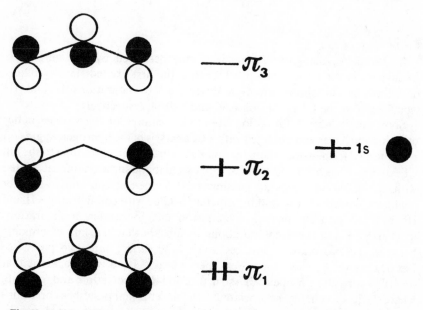

Fig. 10. Valence orbitals for the allyl plus hydrogen system at infinite separation consisting of the π system of allyl and the 1s orbital of hydrogen. (*Reprinted with permission from Bernardi et al.*[69] *Copyright 1984 American Chemical Society.*)

sets. Equilibrium geometries and transition structures have been fully optimized by using MC-SCF gradients[2d] with both basis sets. Each critical point has been characterized by computing the Hessian for the full set of internal coordinates excluding the C–H and C–C stretches but including all the coordinates for the migrating hydrogen.

In this study the three π orbitals of allyl and the 1s orbital of H at infinite separation have been taken as the valence orbitals (see Fig. 10). This choice provides a correct description of the motions investigated here since it allows proper dissociation of both propene and a trimethylene-like species into allyl radical plus a hydrogen atom. An MC-SCF computation of a planar allyl radical plus the hydrogen atom at 20 Å has shown that the ground and first excited states of this system are mainly described in terms of the three Heitler–London configurations shown in Scheme VI:

Scheme VI

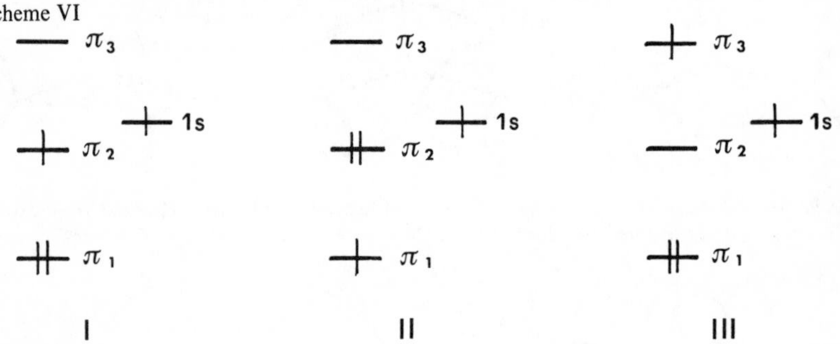

In particular, the ground state is mainly described by configuration I (coefficient in the CI expansion 0.92), while the first excited state is mainly described by an almost equal admixture of configurations II and III (coefficients in the CI expansion 0.66 and -0.64, respectively).

Some preliminary information about the features of the surface under examination have been obtained with a CI analysis of the hydrogen migration over the allyl framework, where the adiabatic surface corresponding to the full CI expansion has been decomposed into two diabatic surfaces, one associated with configuration I plus all possible related one-electron charge-transfer configurations (packet A) and the other associated with configurations II and III plus all possible related one-electron charge-transfer configurations (packet B). From these computations it appears that the line of crossing between the two diabatic surfaces divides the surface into two parts (see Fig. 11): region A, where the lowest energy is associated with packet A that describes correctly a propene-like structure and its dissociation; and region B, where the lowest energy is associated with packet B that describes correctly a trimethylene-type diradical. These results suggest that only part A can be expected to be reasonably described at the SCF level, while part B requires necessarily a CI-type approach.

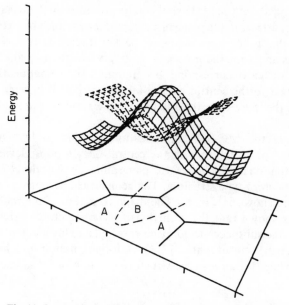

Fig. 11. Intersection of diabatic energy surfaces for hydrogen moving 1.1 A above a rigid allyl radical framework. The full grid corresponds to the closed-shell surface (packet A) and the broken grid to the diradical surface (packet B). (*Reprinted with permission from Bernardi et al.*[69] *Copyright 1984 American Chemical Society.*)

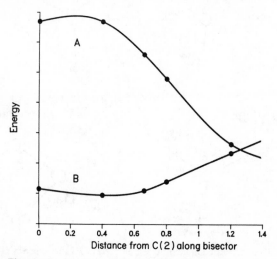

Fig. 12. Cross-section of the diabatic surfaces of Fig. 11 in a plane containing the CCC angle bisector (closed shell A, diradical B). (*Reprinted with permission from Bernardi et al.*[69] *Copyright 1984 American Chemical Society.*)

Information about the [1, 3] shift can be provided by the behaviour of the two diabatic surfaces in the plane bisecting the allyl framework (see Fig. 12). It can be seen that in the bisecting plane, packet B remains at lower energy till very large distances and the crossing occurs very late. In this plane a *supra*-type transition state has to correspond to a minimum and consequently can occur at or very near to the central carbon atom in region B or very far away in region A. In the latter case, the adiabatic surface will be mainly described by packet A.

Accordingly the MC-SCF optimization has led to two structures, a trimethylene-type structure with the central CH_2 group relevantly rotated towards a situation in which the CCC plane almost bisects the HCH angle and the other an *antara*-type structure. The geometrical parameters of these two structures are shown in Fig. 13. The diagonalization of the related Hessian matrices has shown that the trimethylene-type structure corresponds to a minimum in the subspace including co-rotation and inversion of the terminal methylenes, ring closure and [1, 3] shift-like motion of the migrating hydrogen, while the *antara*-type structure is a first-order saddle point for a [1, 3] shift.

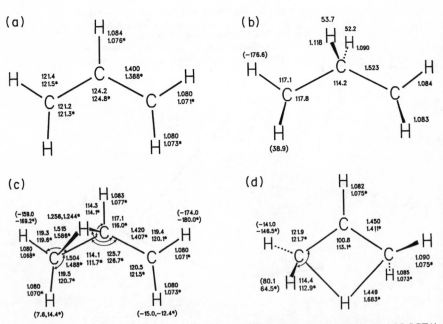

Fig. 13. Optimized geometries computed at MC-SCF/STO-3G (no superscript) and MC-SCF/4-31G (asterisk): (a) allyl radical (C_{2v} symmetry), (b) trimethylene-like structure (C_s symmetry), (c) [1,2] *supra* transition state, (d) [1,3] *antara* transition state (C_2 symmetry). The HCCC dihedral angled are given in parentheses. (*Reprinted with permission from Bernardi et al.*[69] *Copyright 1984 American Chemical Society.*)

When compared with the structure of the allyl radical (see Fig. 13), one can note the following major structural features of these two structures:

1. In the case of the trimethylene 'minimum', there is a significant increase of the C–C distance (from 1.400 Å to 1.523 Å at the STO-3G level), a significant pyramidalization of the methylene centres and a large decrease of the CCC angle (from 124.2° to 114.2°).

2. In the case of the *antara* transition structure, there is a very small change of the C–C distances (from 1.400 Å (1.388 Å) to 1.450 Å (1.411 Å) at the STO-3G (4-31G) level) and a very large bond length between the terminal carbons and the migrating hydrogen (C–H distance of 1.449 Å (1.683 Å) at the STO-3G (4-31G) level, which is 32% (55%) stretched compared to a standard C–H bond length).

As expected on the basis of the previous discussion, the MC-SCF results agree well with the SCF results in the case of the *antara*-type structure, but not in the case of the trimethylene-type structure. In this case, in fact, an SCF treatment leads to a transition structure of C_s symmetry.[28]

Information about the [1, 2] shift can be obtained again from the behaviour of the two diabatic surfaces computed in the plane perpendicular to the allyl plane and containing the CC bond axis (see Fig. 14). In this case the two diabatic curves show an avoided crossing, which occurs approximately at one-

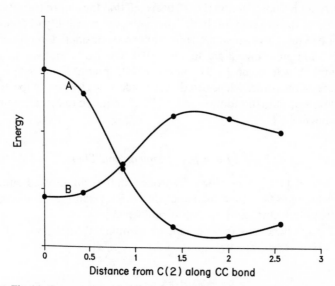

Fig. 14. Cross-section of the diabatic surface of Fig. 11 in a plane containing the C(1)–C(2) bond (closed shell A, diradical B). (*Reprinted with permission from Bernardi* et al.[69] *Copyright 1984 American Chemical Society.*)

half of the C–C bond length and indicates the existence of a transition state for a [1, 2] shift. The MC-SCF geometry optimization has led in this case to a critical point, which, through diagonalization of the Hessian matrix, has been found to be a first-order saddle point. The computed geometrical parameters are shown in Fig. 13 and agree very well with the information provided by the preliminary diabatic surface analysis.

While the origin of the [1, 2] transition state is due to the avoided crossing between the two diabatic surfaces, that of the [1, 3] *antara* transition state is less clear. Information on this problem has been obtained by performing an energy decomposition analysis[8a] of the energy barrier associated with the [1, 3] process. The results of this analysis indicate that the *antara* barrier has a simple interpretation in terms of a large deformation energy of the allyl fragment and a much stronger bond in propene.

To obtain reliable estimates of the energy effects associated with the [1, 2] and [1, 3] hydrogen shifts in propene, the various energy values have been recalculated at the 4-31G level using a size-consistent multi-reference linear coupled-cluster (MR-LCC) approach.[30] It has been found that both the [1, 2] *supra* and the [1, 3] *antara* hydrogen-shift transition states in propene are above allyl radical plus hydrogen atom, the [1, 3] *antara* by 28.3 kcal mol^{-1} and the [1, 2] *supra* by 19.0 kcal mol^{-1}. A similar result for the *antara* transition state has also been obtained by Radom *et al.*,[28] who have suggested, on the basis of coupled electron-pair approximation (CEPA) calculations with a double zeta plus polarization (DZP) basis set, that the *antara* transition state may be above the dissociation limit. Therefore, the possibility of dissociation must not be ignored when sigmatropic shifts are considered in propene. The MR-LCC computations indicate also that the [1, 2] transition state is 9.3 kcal mol^{-1} below the [1, 3] *antara* saddle point. However, once the molecule is on the trimethylene diradical surface, ring closure is expected to be more favourable than the additional [1, 2] shift required to complete the [1, 3] rearrangement.[73–75]

2. [1,5] and [1,7] Sigmatropic Shifts

The thermal [1, 5] sigmatropic hydrogen shift is an allowed suprafacial process, which has been investigated at the SCF level in 1, 3-pentadiene,[67,68] in β-hydroxyacrolein[28] and in cyclopentadiene.[67]

In 1, 3-pentadiene this process can be formulated in the way shown in Scheme VII:

Scheme VII

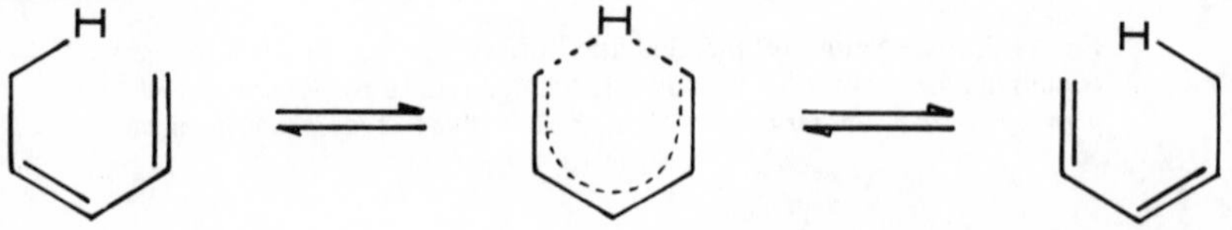

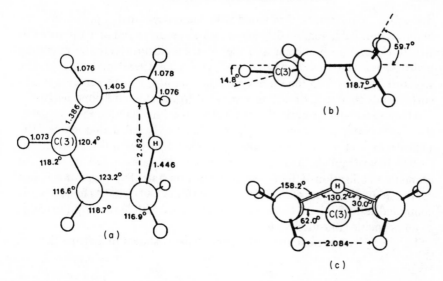

Fig. 15. Geometrical parameters of the transition structure for the [1,5] hydrogen shift of 1,3-pentadiene computed at the 3-21G SCF level. (*Reprinted with permission from Randan and Houk.*[67] *Copyright 1984, Pergamon Press.*)

The SCF geometry optimization has been carried out at the STO-3G[28] and 3-21G[67,68] levels and in all cases has led to a critical point of C_s symmetry, whose geometry is shown in Fig. 15. This structure has been found, through diagonalization of the Hessian matrix, to be a first-order saddle point. As pointed out by Rondan and Houk,[67] this structure is distinctly chair-like and can be described as an aromatic transition state, since the C–C bond lengths are nearly the same as those of benzene. The migrating hydrogen lies 60° out of the C(1)C(2)C(4)C(5) plane and forms a CHC angle of 130° and a C–H bond of 1.45 Å, which is 33% stretched compared to a standard C–H bond length.

The activation energies calculated for the 3-21G fully optimized structure are 55 kcal mol^{-1} (3-21G level[68]), 52.8 kcal mol^{-1} (3-21G level with zero-point energy corrections[67]) and 59 kcal mol^{-1} (6-31G level[68]): all these values are considerably higher than the experimentally estimated activation energy of 35 kcal mol^{-1}.[76]

The [1, 5] hydrogen shift in β-hydroxyacrolein is illustrated in Scheme VIII:

Scheme VIII

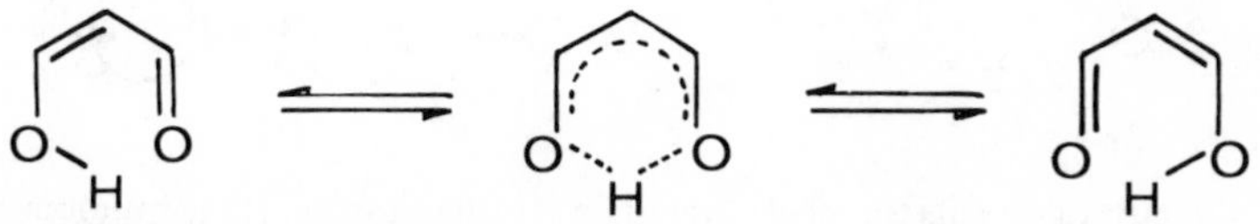

The best *ab initio* calculations include optimizations at the minimal[28] and double-zeta[77] SCF level and inclusion of the effects of electron correlation through a CI treatment. Compared with the process in 1, 3-pentadiene, the main structural differences of the symmetric structures are that in this case this structure seems to be planar and the bond length involving the migrating hydrogen (O–H in this case) varies much less (from 0.998 Å in the asymmetric structure to 1.167 Å in the symmetric structure at the STO-3G level, which corresponds to a 16% stretching compared to the standard bond length).

At both SCF and CI levels the asymmetric structure (A) is found to lie higher than the symmetric structure (S): the barrier is calculated to be 10–11 kcal mol^{-1} at the SCF level and is lowered slightly with the CI treatment (~ 9.8–10 kcal mol^{-1}). This low barrier may be rationalized in terms of aromatic stabilization of the π-electron transition state. This suggestion is also confirmed by the values of the C–C (1.401 Å) and C–O (1.288 Å) bond lengths computed at the STO-3G level, which are in accord with the description of this as an aromatic transition state.

The [1, 5] hydrogen shift in cyclopentadiene is shown in Scheme IX:

Scheme IX

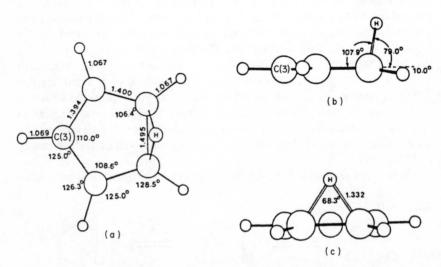

The computations have been performed at the 3-21G[78] level with gradient optimization.[67] The computed transition structure is shown in Fig. 16. It can

Fig. 16. Geometrical parameters of the transition structure for the [1, 5] hydrogen shift of cyclopentadiene computed at the 3-21G SCF level. (*Reprinted with permission from Rondan and Houk.*[67] *Copyright 1984, Pergamon Press.*)

be seen that all five carbons and the hydrogens attached to C(2) and C(4) stay virtually in-plane. The C–C bond lengths, except that between C(1) and C(5), are aromatic bond lengths. The non-migrating hydrogens at C(1) and C(5) are 10° out-of-plane, while the migrating hydrogen is 79° out-of-plane and the related C–H bond length is 1.332 Å, which corresponds to a 22% stretching compared with the standard bond length.

The activation energy is calculated to be $39.7\,\mathrm{kcal\,mol^{-1}}$, which is $18\,\mathrm{kcal\,mol^{-1}}$ higher than the experimental estimates,[79] but $13\,\mathrm{kcal\,mol^{-1}}$ lower than the calculated value of 1, 3-pentadiene, in close agreement with the experimental difference.

For the [1, 7] rearrangement the 'allowed' process would involve an antarafacial hydrogen migration. The process investigated is the hydrogen

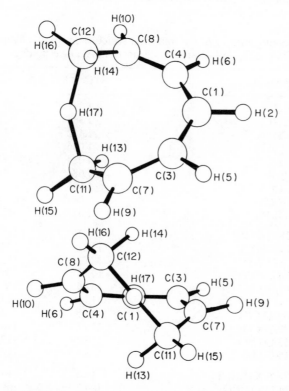

Fig. 17. Drawings of the transition structure (C_2) considered in the [1,7] sigmatropic hydrogen migration of 1,2,3-heptatriene. The values of the related geometrical parameters are given in Table V. (*Reprinted with permission from Hess et al.*[68] *Copyright 1985 American Chemical Society.*)

migration in heptatriene[68] shown in Scheme X:

Scheme X

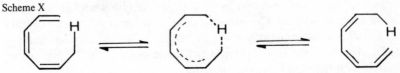

Because of the size of the system, only a transition structure with C_2 symmetry has been considered. The geometry optimization has been performed at the 3-21G level and has led to the structure shown in Fig. 17 and Table V. The size of the system has precluded the calculation of the Hessian matrix at this computational level. Therefore, the structure has been re-optimized with the STO-3G basis set and the Hessian matrix computed at this level. A single negative eigenvalue has been found on diagonalization of this Hessian, thus confirming that the C_s structure is a transition structure.

TABLE V

Optimized geometry of the transition structure in the $1, 7$ hydrogen shift of $1, 3, 5$-heptatriene (distances in Å, angles in deg).

Coordinate	STO-3G	3–21G
C(1)–C(3)	1.397	1.392
C(3)–C(7)	1.384	1.384
C(7)–C(11)	1.412	1.397
C(1)–H(2)	1.086	1.078
C(3)–H(5)	1.084	1.077
C(7)–H(9)	1.084	1.076
C(11)–H(13)	1.085	1.081
C(11)–H(15)	1.083	1.077
C(11)–H(17)	1.327	1.374
⟨ C(3)C(1)C(4)	132.2	132.2
⟨ C(1)C(3)C(7)	130.2	130.4
⟨ C(3)(7)C(11)	125.7	125.6
⟨ C(7)C(11)H(17)	105.4	104.6
⟨ C(11)H(17)C(12)	151.6	149.6
⟨ C(3)C(1)H(2)	113.9	113.7
⟨ H(5)C(3)C(1)	114.5	114.4
⟨ H(9)C(7)C(3)	116.7	116.6
⟨ H(13)C(11)C(7)	116.0	117.0
⟨ H(15)C(11)C(7)	116.5	117.7
⟨ H(5)C(3)C(1)H(2)	14.0	14.4
⟨ C(7)C(3)C(1)H(2)	160.2	159.6
⟨ H(9)C(7)C(3)H(5)	13.5	13.5
⟨ C(11)C(7)C(3)H(5)	168.0	166.8
⟨ H(13)C(11)C(7)H(9)	146.3	150.5
⟨ H(15)C(11)C(7)H(9)	9.2	7.5

From the computed geometrical parameters shown in Table V, it can be seen that all the C–C bond lengths are nearly the same as those of benzene, suggesting a significant degree of aromatic character. It can also be seen that the migrating hydrogen forms a CHC angle of 149.6°, much larger than that in the [1, 3] *antara* transition structure, and a C–H bond of 1.374 Å, which is 26% stretched compared to a standard C–H bond length.

The activation energy for the [1, 7] hydrogen shift has been found to be 44 kcal mol^{-1} at the 3-21G level, a value much smaller than that found for the [1, 3] *antara* shift in propene, estimated to be of the order of 90–100 kcal mol^{-1}.[68] This difference in energy is presumably due to the fact that heptatriene leads to a transition structure with relatively little strain because of the greater flexibility of the larger carbon skeleton.

3. The Cope Rearrangement

The Cope rearrangement is a very interesting example of a [3, 3] sigmatropic shift and can be illustrated in the way shown in Scheme XI:

Scheme XI

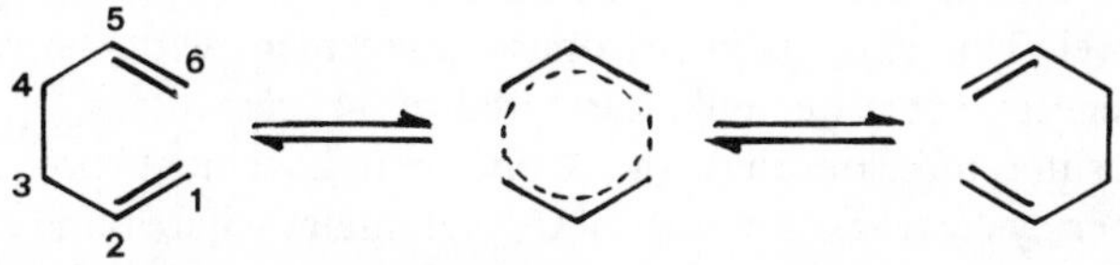

The 'allowed' process would involve a suprafacial migration for both components, a geometrical disposition that can be easily obtained.

The mechanism of the Cope rearrangement has been the subject of numerous experimental studies.[80] Three different pathways are possible *a priori*: (i) a concerted pericyclic pathway; (ii) a non-concerted pathway involving σ-bond formation to give cyclohexane-1, 4-diyl; and (iii) a non-concerted pathway involving σ-bond cleavage to afford two allyl radicals. The latter possibility is ruled out by labelling studies.[81]

This problem has been investigated at the *ab initio* level by Borden *et al.*[70] using two different wavefunctions: a two-configuration wavefunction (TC-SCF) and a wavefunction consisting of 52 ^1A_g spin-adapted configurations (MC-SCF). Calculations have been carried out using both STO-3G and 3-21G basis sets. Because the MC-SCF computations with the 3-21G basis required too much computer memory, the calculations were simulated by using ALIS to find the optimal orbitals. These optimized orbitals have been used to perform a full six-electron, six-orbital CI.

At the 3-21G level, the geometry of the transition structure has been optimized at the SCF level for various fixed values of R, the distance of the two equivalent bonds C(3)–C(4) and C(1)–C(6). Energies have then been calculated at these geometries using the MC-SCF wavefunction and a minimum has been found to occur for $R = 2.062$ Å. The resulting structure is shown in Fig. 18. The energy of this structure is 29.4 kcal mol^{-1} above that of 1, 5-

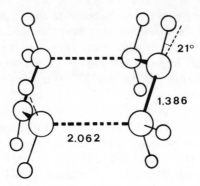

Fig. 18. Geometry of the transition state for the Cope rearrangement optimized at the MC-SCF/3-21G level. (*Reprinted with permission from Osamura* et al.[70] *Copyright 1984 American Chemical Society.*)

hexadiene, whose geometry in the all-transoid conformation was optimized at the SCF level. This value is in reasonable agreement with the value of the activation energy experimentally measured of $34 \, \text{kcal mol}^{-1}$.

These results indicate that the Cope rearrangement proceeds via a concerted pericyclic transition state of C_{2h} symmetry where bond making and bond breaking occur in unison. In fact, both the geometry and the values of the coefficients in the MC-SCF wavefunction (the second largest coefficient is only 0.18) show that the C_{2h} optimized structure is not a diradical. This finding is in disagreement with the result of the STO-3G MC-SCF computations and with those of the STO-3G and 3-21G TC-SCF calculations, which all favour a diradical structure. The reason for this is that the STO-3G basis set and a two-configuration SCF wavefunction each prejudice the calculations by selectively stabilizing cyclohexane-1, 4-diyl.

C. Thermal Cycloaddition of Two Ethylenes

The thermal cycloaddition of two ethylenes is one of the 'textbook' examples used in the illustration of the Woodward–Hoffmann rules[82] of orbital symmetry control in concerted reactions. Therefore, the related potential energy surface can provide various types of information of chemical and theoretical interest. A first question is associated with the mechanistic question of whether this reaction proceeds via diradical or concerted pathways. Since this reaction is an example of a concerted thermally forbidden process, it can be expected that the favoured path be the diradical one. However, it is important to have a detailed description of the structural and energetic features of these different pathways. A second question is associated with the structure of the tetramethylene diradical, which is often invoked as an intermediate in the pyrolysis of cyclobutanes to ethylenes.[83] Though this diradical has never been directly observed, the indirect evidence for its existence is impressive.[83-85] A third question is associated with the *supra–antara* pathway,[82] and in particular with the geometry and energy of the related transition structure.

Because of the sharply avoided crossing in the region of the transition state for the $[2_s + 2_s]$ concerted reaction and of the diradicaloid character of the critical points involved in the non-concerted process, the computation of this surface requires computational methods that transcend the SCF method. Recently this surface, and in particular the transition structure region, has been investigated in detail by *ab initio* molecular-orbital methods.[86] The calculations have been performed at the MC-SCF level with minimal (STO-3G) and extended (4-31G) basis sets. The various critical points have been fully optimized with MC-SCF gradients and characterized by computing the corresponding Hessian matrices.

In previous *ab initio* calculations, Wright and Salem[87] explored the coplanar $[2_s + 2_s]$ rectangular decomposition of cyclobutane to form two ethylenes using an SCF treatment at the STO-3G level followed by a 2×2 CI (HOMO–LUMO), and Segal[88] investigated the non-concerted approaches again using an SCF treatment followed by a 15-dimensional CI. In other *ab initio* computations Doubleday *et al.*[89] studied the structure of the tetramethylene diradicals using a two-configuration and a 20-configuration MC-SCF wavefunction at the 3-21G level. A two-configuration MC-SCF wavefunction was also used by Borden and Davidson[90] in a study of the conformational properties of the tetramethylene diradical. In the following we summarize the MC-SCF results described in Ref. 86 since they should provide an accurate homogeneous description of the whole transition structure region of this surface, and where appropriate we compare these results with those obtained by the other authors.

In the MC-SCF study the molecular orbitals have been constructed from the non-interacting ethylenes, with the valence space formed by the two π orbitals (π and π^*) of each ethylene (see Scheme XII).

Scheme XII

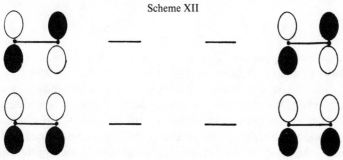

This choice is appropriate for the motions investigated here since it allows proper dissociation of tetramethylene into two ethylenes and also permits the correct description of a diradicaloid intermediate or transition structure. In this valence space, a complete CI for the singlet state has 20 configurations.

The following reaction paths, which are expected to be relevant on the basis of either the Woodward–Hoffmann rules or previous *ab initio* computations, have been investigated: the $[2_s + 2_s]$ *supra–supra* approach; non-concerted approaches (i.e. *gauche* and *trans* approaches); and the $[2_s + 2_a]$ *supra–antara* approach.

The surface can be conveniently divided into two regions: one, the tetramethylene region, containing the *supra–supra*, the *gauche* and the *trans* approaches; and the other, the *supra–antara* region, containing the *supra–antara* approach.

1. *The Tetramethylene Region*

The results obtained in the MC-SCF study of this region, which is chemically more important, are illustrated in Fig. 19 in terms of an R, φ map (for the notation of the internal coordinates, see Scheme XIII).

Scheme XIII

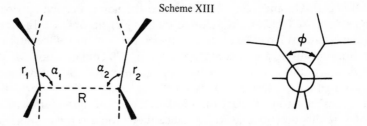

In this map it is possible to include, with some approximations, all the critical points of this region, except that for the $[2_s + 2_s]$ rectangular coplanar approach, which belongs to a different cross-section. For the latter, in fact, the value of α is 90°, while the α values of all the other critical points vary between 110° and 117°. The most important results obtained in the MC-SCF study can be summarized as follows.

a. $[2_s + 2_s]$ *Approach*

Two different types of $[2_s + 2_s]$ approach have been investigated: a rectangular coplanar approach of D_{2h} symmetry in which the two forming C–C bond distances are constrained to be equal, and a coplanar approach of C_{2v} symmetry where the constraint of two equal C–C distances is relaxed. For each approach, a critical point has been found, which has been characterized to be a second-order saddle point. At the STO-3G level, both Hessian matrices have been computed for the full set of internal coordinates, excluding those

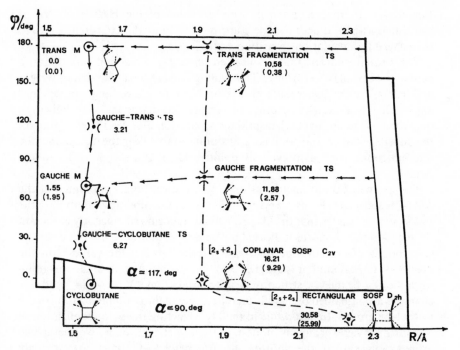

Fig. 19. Critical points of the tetramethylene region of the surface associated with the cycloaddition of two ethylenes (M denotes a minimum, TS a transition state and SOSP a second-order saddle point). The energies are given relative to the energy of the tetramethylene *trans* diradical minimum and are computed at the STO-3G (4-31G) level. (*Reprinted with permission from Bernardi et al.*[64] *Copyright 1986 American Chemical Society.*)

associated with the stretching and bending of the C–H bonds. The lower negative eigenvalue of the Hessian matrix in each second-order saddle point is associated with a transition vector dominated by R (see Scheme XIII), the interfragment distance between the two ethylenes. In contrast, the second negative eigenvalue in the case of the rectangular second-order saddle point is associated with the motion which distorts the symmetry from D_{2h} and clearly leads to the coplanar C_{2v} second-order saddle point (see Scheme XIV).

Scheme XIV

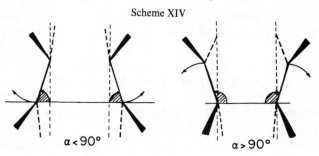

For the latter, the second negative eigenvalue of the Hessian matrix is associated with the dihedral angle φ and motion along this coordinate leads to the *gauche* fragmentation transition state.

These results indicate that a concerted *supra–supra* reaction path does not exist in the case of two ethylenes. This finding is in contrast with that of Wright and Salem,[87] who have found a transition state along a rectangular coplanar approach. However, they obtained this result considering in the optimization only the C–C bonds in ethylene and the separation between the ethylenes. In this subspace this critical point is a transition state, while in the full space of the internal coordinates, as previously pointed out, it is a second-order saddle point.

The geometrical parameters of these two critical points computed at the STO-3G and 4-31G levels are given in Fig. 20. The comparative analysis of the two structures shows that the D_{2h} critical point occurs significantly earlier on the reaction coordinate as indicated by the larger value of R. In both cases the C–C bonds of the ethylene fragments are in between the values[87] in the reactants (1.335 Å) and product (1.550 Å). The variation of the angle α in going from the D_{2h} to the C_{2v} structure is quite large (from 90° to $\sim 118°$) and this change is accompanied by significant differences in the pyramidalization of the methylene groups. In fact, while in the D_{2h} structure the pyramidalization of all four methylene groups is modest, in the C_{2v} structure the two methylene groups adjacent to the shorter interfragment C–C bond are strongly pyramidal while the other two are almost planar.

In both cases the values computed at the minimal level agree well with those computed at the extended level: the only exception is the interfragment distance R in the C_{2v} structure that appears to be overestimated at the minimal level.

From Fig. 19 it can also be seen that the D_{2h} critical point is significantly higher in energy than the C_{2v} one, the energy difference being 14.4 kcal mol^{-1} at the STO-3G level and 16.7 kcal mol^{-1} at the 4-31G level: therefore the energy difference also agrees well at the two computational levels.

b. Gauche and trans approaches

Both the *gauche* and *trans* approaches involve a transition state and a minimum. In the case of the two fragmentation transition states, the diagonalization of either the second derivative matrices computed by finite difference at the STO-3G level or the updated second derivative matrices at the 4-31G level has shown only one negative eigenvalue associated with a transition vector dominated by R.

Therefore the present results indicate that there are two local minima for the tetramethylene diradical corresponding to a *gauche* and a *trans* geometry, in agreement with the CI results of Segal.[88] On the other hand, Doubleday

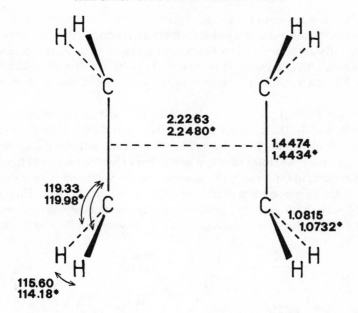

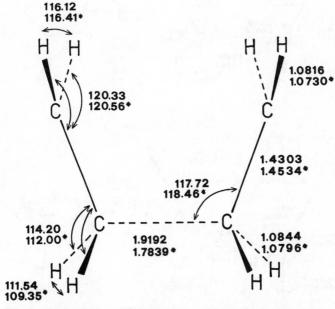

Fig. 20. Optimized geometries computed at MC-SCF/STO-3G (no super-script) and MC-SCF/4-31G (asterisk) for the rectangular (D_{2h}) and the coplanar (C_{2v}) second-order saddle points. (*Reprinted with permission from Bernardi et al.[86] Copyright 1985 American Chemical Society.*)

et al.[89b] find no minima in the region of the tetramethylene diradical. However, because the computed barriers to fragmentation are very small, particularly at the 4-31G level (less than 1 kcal mol^{-1}), one must conclude, in agreement with Doubleday et al., that the high-temperature thermochemistry of such a reaction will not be influenced by such small dips in the potential surface.

The geometrical parameters of these four critical points computed at the STO-3G and 4-31G levels are given in Figs. 21 and 22. The analysis of these results shows that the two minima have very similar inter- and intrafragment C–C distances. A similar situation is also found for the two transition states. Furthermore, in both cases, the geometry of the minimum is very similar to that of the corresponding transition state, in agreement with Hammond's postulate: in both cases the main changes are associated with the interfragment distance R and the intrafragment C–C distance. For the transition states,

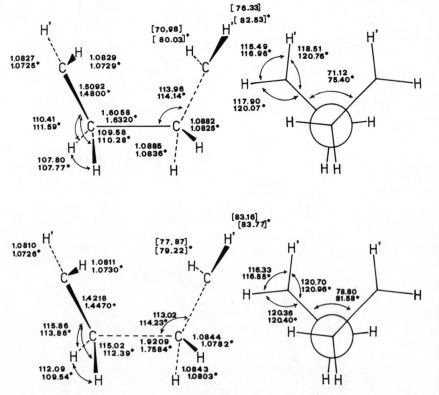

Fig. 21. Optimized geometries computed at MC-SCF/STO-3G (no superscript) and MC-SCF/4-31G (asterisk) for the tetramethylene *gauche* minimum and fragmentation transition state. The values in brackets are the dihedral angles HCCC and H'CCC. (*Reprinted with permission from Bernardi et al.[86] Copyright 1985 American Chemical Society.*)

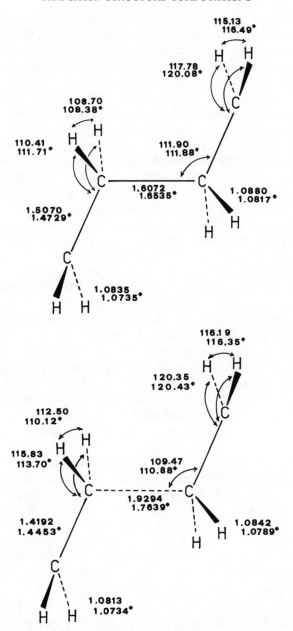

Fig. 22. Optimized geometries computed at MC-SCF/STO-3G (no superscript) and MC-SCF/4-31G (asterisk) for the tetramethylene *trans* minimum and fragmentation transition state. (*Reprinted with permission from Bernardi et al.*[86] *Copyright 1985 American Chemical Society.*)

in fact, the latter values are in between reactants and product and the interfragment distance R is still large, while in the minima the C–C distances have assumed values very near to that of a single C–C bond and the interfragment distance R has become significantly shorter.

For the two minima the geometrical parameters computed at the minimal STO-3G and extended 4-31G levels agree very well. This situation is also found for the two transition states, except for the interfragment distance R, which again appears to be overestimated at the minimal level.

From Fig. 19 it can be seen that the *trans* minimum represents the most stable form of the tetramethylene diradical. However, the energy difference between the two minima is very small, $1.55 \, \text{kcal mol}^{-1}$ at the STO-3G level and $1.95 \, \text{kcal mol}^{-1}$ at the 4-31G level. Therefore, in this case also the energy difference agrees well at the two computational levels. Also the barriers to fragmentation are small. In this case, the basis set effect is more pronounced: in fact, these barriers are reduced from approximately $10 \, \text{kcal mol}^{-1}$ at the STO-3G level to less than $1 \, \text{kcal mol}^{-1}$ at the 4-31G level.

In the tetramethylene surface, there are also two other transition states, one connecting the *trans* and *gauche* minima and the other along the path connecting the *gauche* minimum to cyclobutane. Both these transition states have been optimized only at the STO-3G level. The geometrical parameters are given in Fig. 23. The structures of both these transition states are very similar to that of the *gauche* minimum. The main geometrical change in both cases is associated with the dihedral angle φ. Another interesting feature is the planarization of the two methylene centres and the decrease of the CCC angle in the *gauche*–cyclobutane transition state. Both the barrier between the *trans* and *gauche* minima and that between the *gauche* minimum and cyclobutane are very low (1.69 and $5.08 \, \text{kcal mol}^{-1}$ relative to the *gauche* minimum at the STO-3G level).

Information about the origin and nature of the various critical points can be obtained with a diabatic surface analysis.[8a,8b,91] This analysis has been performed at the STO-3G level, since the main features of the various critical points are well reproduced at this computational level. For such analysis we take as fragments the two ethylene molecules and take as valence orbitals the π and π^* MOs of the two ethylenes. The reactant IFC (configuration I in Fig. 24) involves the singlet states of the two fragments, while the product IFC (configuration II in Fig. 24) corresponds to two ethylenes in a triplet state with the overall spin coupled to a singlet. Thus the two diabatic surfaces are associated with these two IFC plus the related one-electron CTC. In the case of configuration II, where there are four unpaired electrons, there is the possibility of coupling two pairs in two singlets and then coupling the result to an overall singlet or coupling first to two triplets then to a singlet. The packet II contains both possibilities but is dominated by the triplet–triplet coupling and correlates at infinite separation with the triplet states of the two ethylenes.

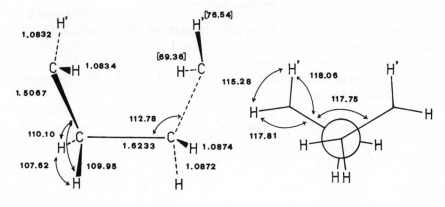

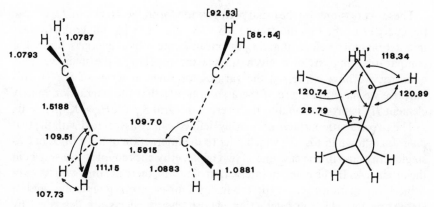

Fig. 23. Optimized geometries computed at MC-SCF/STO-3G for the *gauche–trans* transition state and for the *gauche*–cyclobutane transition state. The values in brackets are the dihedral angles HCCC and H'CCC. (*Reprinted with permission from Bernardi* et al.[86] *Copyright 1985 American Chemical Society.*)

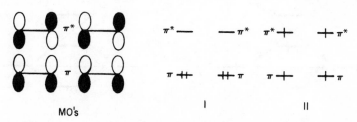

Fig. 24. Valence orbitals for the addition of two ethylenes and IFC associated with reactants (I) and products (II). (*Reprinted with permission from Bernardi* et al.[91] *Copyright 1986 American Chemical Society.*)

In order to satisfy the chemist's need to understand and develop a simple qualitative model, the main features of the diabatic surfaces can be rationalized in terms of very simple MO arguments. In fact, each specific bonding situation can be associated with an MO interaction diagram with specific orbital occupancies, whose energy effects can be discussed in terms of simple MO expressions such as:

$$\Delta E_{ij} = 4\frac{S_{ij}^2\varepsilon_0 - S_{ij}H_{ij}}{1 - S_{ij}^2} \approx K_1 S_{ij}^2 \qquad K_1 = \frac{4(\varepsilon_0 - K)}{1 - S_{ij}^2} > 0 \qquad (24)$$

$$\Delta E_{ij} = 2\frac{(H_{ij} - S_{ij}\varepsilon_i)^2}{\varepsilon_i - \varepsilon_j} \approx K_2 \frac{S_{ij}^2}{\varepsilon_i - \varepsilon_j} \qquad K_2 = 2(K - \varepsilon_i)^2 > 0 \qquad (25)$$

$$\Delta E_{ij} = 2\frac{H_{ij} - \varepsilon_i S_{ij}}{1 + S_{ij}} \approx K_3 S_{ij} \qquad K_3 = \frac{2(K - \varepsilon_i)}{1 + S_{ij}} < 0 \qquad (26)$$

These expressions are the usual perturbation formulae. Here ε_i and ε_j denote the energies of the two interacting MOs, S_{ij} their overlap integral, H_{ij} their interaction matrix element and ε_0 the mean of the orbital energies. Since in a comparative analysis of a given surface the overlap is the main variable quantity, we have expressed the various formulae in terms of S_{ij}. To this purpose we have made use of the assumption[12] that the interaction matrix element H_{ij} is proportional to the overlap integral S_{ij}, i.e. $H_{ij} = KS_{ij}(K < 0)$.

The two diabatic surfaces previously defined can be associated with the two diagrams shown in Fig. 25, which refer to the interaction of the π MOs of two singlet ethylenes in one case and of two triplet ethylenes coupled to a singlet in the other case. In the reactant interaction diagram, there are only two types of orbital interactions, i.e. a destabilizing four-electron interaction and a stabilizing two-electron interaction, whose energy effects are described by Eqs. (24) and (25) respectively. In the product interaction diagram, there are only stabilizing interactions between singly occupied orbitals. The stabiliz-

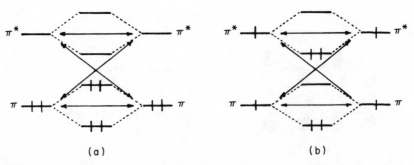

ation energy associated with the interaction of two singly occupied orbitals is described by Eq. (26) when they are degenerate.

We begin our discussion by examining some cross-sections of the diabatic surfaces for the various types of approach. The various types of adiabatic curves along the interfragment distance R are shown in Fig. 26. In these

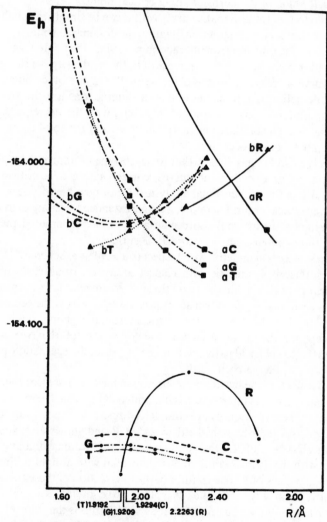

Fig. 26. Reactant (a) and product (b) diabatic curves along R for the rectangular (R) and coplanar (C) *supra–supra, gauche* (G) and *trans* (T) approaches and related adiabatic curves. The numerical values denote the STO-3G optimized R values of the various critical points. (*Reprinted with permission from Bernardi* et al.[91] *Copyright 1986 American Chemical Society.*)

computations, we have used for the ethylene fragments the geometries optimized at the STO-3G level for the corresponding critical points. In each type of approach, with the decrease of R, the reactant diabatic curve increases rapidly and the product diabatic curve decreases. In each case there is an intersection of the two diabatic curves, which occurs very near to the optimized R value of the corresponding critical point.

The behaviour of these diabatic curves with R can be simply rationalized. In each case, in the reactant interaction diagram the dominant orbital interaction is the destabilizing four-electron interaction $\pi-\pi$. Since the related overlap and destabilization increase with the decrease of R, the total energy of the reactant diabatic curve increases with the decrease of R. On the other hand, in the product interaction diagram, the dominant interactions are the stabilizing two-electron interactions $\pi-\pi$ and $\pi^*-\pi^*$. Here, with the decrease of R, the overlap and related stabilization energy increase and the total energy of the product diabatic curve decreases.

The analysis so far has shown that for each type of approach there is an intersection of the two diabatic curves, corresponding to a critical point. However, with a two-dimensional analysis it is not possible to assess the real nature of the critical point. Therefore, we have extended the analysis to include the other important internal coordinate and we have computed the corresponding three-dimensional R, α and R, φ surfaces.

The R, α surface is symmetric with respect to $\alpha = 90°$ (see Scheme XIV), while the R, φ surface is symmetric with respect to $\varphi = 0°$. In all cases the same geometrical parameters for the two ethylenic fragments have been used for each cross-section along R. When appropriate, we have used the geometries of the critical points, while, in the other cases, interpolated geometries have been used, except for the R, φ surface at $\alpha = 90°$ where for the two ethylenic fragments the geometrical parameters of the D_{2h} second-order saddle point for the whole surface have been used.

Let us discuss first the two diabatic surfaces in the R, α space (see Fig. 27); the corresponding curve of intersection is illustrated in Fig. 28a. It can be seen that the shapes of the two surfaces are completely different. The reactant diabatic surface increases rapidly with the decrease of R and shows a maximum for $\alpha = 90°$ at all values of R. On the other hand, the product diabatic surface decreases with the decrease of R, and is almost flat with a slight minimum for $\alpha = 90°$ at all values of R. The features of the intersection of these two surfaces are illustrated in Figs. 27c and 28a. The curve of intersection shows a maximum for $\alpha = 90°$ and a minimum for $\alpha = 117°$ (this is the value of α for the C_{2v} critical point). Therefore, in the R, α space, the D_{2h} approach ($\alpha = 90°$) involves a second-order saddle point and the C_{2v} approach ($\alpha = 117°$) a transition state corresponding to a minimum of the curve of intersection. Thus the shape of the curve of intersection is determined by the shape of the reactant surface.

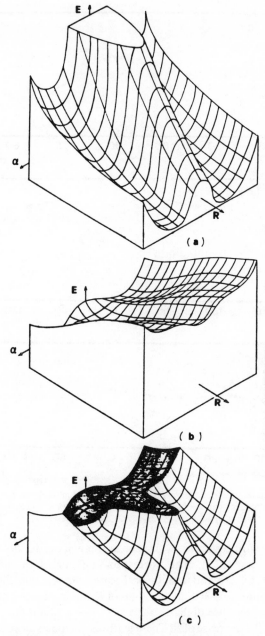

Fig. 27. Diabatic surfaces in the R, α space: (a) reactant surface, (b) product surface and (c) resulting surface. (*Reprinted with permission from Bernardi et al.*[91] *Copyright 1986 American Chemical Society.*)

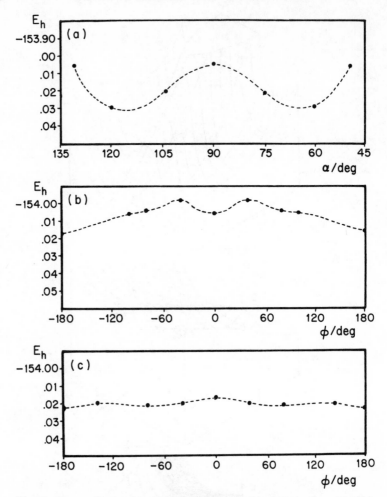

Fig. 28. Curves of intersection in (a) the R, α space and in (b, c) the R, φ space ((b) with $\alpha = 90°$ and (c) with $\alpha = 117°$). (*Reprinted with permission from Bernardi et al.*[91] *Copyright 1986 American Chemical Society.*)

The diabatic surfaces in the R, φ space are shown in Fig. 29 for $\alpha = 90°$ and in Fig. 30 for $\alpha = 117.7°$. The corresponding curves of intersection are illustrated in Figs. 28b and 28c. In both cases the reactant diabatic surface shows a maximum for $\varphi = 0°$ at all values of R and increases with the decrease of R, while the product diabatic surface shows a minimum at $\varphi = 0°$ and decreases with the decrease of R. These features are much more pronounced at $\alpha = 90°$ than at $\alpha = 117°$. The shape of the curve of intersection is dominated by the product surface for $\alpha = 90°$ and by the reactant surface for $\alpha = 117°$: consequently this curve shows a minimum for $\alpha = 90°$ and a maximum for

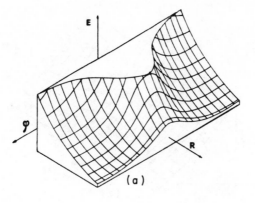

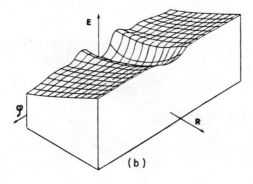

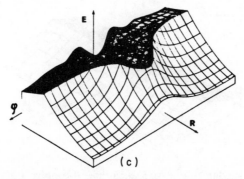

Fig. 29. Diabatic surfaces in the R, φ space ($\alpha = 90°$):
(a) reactant surface, (b) product surface and (c) resulting surface. (*Reprinted with permission from Bernardi et al.*[91] *Copyright 1986 American Chemical Society.*)

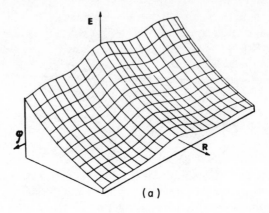

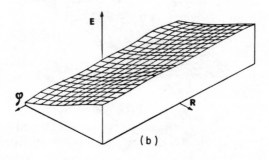

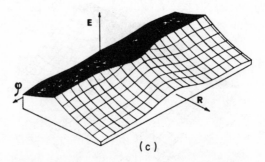

Fig. 30. Diabatic surfaces in the R, φ space ($\alpha =$ 117°): (a) reactant surface, (b) product surface and (c) resulting surface. (*Reprinted with permission from Bernardi* et al.[91] *Copyright 1986 American Chemical Society.*)

$\alpha = 117°$. Therefore, the behaviour of the two component diabatic surfaces in the two different subspaces is very similar. The reactant surface is convex and the product surface is concave, with the maximum in the reactant surface and the minimum in the product surface occurring for the values of the coordinates corresponding to the least-motion path, i.e. for $\alpha = 90°$ in the R, α subspace and for $\varphi = 0°$ in the R, φ subspace.

These different forms of behaviour can be rationalized in terms of simple MO arguments. In all cases the shape of the reactant surface is determined by the energy variation of the four-electron destabilizing interaction. The $\pi-\pi$ overlap and related destabilization decrease when α becomes larger (or smaller) than 90° and when the absolute value of φ becomes larger than 0° and increase with the decrease of R. On the other hand, the shape of the product surface is determined by the energy variation associated with the stabilizing two-electron interactions. The $\pi-\pi$, $\pi^*-\pi^*$ overlaps and the related stabilizations decrease when α becomes larger (or smaller) than 90° and when the absolute value of φ becomes larger than 0° and increase with the decrease of R.

In the R, α space the variation of the destabilization energy is much more pronounced than that of the stabilization energy, because in one case the energy effect is proportional to the square of the overlap while in the other just to the overlap and consequently the crossing is dominated by the reactant surface. In the R, φ space at $\alpha = 90°$ the rate of change of the overlaps with φ is slightly slower than that occurring with α at $\varphi = 0°$. This has the effect of making the reactant surface less repulsive, while the product surface is only slightly affected. In this situation the crossing is dominated by the product surface. At $\alpha = 117°$, the rate of change of the overlaps with φ is even smaller. In this case the product surface is almost flat and the crossing is dominated by the reactant surface.

The other critical points of this surface, i.e. the *gauche* and *trans* minima and the transition states connecting the two minima and the *gauche* minimum to cyclobutane, all lie on the part of the adiabatic surface dominated by the product diabatic surface. Therefore, while the *gauche* and *trans* fragmentation transition states originate from the crossing of the two diabatic surfaces, the *gauche–trans* and the *gauche*–cyclobutane transition states originate from conformational effects.

2. The Supra–Antara Region

The MC-SCF study of this region, performed only at the STO-3G level, has revealed the existence of a transition state, which is 62 kcal mol^{-1} above the tetramethylene *trans* minimum.[86] This result clearly indicates that this approach is very unfavourable.

The STO-3G optimized geometrical parameters of the $[2_s + 2_a]$ transition state are given in Fig. 31. Similar results have also been obtained in a CI study

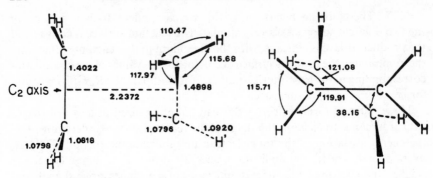

Fig. 31. Optimized geometry computed at MC-SCF/STO-3G for the *supra–antara* transition state (C$_2$). (*Reprinted with permission from Bernardi et al.*[86] *Copyright 1985 American Chemical Society.*)

by Burke and Leroy.[92] The analysis of Fig. 31 shows the following notable features:

1. The two ethylene fragments in the $[2_s + 2_a]$ transition state are very different. One fragment is just an ethylene slightly distorted while the other fragment is a twisted ethylene with a significantly longer C–C bond and methylene groups twisted at almost 90°.
2. The $[2_s + 2_a]$ transition state has a C$_2$ symmetry axis which joins the midpoints of the C–C bonds of the two fragments, with a dihedral angle with respect to the C$_2$ axis of ~ 38°.

The MC-SCF optimized valence orbitals associated with the optimized

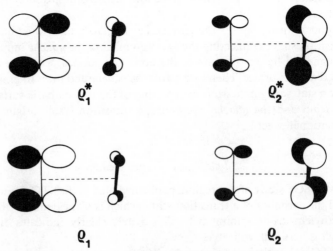

Fig. 32. MC-SCF optimized valence orbitals associated with the optimized geometry of Fig. 31.

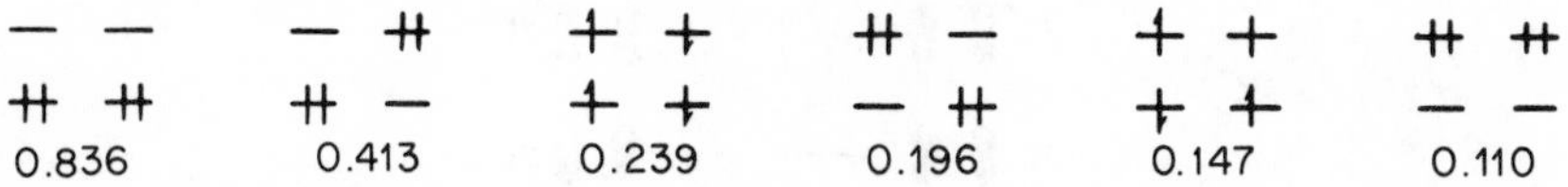

‒ ‒	‒ ⤉⤈	+ +	⤉⤈ ‒	+ +	⤉⤈ ⤉⤈
⤉⤈ ⤉⤈	⤉⤈ ‒	+ +	‒ ⤉⤈	+ +	‒ ‒
0.836	0.413	0.239	0.196	0.147	0.110

Fig. 33. Configuration state functions for the *supra–antara* transition state with coefficients larger than 0.1.

geometry of Fig. 31 are shown in Fig. 32, while the configuration state functions with coefficients larger than 0.1 are given in Fig. 33. It can be seen that in this wavefunction there are various configurations with a significant weight. Furthermore, they correspond to the three MO configurations shown in Scheme XV. This result indicates that the *supra–antara* approach cannot be properly described in the Woodward–Hoffmann model. In fact the Woodward–Hoffman approach can describe satisfactorily only cases described by a monodeterminantal MO wavefunction,[86] while in the present case the wavefunction involves at least three MO configurations, as shown in Scheme XV.

Scheme XV

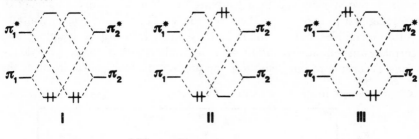

MO configurations

We should point out that the MO configuration I (see Scheme XV) can describe only a *supra–antara* approach where both ethylenes are non-distorted. However, in this region of the *supra–antara* surface there are no critical points. In fact, the only critical point of this surface is that described above, where one fragment is just an ethylene fragment and the other a twisted ethylene. This geometrical situation can only be described by a combination of the MO configurations I and II. In fact, a twisted ethylene in its ground state is described by a wavefunction of the type $(\pi)^2–(\pi^*)^2$. The third MO configuration has a significantly smaller weight and its effect is just that of completing the description of the pyramidalization of the various methylene groups.

A diabatic surface analysis similar to that performed for the other approaches can also provide useful information in this case. For the analysis of the $[2_s + 2_a]$ transition state, where one of the ethylene fragments is twisted, the reactant packet has to involve both the configurations I and II of Fig. 34, as specified above. For completeness we have also included configuration III.

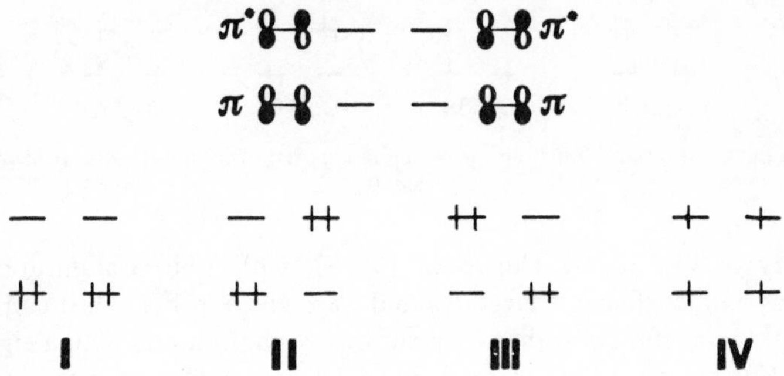

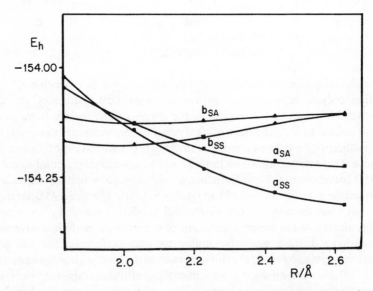

Fig. 34. Fragment orbitals and basic IFC.

Thus the reactant diabatic packet involves these three IFC plus the related one-electron CTC.

In the product diabatic packet, there is only one IFC, i.e. configuration IV of Fig. 34. In the case of this configuration, there is the possibility of coupling two pairs in two singlets and then coupling the result to an overall singlet or coupling first to two triplets then to a singlet. The product packet contains both possibilities but is dominated by the triplet–triplet coupling and correlates at infinite separation with the triplet states of the two ethylenes. Thus, according to our definition, the product diabatic packet involves the

Fig. 35. Behaviour of the reactant (a) and product (b) diabatic curves along R for the *supra–supra* (SS) and *supra–antara* (SA) approaches.

IFC IV plus the related one-electron CTC. All computations presented here[93] have been performed at the STO-3G level.

A comparison of the behaviour of the diabatic curves along the interfragment distance R for the $[2_s + 2_s]$ and the $[2_s + 2_a]$ approaches is shown in Fig. 35. In these computations for the ethylene fragments the geometries optimized at the STO-3G level for the corresponding critical points[86] have been used. In each type of approach, with the decrease of R, the reactant diabatic curve increases rapidly and the product diabatic curve decreases. In each case there is an intersection of the two diabatic curves, which occurs very near to the optimized R values of the corresponding critical points. It can also be seen that the crossing associated with the $[2_s + 2_a]$ approach occurs at a higher energy, mainly because the product diabatic curve is less stabilized.

In Fig. 36 we show the diabatic curves for two $[2_s + 2_a]$ approaches where both ethylenes are kept planar ($\varphi = 180°$) and where the dihedral angle with respect to the C_2 axis (τ) is 38.15° and 90° respectively. These computations allow us to discuss, with reference to the optimized geometry of the $[2_s + 2_a]$ transition state, the effect of increasing the twisting (φ) to 180° and of increasing also the dihedral angle (τ) to 90°. It can be seen that in these geometrical situations not only the reactant but also the product diabatic curve increase rapidly and the crossing, which occurs at a very short R value and at a very large energy when the twisting angle becomes 180°, tends to disappear when the dihedral angle is also increased to 90°.

The behaviour of the diabatic curves along R for the $[2_s + 2_s]$ approach has already been rationalized. In this case, in fact, the behaviour of the total energy of both surfaces is determined by the energy effects associated with the π MOs. In the reactant interaction diagram the dominant orbital interaction is the destabilizing four-electron interaction $\pi_S - \pi_S$. Since the related overlap and destabilization increase with the decrease of R, the total energy of the reactant diabatic curve increases with the decrease of R. On the other hand, in the product interaction diagram, the dominant interactions are the stabilizing two-electron interactions $\pi_S - \pi_S$ and $\pi_A^* - \pi_A^*$. Here, with the decrease of R, the overlap and related stabilization energy increase and the total energy of the product diabatic curve decreases.

For the $[2_s + 2_a]$ approaches where both ethylenes are kept planar the situation is different, since the energy effects associated with the π MOs do not determine the behaviour of the total energy and therefore of the diabatic curves. In this case, in fact, the overlap between the interacting π MOs is very small because of the relative orientation of the π MOs and the behaviour of the diabatic curves is dominated by steric effects, which increase with the decrease of R. Consequently both diabatic curves increase rapidly with the decrease of R. An opposite trend would have been expected if the π-orbital effects were dominant: in fact, in both the reactant and product interaction diagrams, the dominant orbital interactions are stabilizing.

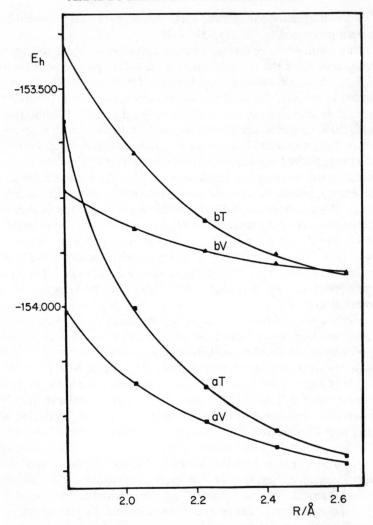

Fig. 36. Behaviour of the reactant (a) and product (b) diabatic curves along R for *supra–antara* approaches where both ethylenes are kept planar and where the dihedral angle with respect to the C_2 axis is 38.15° (T) and 90° (V) respectively.

The situation changes again for the $[2_s + 2_a]$ optimized approach, where one of the two ethylene fragments is twisted by an angle larger than 90°. The twisting has the effect of reducing the steric effects and, mainly, of increasing the overlap between the interacting π MOs. In the reactant interaction diagram there is now a destabilizing contribution arising from the four-electron interaction π_S–π_S. The combined effect of this contribution and of steric effects dominates over the stabilizing effect associated with the $\pi_S-\pi_S^*$ and $\pi_A-\pi_A^*$ two-electron interactions, and consequently the reactant diabatic curve increases with the decrease of R. On the other hand, in the product interaction diagram, the dominant interactions are the stabilizing interactions $\pi_S-\pi_S$ and $\pi_A-\pi_A$. Here, with the decrease of R, the overlap and related stabilization energy increase and the total energy of the product diabatic curve decreases. Since the overlap between the singly occupied MOs is smaller than in the case of the $[2_s + 2_s]$ approach, the product diabatic curve for the $[2_s + 2_a]$ approach is less stabilized.

In summary the following points can be made. The 'textbook' *supra–antara* coordinate with both of the ethylene groups planar is very unfavourable because the overlap of the orbitals (which would lead to an attractive product diabatic surface) is very small and steric effects dominate, as expected. Only for the approach where one ethylene is twisted is the overlap sufficient to stabilize the product diabatic.

D. 1,3 Dipolar Cycloadditions

1, 3 Dipolar cycloadditions are a class of chemical reactions of significant experimental interest. In fact, they provide a versatile method of preparing five-membered heterocyclic compounds.[94] But they are also a class of reactions of theoretical interest, since they represent an example of thermally allowed pericyclic reactions.[65]

The mechanism of these reactions has been the object of controversy between Huisgen[94,95] and Firestone.[96] Huisgen, on the basis of experimental data and theoretical considerations supported by the Woodward–Hoffmann rules, has proposed a concerted mechanism where the two new σ bonds are formed simultaneously (see Scheme XVI):

Scheme XVI

Huisgen classified the 1, 3 dipoles into two categories, the allyl type and the propargyl–allenyl type, and suggested that both types of 1, 3 dipole should follow the concerted mechanism and that both types of 1, 3 dipole and dipolarophiles should approach each other in two parallel planes, forming an envelope-like transition structure.

Firestone, on the contrary, on the basis of experimental results and theoretical considerations supported by the Linnett theory,[97] has suggested that both types of 1, 3 dipole should follow a two-step mechanism, where one σ bond is formed in the first step, leading to a diradical in a *cis* or *trans* conformation, and in a second step this diradical would close to the final five-membered ring, according to Scheme XVII:

Scheme XVII

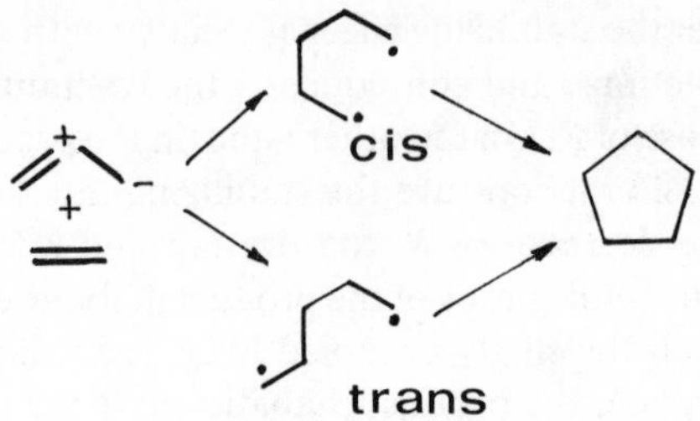

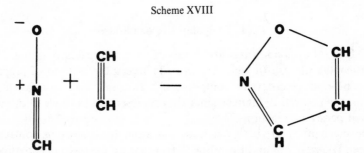

A number of *ab initio* theoretical investigations[98–102] on 1,3 dipolar cycloadditions have appeared in the literature in the last decade with the purpose of providing information on the mechanism of these reactions. Here we review the theoretical results obtained for the reaction between fulminic acid and acetylene (see Scheme XVIII), since this is the 1,3 dipolar cycloaddition most extensively investigated at the theoretical level.

Scheme XVIII

The first *ab initio* study of this reaction is that of Poppinger,[99] who has optimized the critical points at the minimum basis (STO-3G) SCF level of theory and then performed single-point calculations on reactants, product and transition state using the larger 4-31G basis. A more accurate study is that of Komornicki, Goddard and Schaefer,[100] who have used *ab initio* SCF gradient methods at a double-zeta level for the optimizations and then investigated the energetic features of the reaction using a CI treatment. Since in both these studies the transition state region has been investigated at the SCF level, the discovery of any diradical intermediate has been precluded and only quite symmetrical transition states have been found. The diradicaloid region has been recently investigated by Hiberty, Ohanessian and Schlegel[101] using various *ab-initio* computational methods including unrestricted and restricted Hartree–Fock and multireference CI methods. Very recently the transition structure region of this reaction has been investigated also at the MC-SCF level using minimal (STO-3G) and extended (4-31G) basis sets:[102] in this study the various critical points have been fully optimized with MC-SCF gradients and characterized by computing the corresponding Hessian matrices. Following the procedure used in the previous sections, here we concentrate on the MC-SCF results and their analysis.

As specified above, the reaction investigated at the MC-SCF level is the 1, 3 dipolar cycloaddition of fulminic acid to acetylene to give isoxazole. The fulminic acid can be thought of as having two allyl-like systems of three π orbitals, an in-plane π set and an out-of-plane π set. Similarly the acetylene has two sets of π orbitals, an in-plane and an out-of-plane set. These orbitals are

Scheme XIX

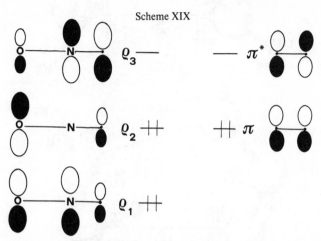

shown in Scheme XIX. The isolated fragments must have the orbital occupancy shown in Scheme XIX for the in-plane and out-of-plane systems. For the product isoxazole, the in-plane π-orbital occupancy must correspond to the promotion of one electron in each of the fragments from HOMO to LUMO. The unpaired electrons must be spin-coupled to a state of triplet spin within a fragment and these two triplet states subsequently spin-coupled to a singlet in order to describe two new σ bonds. For the out-of-plane system of the product isoxazole, the fragment π systems retain the configuration shown in Scheme XIX, since there is no requirement to uncouple and recouple the spins to describe the isoxazole system. From these arguments, it follows that, in terms of the fragment orbitals, one requires a valence space consisting of four in-plane π orbitals (the HOMO and LUMO of each fragment) in order to describe the product, the transition state and the reactants with equal accuracy. Thus the valence space to be used in the MC-SCF calculations should contain these four π orbitals at least. All the geometry optimizations have been carried out with a valence space consisting of these four π orbitals.

In order to improve the energetics, MC-SCF calculations have also been carried out at the previously optimized geometries with a valence space that consisted of the HOMO–LUMO out-of-plane π orbitals as well, corresponding to eight valence orbitals. This calculation should account for some of the dynamic correlation of the delocalized π system.

The calculations have been carried out at the STO-3G and 4-31G levels. At the STO-3G level, each critical point has been characterized by computing the

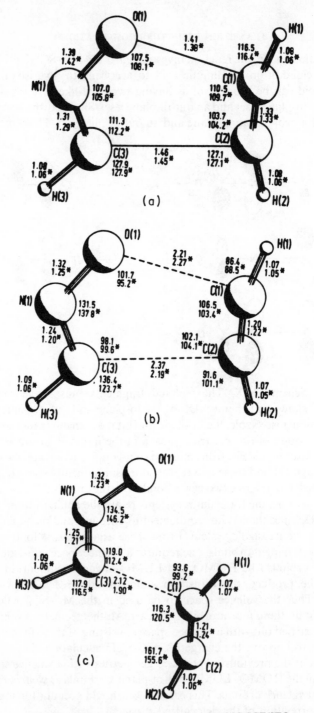

Fig. 37. Optimized geometries computed at MC-SCF/STO-3G (no superscript) and MC-SCF/4–31G (asterisk): (a) iso-xazole, (b) concerted transition state, (c) two-step transition state corresponding to the formation of the first bond,

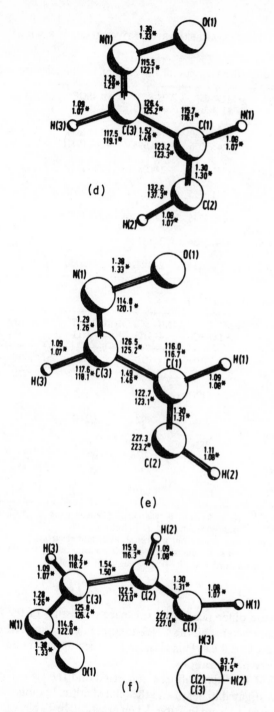

(d) *trans* diradical intermediate, (e) *cis* diradical intermediate and (f) two-step transition state connecting the *cis* diradical intermediate to the product isoxazole. (*from Bernardi* et al.[102] *Reproduced by permission of The Royal Society of Chemistry.*)

TABLE VI
a. Absolute and relative energies for the reaction of fulminic acid with acetylene.

Structure	MC-SCF[a] STO-3G	MC-SCF[a] 4–31G	MC-SCF[b] 4–31G	Ref. 100[c]	Ref. 101[d]
HCCH + HCNO	−241.3130	−244.1360	−244.2394	−244.5444	−244.6181
Cyclic synchronous TS (Fig. 37b)	−241.3056	−244.0952	−244.1941	−244.5062	−244.5973
Asynchronous first TS (Fig. 37c)	−241.2950	−244.0871	−244.1841		−244.6031
trans Diradical intermediate (Fig. 37d)	−241.3715	−244.1214	−244.2005		
cis Diradical intermediate (Fig. 37e)	−241.3746	−244.1246	−244.1977		−244.6081
Asynchronous second TS (Fig. 37f)	−241.3636	−244.1123			
Isoxazole (Fig. 37a)	−241.5211	−244.2426	−244.3150		

b. Relative energies (kcal mol^{-1})

Structure	MC-SCF[a] STO-3G	MC-SCF[a] 4–31G	MC-SCF[b] 4–31G	Ref. 100[c]	Ref. 101[d]
HCCH + HCNO	0.0	0.0	0.0	0.0	0.0
Cyclic synchronous TS (Fig. 37b)	4.6	26.0	28.4	24.0	13.0
Asynchronous first TS (Fig. 37c)	11.3	30.7	34.7		9.39
trans Diradical intermediate (Fig. 37d)	−36.6	9.1	24.4		
cis Diradical intermediate (Fig. 37e)	−38.6	7.2	26.2		8.20
Asynchronous second TS (Fig. 37f)	−31.7	14.9			
Isoxazole (Fig. 37a)	−130.6	−66.9	−47.4	−69.1	−63.4

From Bernardi et al.[102] Reproduced by permission of The Royal Society of Chemistry.
[a]Full CI (CAS-SCF) in four a' orbitals and four electrons.
[b]Full CI (CAS-SCF) in four a' and four a″ orbitals and eight electrons.
[c]CI using SCF orbitals, all single and double excitations from two a' and three a″ orbitals.
[d]CI (two-reference second-order Moller–Plesset) using SCF orbitals.

Hessian by finite difference in the subspace consisting of the interfragment geometrical parameters and the CNO angle of fulminic acid, which is strongly coupled to these variables. This Hessian has been updated numerically in the 4-31G geometry optimizations.

In the MC-SCF study, both the concerted and the two-step mechanisms have been investigated in detail and the related critical points are summarized in Fig. 37 (geometries) and in Table VI (energetics). In Table VI the energies of the critical points are also given relative to the energy of the reactants.

TABLE VII

Optimized geometries computed at MC-SCF/STO-3G and MC-SCF/4–31G for fulminic acid (HCNO) and acetylene (HCCH) (atoms labelled as in Fig. 37).

		Bond length (Å)	
Molecule	Bond	STO-3G	4–31G
HCNO	C(3)–H(3)	1.06	1.05
	C(3)–N(1)	1.17	1.14
	N(1)–O(1)	1.29	1.26
HCCH	C(1)–H(1)	1.07	1.05
	C(1)–C(2)	1.18	1.20

From Bernardi et al.[102] *Reproduced by permission of The Royal Society of Chemistry.*

a. Concerted approach

Along this type of approach, a cyclic transition structure has been found, which connects reactants and product. The STO-3G geometry is in quite good agreement with the 4-31G results, and furthermore the MC-SCF geometries are also in good agreement with the SCF results. The analysis of the geometrical parameters of this transition structure (see Fig. 37) shows that the two incipient interfragment bond distances C–O and C–C are quite large and that the geometries of the two fragments are very similar to those of the reactants (see Table VII). These geometrical features are in agreement with Hammond's postulate, since this reaction is a strongly exothermic reaction (see Table VI).

When the two bonds that are formed in a pericyclic reaction are different (as in the present case, where a C–C and a C–O bond are formed), it is difficult to assess clearly whether a concerted process should be called synchronous or asynchronous. The process might be synchronous with respect to bond length yet asynchronous with respect to bond strength because of different stretching potentials for the two newly formed bonds. However, as discussed in footnote 20 of Ref. 101, the assignment of diagonal force constants for the C–C and the C–O stretch is dependent upon an arbitrary choice of internal coordinates in the case of cyclic molecules. For this reason, it is preferable to use a criterion based on the extent of new bond formation for assessing whether a concerted process is synchronous or asynchronous. If we take r, the bond length in the product, as the reference, and denote by R the bond length in the transition structure, then the extent of new bond formation is given by $S = r/R$. For the concerted transition structure the values of S are 0.64 (0.60) and 0.62 (0.66) for the C–O and C–C bonds respectively at the STO-3G (4-31G) level. According to this criterion, the concerted process is synchronous.

b. Two-step approach

In the diradicaloid region of the surface, the following critical points have been found: (a) an extended diradicaloid transition state for the formation of the first bond; (b) a diradical intermediate, which exists in a *trans* and a *cis* form; and (c) a second diradicaloid transition state which connects the *cis* form of the diradical intermediate with the product isoxazole. On the basis of these results, it seems likely that the transition state for the formation of the first bond connects the reactants with the *cis* form of the diradical intermediate and the *trans* form represents a subsidiary minimum accessible via an in-plane inversion process.

The geometries of all these critical points are shown in Fig. 37. Again, the agreement between the STO-3G and 4-31G results is quite good. The geometrical parameters of the transition state for the formation of the first bond are again in agreement with the expectations based on Hammond's postulate for a strongly exothermic reaction: in fact, the bond distance of the incipient C–C bond is quite long and the geometries of the two fragments are similar to those of the reactants. On the other hand, the geometrical parameters of the *trans* and *cis* intermediates and of the *cis*–isoxazole transition state are already very similar to those of the product.

The description of the diradicaloid region of the surface obtained at the MC-SCF level differs in various aspects from that obtained in Ref. 101. In particular, the geometry of the diradicaloid transition state that connects the reactants and the diradical intermediate is only in qualitative agreement with the transition structures obtained with the same basis set (4-31G) in Ref. 101 either at the restricted Hartree–Fock (RHF) plus 3×3 CI level or using the CIPSI method and interpolating the geometries of the diradicaloid transition state and intermediate optimized at the RHF plus 3×3 CI level. In the case of the structure optimized at the RHF plus 3×3 CI level, the major difference is that the interfragment C–C distance is too long, while in the case of the structure obtained with the interpolating procedure using the CIPSI method the resulting geometry has a *cis* form, rather than a *trans* form. Furthermore, only one form of the diradical intermediate has been found in Ref. 101, while in the MC-SCF study two different structures, a *cis* and a *trans* form, have been found. However, the structure of the intermediate optimized in Ref. 101 at the RHF/4-31G plus 3×3 CI level agrees well with the corresponding *cis* form optimized at the MC-SCF level. Another significant difference is found in the trends of the activation energies. In fact, both second-order Moller–Plesset perturbation calculations and CIPSI calculations favour the two-step mechanism, while the MC-SCF results show the opposite trend. However, it must be stressed that one must be careful in comparing CI calculations for the concerted and diradicaloid regions of the surface. In the diradicaloid region of the surface, SCF orbitals are far from optimum for use in the CI expansion, even if a multi-reference expansion is used. Thus one must be certain that one is describing both regions of the surface with equal accuracy. In the MC-SCF study, the valence space has been chosen so that the region of the concerted transition structure and the region of the diradicaloid structures are described with equivalent accuracy. Thus one has some confidence in the relative energetics of the two different pathways. However, the MC-SCF energetics clearly do not include any true dynamic correlation. Thus the exothermicity and the reaction barriers will not be accurately reproduced.

From the values listed in Table VI it can be seen that the relative energetics for the concerted approach computed at the MC-SCF level are in good agreement with those obtained in Ref. 100 using a CI treatment that included replacements from two á "and three a" orbitals. It can also be seen that the transition state for the concerted process is predicted to be lower than that for the first step of the two-step process at all levels of computations. Furthermore, the second transition state in the two-step process lies at a lower energy than the first and the second barrier is predicted to be quite small.

In summary, the MC-SCF results suggest that for the 1,3 dipolar cycloaddition of fulminic acid to acetylene the concerted synchronous pathway is preferred, in agreement with the mechanism proposed by Huisgen. However, contrary to Huisgen's prediction, the related transition state is found to be planar.

Information about the origin and nature of the various critical points of this surface can be obtained again from a diabatic surface analysis,[102,103] where the fragments are fulminic acid and acetylene. With reference to the orbitals of the reactant IFC (configuration I in Scheme XX) involves the singlet states of the two fragments, while the product IFC (configuration II in Scheme XX) corresponds to the two fragments in a triplet state coupled to a singlet.

Scheme XX

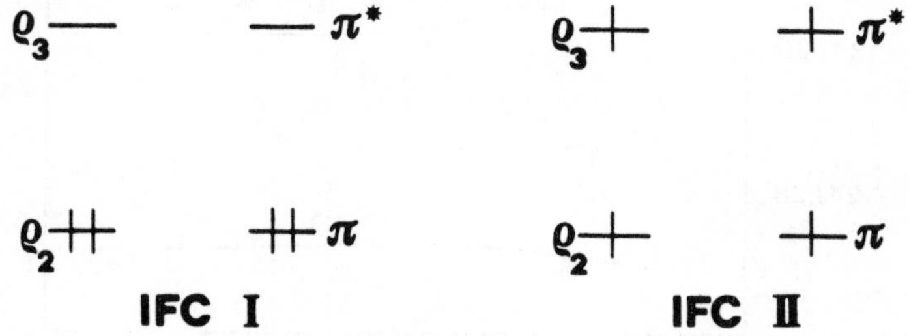

ϱ_3— — π^* ϱ_3+ +π^*

ϱ_2++ ++π ϱ_2+ +π

IFC I   **IFC II**

Only the in-plane orbitals are used to construct these configurations. Again the two diabatic surfaces are associated with these two IFC plus the related one-electron CTC.

The various types of adiabatic and diabatic curves along the interfragment distance R (taken here to be the C(2)–C(3) distance) for the concerted approach and for the first step of the two-step approach computed at the STO-3G level are shown in Figs. 38 and 39 respectively. It can be seen that the diabatic curves have the expected behaviour, i.e. in each type of approach, with the

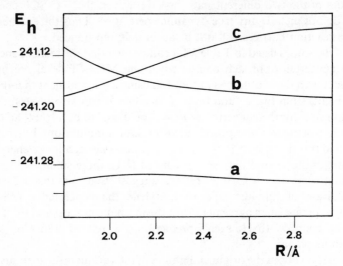

Fig. 38. Behaviour of the reactant (b) and product (c) diabatic curves along R for the concerted approach of fulminic acid and acetylene, and of the related adiabatic curve (a).

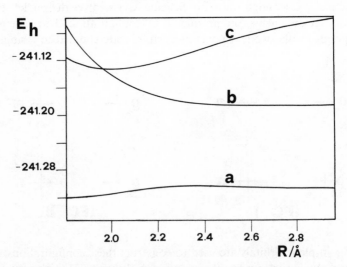

Fig. 39. Behaviour of the reactant (b) and product (c) diabatic curves along R for the two-step approach of fulminic acid and acetylene, and of the related adiabatic curve (a).

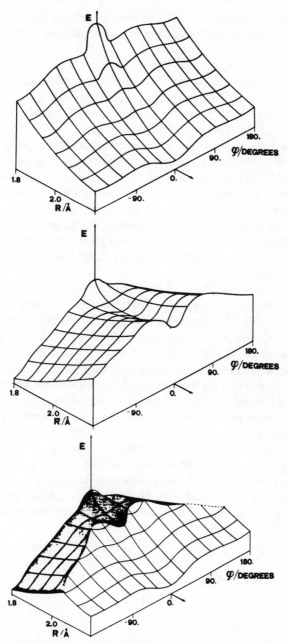

Fig. 40. Diabatic surfaces in the R, φ space. From top to bottom: reactant surface, product surface and resulting surface. (*From Bernardi* et al.[103] *Reproduced by permission of Elsevier Science Publishers.*)

decrease of R, the reactant diabatic curve increases and the product diabatic curve decreases and in each case there is an intersection of the two diabatic curves.

Thus the analysis has shown that these two critical points arise from a crossing of two diabatic curves. The situation is different for the second transition state of the two-step process, which lies in a region of the potential surface dominated by the product diabatic curve. Thus the second barrier originates essentially from conformational effects.

Information about the nature of these critical points can be obtained from the analysis of the diabatic surfaces computed in the R, φ space, where R again denotes the C(2)–C(3) interfragment distance and φ the rotational angle about this incipient bond. The diabatic surfaces are shown in Fig. 40, while the corresponding curve of intersection is given in Fig. 41. These surfaces have been computed at the STO-3G level according to the procedure described in Ref. 102. In these computations, the fragment geometries have been kept fixed along R, while along φ they have been interpolated between the geometries of the critical points.

Also in this case the shape of the two component diabatic surfaces is completely different. The reactant diabatic surface increases rapidly with R and shows a valley with a minimum at $\varphi = 180°$, while around $\varphi = 0°$ the shape of the surface depends on the value of R. In fact, for $R < 1.9$ Å, which also includes the region of the crossing, the reactant diabatic surface shows a maximum for $\varphi = 0°$ while for larger values of R it shows a minimum. On the other hand, the product diabatic surface decreases with the decrease of R and shows minima along φ for $\varphi = 0°$ and $\varphi = 180°$ till very small values of R.

The two component diabatic surfaces have shapes similar to the corresponding diabatic surfaces for the [2 + 2] cycloaddition. However, in spite of the broadly similar shape of the two component surfaces in the two cases, the resulting surfaces (see Figs. 30 and 40) and the curves of intersection (see Figs. 28c and 41) are significantly different. In the case of the [2 + 2] cycloaddition, in fact, the product diabatic surface is almost flat and the curve of intersection is controlled by the reactant diabatic surface, and consequently

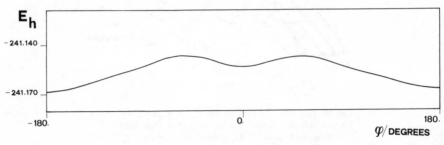

Fig. 41. Curve of intersection in the R, φ space. (*From Bernardi et al.*[103] *Reproduced by permission of Elsevier Science Publishers.*)

shows a maximum for $\varphi = 0°$, corresponding to the coplanar C_{2v} second-order saddle point, and a minimum for $\varphi = 180°$, corresponding to the *trans* transition state. On the other hand, in the case of the 1, 3 dipolar cycloaddition, it is the reactant diabatic surface that is almost flat and therefore the curve of intersection is controlled by the product diabatic surface. Consequently the curve of intersection shows two minima along φ for $\varphi = 0°$ and $\varphi = 180°$, corresponding to the concerted transition state and to the transition state for the first step of the two-step process, respectively.

From this comparative analysis, the concept emerges that even slight modifications of the various energy contributions, such as overlap repulsions, charge transfers and exchange effects, can alter not only the energy barriers of the reaction but also the nature of the critical points.

E. The Diels–Alder Reaction

The Diels–Alder reaction is one of the most important reactions in the history of organic chemistry, both from an experimental[104] and a theoretical[65] point of view. The cycloaddition of 1, 3-butadiene and ethylene is, in fact, one of the 'textbook' examples of a concerted thermally allowed process used in the illustration of the Woodward–Hoffmann rules.[65]

Experimentally,[104] it appears that many [4 + 2] cycloadditions are best described in terms of a symmetry-allowed concerted mechanism of some type. However, also the possibility of a two-step mechanism cannot be ignored. The experimental results suggest, in fact, that the two-step mechanism is in most cases energetically more demanding and can become competitive when simultaneous bond formation may be hindered, for example by steric or electronic effects.

Because of the relatively large number of atoms involved, only a few *ab initio* studies have been performed and only on the simplest Diels–Alder reaction, the addition of ethylene to 1, 3-butadiene. The first *ab initio* studies were those of Burke *et al.*[105] and of Townshend *et al.*[106] Burke *et al.* have carried out extensive calculations on the concerted approach of this reaction at the SCF level with an STO-3G basis set and have also recalculated several points on the hypersurface using a medium-size basis set (7s/3p). Townshend *et al.* have studied not only the concerted but also the two-step approach using an SCF treatment with limited CI at the STO-3G level and have recalculated the energy pathways with an extended 4–31G basis. However, even though these studies already involved a very significant computational effort, they were performed without complete optimization and characterization of the critical points.

Recently the concerted approach has been studied at the SCF level using the STO-3G basis set and gradient techniques with optimization of all variables by Brown and Houk.[107] In this case an authentic transition structure has been obtained, possessing one imaginary vibrational frequency. Ortega *et al.*[108]

have studied the effect of the correlation energy on the mechanism of this reaction by means of Møller–Plesset perturbation theory at the STO-3G SCF level and have found that this energy effect stabilizes preferentially the asynchronous structure. More recently, the transition structure region of this surface has been investigated at the MC-SCF level using STO-3G and 4–31G basis sets,[109] with the critical points fully optimized with MC-SCF gradient techniques and characterized by computing the related Hessian matrices.

Following the procedure adopted in the previous sections, we summarize the MC-SCF results and, where appropriate, we compare these results with those obtained by the other authors.

In this reaction, the relevant orbitals of butadiene and ethylene are the π orbitals (see Scheme XXI).

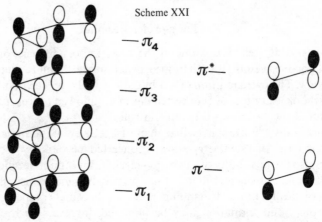

Scheme XXI

In a first series of MC-SCF computations, the various critical points have been fully optimized at the STO-3G level using a valence space consisting of four π orbitals (the HOMO and the LUMO of each fragment). Then the computations at the STO-3G level have been repeated using a valence space consisting of the six π orbitals. The critical points calculated at the minimal level have been used as starting points for searches using the extended basis, with a valence space consisting of four π orbitals. For convenience, the notation CAS1 and CAS2 is used to denote calculations done with a four-electron/four-orbital CAS-SCF wavefunction and with a six-electron/six-orbital CAS-SCF wavefunction, respectively.

a. The concerted approach

A transition state of C_s symmetry has been found at all MC-SCF levels of calculation, i.e. STO-3G CAS1, STO-3G CAS2 and 4–31G CAS1. In all cases the transition state wavefunction is dominated by the SCF wavefunction, with a small contribution of two doubly excited configurations. For instance, at the STO-3G CAS1 level, the coefficient of the SCF configuration is 0.934.

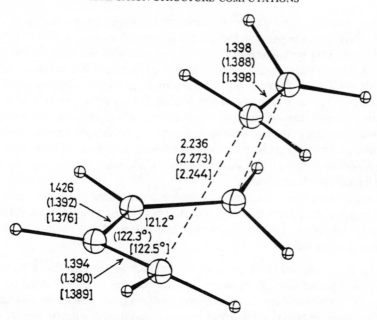

Fig. 42. Calculated transition state structures for the concerted mechanism of the Diels–Alder reaction of butadiene with ethylene. The parameters (in Å and deg) are given for the CAS-SCF expansions, CAS2 and CAS1 (in parentheses), at the STO-3G level. The values in square brackets are for a 4-31G basis and a CAS1 expansion. (*From Bernardi et al.*[109] *Reproduced by permission of The Royal Society of Chemistry.*)

TABLE VIII
Structural parameters for the synchronous 4–31G CAS1 Diels–Alder transition state.

Bond lengths (Å)		Bond angles (deg)		Torsional angles (deg)	
C(1)–C(2)	1.389	C(1)C(2)C(3)	122.5	C(1)C(2)C(3)C(4)	0.0
C(2)–C(3)	1.376	C(3)C(4)C(5)	102.2	C(2)C(3)C(4)C(5)	57.7
C(5)–C(6)	1.398	C(4)C(5)C(6)	109.1	C(3)C(4)C(5)C(6)	49.0
C(6)–C(1)	2.244				
		C(2)C(3)H(10)	118.3	C(1)C(2)C(3)H(10)	169.2
C(1)–H(7)	1.074	C(3)C(4)H(11)	119.3	C(2)C(3)C(4)H(11)	172.1
C(1)–H(8)	1.072	C(3)C(4)H(12)	119.8	C(2)C(3)C(4)H(12)	33.9
C(2)–H(9)	1.075	C(5)C(6)H(15)	119.2	C(3)C(4)C(5)H(13)	73.8
C(5)–H(13)	1.071	C(5)C(6)H(16)	118.9	C(3)C(4)C(5)H(14)	171.4
C(5)–H(14)	1.074			C(4)C(5)C(6)H(15)	106.1
				C(4)C(5)C(6)H(16)	105.8

From Bernardi et al.[109] *Reproduced by permission of The Royal Society of Chemistry.*

The computed geometrical parameters are summarized in Fig. 42 and Table VIII. The various bond distances computed at the different MC-SCF levels agree very well. The agreement is good also with the results of the SCF optimization. The main difference between MC-SCF and SCF results is that at the MC-SCF level the two fragments are more distorted with respect to the reactants. The main features of the MC-SCF results are, in fact, that the forming carbon–carbon distance is quite long (2.2 Å) and that the carbon–carbon distances of the two fragments are strongly distorted: the ethylene and butadiene double bonds have all lengthened by about 0.04 Å and the butadiene single bond has shortened by 0.07 Å, with a tendency towards equalization of these bonds.

The details of these results also reveal some additional remarkable features whic confirm the stereochemical implications discussed by Brown and Houk at the SCF level. Following Brown and Houk, these features can be summarized as follows.

1. The diene and ethylene units approach each other in a distinctly non-parallel-planes fashion. However, the trajectories of bond formation are nearly tetrahedral on the alkene and somewhat smaller on butadiene.
2. The diene terminal methylenes are pyramidal and rotated inwards to facilitate overlap with the termini of ethylene.
3. The hydrogens attached to C(2) and C(3) of butadiene are bent out of the butadiene skeleton even though these hydrogens are in-plane in both reactants and products.
4. There is nearly perfect staggering about the forming C–C bonds, but the steric environment of *exo* and *endo* hydrogens on the ethylene group are different.

The origin of this transition state can be analyzed again in terms of a diabatic analysis, where the fragments are butadiene and ethylene. The reactant IFC (configuration I in Scheme XXII) involves the singlet states of the two fragments, while the product IFC (configuration II in Scheme XXII)

Scheme XXII

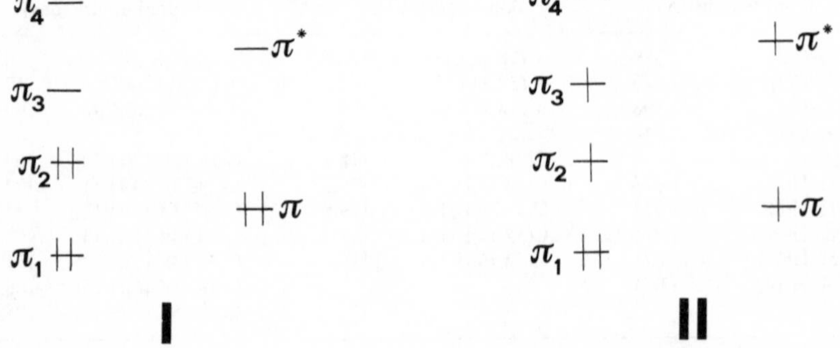

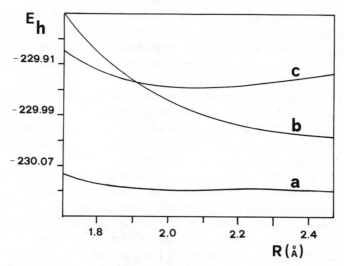

Fig. 43. Behaviour of reactant (b) and product (c) diabatic curves along R for the concerted approach of butadiene and ethylene, and of the related adiabatic curve (a).

corresponds to butadiene and ethylene both in a triplet state with the overall spin coupled to a singlet. Thus the two diabatic surfaces are associated with these two IFC plus the related one-electron CTC. The packet II also contains the configuration with butadiene and ethylene both in a singlet excited state with the overall spin coupled to a singlet.

The behaviour of these diabatic curves along R, the interfragment distance, is shown in Fig. 43, together with that of the corresponding adiabatic curve. The computations have been performed at the STO-3G level using for the two fragments the geometries at the transition state optimized at the STO-3G CAS1 level. The two diabatic curves show the expected behaviour with the decrease of R, with reactant diabatic curve that increases rapidly and the product diabatic curve that decreases. Therefore, there is an intersection of the two diabatic curves, which corresponds to the transition state.

b. The two-step approach

For the two-step approach, three transition states have been located using the STO-3G basis at both the CAS1 and CAS2 levels, corresponding to the formation of the initial carbon–carbon bond for a *syn–gauche*, a *trans* and an *anti–gauche* approach. However, no transition state has been located using the 4–31G basis, in spite of extensive searches.

This latter result is in contrast with the finding of Townshend *et al.*,[106] who have found two distinct two-step pathways, one in which a lone bond is formed without participation by the other non-bonded termini and a second lower pathway in which concertedness or quasi–concertedness is followed by an antisymmetric bond rearrangement. Townshend *et al.* have also found that both these pathways are at higher energy than the concerted approach involving a symmetric transition state.

The 4–31G CAS1 result is in contrast also with the experimental evidence, which shows that the product of addition of ethylene to butadiene consists of cyclohexene (99.98%) and vinylcyclobutane (0.02%), with rate constants in the ratio of 5000 to 1 (7.6 kcal mol^{-1} at 450 K) for the concerted (to cyclohexene) and two-step (to vinylcyclobutane) mechanisms.[106,110]

Thus, while the SCF plus CI results of Townshend *et al.* agree well with the experimental results, the superior 4–31G CAS1 approach provides results in significant disagreement. These discrepancies suggest that this problem has to be investigated in even greater detail, perhaps with the use of basis sets optimized for the transition structure region. The critical importance of basis sets in describing these critical points has also been shown by the results obtained in the MC-SCF study of the ethylene plus ethylene surface.[86] Here it has been found that, although in a STO-3G basis the minimum of the diradicaloid tetramethylene has been well characterized, with a substantial barrier to formation of two ethylenes, in a 4–31G basis this barrier very nearly disappears.

F. Conclusions

The comparative analysis of the results discussed in this section, obtained with MC-SCF and SCF methods and with different basis sets, allows us to draw some conclusions about the computational level to be used for the various types of transition structures, which can be summarized as follows.

1. Saddle points occurring along allowed least-motion paths can be satisfactorily described, in terms of geometry and Hessian matrix, already at the minimal basis SCF level. The geometrical changes associated with the use of an MC-SCF procedure and/or of an extended basis are usually small, while the nature of the critical point remains unchanged.
2. Concerted transition states occurring along non-least-motion paths require more caution: in this case, in fact, as the results obtained for the *supra–antara* approach of two ethylenes show, the SCF level might not be adequate to describe the problem.
3. Diradicaloid saddle points require, as previously pointed out, computational procedures that transcend the SCF method. In particular, the use of the MC-SCF method provides a uniform description of all the transition structure region and therefore a reliable comparison of the various processes. The geometry of diradicaloid saddle points seems also to be

significantly dependent on the basis set used. It appears that an STO-3G basis leads to 'large' interfragment distances and a 4–31G basis to much shorter interfragment distances. Perhaps STO-3G underestimates the long-range interactions and 4–31G overestimates them. These results, and in particular those obtained at the minimal and extended levels for the diradicaloid region of the addition of ethylene to 1, 3-butadiene, suggest the need for basis sets optimized for this type of computation.

Acknowledgements

We are grateful to acknowledge the financial support of NATO under Grant RGO96.81.

References

1. (a) Pulay, P., in *Modern Theoretical Chemistry* (Ed. H. F. Schaefer), Vol. 4, p. 153, Plenum, New York, 1977.
   (b) Bratoz, S., *Colloq. Int. CNRS*, **82**, 287 (1958).
   (c) Bishop, D. M., and Radnic, M., *J. Chem. Phys.*, **44**, 2480 (1966).
   (d) Pulay, P., *Mol. Phys.*, **17**, 197 (1969).
   (e) Moccia, R., *Chem. Phys. Lett.*, **5**, 260 (1970).
   (f) Moccia, R., *Int. J. Quantum Chem.*, **8**, 293 (1974).
   (g) Pople, J. A., Krishnan, R., Schlegel, H. B., and Binkley, J. S., *Int. J. Quantum Chem.*, **S13**, 225 (1979).
   (h) Dupuis, M. and King, H. F., *J. Chem. Phys.*, **68**, 3998 (1978).
2. (a) Goddard, J. D., Handy, N. C., and Schaefer, H. F., *J. Chem. Phys.*, **71**, 1525 (1979).
   (b) Kato, S., and Morokuma, K., *Chem. Phys. Lett.*, **65**, 19 (1979).
   (c) Knowles, P. J., Sexton, G. J., and Handy, N. C., *Chem. Phys.*, **72**, 337 (1982).
   (d) Schlegel, H. B., and Robb, M. A., *Chem. Phys. Lett.*, **93**, 43 (1982).
   (e) Hoffman, M. R., Fox, D. J., Gaw, J. F., Osamura, Y., Yamaguchi, Y., Grev, R. S., Fitzgerald, G., Schaefer, H. F., Knowles, P. J., and Handy, N. C., *J. Chem. Phys.*, **80**, 2660 (1984).
3. (a) Brooks, B. R., Laidig, W. D., Saxe, P., Goddard, J. D., Yamaguchi, Y., and Schaefer, H. F., *J. Chem. Phys.*, **72**, 4652 (1980).
   (b) Krishnan, R., Schlegel, H. B., and Pople, J. A., *J. Chem. Phys.*, **72**, 4654 (1980).
   (c) Jorgensen, P., and Simons, J., *J. Chem. Phys.*, **79**, 334 (1983).
   (d) Fox, D. J., Osamura, Y., Hoffman, M. R., Gaw, J. F., Fitzgerald, G., Yamaguchi, Y. and Schaefer, H. F., *Chem. Phys. Lett.*, **102**, 17 (1983).
4. Fukui, K., Kato, S., and Fujimoto, H., *J. Am. Chem. Soc.*, **97**, 1 (1975).
5. Miller, W. H., Handy, N. C., and Adams, J. E., *J. Chem. Phys.*, **72**, 99 (1980).
6. (a) Woodward, R. B., and Hoffmann, R. B., *The Conservation of Orbital Symmetry*, Academic Press, New York, 1970.
   (b) Fukui, K., *Theory of Orientation and Stereoselection*, Springer-Verlag, Berlin, 1975.
7. (a) Morokuma, K., *J. Chem. Phys.*, **55**, 1236 (1971).
   (b) Morokuma, K., *Int. J. Quantum Chem.*, **10**, 325 (1976).
8. (a) Bernardi, F., and Robb, M. A., *Mol. Phys.*, **48**, 1345 (1983).
   (b) Bernardi, F., and Robb, M. A., *J. Am. Chem. Soc.*, **105**, 54 (1984).

(c) Bernardi, F., Paleolog, S., McDouall, J. J. W., and Robb, M. A., *J. Mol. Struct.* (*Theochem*), **138**, 23 (1986).

9. (a) Murrell, J. N., and Laidler, K. J., *Trans. Faraday Soc.*, **64**, 371 (1968).

   (b) Murrell, J. N., in *Quantum Theory of Chemical Reactions* (Eds. R. Daudel, A. Pullman, L. Salem, and A. Veillard), Vol. 1, Reidel, Dordrecht, 1979.

10. (a) Stanton, R. E., and McIver, J. W., *J. Am. Chem. Soc.*, **97**, 3632 (1975).

    (b) McIver, J. W., *Acc. Chem. Res.*, **7**, 72 (1974).

11. (a) Evans, M. G., and Warhurst, E., *Trans. Faraday Soc.*, **34**, 614 (1938).

    (b) Evans, M. G., *Trans. Faraday Soc.*, **35**, 824 (1939).

12. Dewar, M. J. S., *J. Am. Chem. Soc.*, **106**, 209 (1984).

13. Salem, L., in *The New World of Quantum Chemistry* (Eds. B. Pullman, and R. Parr), Reidel, Dordrecht, 1976.

14. (a) Fukui, K., *J. Phys. Chem.*, **74**, 4161 (1970).

    (b) Fukui, K., Kato, S., and Fuijimoto, H., *J. Am. Chem. Soc.*, **97**, 1 (1975); *ibid.*, **99**, 684 (1977).

    (c) Ishida, K., Morokuma, K., and Komornicki, A., *J. Chem. Phys.*, **66**, 2153 (1977).

    (d) Sana, M., and Reckinger, G., *Theor. Chim. Acta*, **58**, 145 (1981).

15. (a) Wayne, R. P., in *Comprehensive Chemical Kinetics*, (Eds. C. M. Bamford, and C. F. H. Tippen), Vol. 2, p. 189, Elsevier, Amsterdam, 1969.

    (b) Benson, S. W., *Thermochemical Kinetics*, 2nd Edn, p. 194, Wiley, Chichester, 1960.

    (c) Mahan, B. H., *J. Chem. Educ.*, **51**, 709 (1974).

    (d) Pacey, P. D., *J. Chem. Educ.*, **58**, 612 (1981).

    (e) Millar, W. H., *Acc. Chem. Res.*, **9**, 306 (1976).

    (f) Truhlar, D. G., and Garrett, B. G., *Acc. Chem. Res.*, **13**, 440 (1980).

16. Daudel, R., LeRoy, G., Peeters, D., and Sana, M., *Quantum Chemistry*, Wiley, Chichester, 1983.

17. Nagase, S., Fueno, T., and Morokuma, K., *J. Am. Chem. Soc.*, **101**, 5840 (1979).

18. Harding, L. B., Schlegel, H. B., Krishnan, R., and Pople, J. A., in *Potential Energy Surfaces and Dynamics Calculations* (Ed. D. G. Truhlar), p. 164, Plenum, New York, 1981.

19. Morokuma, K., and Kato, S., in *Potential Energy Surfaces and Dynamics Calculations* (Ed. D. G. Truhlar), p. 243, Plenum, New York, 1981.

20. (a) Binkley, J. S., Whiteside, R. A., Krishnan, R., Seeger, R., Defrees, D. J., Schlegel, H. B., Topiol. S. Kahn, L. R., and Pople, J. A., GAUSSIAN 80, *Quantum Chem. Program Exch.*, **13**, 406 (1981).

    (b) Schlegel, H. B., FORCE, *Quantum Chem. Program Exch.*, **13**, 427 (1981).

    (c) Dupuis, M., Rys, J., and King, H. F., HONDO, *Quantum Chem. Program Exch.*, **13**, 401 (1981).

    (d) Komornicki, A., GRADSCF, *NRCC Program* QH04, 1980.

    (e) Pulay, P., TEXAS, *Theor. Chim. Acta*, **50**, 299 (1979).

    (f) Dupuis, M., Spangler, D., and Wendoloski, J. J., GAMMES, *NRCC Program* QG01, 1980.

21. (a) Schlegel, H. B., *J. Comput. Chem.*, **3**, 214 (1985).

    (b) Bell, S., Crighton, J. S., and Fletcher, R., *Chem. Phys. Lett.*, **82**, 122 (1981).

    (c) Bell, S., and Crighton, J. S., *J. Chem. Phys.*, **80**, 2464 (1984).

    (d) Cerjan, C. J., and Miller, W. H., *J. Chem. Phys.*, **75**, 2800 (1981).

22. Halgren, T. A., and Lipscomb, W. N., *Chem. Phys. Lett.*, **49**, 225 (1979).

23. Koga, N., and Morokuma, K., *Chem. Phys. Lett.*, **119**, 371 (1985).

24. Bernardi, F., Robb, M. A., Schlegel, H. B., and Tonachini, G., *J. Am. Chem. Soc.*, **106**, 1198 (1984).

25. Sinanoglu, O., and Bruckner, K. A., *Three Approaches to Electron Correlation in Molecules*, Yale University Press, New Haven, 1970.

26. Iwata, S., in *Quantum Chemistry Literature Data Base* (Eds. K. Ohno and K. Morokuma), Elsevier, Amsterdam, 1982.

27. (a) Harding, L. B., Schlegel, H. B., Krishnan, R., and Pople, *J. Phys. Chem.*, **84**, 3394 (1980).

(b) Goddard, J. D., Yamaguchi, Y., and Schaefer, H. F., *J. Chem. Phys.*, **75**, 3439 (1981).

(c) Dupuis, M., Lester, W. A., Lengsfield, B. H., and Liu, B., *J. Chem. Phys.*, **79**, 6167 (1983).

28. (a) Bouma, W. J., Vincent, M. A., and Radom, L., *Int. J. Quantum Chem.*, **14**, 767 (1978).

(b) Bouma, W. J., and Radom, L., *J. Am. Chem. Soc.*, **101**, 3487 (1979).

(c) Adeny, P. D., Bouma, W. J., Radom, L., and Rodwell, W. R., *J. Am. Chem. Soc.*, **102**, 4069 (1980).

(d) Rodwell, W. R., Bouma, W. J., and Radom, L., *Int. J. Quantum Chem.*, **18**, 107 (1980).

29. Krishnan, R., Frisch, M. J., and Pople, J. A., *J. Chem. Phys.*, **72**, 4244 (1980).

30. Baker, H., and Robb, M. A., *Mol. Phys.*, **50**, 1077 (1983).

31. (a) Handy, N. C., Knowles, P. J., and Somasundram, K., *Theor. Chim. Acta*, **68**, 87 (1985).

(b) Handy, N. C., *Faraday Symp. Chem. Soc.*, **19**, 17 (1985).

32. Saunders, V. R., and van Lenthe, J. H., *Mol. Phys.*, **48**, 923 (1983).

33. (a) Buenker, R. J., and Peyerimhoff, S. D., in *New Horizons of Quantum Chemistry* (Ed. P. O. Lodin), Readel Dordrecht, 1983.

(b) Buenker, R. J., Peyerimhoff, S. D., and Butscher W. *Mol. Phys.*, **35**, 771 (1983).

(c) Buenker, R. J., and Peyerimhoff, S. D., *Theor. Chim Acta*, **35**, 33 (1974); *ibid.*, **39**, 217 (1975).

34. Huron, B., Maleieu, J. P., and Rancuel, P., *J. Chem. Phys.*, **58**, 5745 (1973).

35. Siegbahn, P. E. M., *Int. J. Quantum Chem.*, **18**, 1229 (1980); *ibid.*, **23**, 1869 (1983).

36. Brandow, B., *Adv. Quantum Chem.*, **10**, 187 (1977).

37. (a) Schmidt, M. M., Lam, M. T. B., Elbert, S. T., and Ruedenberg, K., *Theor. Chim. Acta*, **68**, 69 (1985).

(b) Ruedenberg, K., Schmidt, M. W., Gilbert, M. M., and Elbert, S. T., *Chem. Phys.*, **71**, 41 (1982); *ibid.*, **71**, (1982); *ibid.* **71**, 65 (1982).

(c) Lam, B., Schmidt, M. W., and Ruedenberg, K., *J. Phys. Chem.*, **89**, 2221 (1985).

38. Komornicki, A., Goddard, J. D., and Schaefer, H. F., *J. Am. Chem. Soc.*, **102**, 1763 (1980).

39. Metiu, H., Ross, J., and Whitesides, G. M., *Angew. Chem. Int. Edn Engl.*, **18**, 377 (1979).

40. (a) Miller, W. H., and George, T. F., *J. Chem. Phys.*, **56**, 5637 (1972).

(b) Morokuma, K., and George, T. F., *J. Chem. Phys.*, **59**, 1959 (1973).

(c) Lin, Y. W., George, T. F., and Morokuma, K., *Chem. Phys. Lett.*, **22**, 547 (1973).

(d) Preston, R. K., and Tulley, C., *J. Chem. Phys.*, **54**, 4297 (1971); *ibid.*, **55**, 562 (1971).

(e) Chapman, S., and Preston, R. K., *J. Chem. Phys.*, **60**, 650 (1974).

(f) Tully, J. C., in *Dynamics of Molecular Collisions*, Part B (Ed. W. H. Miller), Plenum, New York, 1979.

(g) Nikitin, E. E., and Zulicke, L., *Theory of Chemical Elementary Processes*, Lecture Notes in Chemistry, Springer-Verlag, Berlin, 1978.

41. (a) Epiotis, N. D., and Shaik, S. J., *J. Am. Chem. Soc.*, **99**, 4936 (1977); *ibid.*, **100**, 1, 9, 29 (1978).
    (b) Shaik, S. S., *J. Am. Chem. Soc.*, **103**, 3692 (1981); *ibid.*, **104**, 2708 (1982).
    (c) Shaik, S. S., *Nouv. J. Chim.*, **6**, 159 (1982).
    (d) Epiotis, N. D., *Theory of Organic Reactions*, Springer, Berlin, 1978.
    (e) Epiotis, N. D., *Unifield Valence Bond Theory*, Lecture Notes in Chemistry, Vol. 29, Springer-Verlag, Berlin, 1982.
42. Spiegelmann, F., and Malrieu, J. P., *J. Phys. B: At. Mol. Phys.*, **17**, 1235 (1984); *ibid.*, **17**, 1259 (1984).
43. (a) Van Vleck, J. H., *Phys. Rev.*, **33**, 467 (1929).
    (b) Jordahl, O. M., *Phys. Rev.*, **45**, 87 (1934).
    (c) Kemble, E. C., *Fundamental Principles of Quantum Mechanics*, p. 394, McGraw-Hill, New York, 1937.
44. Shavitt, I., and Redmon, L. T., *J. Chem. Phys.*, **73**, 5711 (1980).
45. (a) Bloch, C., *Nucl. Phys.*, **6**, 329 (1958).
    (b) Lowdin, P.-O., *J. Math. Phys.*, **3**, 969 (1962).
    (c) Jorgensen, F., *Mol. Phys.*, **29**, 1137 (1975).
    (d) Lindgren, I., *J. Phys. B: At Mol. Phys.*, **7**, 2441 (1974).
    (e) Durand, P., *Phys. Rev. A*, **28**, 3184 (1983).
46. Hoffmann, R., Gleiter, R., and Mallory, F. B., *J. Am. Chem. Soc.*, **92**, 1460 (1970).
47. Basch, H., *J. Chem. Phys.*, **55**, 1700 (1971).
48. (a) Cheung, L. M., Sundberg, K. R., and Ruedenberg, K., *Int. J. Quantum Chem.*, **16**, 1103 (1979).
    (b) Ruedenberg, K., Schmidt, M. W., Gilbert, M. M., and Elbert, S. T., *Chem. Phys.*, **71**, 41 (1982).
49. (a) Feller, D., and Davidson, E. R., *J. Phys. Chem.*, **87**, 2721 (1983).
    (b) Ohta, K., Davidson, E. R., and Morokuma, K., *J. Am. Chem. Soc.*, **107**, 3466 (1985).
50. For detailed description of the wavefunctions of these configurations see Ref. 48a.
51. McKellar, A. R. W., Bunker, P. R., Sears, T. J., Evenson, K. M., Saykally, R. J., and Langhoff, S. R., *J. Chem. Phys.*, **79**, 5251 (1983).
52. (a) Bunker, P. R., and Jensen, P. J., *Chem. Phys.*, **79**, 1224 (1983).
    (b) Herzberg, G., and Johns, J. W. C., *Proc. R. Soc. A*, **295**, 107 (1966).
53. Harding, L. B., and Goddard, III, W. A., *Chem. Phys. Lett.*, **55**, 217 (1978).
54. Shih, S., Peyerimhoff, S. D., Buenker, R. J., and Feric, M., *Chem. Phys. Lett.*, **55**, 206 (1978).
55. Meadows, J. J., and Schaefer, III, H. F., *J. Am. Chem. Soc.*, **98**, 4383 (1976).
56. Rice, J. E., and Handy, N. C., *Chem. Phys. Lett.*, **107**, 365 (1984).
57. Zeck, O. F., Su, Y. Y., Gennaro, G. P., and Tang, Y. N., *J. Am. Chem. Soc.*, **96**, 5967 (1974).
58. Dubois, I., Herzberg, G., and Verma, R. D., *J. Chem. Phys.*, **47**, 4262 (1967).
59. Dubois, I., *Can. J. Phys.*, **46**, 2485 (1968).
60. Dubois, I., Duxbury, G., and Dixon, R. N., *J. Chem. Soc., Faraday Trans. 2*, **71**, 799 (1975).
61. Whiteside, R. A., Binkley, J. S., Krishnan, R., DeFrees, D. J., Schlegel, H. B., and Pople, J. A., *The Carnegie–Mellon Quantum Chemistry Archive*, Camegie–Mellon University, Pittsburgh, 1980.
62. Hehre, W. J., Stewart, R. F., and Pople, J. A., *J. Chem. Phys.*, **51**, 2657 (1969).
63. Binkley, J. S., Whiteside, R. A., Hariharan, P. C., Seeger, R., Pople, J. A., Hehre, W. J., and Newton, M. D., GAUSSIAN 76, *Quantum Chem. Program Exch.*, **10**, 368 (1978).

64. Bernardi, F., Olivucci, M., Robb, M. A., and Tonachini, G., *J. Am. Chem. Soc.*, **108**, 1408 (1986).
65. Woodward, R. B., and Hoffmann, R., *Angew. Chem. Int. Edn Engl.*, **8**, 781 (1969).
66. Schaefer, III, H. F., *Acc. Chem. Res.*, **12**, 289 (1979).
67. Rondan, N. G., and Houk, K. N., *Tetrahedron Lett.*, **25**, 2519 (1984).
68. Hess, B. A., Schaad, L. J., and Pancir, J., *J. Am. Chem. Soc.*, **107**, 149 (1985).
69. Bernardi, F., Robb, M. A., Schlegel, H. B., and Tonachini, G., *J. Am. Chem. Soc.*, **106**, 1198 (1984).
70. Osamura, Y., Kato, S., Morokuma, K., Feller, D., Davidson, E. R., and Borden, W. T., *J. Am. Chem. Soc.*, **106**, 3362 (1984).
71. Hehre, W. J., Stewart, R. F., and Pople, J. A., *J. Chem. Phys.*, **51**, 265 (1969).
72. Ditchfield, R., Hehre, W. J., and Pople, J. A., *J. Chem. Phys.*, **54**, 724 (1971).
73. Doubleday, C., Jr, McIver, J. W., Jr, and Page, M., *J. Am. Chem. Soc.*, **104**, 6533 (1982) and references therein.
74. Horsley, J. H., Jean, Y., Moser, C., Salem, L., Steven, R. M., and Wright, J. S., *J. Am. Chem. Soc.*, **94**, 279 (1972).
75. Waage, E. V., and Rabinovitch, B. S., *J. Phys. Chem.*, **76**, 1695 (1972).
76. Roth, W. R., and Konig, J., *Liebigs Ann. Chem.*, **24**, 699 (1966).
77. Isaacson, A. D. and Morokuma, K., *J. Am. Chem. Soc.*, **97**, 4453 (1975).
78. Binkley, J. S., Pople, J. A., and Hehre, W. J., *J. Am. Chem. Soc.*, **102**, 939 (1980).
79. Roth, W. R., *Tetrahedron Lett.*, 1009 (1964).
80. Gajewski, J. J., *Hydrocarbon Thermal Isomerizations*, pp. 166–76, Academic Press, New York, 1981.
81. Humski, K., Molojcic, R., Borcic, S., and Sunko, D. E., *J. Am. Chem. Soc.*, **92**, 6534 (1970).
82. Woodward, R. B., and Hoffmann, R., *Angew. Chem. Int. Edn Engl.*, **8**, 781 (1969).
83. Bartlett, P. D., Montgomery, L. K., and Seidel, B., *J. Am. Chem. Soc.*, **86**, 616 (1964); Montgomery, L. K., Schueller, K., and Bartlett, P. D., *ibid.*, **86**, (1964); Bartlett, P. D., and Montgomery, L. K., *ibid.*, **86**, 628 (1964); Bartlett, P. D., Wallbilich, G. E. H., and Montgomery, L. K., *J. Org. Chem.*, **32**, 1290 (1967); Bartlett, P. D., Wallbilich, G. E. H., Wingrove, A. S., Swenton, J. S., Montgomery, L. K., and Kramer, B. D., *J. Am. Chem. Soc.*, **90**, 2049 (1968); Swenton, J. S., and Bartlett, P. D., *ibid.*, **90**, 2056 (1968); Bartlett, P. D., Wingrove, A. S., and Owyang, R., *ibid.*, **90**, 6067 (1968); Bartlett, P. D., and Schneller, K. E., *ibid.*, **90**, 6071, 6077 (1968); Bartlett, P. D., *Science*, **159**, 833 (1968).
84. (a) Benson, S. W., *J. Chem. Phys.*, **34**, 521 (1961); *ibid.*, **46**, 4920 (1967); Benson, S. W., and Nougia, P. S., *ibid.*, **38**, 18 (1963); O'Neal, H. E., and Benson, S. W., *Int. J. Chem. Kinet.*, **1**, 221 (1969); Benson, S. W., *Thermochemical Kinetics*, Wiley, New York, 1968.
    (b) Frey, H. M., and Walsh, R., *Chem. Rev.*, **69**, 103 (1969); Frey, H. M., *Adv. Phys. Org. Chem.*, **4**, 148 (1966) and references therein.
85. Genaux, C. T., and Walters, W. D., *J. Am. Chem. Soc.*, **73**, 4497 (1951); Carr, R. W., and Walters, W. D., *J. Phys. Chem.*, **67**, 1370 (1963).
86. Bernardi, F., Bottoni, A., Robb, M. A., Schlegel, H. B., and Tonachini, G., *J. Am. Chem. Soc.*, **107**, 2260 (1985).
87. Wright, J. S., and Salem, L., *J. Am. Chem. Soc.*, **94**, 322 (1972).
88. Segal, G., *J. Am. Chem. Soc.*, **96**, 7892 (1974).
89. (a) Doubleday, C. Jr, McIver, J. W., Jr, and Page, M., *J. Am. Chem. Soc.*, **104**, 3768 (1982).
    (b) Doubleday, C., Jr, Camp, R. N., King, H. F., McIver, J. W., Jr, Mullay, D., and Page, M., *J. Am. Chem. Soc.*, **106**, 447 (1984).

90. Borden, W. T., and Davidson, E. R., *J. Am. Chem. Soc.*, **102**, 5409 (1980).
91. Bernardi, F., Olivucci, M., Robb, M. A., and Tonachini, G., *J. Am. Chem. Soc.*, in press.
92. Burke, L. A., and Leroy, G., *Bull. Soc. Chem. Belg.*, **88**, 379 (1979).
93. Bernardi, F., Olivucci, M., and Robb, M. A., submitted for publication.
94. Huisgen, R., *Angew. Chem. Int. Edn Engl.*, **2**, 565 (1963).
95. (a) Huisgen, R., *J. Org. Chem.*, **33**, 2291 (1968).
    (b) Huisgen, R., *ibid.*, **41**, 403 (1976).
96. (a) Firestone, R. A., *J. Org. Chem.*, **33**, 2285 (1968).
    (b) Firestone, R. A., *ibid.*, **37**, 2181 (1972).
    (c) Firestone, R. A., *J. Chem. Soc. A*, 1570 (1970).
    (d) Firestone, R. A., *Tetrahedron*, **33**, 3009 (1977).
97. (a) Linnett, J. W., *J. Am. Chem. Soc.*, **83**, 2643 (1961).
    (b) Linnett, J. W., *The Electronic Structure of Molecules*, Methuen, London, 1964.
98. (a) Leroy, G., Sana, M., Burke, L. A., and Nguyen, M. T., in *Quantum Theory of Chemical Reactions*, Vol. 1, p. 91, Reidel, Dordrecht, 1979.
    (b) Burke, L. A., Leroy, G., and Sana, M., *Theor. Chim. Acta*, **40**, 313 (1975).
    (c) Leroy, G., and Sana, M., *Tetrahedron*, **31**, 2091 (1975).
    (d) Leroy, G., and Sana, M., *ibid.*, **32**, 709 (1976).
    (e) Leroy, G., and Sana, M., *ibid.*, **32**, 1379 (1976).
    (f) Leroy, G., Nguyen, M. T., and Sana, M., *ibid.*, **32**, 1529 (1976).
    (g) Leroy, G., and Sana, M., *ibid.*, **34**, 2459 (1978).
99. (a) Poppinger, D., *J. Am. Chem. Soc.*, **97**, 7486 (1975).
    (b) Poppinger, D., *Aust. J. Chem.*, **29**, 465 (1976).
    (b) Poppinger, D., *Aust. J. Chem.*, **29**, 465 (1976).
100. Komornocki, A., Goddard, J. D., and Schaefer, III, H. F., *J. Am. Chem. Soc.*, **102**, 1763 (1980).
101. Hiberty, P. C., Ohanessian, G., and Schlegel, H. B., *J. Am. Chem. Soc.*, **105**, 719 (1983).
102. Bernardi, F., Bottoni, A., McDouall, J. J. W., Robb, M. A., and Schlegel, H. B., *Faraday Symp. Chem. Soc.*, **19**, 137 (1984).
103. Bernardi, F., Olivucci, M., and Robb, M. A., *J. Mol. Struct. (Theochem)*, **138**, 97 (1986).
104. See Sauer, J., and Sustmann, R., *Angew. Chem. Int. Edn. Engl.*, **19**, 779 (1980) and references therein.
105. (a) Burke, L. A., Leroy, G., and Sana, M., *Theor. Chim. Acta*, **40**, 313 (1975).
    (b) Burke, L. A., and Leroy, G., *ibid.*, **44**, 219 (1977).
106. Townshend, R. F., Ramunni, G., Segal, G., Hehre, W. J., and Salem, L., *J. Am. Chem. Soc.*, **98**, 2190 (1976).
107. Brown, F. K., and Houk, K. N., *Tetrahedron Lett.*, **25**, 4609 (1984).
108. Ortega, M., Oliva, A., Lluch, J. M., and Bertran, J., *Chem. Phys. Lett.*, **102**, 317 (1983).
109. Bernardi, F., Bottoni, A., Robb, M. A., Field, M. J., Hillier, I. H., and Guest, M. F., *J. Chem. Soc., Chem. Commun.*, 1051 (1985).
110. Bartlett, P. D., and Schueller, K. E., *J. Am. Chem. Soc.*, **90**, 6071 (1968).

Ab Initio Methods in Quantum Chemistry—I
Edited by K. P. Lawley
© 1987 John Wiley & Sons Ltd.

OPTIMIZATION OF EQUILIBRIUM GEOMETRIES AND TRANSITION STRUCTURES

H. BERNHARD SCHLEGEL*

Department of Chemistry, Wayne State University, Detroit, Michigan 48202, USA

CONTENTS

*Camille and Henry Dreyfus Teacher–Scholar.

I. INTRODUCTION

Over the past 15 years there have been rapid developments in the area of geometry optimization in *ab initio* molecular-orbital calculations. This progress has come about primarily because of energy gradient techniques. Analytical gradient-based optimization methods are almost an order of magnitude faster than optimization algorithms that do not use gradients. As a result, optimizing equilibrium geometries has become almost routine, and finding transition structures has become tractable. This is amply demonstrated in various bibliographies of *ab initio* calculations.[1-3] For example, the *Carnegie–Mellon Archive*[2] contains over 10 000 optimized structures. Although there had been a number of earlier calculations of energy derivatives,[4-6] Pulay was the first to demonstrate practical energy gradient computations at the self-consistent field (SCF) level.[7] Now, analytical first derivatives are available for most levels of *ab initio* calculations, and second derivatives are also available for a significant number of methods. Energy derivative methods have been surveyed previously,[8-11] and an up-to-date review is given by P. Pulay elsewhere in this series.[12]

With reasonable computational facilities, full geometry optimizations are feasible for molecules as large as 10 non-hydrogen atoms, and perhaps larger if small basis sets are used. For molecules with four or fewer heavy atoms, current practices in theoretical chemistry make it almost mandatory to carry out full geometry optimization at the SCF level with at least a split valence basis set (plus polarization, if possible). For systems with one and two heavy atoms, geometry optimizations at levels of theory that include electron correlation are rapidly becoming the norm. If possible, all of the coordinates of the molecule should be optimized. In the comparison of energies and geometries of equilibrium structures, transition states or points on a reaction path, subtle but significant changes can occur that may be masked by arbitrarily constraining the optimization to a subset of the variables. In some cases, apparent stationary points found by partial optimization may disappear or change drastically on full optimization (particularly for transition states). Fortunately, full optimization with gradient methods is often not much more expensive than partial optimization. For example, if the positions of the heavy atoms are optimized, the coordinates of the hydrogens can be optimized with little additional effort.

The geometry optimization methods discussed in the present chapter are focused primarily on gradient methods for molecular-orbital calculations. Included under optimization methods are algorithms for finding energy minima, locating transition states and higher-order saddle points, and following reaction paths. Beyond the scope of this chapter, but closely related to geometry optimization, is the topic of energy minimization with respect to the variational parameters of the wavefunction. For example, first and second derivative methods have greatly improved the rate of convergence of multi-configuration SCF (MCSCF) calculations. Also not directly addressed by this chapter are optimization methods for empirical force field or molecular mechanics calculations. These usually involve larger molecules with many more variables, but energy functions that can be calculated relatively quickly; the different economics of these calculations requires a somewhat different strategy for efficient geometry optimization.

Section II deals with features of energy surfaces, various definitions and preliminary matters. Section III briefly surveys the calculation of analytical energy derivatives. Section IV discusses some energy minimization algorithms that use the energy only, the first derivatives only, or the first and second derivatives. Section V addresses the more difficult task of locating transition structures. Quite a variety of algorithms can be found, with various ranges of applicability. Section VI examines methods for following reaction paths.

II. FEATURES OF ENERGY SURFACES

The concepts of energy surfaces for molecular motion, equilibrium geometries, transition structures and reaction paths depend on the Born–Oppenheimer approximation to treat the motion of the nuclei separately from the motion of the electrons. Minima on the potential energy surface for the nuclei can then be identified with the classical picture of equilibrium structures of molecules; saddle points can be related to transition states and reaction rates. If the Born–Oppenheimer approximation is not valid, for example in the vicinity of surface crossings, non-adiabatic effects are important and the meaning of classical chemical structures becomes less clear. Non-adiabatic effects are beyond the scope of this chapter and the discussion of energy surfaces and optimization will be restricted to situations where the Born–Oppenheimer approximation is valid.

A. Minima, Maxima and Saddle Points

Minima, maxima and saddle points can be characterized by their first and second derivatives. For a function of several variables, the first derivatives with respect to each of the variables form a vector termed the gradient. The second derivatives form a matrix called the Hessian. In classical mechanics, the first derivative of the potential energy for a particle is minus the force on the particle, and the second derivative (for a quadratic potential) is the force constant. Thus, the negative of the components of the gradient are the forces on the atoms or nuclei in a molecule, and the second derivative matrix, or Hessian, is also called the force-constant matrix.

For a function of one variable, the first derivative of the function is zero at a local minimum or maximum. In more than one dimension, the first derivatives with respect to all of the variables must be zero, i.e. the gradient vector must have zero length. Equivalently, all of the forces on the atoms in a molecule must be zero; hence such a point is also known as a stationary point. In topology, these points are termed critical points. The nature of the stationary point can be determined from the second derivatives. In one dimension, the second derivative is positive for a minimum, negative for a maximum and zero for a point of inflection. In more than one dimension, the eigenvalues of the second derivative matrix, rather than individual elements of the matrix, determine the characteristics of the stationary point. If all the eigenvalues are positive, the point is a local minimum; if all are negative, it is a local maximum. A (first-order) saddle point or col has one negative eigenvalue and all the rest are positive, i.e. a maximum in one direction and a minimum in all perpendicular directions. Thus a first-order saddle point can correspond to a transition structure (see below). An nth order saddle point has n negative eigenvalues, i.e. is a maximum with respect to n mutually perpendicular directions. A zero eigenvalue indicates a point of inflection for motion along the associated eigenvector.

If at all possible, stationary points found by any optimization method should be characterized by computing the Hessian and examining its eigenvalues. This is especially important for high-symmetry structures (where lower-energy, lower-symmetry structures may exist nearby) and for saddle points (which can be higher-order saddle points rather than first-order).

B. Transition Structures and Intrinsic Reaction Coordinates

A minimum on an energy surface represents an equilibrium structure. If there is more than one minimum on a contiguous energy surface, a family of paths can be constructed that connect one minimum to the other. If the highest-energy point on each path is considered, the transition structure can be defined as the lowest of these maxima, i.e. the top of the barrier for the lowest-energy path from one minimum to another. Thus, the transition structure must be a maximum along the reaction path, and a minimum for all displacements perpendicular to the path. This is just the definition of a first-order saddle point. If a higher-order saddle point were chosen (i.e. maximum in more than one direction), it would be possible to move perpendicular to the path to find another path with a lower maximum. The eigenvector corresponding to the single negative eigenvalue of a first-order saddle point is termed the transition vector. At least initially, the steepest-descent path from the saddle point to either of the two minima follows the transition vector.

Since the minima and saddle point are well defined points on the energy surface, it should be possible to define a unique reaction path. The steepest-descent path from the saddle point to the minima can be defined easily, but depends on the particular choice of coordinate system. Cartesian coordinates would yield a different path than internal coordinates. Furthermore, internal coordinates are not unique, since a number of different sets of bond lengths, angles and torsions can represent the same structure.

An intrinsic reaction path[13] can be defined independently of the coordinate system by appealing to classical mechanics. For a given energy surface, the movement of a classical particle must be the same regardless of whether Cartesian coordinates or any of a number of different sets of internal coordinates are used. The intrinsic reaction coordinate (IRC) can be defined as the path traced by a classical particle sliding with infinitesimal velocity from a saddle point down to each of the minima. Since the classical equations of motion can be defined in any coordinate system, and since they must yield the same trajectory, this definition of the intrinsic reaction coordinate is unique. The classical equations of motion are the simplest in mass-weighted Cartesian coordinates, where the effective mass for each coordinate is unity. In this coordinate system, the intrinsic reaction coordinate is the same as the steepest-descent path. An equivalent definition[14] requires the IRC to be the path of minimum distance (geodesic) between the reactants and products, with the metric determined by the potential energy surface.

Intrinsic reaction coordinates are geometrical or mathematical features of the energy surfaces, like minima, maxima and saddle points. Considerable care should be taken not to attribute too much chemical or physical meaning to the reaction coordinate. Since molecules have more than infinitesimal kinetic energy, a classical trajectory will not follow the intrinsic reaction path and may in fact deviate quite widely from it. The intrinsic reaction coordinate is, however, a convenient measure of the progress of a molecule in a reaction. It also plays a central role in the calculation of reaction rates by variational transition state theory[15] and with reaction path Hamiltonians.[16]

C. Symmetry and Topology

Energy surfaces are smooth functions connecting the minima associated with molecular structures. If there are n minima on one contiguous surface, each minimum must be connected to at least one other minimum (otherwise the surface is not contiguous). Since there must be at least one saddle point between connected minima, there are at least $n - 1$ saddle points on a surface with n minima. Other useful relations can also be derived[17] but these either require or yield information about higher-order saddle points or maxima, and are therefore less interesting for chemical reactions.

The energy surface must reflect the symmetry of the molecules it represents. The motion of molecules on energy surfaces has been discussed in terms of non-rigid symmetry groups,[18] and permutation groups have been used to classify reaction paths for isomerization.[19] More directly connected to the present topic are the restrictions that symmetry places on the transition vectors for reactions.[20,21] Some of the more important symmetry requirements for transition vectors discussed by Stanton and McIver[21] are:

1. The transition vector cannot belong to a degenerate representation. Otherwise there would be at least two symmetry-equivalent eigenvectors of the Hessian with negative eigenvalues and the structure would not be a first-order saddle point or transition state. Closely related to this, it can be shown that three valleys cannot meet at a single transition state (except as a numerical accident).
2. The transition vector must be antisymmetric for the symmetry operations that convert the reactants to products.
3. The transition vector must be symmetric for the symmetry operations that leave the reactants and products unchanged.

If an optimization is required to retain the symmetry of a structure, then symmetry-equivalent parameters must remain equal. Thus only displacements belonging to the totally symmetric irreducible representation are permitted; other displacements are forbidden since they would change the symmetry. If it can be determined (e.g. from the Stanton–McIver symmetry rules) that the transition vector does not belong to the totally symmetric representation, then the position of the transition structure along the reaction path is fixed by symmetry, and only the coordinates perpendicular to the reaction path need to be optimized. Since a transition structure must be a minimum with respect to all displacements other than along the reaction path, optimization of the transition structures is reduced to a simple minimization.

By symmetry, the only components of the gradient that are non-zero are those belonging to the totally symmetric representation. Thus, gradient optimization methods (that update the Hessian symmetrically) will not lead to lower-symmetry structures provided the coordinate system reflects the symmetry of the molecule and numerical round-off errors are negligible. Nevertheless, it is quite possible that distortion to a lower symmetry will reduce the energy. This occurs if the higher-symmetry structure is a local maximum or a saddle point with respect to displacements that do not belong to the totally symmetric representation. Calculation of the full second derivative matrix is the best way to test if an optimized structure is stable or unstable with respect to all possible distortions, including those that lower the symmetry.

Sometimes a molecule moves toward a higher symmetry during an optimization. The convergence to the higher symmetry can be relatively slow, and it may be unclear whether the optimized geometry does have the higher symmetry or whether the molecule is slightly distorted and has a lower symmetry. The problem is best resolved by first optimizing the higher-symmetry structure and then testing whether non-totally symmetric distortions lead to a lower energy, preferably by computing the second derivative matrix (numerically or analytically). If the second derivative matrix of the high-symmetry structure has one or more negative eigenvalues, then the structure is not a local minimum and displacement along an eigenvector that corresponds the negative eigenvalue will lead to a lower-energy, lower-symmetry structure. An optimization on the distorted structure then results in a lower-symmetry, lower-energy stationary point (whose stability with respect to further reduction in the symmetry must also be tested).

D. Choice of Coordinates

Before geometry optimization can be carried out, a coordinate system must be chosen to represent the energy surface and the structure of the molecule. Aside from the algorithm for optimization (see Sections IV and V), the choice of the coordinate system is perhaps the single most important factor in determining the ease or difficulty of an optimization. Internal coordinates are normally preferred for geometry optimization, since the total energy is invariant to overall translation and rotation. Usually, several internal coordinate systems are possible. The coordinates must be non-redundant so that the geometry can be specified uniquely (and without extra constraints), and so that the derivatives can be calculated unambiguously. For example, a tetrahedral center has six valence angles, only five of which are independent; a planar six-membered ring has six bond lengths and six angles, but only nine independent coordinates (examples are discussed in Ref. 22).

Difficulties encountered in optimizations can often be traced to problems caused by the coordinate system, especially in cyclic molecules and in transition structures. A different choice of coordinates may significantly improve the behavior of the optimization. Strong coupling degrades the performance of any optimization method on a non-quadratic surface. Coupling between flexible and stiff modes causes special difficulties. This corresponds to a long narrow valley running diagonally across the energy surface. In pathological cases, the valley is also curved, due to non-quadratic terms in the coupling between the coordinates. It may be possible to find a coordinate system that reduces the curvature and/or the coupling. For transition structure optimization, coupling between the transition vector and the other coordinates is also detrimental. A coordinate system should be selected so that the transition vector is dominated by only a few variables (preferably only one), and so that the coupling of these to the remaining coordinates is small.

III. ANALYTICAL ENERGY DERIVATIVES

A comprehensive review of analytical energy methods has been written by P. Pulay and appears elsewhere in this series.[12] The purpose of this section is only to outline some of the computational steps and to supply a few leading references.

An electronic wavefunction can be constructed from a linear combination of many-electron configurations, which can be built from one-electron molecular orbitals, which in turn can be formed from linear combinations of basis functions. To compute the energy, the integrals over the basis functions must be calculated and the coefficients must be determined. The integrals are usually computed analytically. Some or all of the linear coefficients are calculated variationally, others may be obtained by non-variational methods; the particular recipe depends on the specific method. For derivatives of the energy, the corresponding derivatives of the integrals over basis functions are always needed. Derivatives of the linear coefficients may or may not be required, depending on whether the coefficients were determined variationally.

For polyatomic calculations, basis sets usually consist of contracted Gaussians, i.e. fixed linear combinations of Gaussian-type primitives. The necessary integrals can be calculated analytically, and a variety of programs have been developed to compute them efficiently.[23,24] The integral derivatives can also be calculated analytically, although with somewhat greater effort than the integrals themselves, since each integral can have up to 12 first derivatives and 78 second derivatives with respect to the function positions. Efficient algorithms have been devised and coded by several groups.[25-37,49,60] Translational invariance can be used to reduce the number to nine first derivatives and 45 second derivatives. Rotational invariance can further reduce the computational work, especially for third derivatives.[38-40,56] Depending on the energy method and whether first, second or higher derivatives are needed, it may be possible to avoid the explicit storage of the integral derivatives and/or the explicit transformation of the integrals.[49,57,58]

For SCF and MCSCF energy calculations, all of the coefficients— molecular-orbital (MO) and configuration-interaction (CI)—are determined variationally, i.e. by minimizing the expectation value of the energy. The derivative of the energy with respect to any change in these coefficients is zero (provided the molecular orbitals remain orthonormal and the CI coefficients remain normalized). Hence the first derivatives of the coefficients are not required for the first derivative of the SCF and MCSCF energy; only the first derivatives of the integrals are needed. Consequently, first derivatives can be calculated fairly easily at the SCF and MCSCF levels. Efficient programs for the SCF gradients have been developed over the last 15 years (see Refs. 8–12). Typically, the time required for all $3N - 6$ first derivatives of the internal coordinates of an N-atom molecule is equal to or less than the time needed to compute the total energy (integrals plus SCF). First derivatives of the MCSCF energy have appeared more recently[41-45] (and see Refs. 8–12), and are even more economical relative to the time required for the total energy (primarily because of the length of the MCSCF energy computation). For SCF and

MCSCF first derivatives, the integral derivatives do not need to be stored or transformed explicitly.

Analogous to the first derivatives, the second derivatives of the SCF and MCSCF energies do not require the second derivatives of the MO and CI coefficients (provided orthonormality is maintained). However, the first derivatives of the coefficients must be calculated by solving the coupled perturbed Hartree–Fock (CPHF) equations,[6] or the coupled perturbed MCSCF (CPMCSCF) equations. Pople and coworkers[46] developed an efficient method for solving the CPHF equations, which made SCF second derivative calculations practicable for restricted Hartree–Fock (RHF) and unrestricted Hartree–Fock (UHF) wave functions. These methods have been extended to include restricted open-shell SCF calculations.[47,48] More recently, the CPMCSCF equations have been formulated[45,49–52] and MCSCF second derivatives have been implemented by a number of groups.[52–56] The second derivatives of the integrals do not need to be stored; storage of the first derivatives can be avoided for the CPHF equations by constructing the derivatives of the Fock-like matrices as the integral derivatives are computed.[49,57,58] For the CPMCSCF equations, some of the transformed-integral first derivatives appear to be needed.

Third derivatives have also been formulated for the SCF and MCSCF energies by a number of groups[49,59,60] and implemented by at least one group.[60] In addition to the first, second and third derivatives of the integrals, only the first derivatives of the coefficients are needed. This is similar to perturbation theory, where the third-order energy can be computed with the first-order wavefunction.

For energy derivatives at levels of theory other than SCF or MCSCF, the first derivatives of the molecular-orbital coefficients are needed in addition to the derivatives of the integrals over basis functions. The molecular-orbital coefficient derivatives must be determined by solving the appropriate coupled perturbed equations: CPHF for wavefunctions based on a single SCF reference configuration (e.g. configuration-interaction with single and double excitations (CISD), second-order Møller–Plesset theory (MP2), coupled clusters with doubles (CCD), etc.), CPMCSCF for multiple reference based on an MCSCF wavefunction (e.g. multi-reference determinant configuration-interaction (MRDCI)). Because of the special form of the first derivative expression, it is possible to avoid the full CPHF or CPMCSCF equations, and solve only a much smaller set of equations.[61] Energy gradients have been programmed for a number of methods including single-reference CI,[62–64,54] multiple-reference CI,[55] second-order Møller–Plesset theory,[46] third-order Møller–Plesset theory[65] and coupled-cluster methods.[66,67] The latter also requires the solution of the coupled perturbed coupled-cluster equations.

Second derivatives for configuration-interaction, coupled-cluster and many-body perturbation theory require first derivatives of those coefficients determined variationally and the second derivatives of those coefficients not determined variationally. As with the first derivatives of these methods, the solution of the full second-order coupled perturbed equations for the MO coefficients can be reduced to a much smaller set of equations.[61] Second derivatives have been demonstrated for the CI energy[68] and have been formulated for Møller–Plesset theory and for coupled-cluster methods.[50,69]

IV. LOCATING MINIMA ON ENERGY SURFACES

There are quite a number of methods for unconstrained minimization of non-linear functions of many variables[70-74] and it is difficult to assess the goodness of any specific algorithm quantitatively without considering particular applications. Desirable features of an algorithm include: speed of convergence to the minimum, stability and reliability of the method, and the overall cost of the optimization. The choice of the best method for a particular situation depends on the nature of the function to be minimized, the number of variables, the availability of first (and higher) derivatives, and the cost of evaluating the function and its derivatives. Several representative algorithms are examined in this section, with the special features of *ab initio* calculations and geometry optimization in mind.

For geometry optimizations using quantum-mechanical methods, the cost of the function evaluation, i.e. the energy calculation, is much higher than usually assumed for typical minimization problems in numerical analysis. Thus, the total number of evaluations must be kept to a bare minimum, even at the cost of a more complex algorithm. Almost always, the computer time required by the minimization algorithm will be negligible compared to the time needed for the energy evaluation. For example, the cost of inverting or diagonalizing an $N \times N$ matrix, where N is the number of geometrical parameters, is inconsequential compared to the time to calculate the energy. Exceptions to this include determining MCSCF wavefunctions where the energy must be minimized with respect to 10^3 to 10^5 wavefunction parameters, and empirical force field (molecular mechanics) calculations, where the energy, gradient and second derivatives can be calculated relatively easily (neither of these minimization problems are within the scope of this chapter).

A number of representative methods for locating minima are examined more closely. It is convenient to group the algorithms into three categories, depending on the derivative information used: (a) function value only, (b) function plus first derivatives or gradient, and (c) function plus first and second derivatives.

A. Function-only Methods

Algorithms for minimization that require the evaluation of only the function and not its derivatives can be found in a variety of standard books on numerical analysis.[70-74] Because derivatives are not available for all levels of theory, these methods have the widest range of applicability, but the penalty is that they also have the slowest convergence. The simplex method and the pattern search method are generally not used for geometry optimization in *ab initio* calculations, since they require too many steps. The sequential univariate search or axial iteration algorithm is perhaps the most frequently used method for geometry optimization. A modified Fletcher–Powell algorithm[75,76] is somewhat faster and has also seen considerable use. These two algorithms are discussed below and are illustrated in Figs. 1 and 2 for a simple quadratic surface. The optimizations are shown for two variables, but it should be kept in mind that real examples will have about 10–20 variables. There are a number of other, perhaps more efficient, algorithms that depend only on the function evaluation. These are treated in the next section since they in fact use derivatives evaluated numerically.

The sequential univariate search or axial iteration method changes one

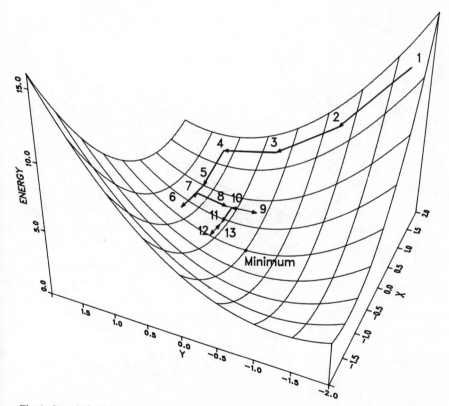

Fig. 1. Steps in finding the minimum on a quadratic surface using the sequential univariate search or axial iteration method.

coordinate at a time (Fig. 1) and cycles over all the coordinates one or more times:

1. Calculate the energy at the initial geometry.
2. Calculate the energy at two displacements along a coordinate.
3. Fit a parabola to the energy at the three geometries (undisplaced and two displaced).
4. Find the minimum of the parabola and calculate the energy at the minimum.
5. Choose the next coordinate and go back to step (2), unless all the coordinates have been processed.
6. If the change in all the coordinates is small enough, then stop; otherwise go back to (2) and cycle through all of the coordinates again.

For the two-dimensional example in Fig. 1, the first cycle gives rise to points (1)–(7). As can be seen, convergence to a minimum is not guaranteed in one

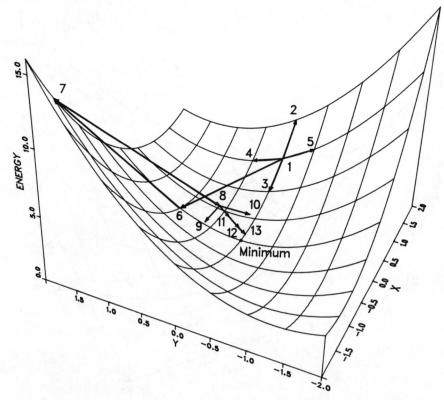

Fig. 2. Steps in finding the minimum on a quadratic surface using a modified Fletcher–Powell algorithm.

cycle, even for a quadratic surface. This is because the example has strong coupling between the two coordinates, which results in a diagonal valley rather than one parallel to the coordinate axes. A second cycle through all of the coordinates (points (8)–(13)) yields a structure closer to the true minimum. For an N-dimensional surface, $3N + 1$ energy calculations are needed for the first cycle and $3N$ for each subsequent cycle, with typically two or three cycles needed for satisfactory convergence. In general, convergence is slow and more cycles are required if the coordinates are strongly coupled and/or the eigenvalues of the second derivative matrix are very different, e.g. a long narrow valley at an angle to the coordinate axes. The only test for the accuracy of the optimization is to carry out one more cycle through all of the coordinates to determine if any change significantly.

A second, more sophisticated, method is a variant of the Fletcher–Powell algorithm[75,76] and is shown in Fig. 2. The method is actually a derivative-based method (fixed metric, using numerical derivatives) but is rather closely related to the axial iteration method. The following steps are involved:

1. Calculate the energy at the starting geometry, $\mathbf{x}_k, k = 0$, and at positive and negative displacements for each of the coordinates (points (1)–(5) in Fig. 2).
2. Fit a parabola for each of the coordinates, i.e. a quadratic surface without interaction terms:

$$E(\mathbf{x}) = E_k + \sum_i \left[g_i^k (x_i - x_i^k) + \tfrac{1}{2} B_{ii} (x_i - x_i^k)^2 \right] \qquad (1)$$

In effect, the gradient, $\mathbf{g}$, and diagonal elements of the Hessian, $\mathbf{B}$, are calculated numerically.
3. Find the minimum on the model surface:

$$dE/dx_i = g_i(\mathbf{x}) = g_i^k + B_{ii}(x_i - x_i^k) = 0$$
$$p_i^k = x_i - x_i^k = - g_i^k / B_{ii} \qquad (2)$$

If $\mathbf{p}_k$, the predicted change in the coordinates, is small enough, stop.
4. Calculate the energy at $\mathbf{x}_k + \mathbf{p}_k$ and $\mathbf{x}_k + 2\mathbf{p}_k$ (points (6) and (7) in Fig. 2).
5. Fit a parabola to the energy at $\mathbf{x}_k$, $\mathbf{x}_k + \mathbf{p}_k$ and $\mathbf{x}_k + 2\mathbf{p}_k$, and find the minimum, $\mathbf{x}_{k+1} = \mathbf{x}_k + \alpha \mathbf{p}_k$.
6. Calculate the energy at $\mathbf{x}_{k+1}$ (point (8) in Fig. 2).
7. Compute $\mathbf{g}_{k+1}$ by calculating the energy at displacements for each coordinate from $\mathbf{x}_{k+1}$ (points (9) and (10)).
8. Set $k = k + 1$, and go back to step 3.

For an N-dimensional surface, $2N + 4 + m(N + 3)$ steps are required, where m is the number of passes through step 7. (typically two or three). A test of the goodness of an individual structure requires $N + 1$ energy computations to evaluate the first derivative numerically. As N increases, the modified Fletcher–Powell method is significantly more efficient than the axial iteration

algorithm, especially for strongly coupled coordinates where several optimization cycles may be required. As can be seen from Fig. 2, this algorithm is also
not exact in one or two cycles for a general quadratic surface, because the off-
diagonal terms in the Hessian are neglected.

Improvements that include the explicit calculation of the off-diagonal
elements of the Hessian, $B_{ij}(i \neq j)$, have been discussed by several authors.[77,78]
Because the B_{ij} are difficult to calculate accurately for a non-quadratic surface,
and because there are $N(N-1)/2$ elements B_{ij}, considerable care is needed to
balance the number of function evaluations with the numerical accuracy of the
B_{ij}, in order to obtain a more efficient algorithm for minimization.

B. Gradient Algorithms

Quasi-Newton, variable metric, conjugate gradient,[79] Fletcher–Powell,[75]
Davidon–Fletcher–Powell,[75]   Murtagh–Sargent,[80]   Broyden–Fletcher–

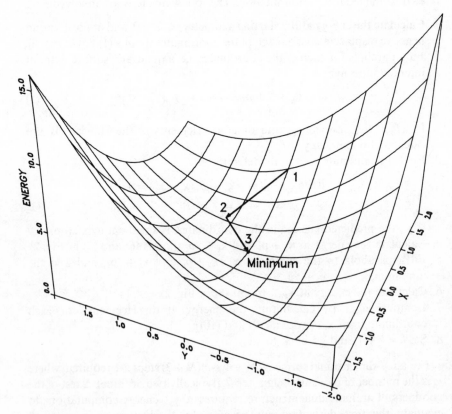

Fig. 3. Steps in finding the minimum on a quadratic surface using a quasi-Newton algorithm with
accurate line searches.

Goldfarb–Shanno,[81] and optimally conditioned[82] are a few of the optimization algorithms that use the first derivative.[70] If analytical derivatives are available, these can be significantly more efficient and can have better convergence properties than the function-only algorithms. If gradients must be calculated numerically, the overall efficiency may not be better than the function-only algorithms discussed above.

The gradient-type optimization algorithms approximate the energy surface at step k by a quadratic expression in terms of the position, x_k, the computed energy, E_k, the computed gradient, g_k, and the approximate Hessian, B_k:

$$E(\mathbf{x}) = E_k + \mathbf{g}_k^T \cdot (\mathbf{x} - \mathbf{x}_k) + \tfrac{1}{2}(\mathbf{x} - \mathbf{x}_k)^T \mathbf{B}_k(\mathbf{x} - \mathbf{x}_k) \tag{3}$$

The initial estimate of the Hessian, $\mathbf{B}_0$ (or its inverse, $\mathbf{H}_0 = \mathbf{B}_0^{-1}$), is improved as the optimization proceeds, and for most methods approaches the true Hessian (or its inverse) if the surface is quadratic. One-dimensional minimization may also be required along each new search direction. A typical search for a two-dimensional quadratic surface is shown in Fig. 3; actual optimization may be in 10–20 dimensions. An optimization can be divided into the following steps:

1. Start with the geometry x_k, $k = 0$, and obtain an estimate of the Hessian, $\mathbf{B}$, or its inverse, $\mathbf{H}$ (either a unit matrix, an empirical estimate, or other, see below).
2. For x_k calculate the energy, E_k, and gradient, $g_k = dE/dx$.
3. Update the Hessian (or inverse Hessian) so that the model surface fits the current energy and gradient as well as those from previous steps (omit for the first point).
4. Find the minimum on the model surface using the gradient and the updated Hessian (note that this requires $\mathbf{H}$, the inverse of the Hessian):

$$dE/d\mathbf{x} = \mathbf{g}(\mathbf{x}) = \mathbf{g}_k + \mathbf{B}_k(\mathbf{x} - \mathbf{x}_k) = 0$$
$$\mathbf{p}_k = \mathbf{x} - \mathbf{x}_k = -\mathbf{B}_k^{-1}\mathbf{g}_k = -\mathbf{H}_k\mathbf{g}_k \tag{4}$$

If the gradient, g_k, is small enough and/or the predicted change in the geometry, p_k, is small enough, stop.
5. Carry out a minimization in the direction of the predicted displacement, i.e. minimize $E(\mathbf{x}_k + \alpha \mathbf{p}_k)$ with respect to α (not used in some methods, i.e. $\alpha = 1$ always).
6. Set $\mathbf{x}_{k+1} = \mathbf{x}_k + \alpha \mathbf{p}_k$, $k = k + 1$, and return to step 2.

This sequence is illustrated for a quadratic surface by points (1)–(3) in Fig. 3 (accurate line search assumed). For an N-dimensional surface, only one gradient calculation is needed to test whether a point is a minimum. For a quadratic surface, gradient methods require $N + 1$ steps or less, and are guaranteed to reach the minimum, regardless of the starting geometry or the magnitude of the coupling between the coordinates. Accurate line searches may be required to prove convergence for some methods. Even for the non-quadratic surfaces encountered in typical geometry optimizations, convergence is usually achieved in about $N + 1$ steps, and often fewer since good estimates of the geometry and the Hessian can be made. Because an energy plus gradient calculation takes about twice as long as an energy calculation alone, each step in a gradient algorithm is more costly than in an energy-only algorithm. However, this is more than compensated by the reduced number of steps and the more stable convergence properties of the gradient optimizations. In practice, as much as an order-of-magnitude increase in efficiency can be obtained with gradient optimization methods compared to energy-only algorithms.

Aside from the nature of the function and the starting coordinates, the overall cost and convergence rate for minimizing a non-quadratic function depend on the updating scheme for the Hessian (or its inverse), the accuracy of the line searches and the initial estimate of the Hessian (or inverse). These topics are considered in the next two subsections. In the numerical analysis literature, much attention has been devoted to the problem of updating the Hessian and the linear searches.[70] For geometry optimization, good initial estimates of the Hessian can often be obtained from general concepts in chemical bonding or from lower levels of theory. An accurate initial estimate of the second derivative matrix can improve the rate of convergence significantly, but will not affect the final, optimized geometry, since the latter depends only on the gradient (where it goes to zero) and not the Hessian.

1. Updating Formulas and Line Searches

The simplest gradient method is the steepest-descent algorithm.[70] The Hessian is taken as the unit matrix (or a constant times the unit matrix) and is not updated. Thus the search is along $\mathbf{p}_k = -\mathbf{q}_k$, i.e. the direction in which the function decreases most rapidly. An accurate linear search is required at each step to achieve convergence. This method reduces the function value quite rapidly at first, but final convergence is slow. Closely related to this algorithm is the fixed metric method[70] in which the Hessian is a more general non-diagonal matrix that is not updated.

At the other end of the spectrum of gradient algorithms is Newton's method.[70] The Hessian is calculated directly at each step (hence it is a second derivative method, and is discussed in Section 4.c below). The most frequently used gradient algorithms fall between the extremes of the steepest-descent/fixed metric methods and Newton's method, and are termed variable metric or quasi-Newton methods. These methods avoid the direct calculation of the Hessian; instead they start with an approximate Hessian and improve it using the gradient information gathered during the course of the optimization. Only the main features are summarized here; a more detailed discussion can be found in books dealing with optimization[70] and in the original literature.

The conjugate gradient (CG) algorithm[79] is one of the older methods and, strictly speaking, is not a quasi-Newton method. However, it is the method of choice for very large problems where the storage of the Hessian is not practicable. In the Fletcher–Reeves approach the search direction is given by

$$\mathbf{p}_k = -\mathbf{g}_k + \mathbf{p}_{k-1}(\mathbf{g}_k^T\cdot\mathbf{g}_k)/(\mathbf{g}_{k-1}^T\cdot\mathbf{g}_{k-1}) \tag{5}$$

To relate the conjugate gradient algorithm to the quasi-Newton methods this formula can be expressed[74] as an updating scheme for the inverse Hessian:

$$\mathbf{H}_k = \mathbf{H}_{k-1} - \frac{\mathbf{H}_{k-1}\Delta\mathbf{g}_k\Delta\mathbf{g}_k^T\mathbf{H}_{k-1}}{\Delta\mathbf{g}_k^T\mathbf{H}_{k-1}\Delta\mathbf{g}_k} \qquad \Delta\mathbf{g}_k = \mathbf{g}_k - \mathbf{g}_{k-1} \quad \text{and} \quad \mathbf{H}_0 = \mathbf{I} \tag{6}$$

If the Hessian can be kept in memory, the quasi-Newton methods provide better convergence to the minimum. Some of the more frequently used schemes for updating the Hessian are as follows.

1. The Davidon–Fletcher–Powell (DFP) algorithm:[75]

$$\mathbf{H}_k = \mathbf{H}_{k-1} + \frac{\Delta\mathbf{x}_k\Delta\mathbf{x}_k^T}{\Delta\mathbf{x}_k^T\cdot\Delta\mathbf{g}_k} - \frac{\mathbf{H}_{k-1}\Delta\mathbf{g}_k\Delta\mathbf{g}_k^T\mathbf{H}_{k-1}}{\Delta\mathbf{g}_k^T\mathbf{H}_{k-1}\Delta\mathbf{g}_k} \qquad \Delta\mathbf{x}_k = \mathbf{x}_k - \mathbf{x}_{k-1} \tag{7}$$

2. The Murtagh–Sargent (MS) algorithm:[80]

$$\mathbf{H}_k = \mathbf{H}_{k-1} + \frac{[\Delta\mathbf{x}_k - \mathbf{H}_{k-1}\Delta\mathbf{g}_k][\Delta\mathbf{x}_k - \mathbf{H}_{k-1}\Delta\mathbf{g}_k]^T}{[\Delta\mathbf{x}_k - \mathbf{H}_{k-1}\Delta\mathbf{g}_k]^T\cdot\Delta\mathbf{g}_k} \tag{8}$$

3. The Broyden–Fletcher–Goldfarb–Shanno (BFGS) algorithm:[81]

$$\mathbf{H}_k = \left[\mathbf{I} - \frac{\Delta\mathbf{x}_k\Delta\mathbf{g}_k^T}{\Delta\mathbf{x}_k^T\cdot\Delta\mathbf{g}_k}\right]\mathbf{H}_{k-1}\left[\mathbf{I} - \frac{\Delta\mathbf{x}_k\Delta\mathbf{g}_k^T}{\Delta\mathbf{x}_k^T\cdot\Delta\mathbf{g}_k}\right]^T + \frac{\Delta\mathbf{x}_k\Delta\mathbf{x}_k^T}{\Delta\mathbf{x}_k^T\cdot\Delta\mathbf{g}_k} \tag{9}$$

The latter can be written in a more general form, which defines the Broyden family of algorithms:[83]

$$\mathbf{H}_k = \mathbf{H}_{k-1} + \frac{\Delta\mathbf{x}_k\Delta\mathbf{x}_k^T}{\Delta\mathbf{x}_k^T\cdot\Delta\mathbf{g}_k} - \frac{\mathbf{H}_{k-1}\Delta\mathbf{g}_k\Delta\mathbf{g}_k^T\mathbf{H}_{k-1}}{\Delta\mathbf{g}_k^T\mathbf{H}_{k-1}\Delta\mathbf{g}_k} + \pi_k\Delta\mathbf{g}_k^T\mathbf{H}_{k-1}\Delta\mathbf{g}_k\mathbf{w}_k\mathbf{w}_k^T \tag{10}$$

$$\mathbf{w}_k = \Delta\mathbf{x}_k/(\Delta\mathbf{x}_k^T\cdot\Delta\mathbf{g}_k) - \mathbf{H}_{k-1}\Delta\mathbf{g}_k/(\Delta\mathbf{g}_k^T\mathbf{H}_{k-1}\Delta\mathbf{g}_k)$$

Note that $\pi_k = 0$ yields the DFP method and $\pi_k = 1$ the BFGS method. The optimally conditioned (OC) method[82] chooses π_k to minimize the condition number of the Hessian (the condition number is the ratio of the largest to the smallest eigenvalue), thereby improving the behavior of the optimization. The CG, MS and DFP methods are also special cases of the Huang family of algorithms.[84] Equations (7)–(10) can also be used to update the Hessian, $\mathbf{B}$, rather than its inverse, $\mathbf{H}$, provided that Δx and Δg are interchanged when $\mathbf{H}$ is replaced by $\mathbf{B}$.[74]

All of the updating formulas have been devised to assure that the Hessian remains symmetric and positive-definite. For non-quadratic functions, the Hessian may have to be reset, if the Hessian is no longer positive-definite, or if Eq. (4) does not yield a search direction leading to a lower value of the function. Convergence can be proven for quadratic functions in $N + 1$ steps; for non-quadratic functions, more steps may be required, but convergence to a stationary point is guaranteed. In numerical tests of the various quasi-Newton methods, the BFGS and OC algorithms appear to be the methods of choice.[70]

For some algorithms, such as conjugate gradient, accurate line searches are necessary. However, for most quasi-Newton methods the line search does not have to be very accurate. Reduction of the directed gradient in the line search by a factor of 0.5 or 0.1 is often sufficient.[70] Frequently this is satisfied by fixing $\alpha = 1$ (i.e no line search). A few more quasi-Newton steps may be needed, but the number of function evaluations in the line search is greatly reduced. The result is a considerable savings in the total number of function and gradient evaluations needed to reach the minimum. On the other hand, if computation of the gradient is expensive (e.g. numerical gradient for a function of many variables), there is some advantage to accurate line searches using only the function value to reduce the number of costly gradient evaluations.

The line searches can be improved with little additional cost by fitting a function to the energy and gradient at $\alpha = 0$ (i.e. $\mathbf{x}_k$) and $\alpha = 1$, and interpolating the energy and gradient at the minimum. A cubic polynomial,[85] a quartic polynomial constrained to have only one minimum[86] and a conic[87] have been used for this purpose. Alternatively, the one-dimensional search need not be along a linear path. McKelvey and Hamilton[88] have suggested searching along a path $\mathbf{p}_k(s)$ given by

$$\frac{d\mathbf{p}_k(s)}{ds} = \mathbf{g}_k + \mathbf{B}_k \mathbf{p}_k(s) \qquad \mathbf{p}_k(s) = \left[\int_0^s \exp(\mathbf{B}_k \sigma)\, d\sigma \right] \mathbf{g}_k \qquad (11)$$

At $s \to 0$, this path follows the steepest-descent direction; at large negative s, the limit is Newton's method.

Two other gradient methods that have been used for geometry optimization do not fall into the schemes outlined above. The algorithm proposed by Császár and Pulay[89] uses a fixed Hessian with a k-dimensional search at the kth iteration. The best linear combination of the current and previous geometries is chosen

$$\mathbf{x}_{k+1} = \sum_{i=0}^{k} c_i \mathbf{x}_i \qquad \sum_{i=0}^{k} c_i = 1 \qquad (12)$$

such that the predicted displacement

$$\mathbf{p}_k = -\mathbf{B}^{-1} \sum_{i=0}^{k} c_i \mathbf{g}_i \qquad (13)$$

is a minimum in a least-squares sense (i.e. $\mathbf{p}_k^T \cdot \mathbf{p}_k$ is a minimum with respect to c_i). Convergence is guaranteed for any positive-definite starting Hessian and appears to be better than some of the quasi-Newton methods in selected geometry optimization problems.[89]

A method devised by Schlegel[86] has seen widespread use as a part of the GAUSSIAN system of MO programs.[30] The Hessian in the full space is updated in the smaller space spanned by the current and previous steps. The available gradients are projected into the small space and elements of the Hessian in the small space are computed according to

$$b_{ij} = \frac{(\mathbf{g}_i - \mathbf{g}_k)^T \cdot \mathbf{r}_j - \sum_{m=i+1}^{k-1} b_{mj}[(\mathbf{x}_i - \mathbf{x}_k)^T \cdot \mathbf{r}_m]}{(\mathbf{x}_i - \mathbf{x}_k)^T \cdot \mathbf{r}_i}$$

$$j \leqslant i, \quad i = k-1, k-2, \ldots \quad (14)$$

where the unit vectors for the small space are given by

$$\tilde{\mathbf{r}}_i = (\mathbf{x}_i - \mathbf{x}_k) - \sum_{m=i+1}^{k-1} \mathbf{r}_m[(\mathbf{x}_i - \mathbf{x}_k)^T \cdot \mathbf{r}_m] \quad (15)$$

$$\mathbf{r}_i = \tilde{\mathbf{r}}_i/|\tilde{\mathbf{r}}_i| \quad i = k-1, k-2, \ldots$$

The correction to the Hessian in the full space is

$$\mathbf{B}_k = \mathbf{B}_{k-1} + \sum_{i \leqslant j} [b_{ij} - \mathbf{r}_i^T \mathbf{B}_{k-1} \mathbf{r}_j][\mathbf{r}_i \mathbf{r}_j^T - (1 - \delta_{ij}) \mathbf{r}_j \mathbf{r}_i^T] \quad (16)$$

At step k, this results in a rank k update to the Hessian, as opposed to the rank 1 and 2 formulas used in the quasi-Newton methods, leading to improved convergence properties. The line search is avoided by using a constrained quartic polynomial to estimate the position of the minimum. The energy and gradient at the line search minimum are obtained by interpolation rather than by recalculation.

2. Estimating the Hessian Matrix

An initial estimate for the Hessian (or its inverse) is required for most gradient methods. The estimate need not be very accurate (e.g. a unit matrix), since the Hessian is updated during the search for the minimum, and gradually approaches the correct second derivative matrix. However, the overall efficiency of the optimization and the rate of convergence to the equilibrium geometry can be improved considerably by a good initial estimate of the Hessian (and the starting geometry). Unlike the general functional minimization problem, the task of geometry optimization is simplified by a hierarchy of theoretical methods that give increasingly more accurate descriptions of the energy surface. A lower level of theory can be used to provide a suitable, inexpensive initial estimate of the Hessian (and the starting geometry) for optimization at a higher level of theory. Several options of increasing complexity (and cost) can be considered.

1. The unit matrix. Although this is an unbiased choice, all the useful structural information about the molecule is discarded, i.e. the nature of the atoms, the bonds between them, etc. Flexible coordinates (i.e. torsion and ring deformation) are not distinguished from stiff modes (i.e. bond stretching) and all coupling between coordinates is ignored. Such information must be accumulated during the course of the optimization at the expense of additional optimization steps. This is particularly detrimental for cyclic molecules whose coordinates are inherently strongly coupled.

2. An empirical guess. For bond stretches, the diagonal elements of the second derivative matrix can be obtained by relating bond length to stretching force constant using Badger's rule;[90] for angle bends and torsions a constant is sufficient. For cyclic molecules, a simple valence force field in redundant internal coordinates, transformed to the non-redundant coordinates used for the optimization, can provide the most important off-diagonal elements.[22] The empirical force fields used in spectroscopy and molecular mechanics are only of limited value for estimating the Hessian, since the parametrizations are usually too specialized to cover the range of molecules encountered in *ab initio* geometry optimizations.

3. A Hessian computed by semi-empirical MO methods. Some empirical adjustment of the second derivatives is usually necessary, since the semi-empirical methods tend to overestimate some terms and underestimate others.

4. Numerical calculation of a few of the more important elements of the second derivate matrix from energy or energy and gradient calculations at small displacements from the initial geometry.

5. An analytically computed second derivative matrix from a lower level of *ab initio* calculation (e.g. from a smaller basis calculation or, in the case of an optimization including correlation energy, from an SCF calculation). Similarly, the approximate Hessian from an optimization at a lower level can provide a good initial estimate of the Hessian.

6. The full second derivative matrix calculated analytically or numerically. This is the most costly and most accurate option, but may be the method of choice for difficult cases.

C. Second Derivatives

If the second derivatives are available, the minimum for a quadratic surface can be found in one step by Newton's method:[70]

$$\mathbf{x}_{min} = \mathbf{x}_k - \mathbf{B}_k^{-1}\mathbf{g}_k = \mathbf{x}_k - \mathbf{H}_k\mathbf{g}_k \qquad (17)$$

For non-quadratic surfaces, more than one step will be needed, but often only the gradient and not the second derivative matrix has to be recalculated on subsequent steps. If necessary, the second derivative matrix can be recalculated every few steps, or can be updated as in the gradient optimization algorithms. The latter is identical to using the exact second derivative matrix as the starting estimate of the Hessian in the gradient method. If second derivatives are not available (e.g. at the MP2 level), those calculated at a lower level of theory (e.g. at the SCF level) may be sufficient to assure rapid convergence to the optimized geometry. For a well behaved minimization, the Hessian must be positive-definite and not be ill-conditioned (i.e. not have a very wide spread in eigenvalues). If these conditions are not fulfilled, the problem can be remedied by adding a constant times the unit matrix to the Hessian (cf. level shifting to solve SCF convergence problems) or by adding a constant to the offending eigenvalue(s).

One of the main drawbacks to optimization methods based on second derivatives is that analytical second derivatives are available only for a few levels of theory, as discussed above. Furthermore, the computer time required for analytical second derivatives can be 5–10 to N times as long as a gradient calculation. Since algorithms using first derivatives often converge in about N steps, there may be no significant cost advantage to using second derivative based algorithms. However, analytical second derivatives may be more cost effective when there are convergence difficulties in the optimization, such as for shallow wells, strongly coupled coordinates and narrow curved valleys, or when second derivatives are only a few times as expensive as the energy and gradient evaluations (e.g. MCSCF calculations).

V. LOCATING TRANSITION STRUCTURES

Both minima and saddle points are stationary points characterized by a zero gradient. However, unlike a minimum, a first-order saddle point must be a maximum along one (and only one) direction. In general, this direction is not known in advance and must be determined during the course of the optimization. Numerous algorithms have been proposed to deal with the problem of locating transition structures.[91–93] In this section a few of the more widely used methods are surveyed. The algebraic details can be found in the original literature.

A. Surface Fitting

The simplest approach to locating a transition structure is to fit a suitable analytical expression to a set of computed energies near the transition structure. The analytical derivatives of the fitted function can then be used to locate the transition structure on the fitted surface, without carrying out additional and expensive molecular-orbital calculations. This method is feasible for transition structures in small molecules.[78] However, surface fitting has several disadvantages. There are no standard, universally applicable methods for fitting multi-dimensional non-quadratic surfaces; each family of reactions is a special case if a large region of the surface is needed. A large number of energy (or energy and gradient) calculations are required to obtain an acceptable fit for the model energy surface in more than two or three variables. Finally, the fitted surface may not be sufficiently good in the region of the barrier to yield an accurate estimate of the geometry of the transition state. However, there are circumstances when the construction of an analytical energy surface may be necessary. An accurate fit to a large portion of the energy surface is needed for the study of reaction dynamics by classical or semiclassical trajectory calculations.[94] Smaller regions near the transition structure are required for variational transition state theory[15] and reaction path Hamiltonian methods.[16]

B. Linear and Quadratic Synchronous Transit

The linear synchronous transit (LST) and quadratic synchronous transit (QST) methods proposed by Halgren and Lipscomb[95] simplify the problem of finding the transition state by making some assumptions about the reaction path. The LST approach assumes the path is a straight line connecting reactants and products. The LST estimate of the transition state, structure 1 in Fig. 4, is the energy maximum along the straight-line path. The energy of 1 is an upper bound to the true saddle point energy, often a rather poor bound if the actual path is curved. The estimate of the transition structure can be improved by minimizing the energy with respect to all of the coordinates perpendicular to the linear reaction path. The resulting point, 2, is a lower bound to the true saddle point. The reaction path can now be approximated by a parabolic path between reactants and products, i.e. a quadratic synchronous transit (QST) path.[95] The maximum on the quadratic synchronous path, 3, yields a much better estimate of the energy and position of the transition state. Under less favorable circumstances (examples are discussed elsewhere[92,93]), the process of minimizing perpendicular to the path and finding a maximum along the quadratic transit path must be repeated a number of times before satisfactory convergence is obtained. If the actual reaction path is very curved, the QST may also be a poor approximation to the true path between reactants and products, and the method may not converge to a satisfactory structure. In part this can be remedied if structures closer to the transition state are used as end points of the LST and QST paths.[95] A number of algorithms have been devised[93,96-100] that alternate between maximizing along a path of predetermined, constrained form and minimizing perpendicular to the path. Methods

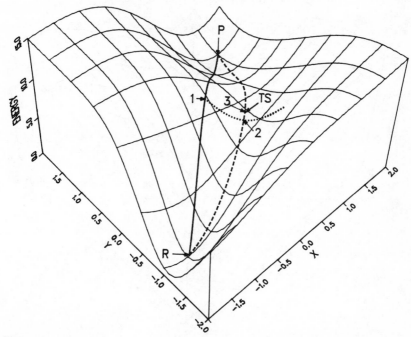

Fig. 4. The linear synchronous transit (LST) and quadratic synchronous transit (QST) methods for finding transition structures: R, reactants; P, products; TS, true transition structure; 1, maximum on LST path (full curve); 2, minimum perpendicular to LST path; 3, maximum on QST path (broken curve). The model surface is constructed from two Gaussians.

using gradient algorithms for the minimization step are considerably more efficient than non-gradient methods.

C. Coordinate Driving

For many reactions, the transformation from reactants to products is dominated by a change in one coordinate. A series of points along the reaction path can be obtained by stepping along the dominant coordinate and minimizing the energy with respect to the remaining $N - 1$ coordinates (Fig. 5). In favorable cases,[101] the path generated passes through the transition structure. Provided the step size is sufficiently small near the saddle point, the energy maximum along the path can be a good approximation to the transition structure. The coordinate driving method can be considered an extension of the QST method because the path is not constrained to be quadratic. However, coordinate driving is rather costly, since an optimization of $N - 1$ coordinates is required for each step taken. Furthermore, if the path is too curved and becomes dominated by one of the coordinates being

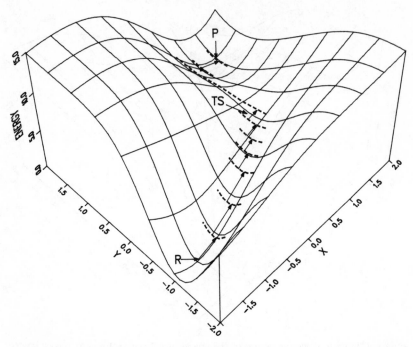

Fig. 5. The coordinate driving method for locating transition structures. A series of points is generated by stepping one coordinate (X) and optimizing the other (Y): R, reactants; P, products; TS, transition structure.

minimized, the method can encounter serious difficulties.[102] If the transition structure is reached, as in Fig. 5, there can be an abrupt change in the geometry just beyond the transition structure, because the minimization is now being carried out along the reaction path rather than perpendicular to it. In other cases (Fig. 6), coordinate driving may miss the transition structure entirely, if the saddle point is off to the side of the main valley. Sometimes (but not always) these problems can be overcome by starting from the products instead of the reactants, and/or using a different driving coordinate.

D. Hill Climbing or Walking Up Valleys

The difficulties encountered by the coordinate driving method when the reaction path is strongly curved can be avoided with a simple change in strategy that does not add appreciably to the computational effort. Instead of a step along a specific coordinate, a step of fixed length is taken in the direction that gives the easiest (least steep) path up the valley,[103-108] as shown in Fig. 7. Near the saddle point the step size can be reduced to locate the stationary point more accurately. Care must be taken that the step length does not

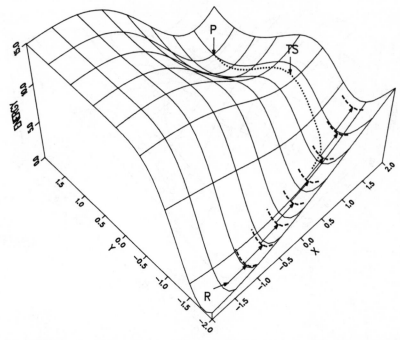

Fig. 6. Failure of coordinate driving method due to a long valley not leading to the transition structure. The dotted curve is the steepest-descent path from the saddle point.

become smaller than the radius of curvature of the contour lines crossing the reaction path (otherwise there will be no minimum-energy uphill step). Alternatively, the step size can be changed similar to a binary search to find the transition state in fewer steps.[103] Like coordinate driving, this algorithm is costly since numerous steps may be needed to approach the transition structure, and each step requires the optimization of $N - 1$ variables. Some applications of this method rely only on the energy[103] or the energy and gradients.[104,105] Other implementations[106-108] calculate the second derivative matrix at each step or every few steps in order to define the uphill path. The direction of shallowest ascent is given by the eigenvector of the Hessian with the smallest eigenvalue. The frequent calculation of the second derivative matrix makes the latter implementation very costly and currently limits it to SCF energy surfaces. In many ways, walking up a valley to the transition structure is the converse of finding the path of steepest descent from the transition structure to the reactants. However, unlike the steepest-descent path, which is guaranteed to find a minimum, a shallowest-ascent path need not find the desired transition state. Like the coordinate driving method, walking up valleys can bypass the transition structure for some surfaces, such as the one shown in Fig. 6.

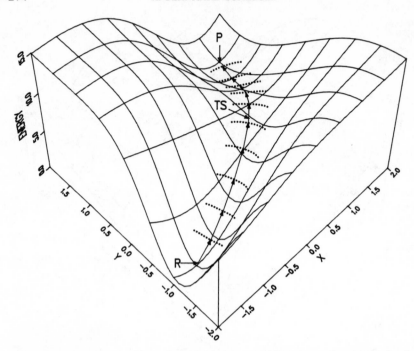

Fig. 7. Hill climbing or walking up valleys to locate transition structures. A series of points is generated by taking a fixed length step and finding the shallowest path uphill: R, reactants; P, products; TS, transition structure.

E. Gradient Norm Method

The gradient at a saddle point is zero, but standard gradient minimization methods cannot be used because the Hessian is not positive-definite. However, a saddle point optimization can be changed into a minimization by recognizing that the norm of the gradient is a minimum at any stationary point (minimum, saddle point or local maximum). In the method proposed by McIver and Komornicki,[109−111] a saddle point is found by minimizing the square of the gradient norm, $|g_k^T \cdot g_k|$. This is a non-linear least-squares problem, and is usually solved using Powell's algorithm[112] or a similar method. The optimization converges directly to the transition structure, provided that the starting point is near the saddle point. At the saddle point the gradient norm must be zero, not just a minimum. As shown in Fig. 8, the gradient norm can be a minimum at a shoulder, in addition to being zero at a stationary point. Care must be taken to avoid converging on such areas of the reaction surface. Figure 9 illustrates the gradient norm squared for the surface used in Figs. 4, 5 and 7. As can be seen, a very good guess for the transition structure may be needed for the gradient norm algorithm to converge to the

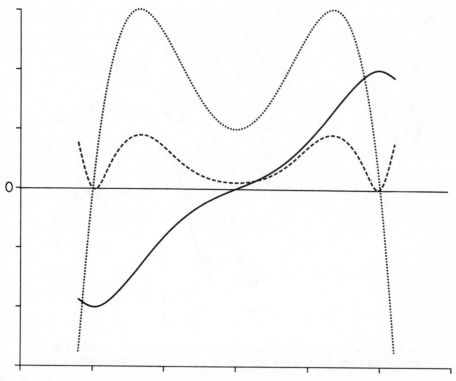

Fig. 8. A potential energy curve (full curve), its gradient (dotted curve) and its gradient norm squared (broken curve). The gradient is zero at the minimum and maximum. The gradient norm squared is a minimum not only at the potential energy minimum and maximum but also at the shoulder.

saddle point, rather than to some other feature on the surface. Furthermore, the region where the Hessian for the gradient norm is positive-definite is smaller than the quadratic region of the saddle point where the Hessian for the energy has one negative eigenvalue.

F. Gradient Algorithms

Gradient methods are very efficient for minimizations; however, to locate transition structures they must be modified or constrained to overcome the problems caused by the Hessian not being positive-definite. One approach is to partition the N-dimensional optimization into a one-dimensional space for maximization, and an $(N - 1)$-dimensional space for minimization.[93,96,99,100] This partitioning in effect chooses a transition vector. The transition vector may be fixed, or may be allowed to vary in a restricted manner (e.g. according to a quadratic synchronous transit path). The search for a maximum in the

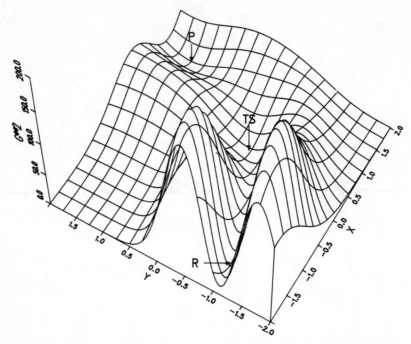

Fig. 9. The gradient norm squared for the model potential energy surface used in Figs. 4, 5 and 7. The gradient norm squared is a local minimum in many places in addition to the minima and saddle point on the potential.

one-dimensional space causes no problem. Minimization in the $(N-1)$-dimensional space can be carried out efficiently using gradient methods as discussed above. The maximization and minimization can be done simultaneously or alternately, but the final structure must be a stationary point in the full space. A drawback of this approach is that the partitioning must be chosen *a priori*. If the one-dimensional space does not contain a significant component of the true transition vector, the energy surface may not have a maximum in the one-dimensional space after several optimization steps.

An alternative approach[86] allows the transition vector to vary without constraint during the optimization. The Hessian is updated as in the minimization procedures, but must be tested to ensure that it has one and only one negative eigenvalue. If not, the offending eigenvalue is changed and the Hessian reconstructed from the eigenvectors and the modified eigenvalues. The initial Hessian must have one negative eigenvalue. An approximate Hessian can be obtained by computing a few rows of the Hessian numerically for the variables expected to dominate the transition vector, or by computing the Hessian at a lower level of theory (options 4 and 5, respectively, in Section IV.B.2).

Gradient methods for optimizing transition structures appear to have wider ranges of convergence than the gradient norm algorithm, do not have the problem of converging onto shoulders, and are considerably more efficient. However, for reliable convergence the starting geometry must be in the quadratic region of the transition state, i.e. where the Hessian has one negative eigenvalue. This may be a significant problem if the barrier is very narrow or the region of the transition state is very flat.

Gradient methods discussed above use a quadratic function (energy, gradient and approximate Hessian) to model the energy surface near the transition state. Distance-weighted interpolants[113] provide a more flexible functional form that can interpolate arbitrarily spaced points with a smooth differentiable function. For a gradient-based optimization, the Shepard interpolation functions[114] seem appropriate

$$E(\mathbf{x}) = \sum W_i(E_i + \mathbf{g}_i^{\mathrm{T}} \cdot (\mathbf{x} - \mathbf{x}_i) + \tfrac{1}{2}(\mathbf{x} - \mathbf{x}_i)^{\mathrm{T}} \mathbf{B}_k(\mathbf{x} - \mathbf{x}_i)) / \sum W_i$$
$$W_i = 1/|\mathbf{x} - \mathbf{x}_i|^2 \tag{18}$$

although other related forms can be constructed. The energies and gradients are interpolated exactly with this function, even for non-quadratic surfaces. The Hessian, $\mathbf{B}_k$, is updated in a manner analogous to the gradient methods. In preliminary applications,[115] this method appears to have a wider range of convergence than the above gradient algorithms. The starting guess for the transition state geometry need not be one structure, but can be a set of structures bracketing the transition state.

G. Second Derivatives

If second derivatives of the energy are available, then Newton's method, Eq. (17), can be used to locate saddle points in a manner analogous to finding minima. However, the Hessian must be tested at each step to insure that it has one and only one negative eigenvalue. For the method to converge without difficulties, the starting structure must be in the quadratic region of the transition state. If the Hessian has the wrong number of negative eigenvalues, either at the starting structure or at any point during optimization, then the Hessian must be adjusted so that it has one negative eigenvalue. This may be done by changing the sign of the offending eigenvalue (usually the smallest one) or by adding/subtracting a constant. The adjusted Hessian is then reconstructed from the eigenvectors and the modified eigenvalues. If the Hessian is initially positive-definite and the smallest eigenvalue is made negative, the result is the same as the algorithm for walking up valleys based on the computation of second derivatives, discussed above. As already cautioned, such procedures can get lost in dead-end valleys. Although the computation of second derivatives is more expensive than gradients, a second derivative-based approach may be the method of choice for difficult problems. Similar to minimization, it may not be necessary to recalculate the second derivatives at every step (the Hessian can be left constant, updated, or recomputed every few steps).

H. Testing Stationary Points

Except for relatively simple surfaces, it is usually necessary to verify that the stationary point found in a transition structure optimization is indeed a first-order saddle point and not a minimum or a higher-order saddle point. This can be done by calculating the full Hessian (analytically or numerically) at the stationary point and checking that there is one and only one negative eigenvalue. Note that the approximate Hessian obtained by an updating procedure in a gradient optimization on a non-quadratic surface is not sufficiently accurate to test the nature of the stationary point. Furthermore, it contains no information about displacements that lead to lower-symmetry, and possibly lower-energy, saddle points. If the full Hessian has a second negative eigenvalue, then one of these two eigenvectors is not a suitable transition vector for the reaction. A lower-energy saddle point can be obtained by displacing along this eigenvector and re-optimizing the geometry.

Once a saddle point has been found and characterized by computing the Hessian, it must still be tested to determine that it is the transition structure for the desired reaction. The transition vector may indicate clearly enough that the reaction path connects the desired reactants and products. If not, it may be necessary to follow the reaction path down from the saddle point (perhaps only for a short distance) to be sure that the path leads to the correct reactants and products.

VI. REACTION PATH FOLLOWING

At the saddle point, the intrinsic reaction coordinate (IRC) in mass-weighted Cartesian coordinates is coincident with the transition vector (i.e. the eigenvector of the Hessian with the negative eigenvalue).[13] On either side of the saddle point, the IRC in mass-weighted Cartesian coordinates is the steepest-descent path from the saddle point to the reactants and to the products.[13] However, geometry optimizations are often more convenient in internal coordinates, and the steepest-descent path in internal coordinates is not the IRC. To obtain the IRC in internal coordinates, the transformation from mass-weighted Cartesian coordinates to internal coordinates must be taken into account.[13,116] The descent direction ($-\mathbf{g}$ in mass-weighted Cartesian coordinates) is then given by $-\mathbf{Gg}$, where $\mathbf{G}$ is Wilson's G matrix (note that $G_{ij} = \delta_{ij}$ for mass-weighted Cartesian coordinates and $G_{ij} = m_i^{-1} d_{ij}$ for Cartesian coordinates, where m_i are the masses of the atoms). In the formulas below, **g** must be replaced by **Gg** if mass-weighted Cartesian coordinates are not used, and the transition vector must be replaced by the normal coordinate for the vibration with imaginary frequency.

The simplest approach for following the IRC is just to take a series of small steps in the descent direction:[117]

$$\mathbf{x}_{k+1}^{IRC} - \mathbf{x}_k^{IRC} = -s\mathbf{g}_k/|\mathbf{g}_k| \tag{19}$$

Unfortunately, except for very small step sizes, this path quickly departs from the IRC and tends to oscillate about it. Equation (19) is just a first

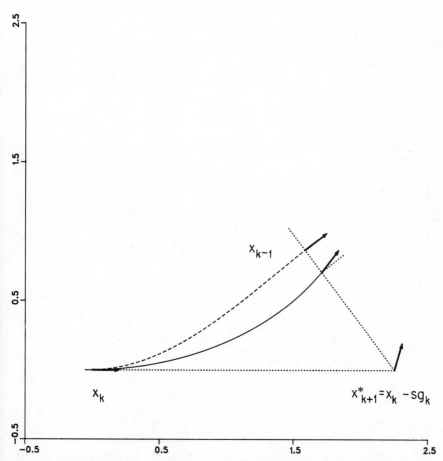

Fig. 10. Reaction path-following algorithm of Ishida, Morokuma and Komornicki (true path (full curve), approximate path (broken curve), vectors point downhill, indicating the orientation of $-\mathbf{g}$). From $\mathbf{x}_k$, a point on the reaction path, a step is taken along $-\mathbf{g}_k$, yielding position $\mathbf{x}_{k+1}^*$ with gradient $\mathbf{g}_{k+1}^*$. Minimization from $\mathbf{x}_{k+1}^*$ along the direction $\mathbf{d}_k = \mathbf{g}_{k+1}^*/|\mathbf{g}_{k+1}^*| - \mathbf{g}_k/|\mathbf{g}_k|$ results in $\mathbf{x}_{k+1}$, the approximation to the next point on the reaction path.

approximation to the solution of the differential equation for the intrinsic reaction path:[13]

$$dx^{IRC}/ds = -g_k/|g_k| \tag{20}$$

A more sophisticated numerical integration of this equation, e.g. a predictor–corrector method, can lead to a much improved reaction path.[118]

An algorithm proposed by Ishida, Morokuma and Komornicki[119] improves on Eq. (19) by adding a one-dimensional search to return the predicted point closer to the IRC (Fig. 10). A step of length s in the direction of $-g_k$

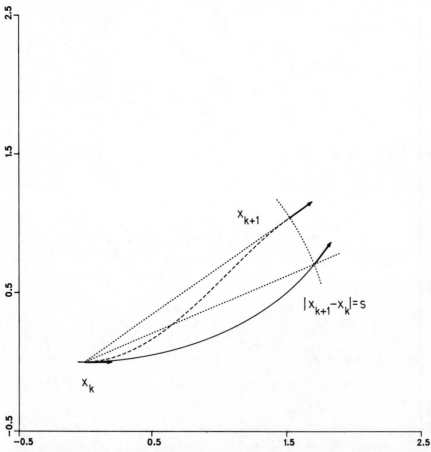

Fig. 11. Reaction path-following algorithm of Müller and Brown (true path (full curve), approximate path (broken curve), vectors point downhill, indicating the orientation of $-g$). From x_k, a point on the reaction path, a step of length s is taken in the downhill direction. The approximation to the next point on the path, x_{k+1}, is found by minimizing the energy with x_{k+1} constrained to be on the hypersphere with radius s centered at x_k.

yields point x_{k+1}^*. A minimization along d_k, the bisector of the angle between $-g_k$ and g_{k+1}^*, results in x_{k+1}, an approximation to the next point on the reaction path. Schmidt, Gordon and Dupuis[120] avoid an explicit search along d_k by computing only one additional point on d_k and using a parabolic fit to interpolate to the minimum. A step size of $|\Delta x/g| = 0.15$ bohr amu$^{1/2}$ was found to be reasonable and 50–200 energy or energy and gradient calculations were typically needed to map out the IRC for small molecules.[119,120]

In the method of Müller and Brown,[103] a point on the steepest-descent path is found by taking a step of fixed length and minimizing the energy with respect

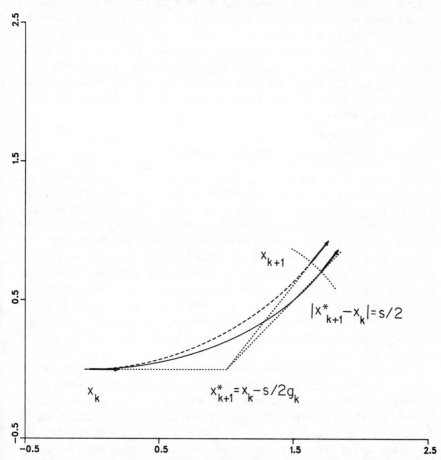

Fig. 12. Reaction path-following algorithm of Schlegel (true path (full curve), approximate path broken curve), vectors point downhill, indicating the orientation of $-g$). From x_k, a point on the reaction path, a step of length $\frac{1}{2}s$. is taken along $-g_k$ yielding x_{k+1}^* (no energy or gradient calculation at x_{k+1}^*). The approximation to the next point on the path, x_{k+1}, is found by minimizing the energy with x_{k+1} constrained to be on the hypersphere with radius $\frac{1}{2}s$ centered at x_{k+1}^*.

to the remaining $N - 1$ degrees of freedom, as shown in Fig. 11. The optimized point x_{k+1} lies on a hypersphere centered at x_k, and the residual gradient, g_{k+1}, is normal to the sphere (i.e. parallel to $x_{k+1} - x_k$). Although each step requires an $(N - 1)$-dimensional optimization, a much larger step size can be used than in the methods described above (typically 10 steps from the saddle point to a minimum). For large step sizes and tightly curved reaction paths, this method has a tendency to underestimate the curvature.[103] This is because the residual gradient is parallel to the straight-line path between x_k and x_{k+1}, rather than tangent to the curved path.

A refinement by Schlegel[121] improves the ability of this approach to follow a curved reaction path, as shown in Fig. 12. The point x_{k+1} is chosen so that the reaction path between x_k and x_{k+1} is an arc of a circle and so that the gradients g_k and g_{k+1} are tangent to this path. Like the Müller and Brown method, this algorithm requires an $(N - 1)$-dimensional optimization on a hypersphere, but about the point $x_{k+1}^* = x_k - \frac{1}{2}sg_k/|g_k|$ rather than x_k. For similar step sizes, this algorithm follows an IRC with large curvature more closely than the Müller and Brown method.

VII. SUMMARY

This survey of methods as well as the accumulated experience by a number of groups suggest several general statements about geometry optimization. Full geometry optimization should be carried out, if possible, rather than partial optimization, and the stationary points found should be characterized by computing the Hessian. Both of these points are particularly important for transition states. For minima and for saddle points, gradient methods appear to be the most efficient techniques for optimization, especially if analytical gradients are available. Second derivative methods are more suitable for very difficult optimization problems, or if analytical second derivatives are relatively inexpensive. Methods for following intrinsic reactions paths have not been tested as widely and it is not yet possible to recommend one method over another.

Beyond the choice of optimization method, the rate of convergence to the stationary point depends significantly on the choice of coordinate system, the starting coordinates and the initial estimate of the Hessian. Coordinate systems (usually internal coordinates) should avoid strong coupling, especially between stiff and flexible modes, and between variables dominating the transition vector and other coordinates. Good starting geometries can be obtained from the structures of similar molecules found in various bibliographies, or from lower-level calculations. Suitable initial estimates of the Hessian can be calculated by empirical means, by semi-empirical MO methods or by *ab initio* MO methods. For transition structure optimizations, an initial estimate of the transition vector is also needed, either to partition the space of variables to be optimized, or to insure that the Hessian has a suitable eigenvector with a negative eigenvalue.

References

1. Richards, W. G., Walker, T. E. H., and Hinkley, R. K., *A Bibliography of ab initio Molecular Wavefunctions*, Oxford University Press, Oxford, 1971, and supplements for 1970–73, 1974–77, 1978–80.
2. Whiteside, R. A., Frisch M. J., and Pople, J. A., *The Carnegie–Mellon Quantum Chemistry Archive*, 3rd Ed, Carnegie–Mellon University, Pittsburgh, 1983.
3. Ohno, K., and Morokuma, K., *Quantum Chemistry Literature Data Base*, Elsevier, Amsterdam, 1982; yearly supplements published in special issues of the journal *Theochem*.
4. Bartož, S., *Colloq. Int. CNRS*, **82**, 287 (1958).
5. Bishop, D. M., and Randic, M., *J. Chem. Phys.*, **44**, 2480 (1966).
6. Gerratt, J., and Mills, I. M., *J. Chem. Phys.*, **49**, 1719 (1968).
7. Pulay, P., *Mol. Phys.*, **17**, 197 (1969).
8. Fogarasi, G., and Pulay, P., *Annu. Rev. Phys. Chem.*, **35**, 191 (1984); Pulay, P., in *The Force Concept in Chemistry* (Ed. B. M. Deb), p. 449, Van Nostrand, New York, 1981; Pulay, P., in *Applications of Electronic Structure Theory* (Ed. H. F. Schaefer III), p. 153, Plenum,, New York, 1977.
9. Gaw, J. F., and Handy, N. C., *Annu. Rep. Prog. Chem. Sec. C*, **81**, 291 (1985).
10. Schlegel, H. B., in *Computational Theoretical Organic Chemistry* (Eds. I. G. Csizmadia, and R. Daudel), p. 129, Reidel, Dordrecht, 1981.
11. Jørgensen, P., and Simons, J. (Eds.), *Geometrical Derivatives of Energy Surfaces and Molecular Properties*, Reidel, Dordrecht, 1986.
12. Pulay, P., *Adv. Chem. Phys.*, (1986).
13. Fukui, K., *Acc. Chem. Res.*, **14**, 363 (1981).
14. Tachibana, A., and Fukui, K., *Theor. Chim. Acta*, **51**, 275 (1979).
15. Truhlar, D. G., and Garrett, B. C., *Acc. Chem. Res.*, **13**, 440 (1980).
16. Miller, W. H., Handy, N. C., and Adams, J. E., *J. Chem. Phys.*, **72**, 99 (1980).
17. For leading references see Mezey, P. G., *Theor. Chim. Acta*, **62**, 133 (1982).
18. Longuet-Higgins, H. C., *Mol. Phys.*, **6**, 445 (1963); for a recent review see Frei, H., Bauder, A., and Gunthard, H. H., *Top. Curr. Chem.*, **81**, 1 (1979).
19. Klemperer, W. G., *J. Chem. Phys.*, **56**, 5475 (1972); for recent references. see McLarnan, T. J., *Theor. Chim. Acta*, **63**, 195 (1983).
20. Murrell, J. H., and Laidler, K. J., *Trans. Faraday Soc.*, **64**, 371 (1968); Murrell, J. H., and Pratt, G. L., *Trans. Faraday Soc.*, **66**, 1680 (1970).
21. Stanton, R. E., and McIver, Jr, J. W., *J. Am. Chem. Soc.*, **97**, 3632 (1975).
22. Schlegel, H. B., *Theor. Chim. Acta*, **66**, 333 (1984).
23. For a recent review see Hegarty, D., and van der Velde, G., *Int. J. Quantum Chem.*, **23**, 1135 (1983).
24. Saunders, V. R., in *Methods in Computational Molecular Physics*, (Eds. G. H. F., Diercksen, and S. Wilson), p. 1, Reidel, Dordrecht, 1983.
25. Schlegel, H. B., *J. Chem. Phys.*, **77**, 3676 (1982).
26. Schlegel, H. B., Binkley, J. S., and Pople, J. A., *J. Chem. Phys.*, **80**, 1976 (1984).
27. Dupuis, M., and King, H. F., *J. Chem. Phys.*, **68**, 3998 (1978).
28. Takada, T., Dupuis, M., and King, H. F., *J. Chem. Phys.*, **75**, 332 (1981).
29. A partial list of integral derivative codes is included in section of the programs

described is references 30–37 and 60 (these codes have also been adapted for numerous other molecular orbital programs).

30. Binkley, J. S., Whiteside, R. A., Krishnan, R., Seeger, R., DeFrees, D. J., Schlegel, H. B., Topiol, S., Kahn, L. R., and Pople, J. A., GAUSSIAN 80, *Quantum Chem. Program Exch.*, **13**, 406 (1981).
31. Dupuis, M., Rys, J., and King, H. F., HONDO 5, *Quantum Chem. Program Exch.*, **13**, 401, 403 (1981).
32. Komornicki, A., GRADSCF, *NRCC Program* QH04, 1980.
33. Meyer, W., and Pulay, P., MOLPRO, Munich and Stuttgart, Germany, 1969.
34. Pulay, P., TEXAS, *Theor. Chim. Acta*, **50**, 299 (1979).
35. Huber, H., Čársky, P., and Zahradnik, R., POLYGRAD, *Theor. Chim. Acta*, **41**, 217 (1976).
36. Schlegel, H. B., FORCE/DRVEXP, *Quantum Chem. Program Exch.*, **13**, 427 (1981).
37. Amos, R. D., CADPAC, *SERC Daresburg Lab. Publication CCP1/84/4* (1984).
38. Kahn, L. R., *J. Chem. Phys.*, **75**, 3962 (1981).
39. Vincent, M. A., Saxe, P., and Schaefer, III, H. F., *Chem. Phys. Lett.*, **94**, 351 (1983); Vincent, M. A., and Schaefer, III, H. F., *Theor. Chim. Acta*, **64**, 21 (1983).
40. Page, M., Saxe, P., Adams, G. F., and Lengsfield, B. H., *Chem. Phys. Lett.*, **104**, 587 (1984); Banerjee, A., Jensen, J. O., and Simons, J., *J. Chem. Phys.*, **82**, 4566 (1985).
41. Kato, S., and Morokuma, K., *Chem. Phys. Lett.*, **65**, 19 (1979).
42. Goddard, J. D., Handy, N. C., and Schaefer, III, H. F., *J. Chem. Phys.*, **71**, 1525 (1979).
43. Dupuis, M., *J. Chem. Phys.*, **74**, 5758 (1981).
44. Schlegel, H. B., and Robb, M. A., *Chem. Phys. Lett.*, **93**, 43 (1982).
45. Osamura, Y., Yamaguchi, Y., and Schaefer, III, H. F., *J. Chem. Phys.*, **77**, 385 (1982).
46. Pople, J. A., Krishnan, R., Schlegel, H. B., and Binkley, J. S., *Int. J. Quantum Chem. Symp.*, **13**, 225 (1979).
47. Saxe, P., Yamaguchi, Y., and Schaefer, III, H. F., *J. Chem. Phys.*, **77**, 5647 (1982).
48. Osamura, Y., Yamaguchi, Y., Saxe, P., Vincent, M. A., Gaw, J. F., and Schaefer, III, H. F., *Chem. Phys.*, **72**, 131 (1982).
49. Pulay, P., *J. Chem. Phys.*, **78**, 5043 (1983).
50. Jørgensen, P., and Simons, J., *J. Chem. Phys.*, **79**, 334 (1983).
51. Almöf, J., and Taylor, P. R., *Int. J. Quantum Chem.*, **27**, 743 (1985).
52. Yamaguchi, Y., Osamura, Y., Fitzgerald, G., and Schaefer, III, H. F., *J. Chem. Phys.*, **78**, 1607 (1983).
53. Camp, R. N., King, H. F., McIver, Jr, J. W., and Mullally, D., *J. Chem. Phys.*, **79**, 1089 (1983).
54. Hoffmann, M. R., Fox, D. J., Gaw, J. F., Osamura, Y., Yamaguchi, Y., Grev, R. S., Fitzgerald, G., Schaefer, III, H. F., Knowles, P. J., and Handy, N. C., *J. Chem. Phys.*, **80**, 2660 (1984).
55. Page, M., Saxe, P., Adams, G. F., and Lengsfield, III, B. H., *J. Chem. Phys.*, **81**, 434 (1984).
56. Banerjee, A., Jensen, J. O., Simons, J., and Shepard, R., *Chem. Phys.*, **87**, 203 (1984).
57. Binkley, J. S., unpublished.
58. Osamura, Y., Yamaguchi, Y., Saxe, P., Fox, D. J., Vincent, M. A., and Schaefer, III, H. F., *J. Mol. Struct.*, **103**, 183 (1983).
59. Simons, J., and Jørgensen, P., *J. Chem. Phys.*, **79**, 3599 (1983).
60. Gaw, J. F., Yamaguchi, Y., and Schaefer, III, H. F., *J. Chem. Phys.*, **81**, 6395 (1984).
61. Handy, N. C., and Schaefer, III, H. F., *J. Chem. Phys.*, **81**, 5031 (1984).

62. Krishnan, R., Schlegel, H. B., and Pople, J. A., *J. Chem. Phys.*, **72**, 4654 (1980).
63. Brooks, B. R., Laidig, W. D., Saxe, P., Goddard, J. D., Yamaguchi, Y., and Schaefer, III, H. F., *J. Chem. Phys.*, **72**, 4652 (1980).
64. Osamura, Y., Yamaguchi, Y., and Schaefer, III, H. F., *J. Chem. Phys.*, **75**, 2919 (1981).
65. Fitzgerald, G., Harrison, R., Ladig, W. D., and Bartlett, R. J., *J. Chem. Phys.*, **82**, 4379 (1985).
66. Fitzgerald, G., Harrison, R., Ladig, W. D., and Bartlett, R. J., *Chem. Phys. Lett.*, **117**, 433 (1985).
67. Pulay, P., *J. Mol. Struct.*, **103**, 57 (1983).
68. Fox, D. J., Osamura, Y., Hoffmann, M. R., Gaw, J. F., Fitzgerald, G., Yamaguchi, Y., and Schaefer, III, H. F., *Chem. Phys. Lett.*, **102**, 17 (1983).
69. Bartlett, R. J., Harrison, R. A., Fitzgerald, G., and Ladig, W. D., to be published.
70. For discussions for non-linear optimization methods see references 71–74 and related books.
71. Fletcher, R., *Practical Methods, of Optimization*, Wiley, Chichester, 1981.
72. Gill, P. E., Murray, W., and Wright, M. H., *Practical Optimization*, Academic Press, New York, 1981.
73. Powell, M. J. D. (Ed.), *Non-linear Optimization, 1981*, Academic Press, New York, 1982.
74. Scales, L. E., *Introduction to Non-linear Optimization*, Macmillan, Basingstoke, 1985.
75. Fletcher, R., and Powell, M. J. D., *Comput. J.*, **6**, 163 (63). Davidon, W., *Argonne National Lab. Report*, ANL-5990.
76. Binkley, J. S., *J. Chem. Phys.*, **64**, 5142 (1976).
77. Payne, P. W., *J. Chem. Phys.*, **65**, 1920 (1976).
78. Sana, M., *Int. J. Quantum Chem.*, **19**, 139 (1981); Comeau, D. C., Zellmer, R. J., and Shavitt, I., in Ref. 11.
79. Fletcher, R., and Reeves, C. M., *Comput. J.*, **7**, 149 (1964).
80. Murtagh, B. A., and Sargent, R. W. H., *Comput. J.*, **13**, 185 (1972).
81. Broyden, C. G., *J. Inst. Math. Appl.*, **6**, 76 (1970); Fletcher, R., *Comput. J.*, **13**, 317 (1970); Goldfarb, D., *Math. Comput.*, **24**, 23 (1970); Shanno, D. F., *Math. Comput.*, **24**, 647 (1970).
82. Davidon, W. C., *Math. Programming*, **9**, 1 (1975).
83. Broyden, C. G., *Math. Comput.*, **19**, 368 (1967).
84. Huang, D. Y., *J. Opt. Theor. Appl.*, **5**, 405 (1970).
85. Biggs, M. C., *J. Inst. Math. Appl.*, **12**, 337 (1973).
86. Schlegel, H. B., *J. Comput. Chem.*, **3**, 214 (1982).
87. Davidon, W. C., in Ref. 73, p. 23.
88. McKelvey, J. M., and Hamilton, Jr, J. F., *J. Chem. Phys.*, **80**, 579 (1984).
89. Császár, P., and Pulay, P., *J. Mol. Struct.*, **114**, 31 (1984).
90. Badger, R. M., *J. Chem. Phys.*, **2**, 128 (1934); *ibid.*, **3**, 227 (1935).
91. For other reviews of transition structure optimization algorithms see Refs. 92 and 93.
92. Müller, K., *Angew. Chem. Int. Edn, Engl.*, **19**, 1 (1980).
93. Bell, S., and Crighton, J. S., *J. Chem. Phys.*, **80**, 2464 (1984).
94. For leading references see Truhlar, D. G., (Ed.), *Potential Energy Surfaces and Dynamics Calculations*, Plenum, New York, 1981.
95. Halgren, T. A., and Lipscomb, W. N., *Chem. Phys. Lett.*, **49**, 225 (1977).
96. Scharfenberger, P. *J. Comput. Chem.*, **3**, 277 (1982).
97. Bálint, I., and Bán, M. I., *Theor. Chim. Acta*, **63**, 255 (1983).

98. Jensen, A., *Theor. Chim. Acta*, **63**, 269 (1983).
99. Tapia, O., and Andrés, J., *Chem. Phys. Lett.*, **109**, 471 (1984).
100. Bell, S., Crighton, J. S., and Fletcher, R., *Chem. Phys. Lett.*, **82**, 122 (1981).
101. Rothman, M. J., and Lohr, Jr, L. L., *Chem. Phys. Lett.*, **70**, 405 (1980).
102. Burkert, U., and Allinger, N. L., *J. Comput. Chem.*, **3**, 40 (1982).
103. Müller, K., and Brown, L. D., *Theor. Chim. Acta*, **53**, 75 (1979).
104. Basilevsky, M. V., and Shamov, A. G., *Chem. Phys.*, **60**, 347 (1981).
105. Dewar, M. J. S., Healy, E. F., and Stewart, J. J. P., *J. Chem. Soc. Faraday Trans. 2*, **80**, 227 (1984).
106. Cerjan, C. J., and Miller, W. H., *J. Chem. Phys.*, **75**, 2800 (1981).
107. Simons, J., Jørgensen, P., Taylor, H., and Ozment, J., *J. Phys. Chem.*, **87**, 2745 (1983).
108. Banerjee, A., Adams, N., Simons, J., and Shepard, R., *J. Phys. Chem.*, **89**, 52 (1985).
109. McIver, Jr, J. W., and Komornicki, A., *J. Am. Chem. Soc.*, **94**, 2625 (1972).
110. Poppinger, D., *Chem. Phys. Lett.*, **35**, 550 (1975).
111. Komornicki, A., Ishida, K., Morokuma, K., Ditchfield, R., and Conrad, M., *Chem. Phys. Lett.*, **45**, 595 (1977).
112. Powell, M. J. D., *Comput. J.*, **7**, 303 (1965).
113. For leading references see Franke, R., *Math. Comput.*, **38**, 181 (1982).
114. Shepard, D., *Proc ACM Natl Conf.*, p. 517 (1968).
115. Schlegel, H. B., to be published.
116. Sana, M., Reckinger, G., and Leroy, G., *Theor. Chim. Acta*, **58**, 145 (1981); Quapp, W., and Heidrich, D., *Theor. Chim. Acta*, **66**, 245 (1984).
117. Baskin, C. P., Bender, C. F., Bauschlicher, Jr, C. W., and Schaefer, III, H. F., *J. Am. Chem. Soc.*, **96**, 2709 (1974).
118. Hase, W. L., unpublished.
119. Ishida, K., Morokuma, K., and Komornicki, A., *J. Chem. Phys.*, **66**, 2153 (1977).
120. Schmidt, M. W., Gordon, M. S., and Dupuis, M., *J. Am. Chem. Soc.*, **107**, 2585 (1985).
121. Schlegel, H. B., to be published.

Ab Initio Methods in Quantum Chemistry—I
Edited by K. P. Lawley
© 1987 John Wiley & Sons Ltd.

RELATIVISTIC QUANTUM CHEMISTRY

K. BALASUBRAMANIAN*

Department of Chemistry, Arizona State University, Tempe, AZ 85287, USA

and

KENNETH S. PITZER

Department of Chemistry and Lawrence Berkeley Laboratory, University of California, Berkeley, CA 94720, USA

CONTENTS

I. INTRODUCTION

Measurements in the classical, the macroscopic, world assume the velocity of the probe (light) to be infinite and the interaction between the probe and the object of measurement to be negligible. The laws governing the motions of particles derived on the above assumptions are Newtonian and lead to classical mechanics. In non-relativistic quantum mechanics, one assumes the velocity of the probe to be infinite but allows interaction between the probe and the particle by way of Heisenberg's uncertainty principle. For most measurements on the lighter elements in the periodic table, non-relativistic quantum mechanics is sufficient, since the velocity of an electron is small compared to that of light. For the heavier elements in the periodic table (Au, Hg, Pb, Tl, etc.) the picture is entirely different. As a result of heavy nuclear

*Alfred P. Sloan fellow.

charge for the heavy atoms, the inner electrons attain such high velocities comparable to that of light that non-relativistic quantum mechanics is far from adequate.

Relativistic quantum mechanics neither assumes infinite probe velocity nor ignores the interaction between the probe and the object of measurement. Thus, the difference between non-relativistic quantum results and relativistic quantum results arises from the true velocity of light. One can define relativistic effects, in general, as the difference in the results obtained with the true velocity of light and infinite velocity of light.

Relativistic effects can be further divided into a number of categories such as the mass–velocity correction, Darwin correction, spin–orbit correction, spin–spin interaction, Breit interaction, etc.[1] The mass–velocity correction is the correction to the kinetic energy of the electron arising from the variation of its mass with velocity. One of the consequences of the finiteness of the velocity of light (as can be shown from the laws of the special theory of relativity) is the variation of the mass of a particle with its speed. This is especially significant as the speed of the particles approaches the speed of light. This variation with velocity in turn affects the kinetic energy of the particle. The spin–orbit correction arises from the strong coupling of the spin of the electron with the orbital angular momentum. This is especially large for electronic states of heavier atoms which arise from open-shell configurations. The Breit interaction is the two-electron counterpart of the spin–orbit interaction. The Darwin correction is a characteristic outcome of the Dirac relativistic equation and there does not seem to be a simple physical explanation for this effect.

Relativistic corrections make significant impact on the electronic properties of heavy atoms and molecules containing heavy atoms. The inner s orbitals are the closest to the nucleus and thus experience the high nuclear charge of the heavy atoms. Thus, the inner s orbitals shrink as a result of mass–velocity correction. This, in turn, shrinks the outer s orbitals as a result of orthogonality. Consequently, the ionization potential is also raised. The p orbitals are also shrunk by mass–velocity correction but to a lesser extent since the angular momentum keeps the electrons away from the nucleus. However, the spin–orbit interaction splits the p shells into $p_{1/2}$ and $p_{3/2}$ subshells and expands the $p_{3/2}$ subshells. The net result is that the mass–velocity and spin–orbit interactions tend to cancel for the $p_{3/2}$ shell but reinforce for the $p_{1/2}$.

Spin–orbit interaction plays an important role in the electronic and spectroscopic properties of states arising from open-shell electronic configurations. Thus, the contribution of spin–orbit interaction to the ground state of Au is small but to Pb is large. Spin–orbit interaction not only splits the electronic state into substates but mixed states which would not mix in the absence of spin–orbit interaction. This is well illustrated by the lead atom. The ground state of the Pb atom would be ^3P in the absence of spin–orbit interaction. In addition, the ^1D and ^1S states arise from the $6p^2$ electronic configuration. However, the spin–orbit interaction splits the ^3P state into $^3P_0$, $^3P_1$ and $^3P_2$ states. The $^3P_0$–$^3P_1$, $^3P_0$–$^3P_2$ splittings are large, 7819 cm^{-1} and 10 650 cm^{-1}, respectively.[2] The spin–orbit interaction also mixes $^3P_0$ with $^1S_0$ among other states. Similarly, $^3P_2$ mixes with $^1D_2$. This mixing is sometimes referred to as spin–orbit contamination and is quite large for heavy atoms. The $^3P_0$–$^3P_1$, $^3P_0$–$^3P_2$ splittings for the carbon atom are only 16 cm^{-1} and 44 cm^{-1}, respectively.[2] Thus one can see the dramatic contribution of relativistic effects for atoms and molecules containing very heavy atoms. In fact, the rare-gas compound RnF is predicted to be ionic, Rn^+F^-, based on the spin–orbit interaction of Rn^+.[3]

Spin–orbit interaction alters the spectroscopic properties of molecules containing heavy atoms to a considerable extent. Even if a molecule has a closed-shell ground state, the excited states may arise from open-shell electronic configurations, in which case the spin–orbit interaction not only splits the excited states but mixes different excited states which would not mix in the absence of spin–orbit interaction. This leads to a number of interesting features in the potential energy curves, such as shoulders, barriers, double minima, etc., which are attributed to relativistic avoided crossings. This aspect of relativistic effects is discussed in the third sections of this chapter. The color of gold is attributed to relativistic effects which split the 5d shell of gold and rise its energy. The golden color results from the 5d–Fermi level transition which contrasts its color in comparison to silver.[8] The Lamb shift, Breit interaction, etc., are more important if one is considering fine-structure calculations. They are normally ignored if one considers electronic and spectroscopic properties in the valence region.

Relativistic effects alter the chemical bonding of molecules containing heavy atoms to a considerable extent. In some cases the bond is strengthened while in other cases it is weakened. Consider the dissociation energies of a few molecules to get some insight into the effect of relativistic corrections on D_e values. The dissociation energy of Au_2 is higher than that for Ag_2, in contrast to the usual trend of a lower D_e for the heavier elements of a group in the periodic table. This anomaly is caused by relativistic contraction and stabilization of the 6s orbital of Au. On the other hand, the relativistic spin–orbit interaction weakens the bound in the case of Pb_2 by 50%. The calculated dissociation energy for Pb_2 with the inclusion of spin–orbit interaction is 50% of the value obtained with spin–orbit interaction.[4]

The lanthanide contraction (the decrease of radii from La to Lu) is usually attributed to incomplete shielding of the 4f shell. However, as pointed out by Pitzer and coworkers,[5] this effect is in part attributable to relativistic effects. If one compares the non-relativistic Hartree–Fock and Dirac–Fock results, one obtains a contribution of about 27% from relativistic effects.

In recent years a number of papers have appeared in the chemical literature dealing with relativistic effects in atoms and molecules. Relativistic atomic energies and other atomic data for heavy atoms are summarized.[6] Some of the earlier developments in this area have been reviewed by Pyykkö,[7] Pyykkö and Desclaux[8] and Pitzer.[9] A NATO conference proceedings edited by Malli[10] lists a number of papers by experts in this area on several topics pertaining to relativistic effects in atoms, molecules and solids. More recently, an issue of the *International Journal of Quantum Chemistry* was devoted to the proceedings of the Finland conference on relativistic effects.[11] Krauss and Stevens[12] have reviewed the use of effective potentials which include relativistic effective potentials. Christiansen, Ermler and Pitzer[13] have recently reviewed relativistic effects in chemical systems. The above reviews have outlined the various developments and applications in relativistic quantum chemistry. Since the appearance of the above reviews, a number of developments and new applications have emerged in this area. Earlier reviews on this topic could not cover all applications in this area in detail as a result of space limitation among other reasons. The present review emphasizes the methods and very recent applications of relativistic quantum chemistry to molecules of spectroscopic interest.

II. METHODS OF RELATIVISTIC QUANTUM CHEMISTRY

The starting point of most relativistic quantum-mechanical methods is the Dirac equation, which is the relativistic analog of the Schrödinger equation. Before Dirac's formulation, an obvious way of starting relativistic quantum mechanics would be the Einstein energy expression

$$E^2 = m_0^2 c^4 + p^2 c^2 \tag{1}$$

One could insert the appropriate quantum-mechanical operators for E and p and obtain a differential equation. The resulting differential equation, known as the Klein–Gordon equation, is

$$(\Box + m_0^2)\psi = 0 \tag{2}$$

where

$$\Box = -\left(\frac{\partial^2}{\partial x^2} + \frac{\partial^2}{\partial y^2} + \frac{\partial^2}{\partial z^2}\right) + \frac{\partial^2}{\partial t^2}$$

and where $\hbar = c = 1$. The resulting equation is covariant to the Lorentz transformation, a fundamental transformation under which all relativistic equations must be invariant. However, $\rho = \psi\psi^*$ obtained from the Klein–Gordon equation can be negative, which leads to difficulties in interpreting ρ, the conventional probability density.

Dirac discovered an equation, now well known as the Dirac equation, in an attempt to overcome the above-mentioned difficulties of the Klein–Gordon equation. The reasoning behind the derivation of this equation is that, in order to prevent the occurrence of negative charge densities, one must avoid time derivatives in the charge density and the resulting equation must be completely symmetric in the treatment of spatial and temporal coordinates. The resulting equation for a single electron in a central Coulombic field is

$$H_D \psi = E \psi \tag{3}$$

where

$$H_D = (\boldsymbol{\alpha} \cdot \mathbf{p} + c^2 \beta - z/r)$$

$$\alpha = \begin{pmatrix} 0 & \sigma_P \\ \sigma_P & 0 \end{pmatrix} \qquad \beta = \begin{pmatrix} I & 0 \\ 0 & -I \end{pmatrix}$$

where the σ_P are the 2×2 Pauli matrics and I is the 2×2 identity matrix.

The Dirac Hamiltonian for a many-electron atom can be written as

$$H_D = \sum_i h_D(i) + \sum_{i<j} \frac{1}{r_{ij}} \tag{4}$$

where $h_D(i)$ is the one-electron Dirac Hamiltonian

$$h_D(i) = \boldsymbol{\alpha}_i \cdot \mathbf{p}_i + \beta_i c^2 - z/r_i$$

Note that the above Hamiltonian ignores the two-electron relativistic Breit interaction. Introduction of the Breit interaction as a perturbation shows that it is very small in the valence region. However, the Breit interaction appears to be more important for the properties of core electrons, for which this makes a significant contribution.

Since the one-particle Dirac Hamiltonian involves 4×4 matrices instead of scalar functions and differential operators, the solution of the Dirac equation is a vector of four components. This is referred to as a four-component spinor. It takes the form

$$\psi_{nkm} = \frac{1}{r} \begin{bmatrix} P_{nk}(r) & \chi_{km}(\theta, \phi) \\ iQ_{nk}(r) & \chi_{-km}(\theta, \phi) \end{bmatrix} \tag{5}$$

where

$$\chi_{km}(\theta, \phi) = \sum_{\sigma = \pm 1/2} C(l\tfrac{1}{2}j; \quad m - \sigma, \sigma) Y_{\lambda}^{m-\sigma}(\theta, \phi) \phi_{1/2}^{\sigma}$$

$Y_{\lambda}^{m-\sigma}$ is a spherical harmonic,

$$\phi_{1/2}^{1/2} = \alpha = \begin{pmatrix} 1 \\ 0 \end{pmatrix} \qquad \phi_{1/2}^{-1/2} = \beta = \begin{pmatrix} 0 \\ 1 \end{pmatrix}$$

are the Pauli spinors, $C(l\tfrac{1}{2}j; m - \sigma, \sigma)$ are the Clebsch–Gordan coefficients, k is the relativistic quantum number, defined as

$$k = \begin{cases} j + \frac{1}{2} & \text{if } j = l - \frac{1}{2} \\ -(j - \frac{1}{2}) & j = l + \frac{1}{2} \end{cases} \tag{6}$$

and λ is defined as

$$\lambda = \begin{cases} k & \text{if } j = l - \frac{1}{2} \\ -(k + 1) & \text{if } j = l + \frac{1}{2} \end{cases}$$

The Q_{nk} are known as the small components and P_{nk} are the large components. They satisfy the following coupled differential equations for a central force field v:

$$\frac{dP_{nk}}{dr} + \frac{kP_{nk}}{r} - \left(\frac{2}{\alpha} + \alpha[v(r) - \varepsilon_{nk}] \right) Q_{nk} = 0$$

$$\frac{dQ_{nk}}{dr} - \frac{kQ_{nk}}{r} + \alpha[v(r) - \varepsilon_{nk}] P_{nk} = 0 \tag{7}$$

Thus one can solve the Dirac–Fock equation to obtain the relativistic energies and four-component spinor wavefunctions.

Desclaux[14] has developed a computer code to solve the many-electron Dirac–Rock equation for atoms in a numerical self-consistent method. In this method, the relativistic Hamiltonian is approximated within the Dirac–Fock method, ignoring the two-electron Breit interaction. The Breit interaction is introduced as a first-order perturbation to energy after self-consistency is achieved. Relativistic wavefunctions and energies calculated this way are available for a number of atoms.[6]

The non-relativistic limit $(c \to \infty)$ of the two coupled radial Dirac equations reduces to the Schrödinger equation if the small components Q_{nk} are eliminated. Thus, the small component is a measure of the magnitude of relativistic effects in these systems. While the small components make a significant contribution in the core region, the effect of these components in the valence region can be ignored. This was illustrated by a comparison of the relativistic and non-relativistic radial solutions for the 6s orbital of Pb by Lee, Ermler and Pitzer.[15] We reproduce their comparison in Fig. 1. As one can see from that figure, the effect of the small component is very small in the valence region.

The effect of small components on the properties of molecules have been studied by Schwarz[16] to a high order using the Foldy–Wouthuysen transformation. Schwarz has demonstrated that the contribution of the small components to chemical properties can be ignored. Thus one can ignore the small components if one is considering chemical properties. Examples of four-component atomic spinors are shown in the review paper by Pitzer.[9]

Relativistic calculations can also be carried out using the Pauli Hamil-

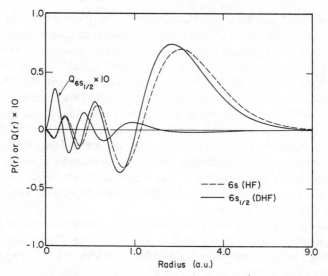

Fig. 1. Relativistic (DHF) and non-relativistic (HF) radial components of 6s wavefunctions of Pb. Only the small component of DHF is labeled.

tonian.[1] The Pauli approximation divides relativistic effects into different categories, thereby facilitating evaluation of different types of relativistic correction. The Pauli Hamiltonian can be written in the absence of a magnetic field as

$$H_{BP} = H_0 + H_D + H_{MV} + H_{s0} \qquad (8)$$

where

$$H_0 = \text{non-relativistic Hamiltonian}$$

$$H_D = +\frac{\alpha^2}{8}(\nabla^2 V) \qquad \text{(Darwin)}$$

$$H_{MV} = -\frac{\alpha^2}{8}\sum_i P_i^4 \qquad \text{(mass–velocity)}$$

$$H_{s0} = \frac{\alpha^2}{2}\left(\sum_i \frac{z}{r_i^3}(\mathbf{L}_i \cdot \mathbf{S}_i) - \sum_{i \neq j}\frac{1}{r_{ij}^3}(\mathbf{r}_{ij} \times \mathbf{P}_i)\cdot(\mathbf{S}_i + 2\mathbf{S}_j)\right) \qquad \text{(spin–orbit)}$$

α is the fine-structure constant, and

$$V = -Z\sum_i \frac{1}{r_i} + \sum_{i<j}\frac{1}{r_{ij}}$$

The Pauli Hamiltonian is ideally suited for carrying out relativistic corrections as a first-order perturbation to a non-relativistic Hamiltonian. In recent years, several authors have considered inclusion of the Pauli terms in variational self-consistent field (SCF) calculations.[17–19] Wadt, Hay and coworkers[20-24] use the Pauli Hamiltonian in deriving relativistic effective core potentials. The choice of appropriate basis sets is crucial in using the Pauli Hamiltonian in polyatomic relativistic calculations. Gaussian basis functions with large exponents are important. It has been noted[17,25] that the mass–velocity term of the Pauli Hamiltonian leads to divergence in the region close to the atomic nucleus. Cowan and Griffin[17] avoid this divergence in their numerical SCF calculations by restricting the wavefunction near the nucleus by a two-term series expansion. For Gaussian basis sets the total Pauli kinetic energy contributions for the s functions with exponents of the order of $1/\alpha^2$ or larger become negative, resulting in unrealistic wavefunctions and energies if corrections above first order in α^2 are included. Nevertheless, this can be avoided if one contracts heavily such high-exponent s functions with functions having much smaller exponents.

Several authors have considered a number of approximate solutions to the Dirac equation. One such method is the use of the Foldy–Wouthuysen transformation[19] (see, for example, Morrison and Moss[25]). Upon application of a unitary transformation of the form

$$H = \exp(is)H_{D}\exp(-is) \qquad (9)$$

where s is an Hermitian operator, to the Dirac Hamiltonian, one can transform the Dirac Hamiltonian to a form for which the solutions have only the two large components. Kuttzelnig[26] has also recently emphasized the importance of the Foldy–Wouthuysen transformation in order to avoid a variational collapse for finite basis sets with the Dirac Hamiltonian. Desclaux and coworkers[27] have formulated a relativistic one-center approach based on the Dirac Hamiltonian; this method, however, is applicable only for hydrides.

Kim[28] has formulated a relativistic Hartree–Fock–Roothaan equation for the ground states of closed-shell atoms using Slater-type orbitals. Relativistic effects in atoms have been reviewed by Grant.[29] Malli and coworkers[30-33] have formulated a relativistic SCF method for molecules. In this method, four-component spinor wavefunctions are obtained variationally in a self-consistent scheme using Gaussian basis sets.

Lee and McLean[34,35] have considered full relativistic all-electron solutions to the Dirac equation for AgH and AuH. In this method, four-component, all-electron spinors are obtained using a LCAS-MS (linear combination of atomic spinor—molecular spinor) method. These authors employ a Slater-type basis for AgH and AuH. However, such relativistic all-electron calculations do not seem to be practicable for molecules other than diatomic hydrides at present.

The Breit two-electron correction[36] arises from the relativistic magnetic retardation between two electrons. The Breit operator, which describes this interaction, is

$$H_{Br} = -\frac{e^2}{2r_{12}}\left(a_1 \cdot a_2 + \frac{(a_1 \cdot r_{12})(a_2 \cdot r_{12})}{r_{12}^2}\right) \qquad (10)$$

where α_1 and α_2 are Dirac matrices, and r_{12} is the distance between the electrons 1 and 2. The eigenfunction of H_{Br} is thus a 16-component spinor, since each electron has a four-component spinor function. The Breit interaction is of the order α^2 (α is the fine-structure constant). Although its contribution is significant in the core, the effect of Breit correction in calculating chemical and valence-level spectroscopic properties appears to be small. The Breit interaction is normally introduced as a first-order perturbation to the Dirac Hamiltonian. Further, inclusion of Breit interaction in the Dirac Hamiltonian would lead to results that are not consistent with the laws of quantum electrodynamics, since the Breit interaction is not Lorentz-invariant. The effects of Breit interaction on chemical properties have not yet been studied in full detail. Stevens and Krauss[37,38] have developed a semi-empirical scaling method which corrects for two-electron screening of the valence spin–orbit by core. The total Breit correction to the spin–orbit splittings are about 15% in the first row and 5% in the second row.

Pyykkö and coworkers[39,40] have formulated a relativistic extended Hückel method for molecules. This method incorporates relativistic effects by a systematic parametrization using Desclaux's atomic relativistic Dirac–Fock calculation.[6] The Dirac–Fock atomic energies and the off-diagonal elements are proportional to the product of the overlap matrix element in the relativistic $|lsjm\rangle$ basis. Pyykkö and coworkers[40,44] have applied this method to a number of molecules.

The Dirac–Slater multiple $X\alpha$ (DS-MS $X\alpha$) method is an approximate way of introducing relativistic corrections. Case and coworkers[42–46] have used this method in a number of studies. This method has been recently reviewed by Case[42] and earlier by Pyykkö.[7] Readers are referred to these two reviews for further details.

We now briefly consider the symmetry aspects of relativistic quantum chemistry. With the introduction of spin–orbit interaction, the appropriate group of the Hamiltonian is the double group of the molecular point group. This arises from the fact that spin is no longer a good quantum number. The double group contains twice the number of operations in the molecular point group. In this group, rotation by an angle $\varepsilon + 2\pi$ is not considered equivalent to the rotation by ε. The group has two sets of irreducible representations. The first set is for the integral spins which correspond to the ordinary irreducible representations of the point group but extended to the double group. For half-integral spins additional representations are generated which are not members of the set of irreducible representations of the point group. The first use of the double group is in correlating non-relativistic electronic states into relativistic states. An equivalent correlation is the correlation of the states without and with spin–orbit interaction. For example, consider the $^3B_1$ state of PbH_2. Since the spin–orbit interaction on Pb is quite large, this state would be split apart into finer relativistic states. The states that result from $^3B_1$ are obtained by first correlating the spin state (triplet, $D^{(1)}$) into the C_{2v}^2 group and then multiplying the resulting spin representations with the spatial symmetry. The correlation of the triplet into C_{2v}^2 gives rise to $A_2 + B_1 + B_2$ representations which upon multiplication with B_1 results in $B_2 + A_1 + A_2$. Thus the $^3B_1$ state of PbH_2 is split into three states in the presence of spin–orbit interaction which are of A_1, A_2 and B_2 symmetry. If one considers the $^2B_1$ state of PbH_2^+, one obtains a different picture. The doublet correlates with the $E_{1/2}$ representation in the C_{2v}^2 group. The overall symmetry is thus $E_{1/2}$, which is not a part of the character table of C_{2v}. Oreg and Malli[47–49] have considered symmetry aspects of the construction of spinors for polyatomics which are symmetry-adapted in the double group of the molecule. These methods have been used in the Dirac–Fock theory of open-shell as well as closed-shell molecules.[50]

An important and reliable method for carrying out relativistic quantum calculations is the relativistic effective potential method. A number of groups have considered the generation of both relativistic and non-relativistic effective potentials. Krauss and Stevens[12] have recently reviewed the use of effective potentials in quantum chemistry. Readers are referred to that review for additional details pertaining to this topic. In this review we consider the methods of relativistic effective potentials.

The objective of the effective potential method is to represent the interaction of the valence electrons with the core electrons by an effective potential, thereby reducing the number of electrons significantly in quantum calculations. The effective potentials must prevent the collapse of the valence electrons into the core. Effective potentials can be relativistic or non-relativistic depending on the nature of the wavefunction from which they are generated. Relativistic effective potentials can be generated by either semi-empirical or *ab initio* methods. We first briefly review the *ab initio* methods.

Many of the effective potentials (relativistic or non-relativistic) are generated using the Phillips–Kleinman transformation.[51] In this method, the explicit core–valence orthogonality constraints are replaced by a modified valence Hamiltonian. If one replaces the potential generated by core electrons by a potential V_c, then one can write the one-electron valence wave equation as

$$(h + V_c)\phi_v = E_v\phi_v \tag{11}$$

where $\langle \phi | \phi_c \rangle = 0$, $H\phi_c = E_c\phi_c$. Phillips and Kleinman[51] suggested that ϕ_v can be written as

$$\phi_v = \chi_v - \sum_c \langle \chi_v | \phi_c \rangle \phi_c \tag{12}$$

It can be easily seen that for any χ_v, ϕ_v is orthogonal to ϕ_c thereby satisfying $\langle \phi_v | \phi_c \rangle = 0$. If one substitutes the above expression for ϕ_v into the one-electron eigenequation, one obtains

$$(h + V_c + V_{Ep})\chi_v = E_v\chi_v$$
$$V_{Ep} = \sum_c (E - E_c)|\phi_c\rangle\langle\phi_c| \tag{13}$$

V_{Ep} thus obtained is often referred to as the Phillips–Kleinman pseudo-potential, while χ_v is known as the pseudo-orbital. $|\phi_c\rangle\langle\phi_c|$ is the projection operator corresponding to the core orbital ϕ_c. Thus the operator $\sum_c |\phi_c\rangle\langle\phi_c|$ is the projection operator of the core orbitals. If the V_{Ep} are derived based on a non-relativistic atomic wavefunction, they are known as non-relativistic effective core potentials. There are also several non-relativistic model potentials which are not strictly derived from *ab initio* methods. For a review of these potentials see Krauss and Stevens.[12]

Ab initio relativistic effective core potentials can be derived from a number of methods. We briefly review these methods. Lee, Ermler and Pitzer[15] have formulated a method for deriving effective potentials from the numerical Dirac–Fock calculations of the atoms. We start with this method.

The solution of the Dirac–Fock equation is a set of four-component spinors. If the spinors are partitioned as core and valence spinors, then one can write the overall many-electron relativistic wavefunction for a single configuration as

$$\psi = A[(\psi_1^c \psi_2^c \cdots \psi_m^c)(\psi_1^v \psi_2^v \cdots \psi_n^v)] \tag{14}$$

where A is the antisymmetrizer, $\psi_1^c, \ldots, \psi_m^c$ are core orbitals, m being the number of core electrons, and $\psi_1^v, \ldots, \psi_n^v$ are the valence orbitals, n being the number of valence electrons. The total energy E_T can be partitioned into core, valence and core–valence interaction energies. In symbols,

$$E_T = E_c + E_v + E_{cv} \tag{15}$$

It can be shown that

$$E_v + E_{cv} = \langle \psi_v^R | H_v^{rel} | \psi_v^R \rangle \tag{16}$$

$$H_v^{rel} = \sum_i \left(h_D(i) + \sum_c (J_c(i) - K_c(i)) \right) + \sum_{i<j} \frac{1}{r_{ij}}$$

where the indices i and j run over valence electrons. For an orthonormal set of valence orbitals, it can be shown that the Dirac–Hartree–Fock (DHF) equation for a single electron is given by

$$\left(h_D + \sum_c (J_c - K_c) \right)\psi_v = \varepsilon_v \psi_v + \sum_c \psi_c \varepsilon_{cv} \tag{17}$$

where the ε_{cv} are the off-diagonal Lagrange multipliers given by

$$\varepsilon_{cv} = \langle \psi_v | h_D + \sum_c (J_c - K_c) | \psi_c \rangle$$

If one defines the core projector and the pseudo-orbital in the same way as done in the Phillips–Kleinman method, for relativistic spinor wavefunctions one obtains relativistic pseudo-orbitals and relativistic effective potentials which are given by

$$\chi_v^R = \psi_v^R + \sum_c a_c \psi_c^R \qquad \psi_v^R = (1 - P)\chi_v^R \qquad P = \sum_c |\psi_c\rangle\langle\psi_c|$$

$$V^{RPK} = -PH_v^{rel} - H_v^{rel}P + PH_v^{rel}P + \varepsilon_v P \tag{18}$$

where

$$(H_v^{rel} + V^{RPK})\chi_v^R = \varepsilon_v \chi_v^R$$

$$(h_D + U^{core})\chi_v^R = \varepsilon_v \chi_v \qquad \text{if } U^{core} = \sum_c (J_c - K_c) + V^{RGPK}$$

The relativistic effective potentials (REP) thus formulated involve four-component spinor projectors. As mentioned earlier, the effect of small components in the valence regions is rather small and one can neglect the small components in considering chemical and spectroscopic properties. Alternatively, one can make the Foldy–Wouthuysen transformation[19] to eliminate the small components and obtain a correction to the large component; for valence spinors this correction is so small that it can be neglected. Thus one can use the non-relativistic kinetic energy operator along with the relativistic large components in an equation from which valence-level, relativistic core potentials are generated. Thus, for a single valence electron

$$(-\tfrac{1}{2}\nabla^2 - z/r + U^{EP})\chi_v' = \varepsilon_v \chi_v' \tag{19}$$

where χ_v' is a two-component pseudowavefunction containing only the large radial components.

With more than one valence electron, this equation becomes

$$[-\tfrac{1}{2}\nabla^2 - z/r + U^{EP} + W(\chi_v', \chi_{v'}')]\chi_v' = \varepsilon_v \chi_v' \tag{20}$$

where $W(\chi_v', \chi_{v'}')$ is the sum of Coulomb and exchange interactions of a given pseudo-orbital χ_v' with all other valence pseudo-orbitals $\chi_{v'}'$.

The effective core potentials U_v^{EP} are not the same for pseudo-orbitals of different symmetry. Thus one can express the REPs as products of angular projectors and radial functions. In the Dirac–Fock approximation, the orbitals with different total j but which have the same l value are not degenerate, and thus the potentials derived from the Dirac–Fock calculations would be j-dependent. The REPs can thus be expressed by introducing the lj-dependent radial potentials U_{lj}^{REP}, as

$$U^{REP} = \sum_{l=0}^{\infty} \sum_{j=|l-1/2|}^{l+1/2} U_{lj}^{REP}(r)|ljm\rangle\langle ljm| \tag{21}$$

where the $|ljm\rangle$ are Pauli two-component spinors.

The expression for the relativistic effective potentials involves an infinite sum over l. This requires calculations of the radial potentials for all the excited states of the atom, which is impractical. However, the radial functions U_{lj}^{REP} cease to change significantly with l and j after these numbers exceed those of electrons in the core. Consequently, it is a good approximation to stop at maximum l and j values denoted by L and J, respectively. The modified relativistic effective potential can be written as

$$U^{\text{REP}} = U_{LJ}^{\text{REP}}(r) + \sum_{l=0}^{L-1} \sum_{j=|l-1/2|}^{l+1/2} \sum_{m=-j}^{j} (U_{lj}^{\text{REP}} - U_{LJ}^{\text{REP}})|ljm\rangle\langle ljm| \qquad (22)$$

One can test alternative values by a series of actual calculations. It is our experience that L should be at least one higher than the maximum l value in the core. Thus, for example, desirable minimal values of L for Sn and Pb are 3 and 4, respectively.

Pitzer and coworkers[15,52-55] have carried out relativistic calculations on a number of diatomics such as Xe_2, Xe_2^+, TlH, Au_2, Au_2^+, PbS, $PbSe^+$, etc. These calculations were carried out with an LCAS-MS (linear combination of atomic spinor—molecular spinor) approach with the relativistic effective potentials. Many of these calculations were at the level of single-configuration SCF. In the earlier calculations, the spin–orbit coupling was ignored at the SCF stage and introduced using a semi-empirical procedure.

These early calculations used effective potentials obtained by the Phillips–Kleinman method[51] (see the review of Krauss and Stevens[12]) wherein the pseudo-orbitals are taken to be linear combinations of the atomic orbitals of the same l and j. This method tends to underestimate the repulsive region of the potential energy curves. There is no reason that the pseudo-orbital must be linear combinations of core and valence orbitals. Christiansen, Lee and Pitzer[56] have proposed a method for constructing pseudo-orbitals in which the pseudo-orbital is represented as

$$P_l^{\text{PS}} = \begin{cases} r\sum_{i}^{N} C_i r^i & \text{for } r \leqslant r_m \\ \phi_l(r) & \text{for } r \geqslant r_m \end{cases} \qquad (23)$$

where r_m is a match radius, and ϕ_l is the all-electron Dirac–Fock orbital. In this method the coefficients C_i are determined by matching the value and the first three derivatives of the P_l and ϕ_l at r_m with the condition that P_l is normalized. The match radius is minimized subject to the condition that P_l can have only one maximum and two inflection points. Pseudo-orbitals derived this way are called shape-consistent because the resulting pseudo-orbital is identical to the Hartree–Fock (Dirac–Fock) orbital in the valence region. In Fig. 2, we reproduce a comparison of this pseudo-orbital and the corresponding orbital obtained using the Phillips–Kleinman method for the chlorine atom as done by Christiansen, Lee and Pitzer (CLP).[56] As one can see from that figure, the maximum of the Phillips–Kleinman orbital is at smaller radius than the all-electron orbital or that of the CLP orbital.

The relativistic effective potentials obtained with the methods described

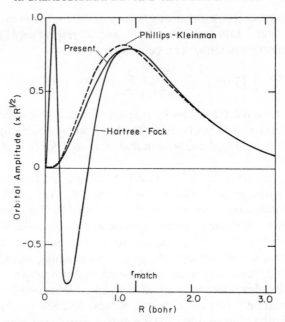

Fig. 2. A comparison of chlorine 3s pseudo-orbitals generated with the Christiansen–Lee–Pitzer method and the Phillips–Kleinman pseudo-orbital and the all-electron Hartree–Fock orbital.

above are numerical potentials. However, for polyatomic calculations a Gaussian analytic fit of such potentials is more desirable since multi-centered integrals over Gaussians can be obtained easily. Kahn, Baybutt and Truhlar[57] have suggested the following Gaussian expansion of numerical potentials:

$$U_{LJ}^{\text{REP}}(r) - U_{lj}^{\text{REP}}(r) = \frac{1}{r^2} \sum_{i=0}^{N} C_i r^{n_i} \exp(-\alpha_i r^2) \tag{24}$$

where the C_i, n_i and α_i are chosen so as to obtain the best fit for the numerical potentials.

The relativistic effective potentials can be averaged with respect to spin. The averaged relativistic effective potentials can be written as

$$U^{\text{AREP}}(r) = U_L^{\text{AREP}}(r) + \sum_{l=0}^{L} \sum_{m=-l}^{l} [U_l^{\text{AREP}}(r) - U_L^{\text{AREP}}(r)] |lm\rangle \langle lm| \tag{25}$$

where

$$U_l^{\text{AREP}} = \frac{1}{2l+1} [lU_{l,l-1/2}^{\text{REP}}(r) + (l+1)U_{l,l+1/2}^{\text{REP}}(r)]$$

The resulting potentials involve the ordinary spherical harmonic projector as opposed to two-component spinor projectors. These potentials include all relativistic effects except the spin–orbit interaction. These potentials resemble non-relativistic effective potentials in their form and can thus be introduced into non-relativistic SCF or multi-configuration SCF (MCSCF) calculations. Schwarz and coworkers[58] and Ermler et al.[59] have suggested that the spin–orbit operator can be obtained as the difference of different j but same l. The resulting spin–orbit operator is simply the difference of $l + \frac{1}{2}$ and $l - \frac{1}{2}$ relativistic effective potentials. The spin–orbit operator thus derived can be written as

$$
H^{so} = \sum_{l=1}^{L-1} \Delta U_l^{REP}(r) \left(\frac{l}{2l+1} \sum_{-l-1/2}^{l+1/2} |l, l + \tfrac{1}{2}, m\rangle \times \langle l, l + \tfrac{1}{2}, m| \right.
$$
$$
\left. - \frac{l+1}{2l+1} \sum_{-l-1/2}^{l-1/2} |l, l - \tfrac{1}{2}, m\rangle \langle l, l - \tfrac{1}{2}, m| \right) \tag{26}
$$

where

$$
\Delta U_l^{REP}(r) = U_{l,l+1/2}^{REP}(r) - U_{l,l-1/2}^{REP}(r)
$$

The spin–orbit operator derived this way can be *ab initio* if it is derived from relativistic *ab initio* potentials. It can be introduced in molecular calculations. Pacios and Christiansen[60] have published Gaussian analytic fits of averaged relativistic effective potentials and spin–orbit operators for Li through Ar. The relativistic potentials of other elements are also being tabulated.[61]

The relativistic effective potentials derived from the numerical Dirac–Fock wavefunctions have been employed in a number of quantum calculations with considerable success. We will review a number of such calculations and results in the next section. We now consider other methods of generating relativistic effective potentials.

Hay, Wadt and coworkers[20–24,62,63] derive relativistic effective potentials from the Pauli Hamiltonians after omitting the spin–orbit term. Cowan and Griffin[17] have suggested a procedure for the inclusion of the mass–velocity and Darwin terms of the Pauli Hamiltonian into a variational SCF calculation. Since Hay and Wadt do not include spin–orbit interaction in their effective potentials, these potentials are analogous to spin-averaged relativistic potentials. Hay and Wadt[22–24] have recently published Gaussian fits of relativistic effective potentials without spin–orbit interaction for all the elements in the periodic table. Also, included in these papers are the optimized Gaussian basis sets of these elements for polyatomic calculations. The results of the Hay–Wadt potentials compare reasonably well with the averaged relativistic effective potentials derived from the Dirac–Fock calculations. However, these potentials do not provide for an *ab initio* spin–orbit operator which can be introduced variationally either in an MCSCF or a configuration-interaction (CI) scheme. The spin–orbit interaction is introduced in the final step as a perturbation using the semi-empirical method.[21]

Schwarz and coworkers[58,64,65] have developed a relativistic model potential method. In this method the relativistic effective potential is expressed as

$$V_{rel}^{eff} = V_{rel} + \sum_{l,j,m_j}^{core} |ljm_j\rangle V_{lj}(r)\langle ljm_j| \tag{27}$$

where $V_{rel}(r)$ and $V_{lj}(r)$ are parametrized as

$$V_{rel}(r) = \frac{-Z_{eff} + A\exp(\alpha r)}{r}$$

$$V_{lj}(r) = B_{lj}\exp(-\beta_{lj}r)$$

where Z_{eff} is the effective charge of the atomic core. The parameters α, β_{lj}, A and B_{lj} are obtained so that the effective Hamiltonian reproduces the valence-electron spectrum of the alkali-like systems. The resulting effective potentials are spin-dependent and they include spin–orbit interaction. Schwarz and coworkers[65] have developed a complex two- and four-index transformation over two-component spinors which enables introduction of these potentials in MCSCF or CI calculations. These authors[66–70] have applied the relativistic model potential method to a number of molecules containing heavy atoms. More recently, Mark, Marian and Schwarz[71] have considered the use of the relativistic Dirac–Breit approach to estimate the fine-structure splittings of F_2 and F_2^+. Schwarz and Chu[72] have considered relativistic contributions to ionization energies and bond lengths with semi-empirical as well as *ab initio* Dirac–Fock calculations. Esser[73] has recently developed a relativistic multi-reference CI (MRCI) method for many-electron wavefunctions. He has presented the unitary group formalism and applications to heavy atoms such as Pb, Hg, etc.

We now briefly review a number of methods of relativistic calculations using effective potentials. As mentioned earlier Pitzer and coworkers carried out their earlier relativistic calculations using a single-configuration SCF spinor scheme.[52–55] Christiansen and Pitzer[74] introduced REPs in MCSCF LCAS-MS spinor calculations. These calculations were carried out by a modification of the BISON MCSCF code to accommodate relativistic two-component spinors. The first calculation along these lines was carried out on the ground state of the TlH molecule.[74] This MCSCF calculation included five spinor configurations. Christiansen and Pitzer[75–78] have carried out such MCSCF calculations on Tl_2 and Tl_2^+ and polarizabilities of the Rb and Cs atoms. Other calculations which compare effective potential and all-electron methods have also been carried out.[79,80] This MCSCF spinor scheme is restricted in that it could accommodate only 10 configurations. This was, however, adequate for the ground-state properties of molecules such as TlH. Christiansen and Pitzer[74] obtained 80% of the experimental dissociation energy using this method.

While the MCSCF spinor approach with up to 10 configurations provides a reasonable picture of bonding for some molecules, such as TlH, for other molecules, like Pb_2, this approach is far from adequate. This is a result of large mixing of configurations arising from both correlation and spin–orbit interaction. Thus an approach that can accommodate a large number of configurations is more desirable. Further, the method should enable calculations of several excited states. Christiansen, Balasubramanian and Pitzer[81] have developed a relativistic configuration-interaction method. In this method the relativistic effective potentials are averaged with respect to the spin at the SCF stage. Thus at this stage relativistic effects such as mass–velocity correction, Darwin correction, etc., are included, but the spin–orbit interaction is not introduced at this stage. However, this spin–orbit operator is obtained as the difference of $l + \frac{1}{2}$ and $l - \frac{1}{2}$ potentials as described earlier. The spin–orbit integrals over MOs with this spin–orbit operator obtained this way are introduced as one-electron integrals at the CI stage. The spin–orbit integrals over real Cartesian basis sets can be imaginary and thus the introduction of these integrals at the SCF or MCSCF stage would lead to a complex Fock operator. Although the CI integrals are imaginary, the CI matrices are diagonalized only once. This procedure could be called the relativistic CI method or the spin–orbit CI (SOCI) method. This provides for a method of introducing spin–orbit interaction variationally rather than through the conventional perturbative scheme. Further, this method allows for mixing of configurations which would not mix in the absence of spin–orbit interaction. The method has been tested on a number of diatomics in the past few years. These results will be reviewed in the next section. R. Pitzer and Winter[82] are considering relativistic CI calculations of UF_6, NpF_6 and PuF_6 with this method.

III. APPLICATIONS TO MOLECULES CONTAINING HEAVY ATOMS

In this section we consider applications of relativistic quantum methods to calculations of properties of molecules containing heavy atoms. In recent years a number of authors have made relativistic calculations of the electronic and spectroscopic properties of a number of molecules using the methods outlined in the earlier section. There are excellent reviews on applications of relativistic calculations to a number of molecules[7–13,42,83,84]. While we review some of these calculations for completeness, additional details on these calculations can be found in these reviews. In the present chapter we review more recent developments in this area. We will divide molecules into several categories and discuss the calculations in each category. Sections III.A and III.B consider homonuclear and heteronuclear diatomics, respectively, while Section III.C considers polyatomics and miscellaneous applications. The heteronuclear diatomics that we consider here include hydrides, halides and oxides.

A. Homonuclear Diatomics Containing Very Heavy Atoms

One of the first relativistic calculations on very heavy diatomics was carried out by Pitzer and coworkers[53] on the ground state of Au_2. For this molecule, the spin–orbit contribution is small in comparison to other relativistic contributions such as mass–velocity and Darwin corrections. The Au_2 molecule exhibits primarily σ_s bounding arising from the overlap of s orbitals. An important result of this calculation is that the Au_2 bond is stronger than the Ag_2 bond as a result of the relativistic contraction of the s orbital. Ermler, Lee and Pitzer[55] carried out calculations of several excited states of Au_2. For the excited states the spin–orbit interaction plays a more important role; it was introduced by a semi-empirical procedure. These calculations employ the Phillips–Kleinman potentials which tend to underestimate the repulsive region of the potential energy curves, thus predicting short bond lengths. Zeigler et al.[85] have carried out Hartree–Fock–Slater calculations of Au_2 as well as other heteronuclear diatomics containing Au and Ag. Ross and Ermler[86] have recently reported calculations on Ag_2, Au_2 and other Ag- and Au-containing molecules. They have carried out SCF, MCSCF and CI calculations on these molecules with the revised Christiansen–Lee–Pitzer potentials. These calculations have shown that the relativistic bond contractions for Au_2 and Ag_2 are about 0.2 Å and 0.05 Å, respectively. These calculations employ a triple-zeta s, double-zeta p and double-zeta d basis, but the f functions which seem to play a significant role for these systems are not included. Further, core–valence correlations, which were not included, may be important for these systems. As a result of these approximations their calculated dissociation energies with MCSF and CI schemes are 1.00 eV and 1.47 eV, for Ag_2 and Au_2, respectively, in comparison to the experimental value of 2.08 eV for Au_2. McLean[87] has also carried out non-relativistic all-electron calculations for Ag_2.

Christiansen and Pitzer[75] carried out MCSCF calculations on Tl_2 and Tl_2^+ in the $\omega-\omega$ coupling scheme. They carried out calculations on three low-lying $\omega-\omega$ states and three $\lambda-s$ states. These calculations revealed that the ground state of TL_2 is the 0_u^- state and is essentially repulsive with only a shallow minimum at long distance. The 0_g^+ and 1_u states are slightly higher in energy. The Tl_2^+ ion, however, has a $1/2_g$ state which is bound by 0.58 eV at this level of calculations. Christiansen[88] carried out relativistic configuration-interaction calculations on Tl_2 with the method described in Ref. 81. These extensive calculations also produced only weak binding for the ground state of Tl_2. Christiansen[88] also recalculated the experimental dissociation energy by correcting the partition function of Tl_2. The revised experimental D_e for Tl_2 is about 0.37 eV(± 0.15). Pitzer[83] has reviewed the earlier calculations on Tl_2 and Au_2.

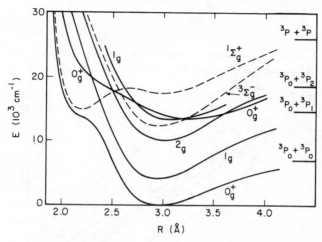

Fig. 3. Calculated potential curves for the g states of Pb_2. The broken curves are computed without the spin–orbit term.

The Pb_2 and Sn_2 molecules are considerably more complex than Au_2 and Tl_2 in that even the atom needs to be treated in an intermediate-coupling scheme. Both correlation and spin–orbit contributions are large for these systems, but the spin–orbit contribution for Pb_2 is much larger than for Sn_2. Balasubramanian and Pitzer[4,89] have carried out relativistic CI calculations on Pb_2 and Sn_2. Results of our calculations on Pb_2 and Sn_2 are shown in Fig. 3–6. Our calculations enabled interpretation of experimentally observed

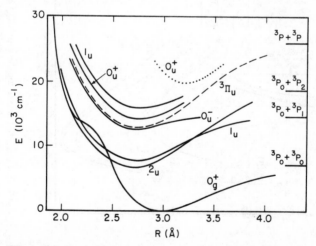

Fig. 4. Calculated potential curves for the u states and the ground 0_g^+ state of Pb_2. An estimated curve at the experimental T_e is also given for the upper 0_u^+ state.

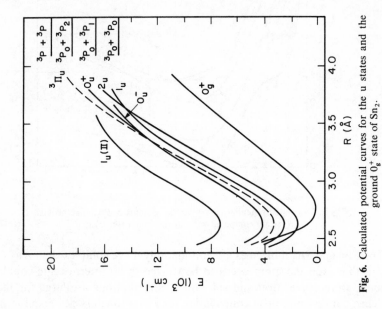

Fig. 6. Calculated potential curves for the u states and the ground 0_g^+ state of Sn_2.

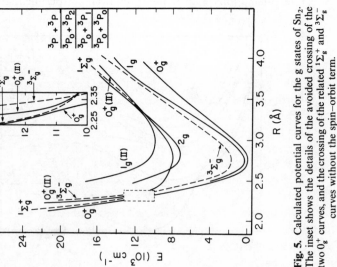

Fig. 5. Calculated potential curves for the g states of Sn_2. The inset shows the details of the avoided crossing of the two 0_g^+ curves, and the crossing of the related $^1\Sigma_g^+$ and $^3\Sigma_g^-$ curves without the spin–orbit term.

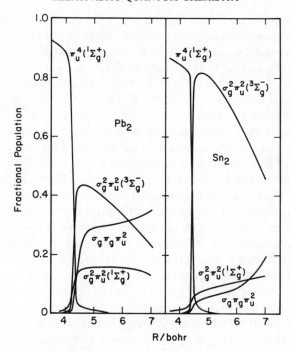

Fig. 7. Fractional populations of various Λ–S configurations for Pb_2 and Sn_2.

laser-induced fluorescence spectra of Pb_2.[91,91] The experimentally observed X, A and C states were assigned to 0_g^+, 1_g and 0_u^+ states. The B state was reassigned to the 0_u^- state. The ground 0_g^+ and low-lying 1_g state are type c analogs of $^3\Sigma_g^-$ ($\sigma_g^2\pi_u^2$). However, the assignment of any state of Pb_2 to any particular spin–orbit component of a λ–s state has limited meaning because of large spin–orbit contamination of several λ–s states. In Fig. 7 we show the fractional CI populations of various λ–s states as a function of internuclear distances for the 0_g^+ ground state. As one can see from these figures, λ–s populations in Pb_2 change dramatically as a function of internuclear distance. It is further interesting to note that at a short distance, the 0_g^+ states of both Pb_2 and Sn_2 exhibit an avoided crossing resulting from π_u^4 (0^+g) with $\sigma_g^2\pi_u^2$ (0_g^+, 1_g) configurations. This results in a shoulder in the 0_g^+ curve at short distance. The Sn_2 is described reasonably well in λ–s coupling. As one can see from Fig. 7, the spin–orbit contamination for Sn_2 is rather small and thus the ground state is $^3\Sigma_g^-$ (0_g^+). The $J = 0$ state of the Pb atom is 88% $^3P_0$ and 12% $^1S_0$. Similarly, the $J = 2$ state of Pb atom is 70% $^3P_2$ and 30% $^1D_2$. The corresponding percentages for the $J = 0$ and $J = 2$ states of Sn are 97% $^3P_0$, 3% $^1S_0$ and 97% $^3P_2$ and 3% $^1D_2$. Thus correlation is more important for Sn than spin–orbit interaction, while both correlation and spin–orbit interaction are important for the Pb compounds. For the Pb_2 dimer than spin–orbit interaction destabilizes the bound by almost 50%. The calculated D_e values for Pb_2 and Sn_2 (0.88 eV and 1.86 eV) are in good agreement with the experimental values[92–94] (0.88 eV and 1.94 eV). After the completion of these calculations, an experimental value of R_e for the ground state of Pb_2 was reported by Sontag and Weber.[95] The calculated value of $R_e = 2.97$ Å agrees very closely with the subsequent experimental value of 2.930 Å.

Pacchioni[96] has recently carried out calculations on the low-lying states of Sn_2 and Pb_2. This author gives the impression that he is the first to carry out a comparative *ab initio* CI calculation on these systems. We would like to clarify this further. First, his calculation starts with the Hafner–Schwarz model potentials in comparison to our relativistic *ab initio* potentials derived from numerical Dirac–Fock solutions of the atoms. Pacchioni's calculations ignore spin–orbit interaction. Our calculations include spin–orbit interaction in a relativistic CI scheme in comparison to the non-relativistic CI of Pacchioni. Thus, he obtains a D_e approximately twice the experimental value which he corrects by a semi-empirical scheme to arrive at a value close to our calculated value with a relativistic CI. Our calculations have clearly demonstrated the need to carry out an intermediate-coupling CI calculation for Pb_2 as a result of large spin–orbit contamination. Calculations without spin–orbit, such as Pacchioni's, have little relationship to the real Pb_2 molecule.

Christiansen[97] has recently carried out relativistic CI calculations on the ground state of Bi_2. His calculations have shown that, although the ground state of Bi_2 is dominantly $^1\Sigma_g^+$ (arising from the $\sigma_g^2 \pi_u^4$ triple bonding configuration), it is significantly contaminated by $^3\Pi_g$ (about 25%). The calculated D_e value of Bi_2 (2.3 eV) is in reasonable agreement with the experimental value of 2.04 eV. The computed R_e value is about 0.16 bohr longer than the experimental value. This seems to be the general trend in a limited relativistic CI scheme which is adequate to describe other properties but somewhat less accurate in calculating bond distances and dissociation energies. It is believed that this discrepancy in calculated bond distances arises from the d correlation. More extensive calculations are warranted to confirm the origin of calculated longer bond lengths.

Celestino and Ermler[98] have carried out calculations on Hg_2 and TlHg. Calculations on Hg_2 were carried out with a full four-electron CI within 16 of the 22 valence and virtual orbitals. The spin–orbit interaction is ignored at the CI stage but introduced after CI using a semi-empirical scheme. For the TlHg molecule, CI calculations included full correlation of five outer electrons with some restrictions. These authors have carried out calculations on a large number of low-lying states.

Stevens, Basch and Krauss[99] have carried out calculations on a number of light diatomics such as P_2 and Cl_2 among other molecules like SiO, CH_4, etc., to test the reliability of effective potentials. They have also generated *ab initio* effective potentials and compatible basis sets for the first- and second-row atoms.[99] Other calculations on homonuclear diatomics which use relativistic methods include Cu_2,[100] and I_2.[101] Relativistic *ab initio* calculations have been carried out on noble-gas dimers and ions such as Xe_2^+, Xe_2, etc.[80,102-105]

B. Heteronuclear Diatomics Containing Very Heavy Atoms

Among the heteronuclear diatomics containing heavy atoms, a number of hydrides have been studied. Desclaux and Pyykkö have studied a number of hydrides using the one-center numerical Dirac–Fock method.[27,106-109] These calculations provide an insight into the magnitude of relativistic effects on the bond lengths; however, they are not useful for dissociation energies as they break down at long distances.

Hay *et al.*[110] have carried out effective potential calculations on AuH which yield the relativistic bond contraction of 0.26 Å. The dissociation energy of AuH is increased by 0.5 eV as a result of relativistic contributions. They also presented calculations for AuCl, HgH and $HgCl_2$. Ziegler, Snijders and Baerends[85] have carried out perturbational relativistic calculations on a number of heteronuclear diatomics such as HgH^+, CdH^+, ZnH^+, AuH, AgH, CuH and CsH.

Lee and JcLean[34,35] have carried out all-electron Dirac four-component spinor LCAS-MS SCF calculations on AgH and AuH. The relativistic effects increase the dissociation energies by 0.08 eV and 0.42 eV in these molecules, while the bond lengths contrast by 0.08 Å and 0.25 Å. These values for AuH confirm the earlier effective potential calculations.[110]

Lee, Ermler and Pitzer[55] have carried out SCF calculations in $\omega-\omega$ coupling on TlH, PbSe and PbS among other diatomics. Christiansen and Pitzer[74] carried out MCSCF LCAS-MS spinor calculations on TlH.

Christiansen, Balasubramanian and Pitzer[81] have carried out relativistic configuration-interaction calculations on the six low-lying states of TlH. In these calculations, spin–orbit interaction is introduced at the CI stage with the spin–orbit operator derived as a difference of $l + \frac{1}{2}$ and $l - \frac{1}{2}$ relativistic effective potentials. The properties of the two lowest 0^+ states are in very good agreement with the experimental results. In addition, these calculations yielded shallow minima in certain excited states of TlH in agreement with provisional interpretation of spectra. The calculated equilibrium bond length (1.99 Å) is somewhat longer than the experimental value of 1.87 Å. This discrepancy was in part attributed to the lack of d correlation. However, this is yet to be confirmed. The calculated dissociation energy (1.81 eV) is in reasonable agreement with the experimental value of 1.97 eV.

Balasubramanian and Pitzer[111,112] also made relativistic CI calculations on a number of low-lying states of PbH and SnH which enabled interpretation of the electronic spectra of these molecules. The low-lying electronic states of PbH (3/2(II), 5/2) exhibit interesting avoided crossings. The 3/2(II) state is $^4\Sigma^-_{3/2}$ at short distances, but at long distances it becomes a mixture of $^2\Pi_{3/2}$, $^2\Delta_{3/2}$ and other λ–s states. The 5/2 state is $^2\Delta_{5/2}$ at short distances, but it becomes $^4\Pi_{5/2}$ at very long distances.

Balasubramanian[113] has recently reported relativistic CI calculations on four low-lying ω–ω states of BiH arising from the $\sigma^2\pi^2$ electronic configur-

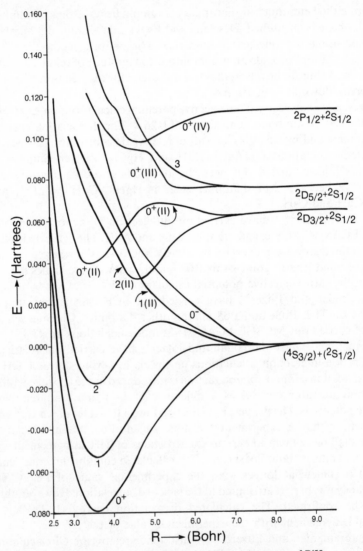

Fig. 8. Potential energy curves of 10 low-lying states of BiH.

ation. These calculations enabled the assignment of the X(0^+), A(1) and B(0^+) states which were observed experimentally. The potential energy curve of the 2 state exhibits a barrier resulting from the avoided crossing of $^1\Delta_2$ with $^5\Sigma_2^-$. The experimental D_e value of this molecule is uncertain since it is estimated from the predissociation of the E(0^+) state into $^2D_{3/2} + {}^2S_{1/2}$ atoms. In another investigation more extensive relativistic CI calculations have been carried out[114] on 10 low-lying ω–ω states of BiH (0^+, 1, 2, 0^+(II), 1(II), 2(II), 0^-, 0^+(III), 3 and 0^+(IV)). Potential energy curves of the excited states of this

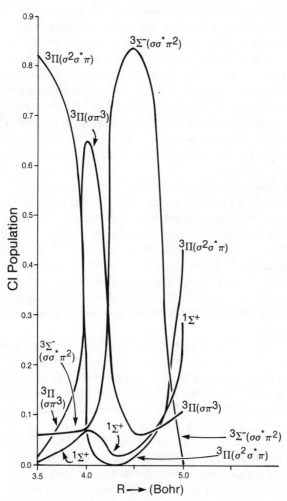

Fig. 9. CI population of the 0^+(IV) (E(0^+)) states of BiH as a function of internuclear distance.

molecule exhibit very interesting properties such as shoulders, barriers, etc. Figure 8 shows the potential energy curves of some of the low-lying states of BiH. In Fig. 9 we show the relativistic CI population of the 0^+(IV)(E(0^+)) state of BiH. As one can see from Fig. 9, this state exhibits a number of avoided crossings. The spin–orbit contaminations of these states are quite large. In another investigation Balasubramanian[115] has carried out relativistic CI calculations of eight low-lying states of BiH$^+$ (1/2, 3/2, 1/2(II), 1/2(III), 3/2(II), 3/2(III), 5/2 and 5/2(II)). The ionization potential of BiH$^+$ was calculated to be 8.08 eV. The excited states of BiH$^+$ show interesting relativistic avoided crossings. The calculated D_e of BiH$^+$ is about 1.05 eV in comparison to the neutral molecule whose D_e value is about 2.2 eV. Since the π orbital of BiH is essentially non-bonding, one might expect the D_e values of BiH and BiH$^+$ to be about the same. However, at the dissociation limits, the spin–orbit stabilization of the $^4S_{3/2}$ of the neutral atom is substantially smaller than that of the $^3P_0$ state of the ion. The ground state of the Bi neutral atom ($^4S_{3/2}$) cannot be split but is lowered by 0.032 66 hartree atomic units when the spin–orbit operator is included. The ground state of Bi$^+$ ($^3P_0$) is stabilized by 0.0859 atomic units in comparison to ^3P without the spin–orbit term. The D_e values of BiH and BiH$^+$ obtained without spin–orbit interaction are 2.06 eV and 2.17 eV, respectively. Thus the D_e value of BiH$^+$ is lowered considerably by spin–orbit interaction.

Chapman, Balasubramanian and Lin[116] have carried out relativistic CI calculations on the six low-lying states of HI. These calculations enabled the assignment of absorption continua near 46 000 cm^{-1}, 23 000 cm^{-1} as well as bands which extend from 55 000 cm^{-1}. Calculations of a number of low-lying states of HBr (0^+, 0^+(II), 0^+(III), 1, 1(II), 0^-, 0^-(II), 2, 2(II) and 0^-(III)) have just been completed.[117] These calculations were carried out with an extended triple-zeta basis set since the 0^+(III) and 2(II) states dissociate into Rydberg atoms. These calculations have not only enabled the assignment of experimentally observed bands but also predicted several new transitions which are yet to be observed.

Wang and Pitzer[118] have carried out relativistic CI calculations on the five low-lying states of PtH as well as PtH$^+$. The calculated dissociation energy is above 2.45 eV in comparison to the experimental value of 3.44 eV. There are three states of nearly equal energy. In each case there is a σ bond and a single vacancy in the 5d shell of Pt. The calculations indicate the $^2\Delta_{5/2}$ state to be lowest with the $^2\Sigma_{1/2}$ higher by 1008 cm^{-1} and the $^2(\Pi + \Delta)_{3/2}$ above $^2\Delta_{5/2}$ by 2742 cm^{-1}. But these differences are so small that further refinement of the calculations might change the order. All three states are known experimentally,[119] but the present measurements leave many uncertainties. A tentative value of 1300 cm^{-1} has been given for T_e of the 3/2 state with the 5/2 state at zero.

Balasubramanian and Pitzer[120,121] have carried out relativistic CI calculations on 11 low-lying states of PbO and SnO with the objective of interpreting chemiluminescent spectra resulting from $Pb + O_3$, $Sn + N_2O$ and other reactions. These calculations enabled the assignment of a number of experimentally observed bands. Calculations of low-lying states of iso-electronic PbS and SnS have also been carried out.[122] Balasubramanian[123] reported relativistic CI calculations on the low-lying states of PbO^+, SnO^+, PbS^+ and $PbSe^+$. These calculations confirmed the breakdown of Koopman's theorem for SnO and PbO which had previously been noted by Dyke et al.[124] For SnO and PbO, Koopman's theorem predicts the $^2\Sigma^+$ state to be the ground state, while the actual ground state of these molecules is $^2\Pi_{3/2}$. Actually, at very short distances the $^2\Sigma^+$ state is lower than the $^2\Pi$ state. At near-equilibrium geometries this ordering is reversed, which results in an avoided crossing in the 1/2 state of PbO^+ and SnO^+. The breakdown of Koopman's theorem was explained based on orbital relaxation effects. However, for $PbSe^+$, it was noted that the orbital relaxation effects are not large enough to effect the ordering. This Koopman's theorem correctly predicts the $^2\Pi$ state to be the ground state of $PbSe^+$.

Balasubramanian[125] carried out relativistic CI calculations on the low-lying states of TlF. There is considerable experimental interest in the photoionization of vapors of relatively non-volatile materials such as the halides. The thermodynamic and spectroscopic dissociation energies of this molecule do not agree, which seems to suggest the existence of barriers in the excited states. Relativistic CI calculations of the potential energy curves of nine low-lying states of TlF arising from $^1\Sigma^+$, $^3\Pi$, $^1\Pi$, $^3\Sigma^-$, $^3\Sigma^+$ and $^3\Delta$ were carried out.[125] These calculations confirmed the existence of barriers in the excited states of TlF which arise from relativistic avoided crossings. Calculations on a number of low-lying states of ICl and ICl^+ have also been carried out.[126,127] These calculations enabled interpretation of the electronic spectra of these molecules.

Relativistic CI calculations of the low-lying states of PbF were recently carried out.[128] The assignments of the experimentally observed A and B states were ambiguous as a result of the existence of a number of low-lying states. Relativistic CI calculations of these states enabled the assignment of A and B states to $^2\Sigma^+_{1/2}$ and $^2\Sigma^+_{1/2}(II)$. The calculations of the B state, which is a Rydberg state, were carried out with an extended triple-zeta basis set. The ground configuration $^2\Pi$ is split into $^2\Pi_{1/2}$ and $^2\Pi_{3/2}$ states with a spin–orbit splitting of about 7895 cm^{-1}. Basch and Topiol[129] have carried out calculations on AuCl and PtH in addition to $HgCl_2$.

Laskowski and Langhoff[130] have carried out calculations on CrI using averaged relativistic effective potentials. Similar calculations have been carried out on CsO[131] as well as CsH[132]. Krauss and Stevens[133,134] carried out SCF calculations on UO, UH, UF and their ions. Krauss and Stevens[135] have investigated the electronic structure of FeO and RuO using relativistic effective potentials. Relativistic configuration-interaction calculations of low-lying states of BiF have been completed.[136]

C. Relativistic Calculations of Polyatomics and Miscellaneous Applications

Relativistic calculations of a number of polyatomic molecules have been carried out by both semi-empirical and *ab initio* methods. Many of the earlier calculations have been reviewed by Pyykkö[7] as well as by Christiansen, Ermler and Pitzer.[13] The semi-empirical methods include the relativistic extended Hückel method, relativistic $X\alpha$ method, etc. Most of the *ab initio* calculations on polyatomics initially omit the spin–orbit interaction. This is introduced at a later stage as a perturbation or by a semi-empirical scheme.

One of the very interesting earlier calculations on polyatomics was carried out by Wadt[137] on ThO_2 and UO_2^{2+}. While the UO_2^{2+} ion is linear, ThO_2 is bent. There is a greater 5f participation in UO_2^{2+} in contrast to ThO_2, and this seems to be the reason for the change in structure. Wadt notes that earlier semi-empirical calculations on these molecules yielded results which do not agree with his accurate *ab initio* calculations.

Another interesting molecule is the cyclooctatetraene sandwich complex, $U(C_8H_8)_2$. The 5f contribution is important to the stability of uranium-containing molecules. Semi-empirical calculations have been carried out on this molecule as well as other actinocene compounds.[138,139] Relativistic calculations on UF_6, NpF_6 and PuF_6 have been reported.[140–144]

Case and coworkers[45,46] have reported calculations on Pt clusters and their reactivity with CO using relativistic $X\alpha$ methods. Noell and Hay[145] have carried out SCF calculations on Pt complexes such as $Pt(NH_3)$ using the REPs generated from Pauli Hamiltonians. Hay[146] has also studied the Pt–ethylene complex. Similar calculations have been carried out by Basch and Cohen[147] on PtCO.

Collignon and Schwarz[148] have studied the change of molecular structures as a result of relativistic effects. They have carried out calculations on $PbCl_2$, PbH_2, Pb_2H_4, TeH_2, PoH_2 and $^{114}EH_2$ using a relativistic SCF pseudo-potential method. Spin–orbit interaction changes the bond angle by less than a degree in these systems with the exception of $^{114}EH_2$ where the bond angle is increased by $2.3°$.

Relativistic calculations have made an impact on biological compounds recently. Miller *et al.*[149] have recently investigated the binding of *cis*-$Pt(NH_3)_2^{2+}$ to the bases of DNA (guanine (G), cytocine (C), adenine (A) and thymine (T)). Electronic and geometrical structures of $Pt(NH_3)_2Cl_2$, $Pt(NH_3)_3X$ and $Pt(NH_3)_2XY$ (X, Y = H_2O, OH^-) have also been investigated recently by Basch, Krauss and Stevens.[150]

Pitzer and Winter[144] have developed a computer code for including the *ab inito* spin–orbit operator in polyatomic calculations. In this method the spin–orbit integrals are evaluated using the difference of $l + \frac{1}{2}$ and $l - \frac{1}{2}$

potentials over Gaussian basis sets. The procedure is being tested on UF_6, NpF_6 and PuF_6.

Ross, Ermler and Christiansen[151] have recently carried out *ab initio* EP calculations of spin–orbit coupling in the group III A and group VII A atoms. The spin–orbit splittings are computed with the operator represented as the difference of $l + \frac{1}{2}$ and $l - \frac{1}{2}$ effective potentials. Comparisons were also made with the all-electron Dirac–Fock results as well as the first-order perturbation calculations. These authors have shown that the first-order perturbation results could be in error by 9% in comparison to the Dirac–Fock results and that the EP spin–orbit operator yields accurate results.

Self-consistent Dirac–Slater calculations of molecules and embedded clusters have been recently reviewed by Ellis and Goodman.[152] Relativistic band structure calculations have also been carried out.[153] Dirac scattered-wave calculations have been carried out on a number of inorganic complexes such as $W(CO)_6$[154] and $W_2Cl_8^{4-}$.[155] The electronic structure and geometries of X_2H_2 ($X = O, S, Se$ and Te) have also been investigated recently.[156]

Malli[157] has investigated recently the use of Dirac-type functions (DTF) as basis sets for relativistic Dirac–Fock–Roothaan calculations for atoms. The well known Slater-type functions with non-integral principal quantum numbers are special cases of DTFs. Recently, self-consistent relativistic Thomas–Fermi equations for heavy atomic ions have been formulated.[158] Matsuhita *et al.*[159] have reported *ab initio* spin–orbit splittings of the ^3P ground state of the Se atom using the Pauli spin–orbit term. They obtain spin–orbit splitting as a perturbation to multi-reference singles and doubles CI (MRSDCI) wavefunction. The relativistic Pauli integrals over Gaussian basis sets are evaluated by the procedure proposed by Chandra and Buenker.[160,161] Similar calculations have also been made on CBr.[162]

Balasubramanian and coworkers[163] have carried out relativistic CI calculations on the collision of Kr with Br^+. The collisions of rare-gas atoms with heavy halogen ions are the topics of a number of investigations. (For a brief review of recent progress in experimental and theoretical works in this area see Ref. 164.) Krauss, Stevens and Basch[165] have very recently carried out EP calculations on AgH and AuH. Their results are in very good agreement with all-electron Dirac–Fock calculations.

Acknowledgements

K. Balasubramanian would like to thank the Arizona State University for the Institutional Biomedical Research Grant, Faculty Grant-in-Aid and the Alfred P. Sloan Foundation for their support. The work at Berkeley was supported by the Director, Office of Energy Research, Office of Basic Energy Sciences, Chemical Sciences Division of the US Department of Energy under Contract no. DE-A3O3-765F0098.

References

1. Bethe, H. A., and Salpeter, E. E., *Quantum Theory of One and Two Electron Atoms*, Plenum, New York, 1977.
2. Moore, C. E., *Atomic Energy Levels*, National Bureau of Standards, 1971.
3. Pitzer, K. S., *J. Chem. Soc., Chem. Commun.*, 760 (1975).
4. Balasubramanian, K., and Pitzer, K. S., *J. Chem. Phys.*, **78**, 321 (1983).
5. Bagus, P. S., Lee, Y., and Pitzer, K. S., *Chem. Phys. Lett.*, **33**, 408 (1975).
6. Desclaux, J. P., *At. Data Nucl. Data*, **12**, 311 (1973).
7. Pyykkö, P., *Adv. Quantum Chem.*, **11**, 353 (1978).
8. Pyykkö, P., and Desclaux, J. P., *Acc. Chem. Res.*, **12**, 276 (1979).
9. Pitzer, K. S., *Acc. Chem. Res.*, **12**, 271 (1979).
10. Malli, G. L. (Ed.), *Relativistic Effects in Atoms, Molecules and Solids*, Plenum, New York, 1982.
11. Pyykkö, P. (Ed.), Proceedings of the symposium on relativistic effects in quantum chemistry. *Int. J. Quantum Chem.*, **25**, 1984.
12. Krauss, M., and Stevens, W. J., *Annu. Rev. Phys. Chem.*, **35**, 357 (1984).
13. Christiansen, P. A., Ermler, W. C., and Pitzer, K. S., *Annu. Rev. Phys. Chem.*, **36**, 407 (1985).
14. Desclaux, J. P., *Comput. Phys. Commun.*, **9**, 31 (1975).
15. Lee, Y. S., Ermler, W. C., and Pitzer, K. S., *J. Chem. Phys.*, **15**, 5861 (1977).
16. Schwarz, W. H. E., in *Relativistic Effects in Atoms, Molecules and Solids* (Ed. G. Malli), p. 518, Plenum, New York, 1982; also private communications, 1982.
17. Cowan, R. D., and Griffin, D. C., *J. Opt. Soc. Am.*, **66**, 1010 (1976).
18. Davidson, E. R., Feller, D., and Phillips, P., *Chem. Phys. Lett.*, **76**, 416 (1980).
19. Foldy, L. L., and Wouthuysen, S. A., *Phys. Rev.*, **78**, 29 (1980).
20. Cohen, J. S., Wadt, W. R., and Hay, P. J., *J. Chem. Phys.*, **71**, 2955 (1979).
21. Hay, P. J., Wadt, W. P., Kahn, L. R., Raffenetti, R. C., and Phillips, D. H., *J. Chem. Phys.*, **71**, 1767 (1979).
22. Hay, P. J., and Wadt, W. R., *J. Chem. Phys.*, **82**, 270 (1985).
23. Wadt, W. R., and Hay, P. J., *J. Chem. Phys.*, **82**, 284 (1985).
24. Hay, P. J., and Wadt, W. R., *J. Chem. Phys.*, **82**, 299 (1985).
25. Morrison, J. D., and Moss, R. E., *Mol. Phys.*, **41**, 191 (1980).
26. Kuttzelnig, W., *Int. J. Quantum Chem.*, **25**, 107 (1984).
27. Desclaux, J. P., and Pyykkö, P., *Chem. Phys. Lett.*, **29**, 534 (1974).
28. Kim, Y. K., *Phys. Rev.*, **154**, 17 (1967).
29. Grant, I. P., *Adv. Phys.*, **19**, 747 (1970).
30. Malli, G., *Stud. Phys. Theor. Chem.*, **21**, 199 (1982).
31. Malli, G., and Oreg, J., *J. Chem. Phys.*, **63**, 830 (1975).
32. Matsuoka, O., Suzui, N., Aoyama, T., and Malli, G., *J. Chem. Phys.*, **73**, 1320 (1980).
33. Aoyama, T., Yamakawa, H., and Matsuoka, O., *J. Chem. Phys.*, **73**, 1329 (1980).
34. Lee, Y. S., and McLean, A. D., *J. Chem. Phys.*, **76**, 735 (1982).
35. McLean, A. D., and Lee, Y. S., *Stud. Phys. Theor. Chem.*, **21**, 219 (1982).
36. Breit, G., *Phys. Rev.*, **34**, 553 (1929).
37. Stevens, W. J., and Krauss, M., *Chem. Phys. Lett.*, **86**, 320 (1982).
38. Stevens, W. J., and Krauss, M., *J. Chem. Phys.*, **76**, 3834 (1982).
39. Lohr, L. L., and Pyykkö, P., *Chem. Phys. Lett.*, **62**, 333 (1979).
40. Pyykkö, P., and Lohr, L. L., *Inorg. Chem.*, **20**, 1950 (1981).
41. Pyykkö, P., and Laaksonen, L., *J. Phys.*, **88**, 4892 (1984).
42. Case, D. A., *Annu. Rev. Phys. Chem.*, **33**, 151 (1982).

43. Case, D. A., and Lopez, J. P., *J. Chem. Phys.*, **80**, 3270 (1984).
44. Lopez, J. P., and Case, D. A., *J. Chem. Phys.*, **81**, 4554 (1984).
45. Yang, C. Y., Yu, H.-L., and Case, D. A., *Chem. Phys. Lett.*, **81**, 170 (1981).
46. Yang, C. Y., and Case, D. A., *Surface Sci.*, **106**, 523 (1981).
47. Oreg, J., and Malli, G., *J. Chem. Phys.*, **65**, 1746 (1976).
48. Oreg, J., and Malli, G., *J. Chem. Phys.*, **65**, 1755 (1976).
49. Malli, G., and Oreg, J., *Chem. Phys. Lett.*, **69**, 313 (1980).
50. Malli, G., *Chem. Phys. Lett.*, **73**, 510 (1980).
51. Phillips, J. C., and Kleinman, L., *Phys. Rev.*, **116**, 287 (1959).
52. Ermler, W. C., Lee, Y. S., Pitzer, K. S., and Winter, N. W., *J. Chem. Phys.*, **69**, 976 (1978).
53. Lee, Y. S., Ermler, W. C., Pitzer, K. S., and McLean, A. D., *J. Chem. Phys.*, **70**, 288 (1979).
54. Ermler, W. C., Lee, Y. S., and Pitzer, K. S., *J. Chem. Phys.*, **70**, 293 (1979).
55. Lee, Y. S., Ermler, W. C., and Pitzer, K. S., *J. Chem. Phys.*, **73**, 360 (1980).
56. Christiansen, P. A., Lee, Y. S., and Pitzer, K. S., *J. Chem. Phys.*, **71**, 4445 (1979).
57. Kahn, L., Baybutt, P., and Truhlar, D. G., *J. Chem. Phys.*, **65**, 3826 (1976).
58. Hafner, P., and Schwarz, W. H. E., *Chem. Phys. Lett.*, **65**, 537 (1979).
59. Ermler, W. C., Lee, Y. S., Christiansen, P. A., and Pitzer, K. S., *Chem. Phys. Lett.*, **81**, 70 (1981).
60. Pacios, L. F., and Christiansen, P. A., *J. Chem. Phys.*, **82**, 2664 (1985).
61. Christiansen, P. A., and Ermler, W. C., to be published.
62. Hay, P. J., Wadt, W. R., and Kahn, L., *J. Chem. Phys.*, **68**, 3059 (1978).
63. Kahn, L. R., Hay, P. J., and Cowan, R. D., *J. Chem. Phys.*, **68**, 2836 (1979).
64. Hafner, P., and Schwarz, W. H. E., *J. Phys. B: At. Mol. Phys.*, **11**, 217 (1978).
65. Esser, M., Butscher, W., and Schwarz, W. H. E., *Chem. Phys. Lett.*, **71**, 539 (1981).
66. Hafner, P., and Schwarz, W. H. E., *Chem. Phys. Lett.*, **65**, 537 (1979).
67. Hafner, P., Habitz, P., Ishikawa, Y., Wechsel-Trakowski, E., and Schwarz, W. H. E., *Chem. Phys. Lett.*, **80**, 311 (1981).
68. Schwarz, W. H. E., and Wallmeier, H., *Mol. Phys.*, **46**, 1045 (1982).
69. Schwarz, W. H. E., and Wechsel-Trakowski, E., *Chem. Phys. Lett.*, **85**, 94 (1982).
70. Mark, F., and Schwarz, W. H. E., *Phys. Rev. Lett.*, **48**, 673 (1982).
71. Mark, F., Marian, C., and Schwarz, W. H. E., *Mol. Phys.*, **53**, 535 (1984).
72. Schwarz, W. H. E., and Chu, S. Y., *Mol. Phys.*, **50**, 603 (1983).
73. Esser, M., *Int. J. Quantum Chem.*, **26**, 313 (1984).
74. Christiansen, P. A., and Pitzer, K. S., *J. Chem. Phys.*, **73**, 5160 (1980).
75. Christiansen, P. A., and Pitzer, K. S., *J. Chem. Phys.*, **74**, 1162 (1981).
76. Christiansen, P. A., and Pitzer, K. S., *Chem. Phys. Lett.*, **85**, 434 (1981).
77. Christiansen, P. A., and Pitzer, K. S., *J. Chem. Phys.*, **73**, 5148 (1980).
78. Pitzer, K. S., and Christiansen, P. A., *Chem. Phys. Lett.*, **77**, 589 (1981).
79. Christiansen, P. A., Pitzer, K. S., Lee, Y., Yates, J. H., Ermler, W. C., and Winter, N. W., *J. Chem. Phys.*, **75**, 5410 (1981).
80. Yates, J. H., Ermler, W. C., Winter, N. W., Christiansen, P. A., Lee, Y., and Pitzer, K. S., *J. Chem. Phys.*, **79**, 6145 (1983).
81. Christiansen, P. A., Balasubramanian, K., and Pitzer, K. S., *J. Chem. Phys.*, **76**, 5087 (1982).
82. Pitzer, R. M., and Winter, N. W., to be published.
83. Pitzer, K. S., *Int. J. Quantum Chem.*, **25**, 131 (1984).
84. Kahn, L., *Int. J. Quantum Chem.*, **25**, 149 (1984).
85. Ziegler, T., Snjders, J. G., and Baerends, E. J., *J. Chem. Phys.*, **74**, 1271 (1981).
86. Ross, R. B., and Ermler, W. C., *J. Phys. Chem.*, in press.

87. McLean, A. D., *J. Chem. Phys.*, **79**, 7 (1983).
88. Christiansen, P. A., *J. Chem. Phys.*, **79**, 2928 (1983).
89. Pitzer, K. S., and Balasubramanian, K., *J. Phys. Chem.*, **86**, 3068 (1982).
90. Bondybey, V. E., and English, J. H., *J. Chem. Phys.*, **76**, 2165 (1982); *ibid.*, **67**, 3405 (1977).
91. Teichman, R. A., and Nixon, E. R., *J. Mol. Spectrosc.*, **59**, 299 (1976).
92. Gingerich, K. A., Cooke, D. L., and Miller, F., *J. Chem. Phys.*, **64**, 4027 (1976).
93. Pitzer, K. S., *J. Chem. Phys.*, **74**, 3078 (1981).
94. Ackerman, M., Droward, J., Stafford, F. E., and Verhaegen, G., *J. Chem. Phys.*, **36**, 1557 (1962).
95. Sontag, H., and Weber, R., *J. Mol. Spectrosc.*, **100**, 75 (1983).
96. Pacchioni, G., *Mol. Phys.*, **55**, 211 (1985).
97. Christiansen, P. A., *Chem. Phys. Lett.*, **109**, 145 (1984).
98. Celestino, K.-C., and Ermler, W. C., *J. Chem. Phys.*, **81**, 1872 (1984).
99. Stevens, W. J., Basch, H., and Krauss, M. J., *J. Chem. Phys.*, **81**, 6026 (1984).
100. Pelissier, M., *J. Chem. Phys.*, **79**, 2099 (1983).
101. Adachi, H., Rosen, A., and Ellis, D. E., *Mol. Phys.*, **33**, 199 (1977).
102. Christiansen, P. A., and Ermler, W. C., *Mol. Phys.*, to be published.
103. Teichtal, C., and Spiegelmann, F., *J. Chem. Phys.*, **81**, 273 (1983).
104. Flannery, M. R., McCann, K. J., and Winter, N. W., *J. Phys. B: At. Mol. Phys.*, **14**, 3789 (1981).
105. Morgan, W. L., Winter, N. W., and Kulander, K. C., *J. Appl. Phys.*, **54**, 4275 (1983).
106. Desclaux, J. P., and Pyykkö, P., *Chem. Phys. Lett.*, **39**, 300 (1976).
107. Pyykkö, P., and Desclaux, J. P., *Nature*, **266**, 336 (1977).
108. Pyykkö, P., and Desclaux, J. P., *Chem. Phys.*, **34**, 261 (1978).
109. Pyykkö, P., and Desclaux, J. P., *Chem. Phys. Lett.*, **50**, 503 (1977).
110. Hay, P. J., Wadt, W. R., Kahn, L. R., and Bobrowicz, I. W., *J. Chem. Phys.*, **69**, 984 (1978).
111. Balasubramanian, K., and Pitzer, K. S., *J. Phys. Chem.*, **88**, 1146 (1984).
112. Balasubramanian, K., and Pitzer, K. S., *J. Mol. Spectrosc.*, **103**, 105 (1984).
113. Balasubramanian, K., *Chem. Phys. Lett.*, **114**, 201 (1985).
114. Balasubramanian, K., *J. Mol. Spectrosc.*, **115**, 258 (1986).
115. Balasubramanian, K., *J. Phys. Chem.*, **90**, 1037 (1986).
116. Chapman, D. A., Balasubramanian, K., and Lin, S. H., *Chem. Phys. Lett.*, **118**, 192 (1985).
117. Chapman, D. A., Balasubramanian, K., and Lin, S. H., manuscript in preparation.
118. Wang, S. W., and Pitzer, K. S., *J. Chem. Phys.*, **79**, 3851 (1983).
119. Huber, K. P., and Herzberg, G., *Molecular Spectra and Molecular Structure*, Vol. IV, *Constants of Diatomic Molecules*, Van Nostrand Reinhold, New York, 1979.
120. Balasubramanian, K., and Pitzer, K. S., *J. Phys. Chem.*, **87**, 4857 (1983).
121. Balasubramanian, K., and Pitzer, K. S., *Chem. Phys. Lett.*, **100**, 273 (1983).
122. Balasubramanian, K., *J. Chem. Phys.*, in press.
123. Balasubramanian, K., *J. Phys. Chem.*, **88**, 5759 (1984).
124. Dyke, J. M., Morris, A., Ridha, A. M., and Snijders, J. G., *Chem. Phys.*, **67**, 245 (1982).
125. Balasubramanian, K., *J. Chem. Phys.*, **82**, 3741 (1985).
126. Balasubramanian, K., *J. Mol. Spectrosc.*, **110**, 339 (1985).
127. Balasubramanian, K., *Chem. Phys.*, **95**, 225 (1985).

128. Balasubramanian, K., *J. Chem. Phys.*, **83**, 2311 (1985).
129. Basch, H., and Topiol, S., *J. Chem. Phys.*, **71**, 802 (1979).
130. Laskowski, B. C., and Langhoff, S. R., *Chem. Phys. Lett.*, **92**, 49 (1982).
131. Laskowski, B. C., Langhoff, S. R., and Seighbahn, P. E., *Int. J. Quantum Chem.*, **23**, 483 (1983).
132. Laskowski, B. C., Walch, S. P., and Christiansen, P. A., *J. Chem. Phys.*, **78**, 6824 (1983).
133. Krauss, M., and Stevens, W. J., *Chem. Phys. Lett.*, **99**, 417 (1983).
134. Krauss, M., and Stevens, W. J., *J. Comput. Chem.*, **4**, 127 (1983).
135. Krauss, M., and Stevens, W. J., *J. Chem. Phys.*, **82**, 5584 (1985).
136. Balasubramanian, K., *Chem. Phys. Lett.*, in press.
137. Wadt, W. R., *J. Chem. Soc.*, **103**, 6053 (1981).
138. Streitweiser, A., Jr, and Muller-Westerhoff, V., *J. Chem. Soc.*, **90**, 7264 (1968).
139. Hayes, R. G., and Edelstein, N., *J. Am. Chem. Soc.*, **94**, 8688 (1972).
140. Koelling, D. D., Ellis, D. E., and Bartlett, R. J., *J. Chem. Phys.*, **65**, 3331 (1976).
141. Rosen, A., *Chem. Phys. Lett.*, **55**, 311 (1978).
142. Hay, P. J., Wadt, W. R., Kahn, L. R., Raffennetti, R. C., and Phillips, D. H., *J. Chem. Phys.*, **71**, 1767 (1979).
143. Hay, P. J., *J. Chem. Phys.*, **79**, 5469 (1983).
144. Pitzer, R. M., and Winter, N. W., 5th Int. Congr. on Quantum Chemistry, Montreal, Abstract, 1985, full paper to be published.
145. Moell, J. O., and Hay, P. J., *Inorg. Chem.*, **21**, 14 (1982).
146. Hay, P. J., *J. Am. Chem. Soc.*, **104**, 7007 (1982).
147. Basch, H., and Cohen, D., *J. Am. Chem. Soc.*, **105**, 3856 (1983).
148. Collignon, G., and Schwarz, W. H. E., 5th Int. Congr. on Quantum Chemistry, Montreal, Poster, 1985.
149. Miller, K. J., Taylor, E. R., Basch, H., Krauss, M., and Stevens, W. J., *J. Biomol. Struct. Dyn.*, **2**, 1157 (1985).
150. Basch, H., Krauss, M., and Stevens, W. J., *Inorg. Chem.*, in press.
151. Ross, R. B., Ermler, W. C., and Christiansen, P. A. submitted.
152. Ellis, D. E., and Goodman, G. L., *Int. J. Quantum Chem.*, **25**, 185 (1984).
153. Christiensen, N. E., *Int. J. Quantum Chem.*, **25**, 233 (1984).
154. Yang, C. Y., Arratia-Perez, R., and Lopez, J. P., *Chem. Phys. Lett.*, **107**, 112 (1984).
155. Arratia-Perez, R., and Case, D. A., *Inorg. Chem.*, **23**, 3271 (1984).
156. Block, P., and Jansen, L., *J. Chem. Phys.*, **82**, 3322 (1985).
157. Malli, G., *J. Chem. Phys.*, **80**, 2060 (1984).
158. March, N. H., *Int. J. Quantum Chem.*, **27**, 595 (1985).
159. Matsuhita, T., Marian, C. M., Klotz, R., Hess, B. H., and Peyerimhoff, S. D., *Chem. Phys.*, **96**, 371 (1985).
160. Chandra, P., and Buenker, R. J., *J. Chem. Phys.*, **79**, 358 (1983).
161. Chandra, P., and Buenker, R. J., *J. Chem. Phys.*, **79**, 366 (1983).
162. Hess, B. A., Chandra, P., and Buenker, R. J., *Chem. Phys. Lett.*, **119**, 403 (1985).
163. Balasubramanian, K., Hariharan, P. C., Kaufman, J. J., and Koski, W. S., *Chem. Phys. Lett.*, submitted for publication.
164. Hotoka, M., Roos, B., Balasubramanian, K., Semo, N., Sharma, R. B., and Koski, W. S., in *Advances in Chemical Reaction Dynamics* (Eds. P. Renczepis, and C. Capellos), Reidel, Dordrecht (in press).
165. Krauss, M., Stevens, W. J., and Basch, H., *J. Comput. Chem.*, **4**, 287 (1985).

Ab Initio Methods in Quantum Chemistry—I
Edited by K. P. Lawley
© 1987 John Wiley & Sons Ltd.

EFFECTIVE HAMILTONIANS AND PSEUDO-OPERATORS AS TOOLS FOR RIGOROUS MODELLING

PHILIPPE DURAND and JEAN-PAUL MALRIEU

Laboratorie de Physique Quantique, Unité Associée au CNRS no. 505, Université Paul Sabatier, 118 route de Narbonne, 31062 Toulouse Cedex, France

CONTENTS

I. INTRODUCTION

Progress in the reliability of quantum-chemical *ab initio* techniques has been so impressive that the quantum-chemical calculations are sometimes proposed to chemists as a 'new spectroscopy'; in many cases they can actually provide information more direct than (and almost as reliable as) the experimental spectroscopies. This is especially true for problems concerning the possible existence and structure of transient polyatomic small molecules, the order of magnitude of activation energies, and so on. This increasing efficiency in the prediction ability of these instruments is largely due to technological progress in computers (i.e. to an exogenic factor), but it has proceeded through the development of mathematically efficient and physically relevant algorithms, which required a new scientific profile for quantum chemists. Their pride and hopeful prospects are thus partly grounded. One should, however, point out two types of problems.

A. Quantum Chemistry as a Science or a Technology?

The numerical efficiency of black boxes giving more and more precise numbers to experimentalists is a technological goal; it actually exhibits a tendency to transform this field into a numerical spectrometer, but it defines neither a scientific field nor a scientific demand. The physical sciences are (or were?) essentially deductive, i.e. they started from general principles and, through a series of simplifying and reasonable assumptions, they were able to derive—mostly through analytical models—some laws, trends, orders of magnitude, etc., in an explicit way. Computational black boxes deliver the desired energy, and as a by-product, a wavefunction spread on thousands of determinants, which can neither be read nor understood, and does not even offer a possible way for an *a posteriori* rationalization. The information becomes so vast that it becomes useless. If one still believes that science must bring some explanation or derivation, reduction of information must be considered as a desirable task. The present contribution assumes that the rational reduction of information represents an essential goal for *understanding* physics and chemistry, and this must proceed through physically grounded simplified schemes.

B. Reduction of Information: Two Essential Tools

The present review is devoted to two main approaches that may lead in a controlled way from the exact Hamiltonian to simplified Hamiltonians, which are more easy to handle and on which deductive derivations may be easier to draw. One approach proceeds through *projections* of some exact wavefunctions into a relevant reduced subspace and leads to the *effective Hamiltonian* methodology. The techniques will be described in Section II.A, and their applications in Section III. The other procedure may be considered as a *simulation* of the considered exact Hamiltonian by a simpler Hamiltonian, the efficiency of the simulation being measured through a reduced distance, taken on a small subspace. This procedure, described in Section II.B, leads to the definition of pseudo-Hamiltonians or pseudo-operators. The corresponding applications are reviewed in Section IV.

C. Terminology: a Proposal

Despite the large confusion in the terminology existing in this field (to which the authors have also contributed), we propose for the future the following vocabulary:

1. The term 'effective Hamiltonian' should be used for Hamiltonians obtained

by projections of some exact wavefunctions onto a finite model space. The corresponding theory is well established and the practical (perturbative or not) procedures are numerous and well documented.

2. The term 'pseudo-Hamiltonian' should be used when they are obtained through simulation techniques, i.e. minimization of the distance between the exact and pseudo-operators in a reduced subspace.

The term 'model Hamiltonian' is more general. From this point of view the 'effective Hamiltonians' and 'pseudo-Hamiltonians' could also be considered as 'model Hamiltonians'. However, for clarity, we suggest that the term 'model Hamiltonian' should be used when some simplified form of an approximate Hamiltonian has been guessed from a preliminary physical analysis. In contrast with the 'effective Hamiltonians' and the 'pseudo-Hamiltonians' that can be obtained by means of well defined mathematical procedures, the 'model Hamiltonians' are generally parametrized from experiment. They would involve the semi-empirical Hamiltonians of quantum chemistry and solid-state physics (Hückel, Hubbard, Pariser–Parr–Pople (PPP),...).

D. Desirable Simplifications in Quantum Chemistry

Modelling involves a reduction or a simplification of the problem. This reduction may concern:

1. The number of particles of the problem. This is the scope of π theories for conjugated systems and valence Hamiltonians more generally, but other reductions are conceivable.

2. The basis set of atomic orbitals (AO) in which the problem is supposed to be treated. All *ab initio* calculations use projected Hamiltonians since they work in finite basis sets, of course, solving

$$P_S H P_S \psi_S^m = E_S \psi_S^m$$

where P_S is the projector onto the space of determinants built in the considered AO basis set. But one may also try to build 'effective' Hamiltonians, different from the exact one, working only in a small basis set, but with modified operators giving better energies and wavefunctions than the previously defined E_S^m and ψ_S^m. This strategy is the one followed by K. Freed and coworkers in a series of papers, discussed later, which also try to achieve the previously mentioned goal (reduction of the number of particles).

3. The number of N-electron configurations (or determinants) in which one would like to treat the problem. Instead of handling the huge number of configurations which span the Hamiltonian in a finite basis set (full configuration interaction (CI)), one may wish to explain the behaviour of a few solutions in terms of a few leading configurations only, without losing the quality of the energetic information. There exist two main examples of this strategy,

a. one is the interpretation of curve crossings between molecular potential energy surfaces and their diabatic description in terms of dominant configurations,

b. the other concerns the Heisenberg Hamiltonians, which only treat neutral situations in minimal basis sets in such a way that they deliver correct energies, despite the lack of inclusion of ionic states.

One should notice that point 2. implies a reduction of the number of determinants, but point 3 is not reducible to point 2.

4. The relativistic molecular calculations are very difficult as long as they keep four-component wavefunctions according to the Dirac theory. The reduction to a two-component Pauli-like formalism within the effective Hamiltonian theory allows one to perform standard relativistic variational calculations.

5. As a basically different approach, one may consider the simplification of the Hamiltonian, for instance its analytical form. One may try to build a purely monoelectronic Hamiltonian that is 'as close as possible' in some aspects to the exact (bielectronic) Hamiltonian; for instance, which reproduces as closely as possible the Fock monoelectronic energies obtained from the exact Hamiltonian, or the total energy and its changes.

E. Simplification and Efficiency Overlap

As a concluding remark, we would like to stress the fact that effort to find efficient and physically grounded simplified Hamiltonians is not only the answer to a desire for interpretation, it is also a valuable technical tool for the treatment of large systems, since the N dependences of *ab initio* quantum-chemical algorithms are still prohibitive and forbid the treatment of large polyatomic systems, despite the expected progress in computational facilities. Happily enough, the desire for simplification of the information, the desire for understanding, is not at odds with research into numerical efficiency. This is clear for the treatment of large systems, but it is even true for the heavy *ab initio* techniques concerning small systems: techniques that express the CI results in terms of more convenient nearly diabatic pictures rest on the effective Hamiltonian theory. They are less expensive than the usual adiabatic approaches and they may help to solve numerical accuracy problems in the treatment of excited states in large CI approaches, as shown in Section III.B.

II. MATHEMATICAL TOOLS

A. Effective Hamiltonians by Projection Techniques[1-3]

1. The Model Space

The concept of *model space* plays a central role in the theory of effective Hamiltonians.[2] It is a finite N_m-dimensional subspace S_0 of the entire Hilbert space. Physics will further be projected in this model space. Its orthogonal complement is the *outer space* $S_0^\perp$. The orthogonal projection operators associated with S_0 and $S_0^\perp$ are P_0 and Q_0, respectively:

$$P_0 = \sum_{m=1}^{N_m} |m\rangle\langle m| \qquad Q_0 = \sum_\alpha |\alpha\rangle\langle\alpha| \qquad P_0 + Q_0 = 1 \qquad (1)$$

The Latin letters $m, n, \ldots$ and Greek letters $\alpha, \beta, \ldots$ will label functions of the model space and the outer space, respectively.

It is also useful to consider an N_m-dimensional subspace S spanned by N_m exact solutions ψ_m of the exact Hamiltonian H. These solutions will correspond to the part of the spectrum in which we are interested. The orthogonal complement of S is denoted $S^\perp$. The orthogonal projectors associated with S and $S^\perp$ are P and Q:

$$P = \sum_{m=1}^{N_m} |\psi_m\rangle\langle\psi_m| \qquad Q = \sum_\alpha |\psi_\alpha\rangle\langle\psi_\alpha| \qquad P + Q = 1 \qquad (2)$$

The ψ_α are the exact solutions of H belonging to $S^\perp$. The projections Φ_m of the exact solutions ψ_m into the model space play a central role in the theory:

$$\psi_m \in S \rightleftharpoons \Phi_m = P_0\psi_m \in S_0 \qquad (3)$$

Relation (3) establishes a one-to-one correspondence between S and S_0. The projected wavefunctions Φ_m are usually not orthogonal and the corresponding bi-orthogonal states in S_0, noted $\Phi_m^\perp$, have the usual properties:

$$\langle\Phi_m^\perp|\Phi_n\rangle = \delta_{mn} \qquad \sum_{m=1}^{N_m} |\Phi_m\rangle\langle\Phi_m^\perp| = P_0 \qquad (4)$$

2. The Feschbach–Löwdin Hamiltonian[4-6]

Let us consider an exact eigenstate of energy E:

$$H\psi = E\psi \qquad (5)$$

This state may be non-degenerate or degenerate. Using the projections operators P_0 and Q_0, Eq. (5) can be partitioned according to (partitioning technique)

$$P_0 H P_0 \psi + P_0 H Q_0 \psi = E P_0 \psi$$
$$Q_0 H P_0 \psi + Q_0 H Q_0 \psi = E Q_0 \psi \qquad (6)$$

It can easily be seen from (6) that the projection $\Phi = P_0\psi$ of ψ in the model space is a solution of

$$H^{\text{eff}}\Phi = E\Phi \qquad (7)$$

where

$$H^{\text{eff}} = P_0 H P_0 + P_0 H \frac{Q_0}{E - H} H P_0 \qquad (8)$$

$Q_0/(E - H)$ is a reduced resolvent defined in the outer space. H^{eff} can also be written in the form

$$H^{\text{eff}} = P_0 H P_0 + P_0 V \frac{Q_0}{E - H} V P_0 \qquad (9)$$

where the coupling operators $P_0 H Q_0$ and $Q_0 H P_0$ have been denoted $P_0 V Q_0$ and $Q_0 V P_0$, respectively.

H^{eff} given by (8) or (9) is an effective Hamiltonian which possesses the exact eigenenergy E and the corresponding wavefunction $\Phi = P_0 \psi$. It can also be written in the form

$$H^{\text{eff}} = P_0 H \Omega_E \qquad (10)$$

where

$$\Omega_E = P_0 + \frac{Q_0}{E - H} V P_0 \qquad (11)$$

Ω_E is a *wave operator*, parametrized by the exact energy E, which has the basic property of generating the exact wavefunction when acting on its projection in the model space:

$$\Omega_E \Phi = \Omega_E P_0 \psi = \psi \qquad (12)$$

Figure 1 gives an illustration of the correspondence between ψ and Φ by means of P_0 and Ω_E.

Despite its apparent simplicity the Feschbach–Löwdin Hamiltonian suffers from a severe limitation: it is relevant for only one energy level and there are as many H^{eff} as energies. This limitation pleads in favour of a energy-independent formalism that will now be presented throughout this section.

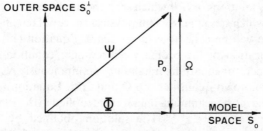

Fig. 1. From the exact wavefunction ψ to the projected wavefunction Φ and back from Φ to ψ by the wave operator Ω.

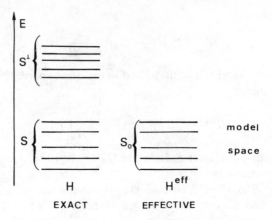

Fig. 2. The exact Hamiltonian H and the effective Hamiltonian have the same eigenenergies in the subspaces S and S_0, respectively.

3. The Bloch and des Cloizeaux Effective Hamiltonians[7,8]

The basic idea is to pass from the exact Hamiltonian H to an effective Hamiltonian whose eigenenergies coincide within a subset of the eigenenergies of H (Fig. 2). In this approach all the information on the other eigenenergies of H is lost. The theory can easily be derived from the basic equation[9]

$$H\Omega = \Omega H^{\text{eff}} \tag{13}$$

where H^{eff} is an effective Hamiltonian defined in the N_m-dimensional model space S_0 and Ω is the associated *wave operator* acting in the model space:

$$\Omega = \Omega P_0 \tag{14}$$

Equation (13) shows immediately that if Φ_m is an eigensolution of energy E_m then

$$\psi_m = \Omega \Phi_m \tag{15}$$

is an exact solution of H. Equation (15) expresses a one-to-one correspondence between the N_m solutions of H^{eff} which span the model space S_0 and N_m exact solutions of H which span the relevant physical space S. Hereafter this *a priori* unknown space will be called the *target space*. Equation (13) appears as a simple generalization of the ordinary Schrödinger equation for one state (Eq. (5)) to an operator equation for handling simultaneously N_m states. In this equation the unknown quantities are Ω and H^{eff}. Equation (13) has many solutions and the most meaningful ones were identified in the 1950s by Okubo, Bloch and des Cloizeaux.[7,8,10] The solution obtained by Bloch in the framework of perturbation theory now appears as the most fundamental in the theory of effective Hamiltonians.

The Bloch formalism is determined in a natural way by choosing that the N_m solutions of H^{eff} in the model space must be the projections in the model space of N_m exact solutions ψ_m spanning the target space S:

$$H^{\text{eff}}\Phi_m = E_m\Phi_m \qquad \Phi_m = P_0\psi_m \qquad m = 1, 2, \ldots, N_m \qquad (16)$$

The wave operator associated with the solutions can be written in compact form as[11,12]

$$\Omega = P(P_0PP_0)^{-1} \qquad (17)$$

It is worth noting that Ω depends only on the projectors P and P_0. From (17) it can immediately be checked that Ω has the following properties:

$$P_0\Omega = P_0 \qquad \Omega^2 = \Omega \qquad \Omega^\dagger \neq \Omega \qquad (18)$$

The first one is associated with the so-called intermediate normalization and the second one means that Ω is a non-orthogonal projection operator. From Ω one can obtain the expression for the projector P:

$$P = \Omega(\Omega^\dagger\Omega)^{-1}\Omega^\dagger \qquad (19)$$

Multiplying both sides of Eq. (13) on the left by P_0 and using the intermediate normalization property leads to the Bloch effective Hamiltonian:

$$H^{\text{eff}} = P_0H\Omega \qquad (20)$$

the spectral decomposition of which is

$$H^{\text{eff}} = \sum_{m=1}^{N_m} E_m|\Phi_m\rangle\langle\Phi_m| \qquad (21)$$

Introducing the above expression for H^{eff} in (13) leads to the basic wave operator equation:[13]

$$H\Omega = \Omega H\Omega \qquad (22)$$

This equation is a generalization of an equation previously found by Bloch and generalized by Lindgren.[14] Solving the operator equation (22) appears as the main task in the Bloch theory of effective Hamiltonians since H^{eff} can immediately be deduced from Ω by means of (20). It will also be shown below that the wave operator Ω plays a fundamental role in the whole theory of effective Hamiltonians.

The Bloch Hamiltonian (20) is non-Hermitian since its solutions are the projections in the model space of exact solutions of the exact Hamiltonian. This can also be seen directly from expressions (17) and (20). The hermitization of the Bloch Hamiltonian can easily be obtained by requiring that the solutions of the new effective Hamiltonian are the symmetrically ortho-gonalized solutions of the Bloch Hamiltonian:[3]

$$H^{\text{eff}}(\text{des Cloizeaux}) = (\Omega^\dagger\Omega)^{1/2}H^{\text{eff}}(\text{Bloch})(\Omega^\dagger\Omega)^{-1/2} \qquad (23)$$

This expression does not appear to be Hermitian, but in fact it is. From (17) one obtains

$$\Omega^{\dagger}\Omega = (P_0 P P_0)^{-1} \tag{24}$$

and (23) becomes

$$H^{\text{eff}}(\text{des Cloizeaux}) = (P_0 P P_0)^{-1/2} P H P (P_0 P P_0)^{-1/2} \tag{25}$$

This last expression is obviously Hermitian and its spectral decomposition is

$$H^{\text{eff}}(\text{des Cloizeaux}) = \sum_{m=1}^{N_m} E_m |\Phi'_m\rangle\langle\Phi'_m| \tag{26}$$

where Φ'_m is defined in eq. (28).

a. The non-orthogonality problem

For a deeper understanding of the non-Hermitian Bloch and Hermitian des Cloizeaux formalisms, it is useful to consider the expressions of Φ_m and Φ'_m appearing in (21) and (26):

$$\Phi_m^{\perp} = \sum_{n=1}^{N_m} (S^{-1})_{mn} \Phi_n \tag{27}$$

$$\Phi'_m = \sum_{n=1}^{N_m} (S^{-1/2})_{mn} \Phi_n \tag{28}$$

S being the overlap matrix between the Φ_m, and $(S^{-1})_{mn}$ and $(S^{-1/2})_{mn}$ represent the (m, n) matrix element of the matrices S^{-1} and $S^{-1/2}$, respectively. When this matrix is almost unity, the functions $\Phi_m, \Phi_m^{\perp}$ and Φ'_m look like the exact wavefunction ψ_m and the Bloch effective Hamiltonian is almost Hermitian. In contrast, when the physical content of at least one projected wavefunction $\Phi_m \equiv P_0\psi_m$ is far from the exact wavefunction ψ_m or, expressed in mathematical terms, when at least one diagonal matrix element of the overlap matrix S is small with respect to unity, then the matrices $S^{-1/2}$ and S^{-1} become singular. The bi-orthogonal wavefunctions Φ_m become meaningless and the Bloch Hamiltonian (21), the spectral decomposition of which implies the $\Phi_m^{\perp}$, becomes strongly non-orthogonal. Many possibilities can be considered for facing up to these difficulties.

First one can build up other effective Hamiltonians based on hierarchized orthogonalization procedures. The Gram–Schmidt procedure is recommended if one starts from the best projected wavefunctions of the bottom of the spectrum. Thus one can obtain a quite reliable effective Hamiltonian with well behaved wavefunctions and good transferability properties (see Section III.D.2). The main drawback of this approach is that the Gram–Schmidt method, which involves triangular matrices, does not lead to simple analytical expressions for perturbation expansions. A partial solution to these limitations is brought about by the new concept of intermediate Hamiltonian, which will be presented in Section III.A.6. One can easily obtain almost Hermitian Hamiltonians that can easily be expanded by standard perturbation theory but with the new limitation that only one part of their roots are exact eigenenergies of the full Hamiltonian.

Another way of handling the non-orthogonality problem was suggested by Kato more than 30 years ago.[15] By means of (17) and (20), the eigenvalue equation (16) can be transformed into[16]

$$(\mathcal{H} - E\mathcal{I})\Phi_m^{\perp} = 0 \qquad m = 1, 2, \ldots, N_m \tag{29}$$

where

$$\mathcal{H} = P_0 P H P_0 \qquad \text{and} \qquad \mathcal{I} = P_0 P P_0 \tag{30}$$

The 'effective' Hamiltonian $\mathcal{H}$ and the metric (or overlap) operator $\mathcal{I}$ are Hermitian. The equation looks like those frequently used for solving the Schrödinger equation by means of a non-orthogonal basis set, as for instance in valence-bond (VB) theory. The operators $\mathcal{H}$ and $\mathcal{I}$ seem to have interesting transferability potentialities but up to now they do not seem to have been used for practical investigations. A more general approach has recently been given by Suzuki.[17,18]

4. Solutions of the Bloch Equation

Im most cases the wave operator cannot be determined from expression (17) since the projector P on the target space is generally unknown. Then one has to solve directly the wave operator equation (22). Let us first introduce a *reduced wave operator* X by using the intermediate normalization and writing Ω as

$$\Omega = (P_0 + Q_0)\Omega = P_0 + X \tag{31}$$

X is a transition operator that couples the model space and the outer space:

$$X = Q_0 X P_0 \tag{32}$$

This operator had previously been denoted χ by Lindgren[2] and ω by Suzuki.[17] A lower-case letter seems a good choice for the reduced operator, keeping the capital letter Ω for the full operator. However, we prefer to use the notation X, which clearly indicates that X is the unknown quantity.[13] We will see below that an operator equation for X can be put in the form $F(X) = 0$, which suggests that this equation could be solved in close analogy with the standard methods of resolution of an ordinary algebraic equation $f(x) = 0$.

An equation for X, originally given by Okubo, can easily be obtained by multiplying both sides of Eq. (22) on the left by Q_0:

$$Q_0 H (P_0 + X) = X H (P_0 + X) \tag{33}$$

This equation can be written in the form

$$F(X) = 0 \qquad \text{with} \qquad F(X) \equiv Q_0 (1 - X) H (1 + X) \tag{34}$$

Equation (34) can be solved by the standard iterative methods, which can be classified according to their rates of convergence: linear, quadratic or quasi-quadratic.[19] The true quadratic methods, which are analogous to the Newton–Raphson procedure for ordinary algebraic equations, would involve the inversion of huge superoperators acting in the linear space of all transition operators coupling the model space and the outer space. These inversions would be as difficult as the direct resolution of the Schrödinger equation for the exact Hamiltonian. For many-body applications the inversion of these operators is performed in an approximate way by methods similar to the partial infinite summation techniques of many-body theory. The linear methods are also attractive for a first approach since they are closely associated with the standard perturbation theory which finds in this approach its rigorous foundation.[13,20,21]

The simplest way for deriving approximate compact expressions for X is to introduce an energy parameter E_0 approximately equal to the mean value of the eigenenergies of H^{eff}. Subtracting $E_0 X$ from both sides of Eq. (33) gives.

$$Q_0(E_0 - H)X = Q_0 H P_0 - XH(P_0 + X) + E_0 X \tag{35}$$

which can be put in the form

$$X = f(X) \qquad \text{with}$$

$$f(X) \equiv \frac{Q_0}{E_0 - H} V P_0 + \frac{Q_0}{E_0 - H}[X(E_0 - H) - XVX]P_0 \tag{36}$$

In this last expression we have again used the perturbation notation $Q_0 V P_0 \equiv Q_0 H P_0$ and $P_0 V Q_0 \equiv P_0 H Q_0$. The solution of Eq. (36) can immediately be found by iteration:

$$X^{(0)} = 0 \tag{37a}$$

$$X^{(1)} = f(X^{(0)}) = \frac{Q_0}{E_0 - H} V P_0 \tag{37b}$$

$$\vdots \qquad \vdots$$

$$X^{(n+1)} = f(X^n) \tag{37c}$$

The expression (36) of $f(X)$ indicates that the convergence properties of the iterative procedure will depend on the smallness of the coupling operator and on $P_0(E_0 - H)P_0$ defined in the model space with respect to the reduced Green's operator $Q_0/E_0 - H$. Keeping terms up to the second iteration gives

$$X = \frac{Q_0}{E_0 - H} V P_0 + \frac{Q_0}{(E_0 - H)^2} V P_0(E_0 - H)P_0$$

$$- \frac{Q_0}{(E_0 - H)^2} V P_0 V \frac{Q_0}{E_0 - H} V P_0 + \cdots \tag{38}$$

This expression is more useful for general discussions than for practical applications since it would require knowledge of the reduced Green's operator $Q_0/E_0 - H$. However, it must be noted that in relativity the very particular structure of the Dirac equation allows direct determination of $Q_0/E_0 - H$ (see Section III.A).

a. Solutions by perturbation

The exact Hamiltonian is split into an unperturbed zero-order Hamiltonian and a perturbation:

$$H = H_0 + V \tag{39}$$

where

$$H_0 = \sum_{m=1}^{N_m} E_m^0 |m\rangle\langle m| + \sum_{\alpha} E_\alpha^0 |\alpha\rangle\langle\alpha| \tag{40}$$

With (39) and (40), Eq. (33) can be transformed into the commutator equation

$$[X, H_0] = Q_0(1 - X)V(1 + X)P_0 \tag{41}$$

which can be put in the form of an implicit equation for X:

$$X = \sum_{m=1}^{N_m} \frac{Q_0}{E_m^0 - H_0}(1 - X)V(1 + X)P_m \tag{42}$$

In Eq. (42) $P_m = |m\rangle\langle m|$ is the projector associated with the unperturbed eigensolution $|m\rangle$ of H_0. Equation (42) can immediately be solved by successive iterations, which provide the perturbation expansion of X in powers of V:

$$X = \sum_{k=1}^{\infty} X^{(k)} \tag{43}$$

$$X^{(1)} = \sum_{m=1}^{N_m} \frac{Q_0}{E_m^0 - H_0} V P_m \tag{44}$$

$$X^{(2)} = \sum_{m=1}^{N_m} \frac{Q_0}{E_m^0 - H_0} V \frac{Q_0}{E_m^0 - H_0} V P_m$$

$$- \sum_{m,n=1}^{N_m} \frac{Q_0}{(E_m^0 - H_0)(E_n^0 - H_0)} V P_m V P_n + \cdots \tag{45}$$

The choice of a degenerate unperturbed Hamiltonian in the model space greatly simplifies the perturbation expansion of X. With the assumption that $E_1^0 = E_2^0 = \cdots = E_m^0 = E^0$, expressions (44) and (45) reduce to

$$X^{(1)} = \frac{Q_0}{E_0 - H_0} V P_0 \tag{46}$$

$$X^{(2)} = \frac{Q_0}{E_0 - H_0} V \frac{Q_0}{E_0 - H_0} VP_0 - \frac{Q_0}{(E_0 - H_0)^2} VP_0 VP_0 \qquad (47)$$

Let us now introduce a more compact notation, which leads to the expansion

$$\Omega = P_0 + X = \sum_{n=0}^{\infty} \Omega^{(n)}$$

$$\Omega^{(0)} = P_0$$

$$\Omega^{(1)} = gVP_0$$

$$\vdots \qquad \vdots$$

$$\Omega^{(n)} = g\left(V\Omega^{(n-1)} - \sum_{k=1}^{n-1} \Omega^{(k)} V\Omega^{(n-k-1)} \right) \qquad (48)$$

where

$$g = \frac{Q_0}{E_0 - H_0} = \sum_{\alpha} \frac{|\alpha\rangle\langle\alpha|}{E_0 - E_\alpha^0} \qquad (49)$$

The expansion of Ω immediately generates the perturbation expansion of the Bloch effective Hamiltonian:

$$H^{\text{eff}} = \sum_{n=0}^{\infty} H_{\text{eff}}^{(n)}$$

$$H_{\text{eff}}^{(0)} = P_0 H_0$$

$$H_{\text{eff}}^{(1)} = P_0 VP_0$$

$$H_{\text{eff}}^{(2)} = P_0 VgVP_0$$

$$\vdots \qquad \vdots$$

$$H_{\text{eff}}^{(n)} = P_0 V\Omega^{(n-1)} \qquad (50)$$

Expansion of H^{eff} up to third order can be found in Lindgren.[2] Very often the perturbation expansion diverges and the first few terms give only an approximation of the true solution (asymptotic convergence). Compact perturbation expansions of the wave operator have recently been given by Suzuki for both the degenerate and quasi-degenerate cases.[22]

b. The choice of the target space

The above theory of the Bloch effective Hamiltonian was mainly based on two finite N_m-dimensional subspaces: the model space S_0 and the target space S, with projectors P_0 and P respectively. There is no difficulty in choosing the model space on which the reduced quantum information will be projected. The definition of the target space is not so obvious. If the energy levels associated

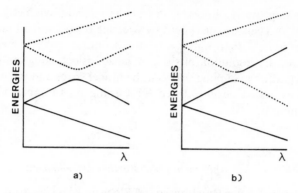

Fig. 3. Eigenenergies of a four-dimensional Hamiltonian (full curves and dotted curves). The two eigenenergies of H^{eff} (full curves) are represented as a function of an arbitrary parameter λ (for example the internuclear distance in a diatomic molecule). Cases (a) and (b) correspond to the adiabatic and diabatic definitions of H^{eff}, respectively.

with the model space are well separated from all the other energies of the spectrum, ore more precisely, in terms of perturbation, if there is a small coupling between the states of the model space S_0 and those of the outer space $S_0^\perp$ for the exact Hamiltonian, then there is no ambiguity in identifying the target space. The resolution of the generalized Bloch equation (22) by means of the techniques of Section II.A.4 will always lead to a stable subspace S. The definition of S is not so obvious when there is a strong coupling between S_0 and $S_0^\perp$. This occurs for example in the presence of an ionic molecular intruder state in the model space of neutral states[13] (see Sections III.C and III.D). This situation is illustrated in Fig. 3 where one considers two possible ways of defining S from S_0 for a two-dimensional effective Hamiltonian. In case (a) there is a rapid change in the second eigenvector, the component of which in the model space vanishes asymptotically, making the wave operator undefined. (The operator $P_0 P P_0$ is singular and the inversion occurring in equation (2.17) becomes impossible.) In contrast, in case (b), the eigenvalues and eigenvectors are discontinuous whereas the physical content of S remains stable. This means that, for all values of parameter λ, the subspace S remains as similar as possible to the model subspace S_0. In both cases we are faced with discontinuities in the effective Hamiltonian either in the physical content of the solutions (case (a)) or in the energies (case (b)). These difficulties are partially solved by introducing the concept of intermediate Hamiltonian (Section II.A.6). From a mathematical point of view, any subspace made up of N_m exact solutions of H can be chosen as the target space but for physical purposes only two main criteria seem to be useful: the subspace S can be chosen from an energetic criterion, taking the N_m lowest states of the relevant symmetry (adiabatic definition of S); S can also be chosen by selecting the N_m states that maximize their occupation $\langle \psi | P_0 | \psi \rangle$ in S_0 (diabatic definition of S). From a practical point of view, the resolution of the generalized Bloch equation is unstable when there is strong coupling between the model space and the outer space. Special devices must be introduced for turning on to either adiabatic or diabatic solutions.

5. Other Approaches and Further Developments

The two basic effective Hamiltonians of Bloch and des Cloizeaux can also be obtained by means of similarity transformations, which reveals with deeper details the mathematical structure of the theory.[23-28] The exact Hamiltonian is first transformed by means of a similarity transformation:

$$\mathcal{H} = U^{-1}HU \tag{51}$$

U has an inverse but it is not required to be unitary. Various effective Hamiltonians can be obtained under the assumption that the transformation U decouples H within the model space and the outer space (Fig. 4):

$$\mathcal{H} = P_0\mathcal{H}P_0 + Q_0\mathcal{H}Q_0 \tag{52}$$
$$H^{\text{eff}} \qquad H^{\text{eff}\perp}$$

The most basic transformation can be written as

$$U = P(P_0PP_0)^{-\nu} + Q(Q_0QQ_0)^{-\nu}$$
$$U^{-1} = (P_0PP_0)^{\nu-1}P + (Q_0QQ_0)^{\nu-1}Q \tag{53a}$$

where ν is a non-negative index.[13,26] It can easily be checked that the values

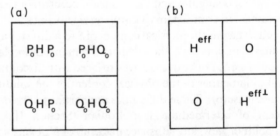

Fig. 4. (a) Matrix representation of H. (b) Matrix representation of H^{eff} and $H^{\text{eff}\perp}$. The projectors onto the model space S_0 and its orthogonal complement are P_0 and Q_0, respectively.

$v = 1$ and $v = \frac{1}{2}$ generate the Bloch and des Cloizeaux Hamiltonians. For $v = 1$

$$U(\text{Bloch}) = P(P_0PP_0)^{-1} + Q(Q_0QQ_0)^{-1} \tag{54a}$$

$$U^{-1}(\text{Bloch}) = P_0P + Q_0Q \tag{54b}$$

The Bloch U operator can also be written in the form

$$U(\text{Bloch}) = 1 + u \tag{55}$$

where

$$u = X - X^\dagger \tag{56}$$

u is an anti-Hermitian operator ($u^\dagger = -u$). Expression (56) emphasizes again the central role played by the reduced wave operator X in the theory of effective Hamiltonians. The value $v = \frac{1}{2}$ leads to the des Cloizeaux formalism

$$U(\text{des Cloizeaux}) = P(P_0PP_0)^{-1/2} + Q(Q_0QQ_0)^{-1/2} \tag{57}$$

This operator is obviously Hermitian:

$$U^{-1}(\text{des Cloizeaux}) = U^\dagger(\text{des Cloizeaux})$$
$$= (P_0PP_0)^{-1/2}P + (Q_0QQ_0)^{-1/2}Q \tag{58}$$

U of expression (57) can also formally be written as

$$U(\text{des Cloizeaux}) = (1 + u)(1 - u^2)^{-1/2} = \left(\frac{1+u}{1-u}\right)^{1/2} \tag{59}$$

It is sometimes useful for separability problems (i.e. size consistency) to introduce an exponential operator G by

$$U(\text{des Cloizeaux}) = e^G \tag{60}$$

G can be expressed as a simple analytical function of u[24,25,27]

$$G = \tanh^{-1} u \tag{61}$$

In the last few years the general theory of effective Hamiltonians has been reformulated by Kutzelnigg in the Fock space.[29-33] The use of creation and annihilation operators is supposed to simplify the calculation of quantities involving variation of the number of particles, such as ionization potentials or electron affinities. Up to now no specific applications have been published with this formalism, the practical efficiency of which is still to be established.

6. Intermediate Hamiltonians[34]

The discussions concerning the non-orthogonality problem (Section II.A.3) and the choice of the target space (Section II.A.4) have shown that the Bloch and des Cloizeaux effective Hamiltonians suffer from convergence difficulties especially in the presence of molecular intruder states that lead to discontinu-

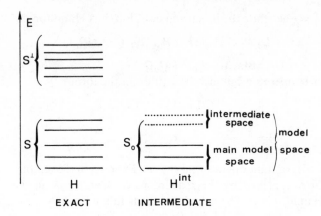

Fig. 5. The exact Hamiltonian H and the intermediate Hamiltonian H^{int} have the same eigenenergies in a subspace of S and in the main model subspace. Moreover, H^{int} also has approximate eigenenergies (dotted lines) in the intermediate subspace.

ities of their matrix elements as a function of internuclear distances. To combat these difficulties, the concept of an intermediate Hamiltonian has recently been proposed (Fig. 5).

The model space is split into an N_m-dimensional *main model space* and an N_i-dimensional *intermediate space* with projectors

$$P_m = \sum_{m=1}^{N_m} |m\rangle\langle m| \qquad P_i = \sum_{i=1}^{N_i} |i\rangle\langle i| \qquad P_m + P_i = P_0 \qquad (62)$$

The *intermediate Hamiltonian*, denoted H^{int}, is defined in the $(N_m + N_i)$-dimensional model space. Among the $N_m + N_i$ eigenenergies of H^{int}, only N_m are exact eigenvalues of H. This loss of information on the energies is the price one has to pay for improving the hermiticity and the convergence properties of an intermediate Hamiltonian with respect to the previously described effective Hamiltonians.

The theory of intermediate Hamiltonians is based on the introduction of a new wave-like operator R. Then a large class of intermediate Hamiltonians can be written as

$$H^{int} = P_0 H R \qquad (63)$$

with R fulfilling the condition

$$R\Omega = \Omega \qquad (64)$$

Ω is the Bloch operator associated with the main model space. In close analogy with the Bloch theory, an equation for R can be chosen in the form

$$Q_0 H R = Q_0 \Omega H R \qquad (65)$$

or more generally one can take

$$Q_0(E_0 - H)R = Q_0\Omega(E_0 - H)R \tag{66}$$

where E_0 is an unperturbed energy typically of the order of magnitude of the exact eigenenergies in the main model space. In Eq. (66) the operator R is weakly E_0-dependent. Equation (66) can be solved by perturbation theory. Assuming that H_0 is degenerate in the main model space:

$$H_0 = \sum_{m=1}^{N_m} E_0 |m\rangle\langle m| + \sum_{i=1}^{N_i} E_i^0 |i\rangle\langle i| + \sum_\alpha E_\alpha^0 |\alpha\rangle\langle\alpha| \tag{67}$$

Eq. (66) can be written in the form

$$Q_0 R = g(1 - \Omega)VR \qquad g = \sum_\alpha \frac{|\alpha\rangle\langle\alpha|}{E_0 - E_\alpha^0} \tag{68}$$

Assuming that R still obeys the intermediate normalization, Eq. (68) leads to the perturbation expansion:

$$R = P_0 + Q_0 R = \sum_{n=0}^{\infty} R^{(n)} \tag{69}$$

$$R^{(0)} = P_0$$

$$R^{(1)} = gVP_0$$

$$\vdots \qquad \vdots$$

$$R^{(n)} = g\left(VR^{(n-1)} - \sum_{k=1}^{n-1} \Omega^{(k)} VR^{(n-k-1)} \right)$$

The expansion of the intermediate Hamiltonian is obtained order by order from the expansion of R. Note that the expansions of H^{int} and H^{eff}(Bloch) coincide up to second order.

7. Size-consistency Problems[35-44]

The present review does not enter into the famous and fundamental question of size consistency in many-body problems. This problem is well clarified in non-degenerate perturbation theory through the linked cluster theorem and for other approximate algorithms for the calculation of the correlation energy of an N-electron system. The size-consistency properties of the (nearly) degenerate perturbation theories are not as easy to establish. Brandow[39] has demonstrated a generalized linked cluster theorem when the model space is a complete active space, i.e. when it involves all the determinants in which p molecular orbitals (MO) (the core) are always doubly occupied while q (active) MOs receive k electrons with all possible distributions of these k electrons among the q MOs. This is a strong limitation. Diagrammatic expansions have also been proposed for general model spaces[41,42] (i.e. leaving the preceding restriction) but the diagrammatic factorization ensuring size consistency is not perfect. One should notice here a fundamental contradiction: the complete active space requirement frequently leads to very large and broad model spaces (spread in a large range of energies), i.e. to convergence problems and difficulties in defining a target space while it ensures the size consistency; in contrast, smaller (non-complete) model spaces may face size-consistency problems. One should notice, however, that the Heisenberg Hamiltonians (cf. Section III.D) are based on a very incomplete model space, and are perfectly size-consistent. The complete active space condition thus appears as a sufficient but non-necessary condition. Much work remains to be done in this field.

B. Pseudo-Hamiltonians by Simulation Techniques

The previously described effective Hamiltonians cannot give a direct solution to the general problem of modelling in quantum chemistry and atomic physics. They only provide finite matrices of numbers that can be diagonalized to obtain a finite number of exact eigenenergies. In a rather unphysical way, all the other eigenenergies are equal to zero. On the contrary, the empirical or semi-empirical Hamiltonians (Hückel, PPP) generally involve one- and two-body interactions which have a clear physical meaning. The spectrum of these model Hamiltonians is extended but most often only the lowest states are significant. For example, a CNDO-type (complete neglect of differential overlap) Hamiltonian describes only the ground and the first few valence excited states of a molecule. An attractive characteristic of these Hamiltonians is that they contain transferable potentials and interactions, such as, for example, the β parameter in the Hückel theory which characterizes the interaction between any two π-bonded carbon atoms. A severe limitation of these models, generally parametrized from experiment, is that their theoretical status and their range of applicability are not well defined. The purpose of this section is to show that theoretical approximate Hamiltonians or *pseudo-Hamiltonians* can easily be derived from first principles by rigorous simulation techniques. This technique will be presented in Section II.B.1. It will also be shown in a second step that the practical determination of pseudo-operators and pseudo-Hamiltonians generally requires an intermediate step with the knowledge of a *truncated Hamiltonian* that contains less information than the original exact Hamiltonian.

1. Simulation Techniques[45]

Starting from an exact Hamiltonian H, one looks for an approximate pseudo-Hamiltonian H^{ps} as close as possible to H. This can be achieved by minimizing the distance between H^{ps} and H:

$$\| H^{ps} - H \|_{minimum} \tag{70}$$

In the following it is assumed that H and H^{ps} are Hermitian. Let us suppose that H is acting in a basis of orthonormalized states denoted $|I\rangle$. The simplest definition of the distance between H^{ps} and H is

$$\| H^{ps} - H \| = \left(\sum_{I,J} |\langle I|H^{ps} - H|J\rangle|^2 \right)^{1/2} \tag{71}$$

For most applications this definition of the distance is not relevant; if the distance was taken in $\mathscr{L}^2$, the distance would be infinite and since the determination of H^{ps} is not required to have the same quality for all states $|I\rangle$, a better definition of the distance requires a *reduced distance*:

$$\| H^{ps} - H \| = \left(\sum_{I,J} |\langle I|H^{ps} - H|J\rangle|^2 w_{IJ} \right)^{1/2} \tag{72}$$

where the w_{IJ} are real positive (or zero) weights. For the following, it is useful to introduce a scalar-product notation. The Hermitian scalar product between two operators A and B belonging to the vectorial space of all operators acting in the space of the states $|I\rangle$ is defined by

$$(A|B) = \sum_{I,J} \langle I|A^\dagger|J\rangle\langle J|B|I\rangle w_{IJ} \tag{73a}$$

$$(A|B) = (B|A)^* \tag{73b}$$

With this notation the reduced distance (72) can be expressed as the square root of the scalar product of $H^{ps} - H$ with itself:

$$\| H^{ps} - H \| = (H^{ps} - H|H^{ps} - H)^{1/2} \tag{74}$$

and condition (70) becomes

$$(H^{ps} - H|H^{ps} - H)_{minimum} \tag{75}$$

Very often the pseudo-Hamiltonian can be written in the form of a linear combination of *a priori* known operators:

$$H^{ps} = \sum_i C_i A_i \tag{76}$$

The operators A_i may also depend on non-linear coefficients. In all cases the best choice for the non-linear and linear coefficients C_i is obtained by minimizing the distance between H and H^{ps}. According to the usual Fourier techniques, the minimization in (75) with respect to the coefficients C_i, assumed to be real, leads to the linear system of equations:

$$(A_i|H) = \sum_j (A_i|A_j)C_i \tag{77}$$

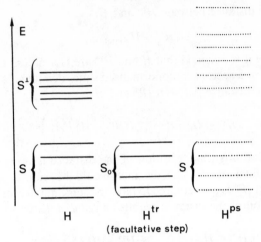

Fig. 6. The exact Hamiltonian H and the pseudo-Hamiltonian have approximately the same eigenenergies in the subspaces S and S', respectively H^{tr} having exact energies and truncated wave functions.

If the operators A_i are orthonormal,

$$(A_i | A_j) = \delta_{ij} \tag{78}$$

the resolution of (77) gives immediately

$$C_i = (A_i | H) \tag{79}$$

In most applications one is only interested in an accurate description of the bottom of the spectrum (Fig. 6). The previously undetermined states $|I\rangle$ become the exact solutions of H and the N_m lowest solutions define a subspace S previously called the target space. The projectors associated with S and its orthogonal complement are denoted P and Q, respectively. H^{ps} is now characterized by the requirement that H^{ps} and H have almost the same matrix elements in S and that there is almost no coupling between S and $S^\perp$. This implies that

$$w_{IJ} = \begin{cases} 1 & \text{if } I, J \in S \\ 0 & \text{if } J \in S^\perp \end{cases} \tag{80}$$

The reduced distance (72) can now be written as a partial trace:

$$\| H^{ps} - H \| = \left(\sum_{\substack{m,i \\ (m \in S)}} |\langle \psi_m | H^{ps} - H | \psi_i \rangle|^2 \right)^{1/2} \tag{81}$$

$$= \{ \text{Tr}[P(H^{ps} - H)^2] \}^{1/2}.$$

From a practical point of view, the determination of a pseudo-Hamiltonian by minimizing the distance between H^{ps} and H has the decisive advantage of keeping simple linear mathematics. However, it has the drawback that there is no guarantee concerning the energies of H^{ps} belonging to $S^{\perp}$. They can enter the desired region of energies or even go below the energies of H^{ps} obtained by simulation in S. To eliminate these spurious unphysical intruder states, one can proceed as follows. First one diagonalizes H^{ps} characterized by an initial choice of parameters. The N_m lowest solutions ψ'_m with energies E'_m define a subspace S', the projector of which is P'. In a second step one minimizes the distance between $P'H^{ps}P'$ and PHP. The reduced distance can be taken either in S or in S':

$$\| H^{ps} - H \|_S = \{\operatorname{Tr}[P(P'H^{ps}P' - PHP)]^2\}^{1/2}$$

$$= \left(\sum_{m,n=1}^{N_m} \left| \sum_{p=1}^{N_m} (E'_p S_{mp'} S_{p'n} - E_p \delta_{mp} \delta_{pn}) \right|^2 \right)^{1/2} \tag{82}$$

$$\| H^{ps} - H \|_{S'} = \{\operatorname{Tr}|P'(P'H^{ps}P' - PHP)|^2\}^{1/2}$$

$$= \left(\sum_{m,n=1}^{N_m} \left| \sum_{p=1}^{N_m} (E'_p \delta_{mp} \delta_{pn} - E_p S_{m'p} S_{pn'}) \right|^2 \right)^{1/2} \tag{83}$$

where $S_{mp'}$ and $S_{p'n}$ denote the overlaps $\langle \psi_m | \psi'_p \rangle$ and $\langle \psi'_p | \psi_m \rangle$, respectively. Expressions (82) and (83) clearly indicate that their minimization implies that the N_m lowest solutions of H^{ps} (wavefunction and energies) are as close as possible to the N_m solutions of H. The minimization of (82) or (83) provides new parameters for H^{ps} and the above two-step procedure must be repeated until the determination of a fixed parametrization for H^{ps}. The drawback of the method is that it is highly non-linear and that the self-consistent procedure may become very computer-time-consuming for many-body problems.

2. The Concept of Truncated Hamiltonian

Up to now it has been considered that the pseudo-Hamiltonian H^{ps} should simulate the exact Hamiltonian H. This means that, at least for a part of the spectrum, the solutions of H^{ps} (energies and wavefunctions) look like those of H. This is not always the best way to proceed since the main purpose of theoretical *ab initio* modelling is to derive pseudo-Hamiltonians that possess for part of the spectrum energies as close as possible to the exact energies and *simplified* truncated wavefunctions belonging to some well characterized predetermined model space. For instance, one may wish to project the physics in the space of all Slater determinants arising from a minimal basis set of atomic orbitals. For that purpose, it may be convenient to introduce some truncated Hamiltonian defined only in a restricted space. Figure 6 gives a schematic illustration of the intermediate role played by such a truncated Hamiltonian, denoted H^{tr}, which appears as a new intermediate step for deriving a pseudo-Hamiltonian from an exact Hamiltonian. H^{tr} is defined in some significant model space S_0, the projector of which is P_0. Its spectral decomposition can be written as

$$H^{tr} = \sum_{m=1}^{N_m} E_m |\psi_m^0\rangle\langle\psi_m^0| \tag{84}$$

Expression (84) defines a Hermitian Hamiltonian. A possible choice for H^{tr} is obviously the des Cloizeaux effective Hamiltonian or some of its generalizations investigated in Section II.A. With this choice, the E_m of (84) remain exact energies of the original exact Hamiltonian H. But other choices are possible for H^{tr} that also may be characterized by its matrix elements. The best pseudo-Hamiltonian is now obtained by minimizing the distance

$$\| H^{ps} - H^{tr} \| = \{ |\text{Tr}[P_0(H^{ps} - H^{tr})]|^2 \}^{1/2} \tag{85}$$

or equivalently by requiring

$$(H^{ps} - H^{tr} | H^{ps} - H^{tr})_{\text{minimum}} \tag{86}$$

Up to now the conceptual importance of this truncated Hamiltonian does not seem to have been clearly recognized. According to our experience in the field, it is most often a necessary intermediate step for deriving pseudo-operators and pseudo-Hamiltonians from exact Hamiltonians by first principles. This will be clearly indicated by applications in Section IV.

III. APPLICATIONS OF EFFECTIVE HAMILTONIANS

A. Relativity

For heavy atoms, the instantaneous velocities of the electrons near the nuclei cannot be neglected with respect to the velocity of light. These electrons must be described within the Dirac relativistic theory. For the sake of simplicity, let us consider a one-electron system in a central field. The Dirac Hamiltonian, shifted for the energy by c^2, can be written in atomic units (a.u.) as

$$H = (\beta - 1)c^2 + c\boldsymbol{\alpha}\cdot\mathbf{p} + V \tag{87}$$

Here $\boldsymbol{\alpha}$ and β are the four-component matrices

$$\alpha = \begin{pmatrix} 0 & \sigma \\ \sigma & 0 \end{pmatrix} \qquad \beta = \begin{pmatrix} I & 0 \\ 0 & I \end{pmatrix} \tag{88}$$

where σ are the Pauli matrices in their standard representation and I is the 2×2 unit matrix; c is the velocity of light ($c = 137$ in atomic units) and $V = V(r)$ depends only on the distance r between the electron and the nuclei.

Most often we are interested in solutions corresponding to the negatively charged electrons of chemistry and physics, the so-called positive-energy solutions.[46] For these solutions the first two components ψ_1 and ψ_2 of ψ are large with respect to the small components ψ_3 and ψ_4. Thus it seems obvious that one has to look for two-component formalisms, which were first investigated by Pauli. Foldy and Wouthuysen[47] have shown that, by means of successive unitary transformations, it is possible to expand in a systematic way the Pauli Hamiltonian in powers of $1/c^2$. Unfortunately, this series is highly singular for a Coulomb potential and it results that the standard two-component Pauli Hamiltonian cannot be used for variational calculations.

The purpose of this section is to show how the problem of passing from the four-component Dirac equation to two-component Pauli-like equations can be systematically investigated within the framework of the theory of effective Hamiltonians.[10,11] Beyond the above-mentioned difficulties, we will be able to derive energy-independent two-component effective Hamiltonians that can be used for variational atomic and molecular calculations. To introduce the subject and the notation, let us first consider the simple case of a free electron.

1. Free Electron (V = 0)

Hamiltonian (87) reduces to

$$H = (\beta - 1)c^2 + c\boldsymbol{\alpha}\cdot\mathbf{p} \tag{89}$$

Within the energy-dependent Feschbach–Löwdin approach (Section II.A.2) one obtains

$$\Omega_E = P_0 + \frac{1}{1 + E/2c^2}\frac{\boldsymbol{\alpha}\cdot\mathbf{p}}{2c}P_0 \tag{90}$$

and

$$H_E^{\text{eff}} = \frac{1}{1 + E/2c^2}\frac{p^2}{2}P_0 \qquad p^2 \equiv -\Delta \tag{91}$$

Within the energy-independent Bloch formalism (Section II.A.3), the reduced wave operator (33) can be written as

$$X = \frac{\boldsymbol{\sigma}\cdot\mathbf{p}}{2c}P_0 - X\frac{\boldsymbol{\sigma}\cdot\mathbf{p}}{2c}X \tag{92}$$

The exact solution of this equation is[10]

$$X = \frac{\boldsymbol{\alpha}\cdot\mathbf{p}}{2c}\hat{f} \qquad \hat{f} = \frac{2}{1 + (1 + p^2/c^2)^{1/2}} \tag{93}$$

where $\hat{f}$ is a kinetic cut-off operator, which tends towards $2c/p$ for high values of p. From X, the Bloch and des Cloizeaux Hamiltonians can easily be obtained. In the particular case of a free electron, they have the same expression:

$$H^{\text{eff}}(\text{Bloch}) = H^{\text{eff}}(\text{des Cloizeaux}) = \tfrac{1}{2}p^2 \hat{f} \tag{94}$$

Note that by adding the energy c^2 (in a.u.) these effective Hamiltonians can be put in the usual form:

$$H^{\text{eff}} = c^2 + \tfrac{1}{2}p^2 \hat{f} = c^2(1 + p^2/c^2)^{1/2} \tag{95}$$

2. Central-field Potential

In the presence of V, the Feschbach–Löwdin wave operator becomes

$$\Omega_E = P_0 + \frac{1}{1+(E-V)/2c^2} \frac{\boldsymbol{\alpha}\cdot\mathbf{p}}{2c} P_0 \tag{96}$$

and

$$H_E^{\text{eff}} = \boldsymbol{\sigma}\cdot\mathbf{p} f_E \boldsymbol{\sigma}\cdot\mathbf{p} + V \tag{97}$$

where

$$f_E = \frac{1}{1+(E-V)/2c^2} \tag{98}$$

is a cut-off function of r which for a Coulomb potential $(V = -Z/r)$ tends towards zero with r (Fig. 7).[48,49]

Within the energy-independent Bloch approach, the reduced wave operator becomes

$$X = \frac{\boldsymbol{\alpha}\cdot\mathbf{p}}{2c} P_0 + \frac{1}{2c^2}[V, X] - X \frac{\boldsymbol{\alpha}\cdot\mathbf{p}}{2c} X \tag{99}$$

$[V, X]$ is the commutator of V and X. The iterative solution of (99) leads to the perturbation expansion

$$X = \frac{\boldsymbol{\alpha}\cdot\mathbf{p}}{2c} P_0 + \frac{1}{4c^2}[V, \boldsymbol{\alpha}\cdot\mathbf{p}]P_0 - \frac{\boldsymbol{\alpha}\cdot\mathbf{p}}{8c^3} p^2 P_0 + \cdots \tag{100}$$

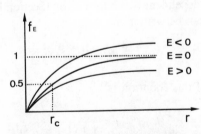

Fig. 7. The cut-off function f_E as a function of the distance r between the electron and the nucleus. For a Coulomb potential, the cut-off radius is $r_c = Z/2c^2$, where Z is the atomic number and c is the velocity of light.

From knowledge of X one can easily derive the Bloch and des Cloizeaux effective operators:

$$H^{\text{eff}}(\text{Bloch}) = -\frac{\Delta}{2} + V - \frac{\Delta^2}{8c^2} - \frac{1}{4c^2}\left(\frac{d}{dr}\right)^\dagger \frac{dV}{dr} + \frac{1}{4c^2}\frac{1}{r}\frac{dV}{dr}\boldsymbol{\sigma}\cdot\mathbf{L} + \cdots \quad (101)$$

$$H^{\text{eff}}(\text{des Cloizeaux}) = -\frac{\Delta}{2} + V$$

$$-\underbrace{\frac{\Delta^2}{8c^2}}_{\text{mass}} + \underbrace{\frac{\Delta V}{8c^2}}_{\text{Darwin}} + \underbrace{\frac{1}{4c^2}\frac{1}{r}\frac{dV}{dr}\boldsymbol{\sigma}\cdot\mathbf{L}}_{\text{spin–orbit}} + \cdots$$

$$\text{correction} \quad (102)$$

The Hermitian des Cloizeaux effective Hamiltonian is identical to the Pauli Hamiltonian obtained by means of the Foldy–Wouthuysen transformation. It is highly singular for a Coulomb potential, since the Darwin term becomes proportional to the Dirac function $\delta(\mathbf{r})$ and since the spin–orbit term behaves as r^{-3} near the nucleus. Even worse, the series diverge for higher terms in $1/c^4$, which would imply the meaningless product of Dirac functions. Moss et al.[50–52] have studied these singularities and searched for methods to avoid these divergences and eliminate the singularities. However, it is easier to notice that the perturbation expansion of X from (99) is incorrect near the nucleus where V becomes infinite and for instantaneous classical velocities of the electron greater than c. To isolate such a strong singularity one has to look for some infinite summation in (99). This can easily be performed if (99) is written in the form

$$X = f\frac{\boldsymbol{\alpha}\cdot\mathbf{p}}{2c}P_0 - \frac{f}{2c^2}XV - fX\frac{\boldsymbol{\alpha}\cdot\mathbf{p}}{2c}X \quad (103)$$

where

$$f = \frac{1}{1 - V/2c^2} \quad (104)$$

For a Coulomb potential the cut-off function f varies from 0 to 1 as illustrated in Fig. 7. The iterative solution of (103) provides regular terms that finally lead to a regular expansion of the Bloch and des Cloizeaux effective Hamiltonians. The first terms are:

$$H^{\text{eff}}(\text{Bloch}) = -\tfrac{1}{2}\Delta_1 + V - \frac{1}{8c^2}\Delta_2\Delta_1 + \frac{1}{4c^2}\Delta_2 V + \cdots \quad (105)$$

$$H^{\text{eff}}(\text{des Cloizeaux}) = -\tfrac{1}{2}\Delta_1 + V - \frac{1}{16c^2}(\Delta_2\Delta_1 + \Delta_1\Delta_2)$$

$$+ \frac{1}{8c^2}(\Delta_2 V + V\Delta_2) + \cdots \quad (106)$$

where

$$\Delta_k = \boldsymbol{\sigma} \cdot \mathbf{p} f^k \boldsymbol{\sigma} \cdot \mathbf{p} \tag{107}$$

These effective Hamiltonians are regular near the nuclei. They provide accurate results for atoms and open the way for relativistic variational two-components calculations in molecules.[53]

B. Research on Nearly Diabatic Potential Energy Surfaces

1. Difficulties of Adiabatic Approaches; Interest in Nearly Diabatic Treatments

It is well known that the solutions of the electronic Hamiltonian H^{el} for a molecule in the Born–Oppenheimer approximation define potential energy curves or surfaces that do not cross when they belong to the same symmetry (except for a few nuclear configurations for which essential or accidental degeneracies may occur). It is also well known that in most cases the molecular wavefunctions keep a well defined dominant character, at least in some regions of the nuclear configuration space. They are for instance valence or Rydberg, ionic or neutral. When following the ith root of the electronic Hamiltonian, the dominant character may change when the nuclear configuration changes. For instance in a diatom, like Na_2 (Fig. 8), the second $2^1\Sigma_u^+$ root[54,55] is essentially 'Rydberg' $Na(4s) + Na(3s)$ at long interatomic distances and it generates a rather flat potential curve. Then at shorter distances, it becomes essentially ionic, dominated by a $Na^+ + Na^-$ $(3s^2)$ character, and the corresponding potential curve is attractive, due to r^{-1} electrostatic interaction. At still shorter interatomic distances the same root changes its physical content to become essentially of $Na(3s) + Na(3p)$ valence-promoted character, and the curve becomes repulsive, defining a long-range minimum near 12 bohr. At still shorter interatomic distances ($r < 10$ bohr), the dominant character changes once more, becoming dominated by a Rydberg character and attractive (as is the Na_2^+ potential curve) (Fig. 8). This adiabatic description, in which the potential curves do not cross (no electronic coupling between the adiabatic states), leads to wavefunctions that change their electronic content from one region of the configurational space to another. It is well known[56,57] that one might be tempted to propose an alternative picture where the wavefunctions would keep an invariant content but would of course be electronically coupled. If the derivative $\partial \psi / \partial r$ of the wavefunctions were strictly zero, the representation would be said to be diabatic. The dynamic (collisional or vibronic) treatments which follow the calculation of potential energy curves require computation of radial couplings, i.e. $\langle \psi_i | \partial \psi_j / \partial r \rangle$ matrix elements between the adiabatic wavefunctions, and if the function ψ_i changes its character rapidly the derivative is a sharp function of the nuclear coordi-

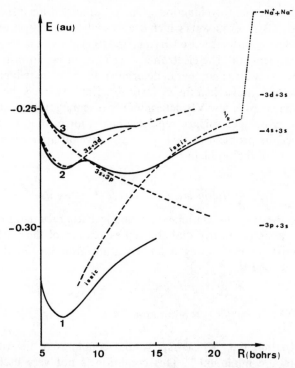

Fig. 8. Illustration of the changes in the content of an adiabatic wavefunction and of a qualitative diabatic reading. Case of the $2\ ^2\Sigma_u^+$ state of Na_2.

nates.[58] Owing to this rapid variation, the integration of the matrix element $\langle \psi_i | \partial \psi_i / \partial r \rangle$ requires the calculation of the wavefunctions for a very dense series of nuclear geometries, and it is therefore either very costly or inaccurate.

One has thus in principle a choice between adiabatic and diabatic approaches

Adiabatic	*Diabatic*
Solve $H^{el}\psi = E\psi$	Use diabatic wavefunctions
No electronic coupling	No radial coupling
Radial coupling difficult to calculate	Large and numerous electronic couplings

depending on the privilege given to a preliminary research of the electronic Hamiltonian solutions. The adiabatic approach has the advantage of reducing the dynamic or vibronic problem to a few adiabatic states close in energy, since the radial coupling cannot mix states lying too far apart in energy. The diabatic approach faces two difficulties: (i) It is general impossible[59] to find or to use strictly diabatic wavefunctions. One is compelled for instance to use valence-bond (VB) determinants which keep a well defined physical character when the nuclear coordinates are varying, and to neglect their derivatives (or to interpolate them). (ii) The electronic wavefunction has in principle to be developed on a very large number of determinants to obtain reliable energies.

Both approaches have then their practical defects. As far as the wavefunction is dominated by a few VB determinants or configurations (a Rydberg configuration crossing a valence repulsive curve, for instance), one may of course restrict the problem to this space and give an approximate elegant diabatic picture of the problem.[57]

2. Ab Initio Nearly Diabatic Treatments

There exist two ways to combine the numerical accuracy requirement with the desire of a simple nearly diabatic representation of the problem. One solution consists of performing a unitary transformation on the set of n eigenvectors ψ_k^a of H^{el}:

$$\{\psi_i^d\} = U\{\psi_k^a\} \qquad i, k = 1, 2, \ldots, n \qquad (108$$

$$\sum_{i,j} |\langle \psi_i^d | \partial \psi_j^d / \partial r \rangle| = \text{minimum} \qquad (109)$$

U being such that the $\langle \psi_i^d | \partial \psi_j^d / \partial r \rangle$ radial couplings of the transformed wavefunctions are minimum.[60] This condition is not very useful from a practical point of view since it requires the knowledge of the radial couplings in the adiabatic basis, which are difficult to calculate. A more practical solution[61] consists of maximizing the overlap of the transformed vectors ψ_k^d with unvariant asymptotic vectors ψ_k^0:

$$\sum_k |\langle \psi_k^d | \psi_k^0 \rangle| = \text{maximum} \qquad (110)$$

This procedure gives nearly diabatic vectors; their radial couplings may be either neglected or interpolated, since the ψ_k^d vary slowly with nuclear coordinate changes (see also Ref. 62).

Instead of working in a stable subspace of H^{el} (i.e. in a basis of 10p determinants), one may work in a basis of a rather limited number of configurations, those which play a major role in the adiabatic eigenfunctions of interest, and build an effective electronic Hamiltonian in this model space. For instance, for the curve crossing between the ionic and neutral configuration in NaCl, one may define as a model space the two leading configurations

$$\phi_n = |4s_{Na}^1 3s_{Cl}^2 3p_{Cl}^5| \qquad \text{Na·Cl·} \qquad \text{neutral}$$

$$\phi_i = |3s_{Cl}^2 3p_{Cl}^6-| \qquad \text{Na}^+\text{Cl}^- \qquad \text{ionic}$$

If the treatment was limited to these two determinants with a common set of $3s_{Cl}3p_{Cl}$ atomic orbitals (AO) taken from either the Cl· or Cl⁻ Hartree–Fock (HF) calculation, the treatment would be quite incorrect and would predict an erroneous curve crossing distance and avoidance. But one may use these two determinants to define a 2×2 model space and apply the theory of effective Hamiltonians, as suggested by Levy[63,64] (with a slightly non-orthodox definition of the effective Hamiltonian). One may use either the Bloch or des Cloizeaux definition of H^{eff} as a 2×2 matrix, the eigenvalues of which are the exact adiabatic eigenvalues

$$\begin{bmatrix} H_{ii}^{eff} & H_{in}^{eff} \\ H_{ni}^{eff} & H_{nn}^{eff} \end{bmatrix} \Phi_{m,0} = \varepsilon_m \Phi_{m,0} \tag{111}$$

where

$$H^{el}\psi_m = \varepsilon_m \psi_m$$

$$\Phi_{m,0} = \langle \phi^n | \psi_m \rangle \phi_n + \langle \phi^i | \psi_m \rangle \phi_i \tag{112}$$

The matrix elements of H^{eff} may be obtained from a former large CI diagonalization, by perturbative expansions, or by iterative techniques.

Spiegelmann and Malrieu have proposed improved versions of this procedure,[65] where the model space is spanned by several multi-configurational zeroth-order descriptions of the various diabatic eigenfunctions. This proposal, which fits very well the architecture of the CIPSI algorithm,[66,67] has received applications on Ar_2^* excited states, NaCl curve crossing,[65] $HeNe^{2+}$[64] and the $Cs^*(7p) + H_2(X^1\Sigma_g^+) \rightarrow CsH(X^1\Sigma_g^+) + H$ reactive collision.[68]

Other refinements, combining unitary transformation theory, have been proposed by Persico et al.[61] and now seem to be the most rational and efficient procedure to convert the large CI calculations (keeping their energetic reliability) into a simple nearly diabatic picture through a small-sized Hamiltonian. The diagonal (and off-diagonal) terms of this effective Hamiltonian have regular behaviour, since they reflect the variation of the energy of (and coupling between) physically nearly invariant wavefunctions. They may be interpolated easily without loss of accuracy. This solves two types of problems: (i) the calculation of the radial couplings for vibronic problems becomes easy, and (ii) the quantum semiclassical or classical calculations of the dynamical cross-sections requires the knowledge of the energy at a very large number of nuclear configurations; the use of adiabatic potential surfaces usually requires difficult analytic fittings.[69] The analytic fitting of well behaved diabatic potential surfaces and electronic couplings is much easier. One must simply diagonalize a small interpolated effective Hamiltonian when energies are needed in trajectory calculations; this is the philosophy of the use of the DIM method[70] in dynamical calculations, and the *ab initio* nearly diabatic effective Hamiltonian approach introduces three- (or more) body effects, and ensures a much better reliability of the energies while keeping the same advantage in the dynamics calculations.

3. A Special Application to Research on Resonant States in e^--Molecule Collisions

As a special application of the diabatization techniques of *ab initio* MO-CI calculations, one should mention the research on resonant states in electron–molecule collisions.[71,72] The problem concerns the existence of a discrete (usually valence) state of the molecular anion, embedded in a continuum of diffuse states. For a molecule like H_2, the electronic Hamiltonian of H_2^- has a $^2\Sigma_u^+$ bound state at large interatomic distance (since H^- is stable), while it has no bound state for interatomic distances shorter than 1.4 Å. Collisional properties, however, suggest the existence of a broad resonance of $^2\Sigma_u^+$ character, which may be seen to be due to the coupling of a discrete state of essentially $\sigma_g^2\sigma_u$ valence character with a continuum of diffuse states associated with the $^1\Sigma_g^+$ (σ_g^2) ground state of the molecule, the outer electron being unbound. This problem may be treated in various modes, but it appears as a challenge to quantum chemists, whose finite basis sets seem to forbid the examination of such a problem. Nevertheless some approximate methods have been proposed (known as 'stabilization techniques'[73–76]) to find the

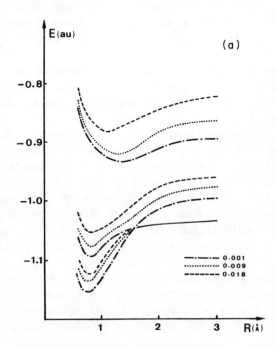

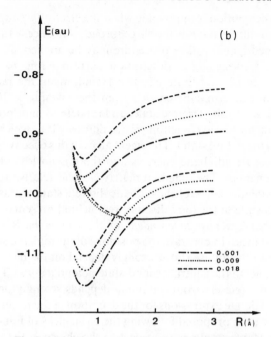

Fig. 9. (a) Adiabatic potential curves of the H_2^- problem in various finite basis set CI calculations.[71] (The numbers refer to the exponent of a diffuse p AO in the centre of the molecule). The full curve refers to the bound-state region. (b) Diabatization of the preceding potential curves;[71] (same comments as in Fig. 8a).

position of the so-called discrete state; for weak couplings between the resonant state and the continuum, the method simply consists in a qualitative diabatic reading of the resonant state potential curve through the avoided crossings with potential curves which are basis set dependent. For strong couplings (broad resonances) some constraints must be added. It may be noticed that finite basis set MO-CI calculations give a sampling of the continuum states through wavefunctions of the type $\phi(X(^1\Sigma_g^+)) \times d$ where d is a diffuse MO. The more diffuse is the d MO, the closer the energy of this unbound state to that of the neutral ground state; the coupling between the hypothetical discrete state, the potential curve of which should be repulsive, and the sampling of diffuse states (with potential curves parallel to the ground-state potential curve) should result in a series of curve crossings, according to Figs. 19a and b.

The usual MO-CI calculations provide adiabatic potential curves. In some cases of weak coupling between the discrete and diffuse states, it has been possible to recognize the place of the discrete potential curve under the net of adiabatic potential curves calculated from various basis sets (see for instance

Ref. 75). This recognition is impossible when the electronic coupling between the discrete and diffuse states is too large (broad resonances). In that case the above-mentioned diabatization procedure may be applied, as suggested by Komiha et al.,[71] furnishing a description corresponding to Fig. 8b. The problem is presented as a finite effective Hamiltonian; the model space is spanned by (a) a determinant derived from the asymptotic $H + H^-$ limit, which corresponds to the discrete state, and (b) determinant products of the neutral ground-state wavefunction by a diffuse MO, which have large components in the diffuse states. The eigenvalues of this effective Hamiltonian reproduce the exact adiabatic energies and there is no loss of information regarding the energies. The diagonal matrix element relative to the discrete state, i.e. the potential curve of the diabatic discrete state, fits asymptotically into the bound state in the long-distance region, and presents a well defined and specific shape, *almost independent of the chosen basis set*, for short interatomic distances. In contrast, the positions of the diabatic state potential curves relative to the diffuse states strongly depend on the basis set, but they are always parallel to the neutral ground-state potential curve. The amplitude of the diffuse–discrete electronic coupling depends strongly on interatomic distance. Approximate expressions for the definition of the resonance position and width have been proposed following the basic ideas of Fano,[77] and they appear to give stable results with respect to the choice of the basis set.

C. Reduction of the Number of Particles and/or of the Atomic-orbital Basis Set Through Effective Hamiltonian Approaches: Construction of Valence-only Effective Hamiltonians

1. Preliminary Remarks

The present section essentially discusses the attempts to build effective valence-only Hamiltonians spanned by a valence minimal basis set, which has been the subject of several tens of papers by Freed et al. (reviewed in Ref. 78; see also Westhaus[79] and Mukherjee[80]). One should notice first that the purpose of such attempts is two-fold: elimination of the core electrons from the explicit treatment; and reduction of the basis set for the treatment of the valence states of the molecule to a minimal valence basis set. The effective Hamiltonian should reproduce the energetic results of a calculation including the core electrons, their exclusion, polarization and correlation effects, and performed in a large basis set (for both the concentrated core distribution and the more diffuse valence cloud).

From a basic point of view, the two purposes are different and should be distinguished; one may be tempted to eliminate the core electrons while keeping a non-minimal valence basis set, as occurs in pseudopotential techniques, on the one hand, while on the other hand, one may try to treat a coreless problem such as a cluster of hydrogen atoms in a minimal basis set, or treat both the core and the valence shells in a minimal basis set. The present section will try to analyse the two purposes independently, in order to simplify the discussion.

As a second important point, one should note that the reduction of the number of electrons or of the basis set, which may be performed through the effective Hamiltonian theory on the atom or on a diatom, is of little interest if it is not used in the treatment of *larger* systems. The reduction of the information on a diatomic problem is not an interesting problem *per se*, it is only worth while if it results in *transferability* of effective interactions from small to large systems. The present section will always discuss the effective Hamiltonian reliability from this point of view, which initiated the work of Freed when he realized that effective Hamiltonian theory might give a strong ground and a practical *ab initio* derivation to the very cheap and quite inelegant (since resting on trial and error in the determination of the parameters) semi-empirical Hamiltonians of quantum chemistry. We shall not discuss the practical successes of the attempts of transfer, which are quite limited, but its theoretical legitimacy, from the basic content of the model spaces.

The present section will be organized as follows. It first analyses briefly the effect of the core electrons on the valence energies, and which are the main impacts of enlargement of the valence basis set for the molecular calculation. This will be done on an elementary problem, an alkali diatom, and the analysis will be performed through an appropriate orthogonal valence-bond (OVB) approach. The strategy followed by Freed and coworkers will then be briefly reported. It rests on a low-order degenerate perturbative expansion which faces difficult problems, some of them being mentioned there. The next subsections move to the basic definition of H^{eff}. In a following subsection, the chances of eliminating the core electrons are briefly discussed, in order to compare with the pseudopotential techniques discussed elsewhere. The last subsection discusses a few difficulties for deriving a minimal basis set effective Hamiltonian, namely intruder state problems, especially at large interatomic distances and the occurrence of intrinsically new valence situations in a polyatomic molecules valence problem, which cannot be derived from diatomic interactions.

This section concludes that, despite its advantages, the attempts to define complete valence effective Hamiltonians will probably face tremendous problems, owing to the large heterogeneity of the valence spectrum, which involves low-lying neutral states and up to multiply ionic states embedded in a far continuum (autoionizing states).

2. Role of Core Electrons and of Basis Set Extension. Analysed for an Alkali Diatom Problem

The simplest typical problem concerns the alkali diatom (Li_2) or the ethylene molecule. The active electrons are the two valence 2s electrons or the two π electrons, and the effective space is generated by two 2s or two $2p_z$ AOs a and b supposed to be properly orthogonalized (*vide infra*). The core electrons represent either the 1s pairs or the σ frame. For the sake of simplicity, we reduce the core to one electron pair, for instance the localized C–C bond pair, in one σ_g MO. The model space is spanned by four determinants, two of them being neutral, two of them being ionic in the sense of VB theory

$$P_0 = \sum_{i=1}^{n} |\phi_i\rangle\langle\phi_i|$$

$$
\begin{aligned}
\phi_1 &= |\sigma\bar{\sigma}a\bar{b}| \\
\phi_2 &= |\sigma\bar{\sigma}b\bar{a}|
\end{aligned}
\left.\right\} \text{A·B·} \quad \text{neutral}
$$

$$
\begin{aligned}
\phi_3 &= |\sigma\bar{\sigma}a\bar{a}| & \quad \text{A}^-\text{B}^+ \\
\phi_4 &= |\sigma\bar{\sigma}b\bar{b}| & \quad \text{A}^+\text{B}^-
\end{aligned}
\left.\right\} \text{ionic} \tag{113}
$$

The first-order Hamiltonian keeps the form

$$
\begin{array}{cccc}
\phi_1 & \phi_2 & \phi_3 & \phi_4 \\
\hline
0 & K_{ab} & F_{ab} & F_{ab} \\
 & 0 & F_{ab} & F_{ab} \\
 & & \Delta E & K_{ab} \\
 & & & \Delta E
\end{array} \tag{114}
$$

where K_{ab} is an interatomic exchange integral, F_{ab} is a hopping integral and ΔE, the transition energy from the neutral to the ionic determinants, is positive. If these four determinants define the model space, the corresponding effective Hamiltonian will keep the same structure, imposed by symmetry. In Hermitian formalisms the outer space will change the matrix into

$$
\begin{array}{cccc}
-\delta E & K_{ab} + \delta K & F_{ab} + \delta F & F_{ab} + \delta F \\
 & -\delta E & F_{ab} + \delta F & F_{ab} + \delta F \\
 & & \Delta E - \delta E' & K_{ab} + \delta K' \\
 & & & \Delta E - \delta E'
\end{array} \tag{115}
$$

and the question under analysis is the understanding of the main differences between this effective OVB matrix and the first-order one.

This matrix may be factorized by spin and space symmetries into a one-dimensional triplet antisymmetric, a one-dimensional singlet antisymmetric and a two-dimensional singlet symmetric subspaces

$$
\begin{aligned}
{}^3\text{A} &= (\phi_1 - \phi_2)/\sqrt{2} \\
{}^1\text{A} &= (\phi_3 - \phi_4)/\sqrt{2} \\
{}^1\text{S}_1 &= (\phi_1 + \phi_2)/\sqrt{2} \\
{}^1\text{S}_2 &= (\phi_3 + \phi_4)/\sqrt{2}
\end{aligned}
$$

The outer space will involve two types of determinants: (i) those which keep the frozen σ core, and (ii) those involving excitations from the σ core. The former represent the effect of an extension of the valence basis set, and their effect may be analysed first.

We may assume first that the a and b orbitals have been determined self-consistently for the lowest triplet state ^3A which is neutral in character since the $S_z = 1$ solution may be written unambiguously by a localization of the symmetry-adapted singly occupied MOs, into a localized picture[81]

$$^3A = |\sigma\bar{\sigma}ab| \tag{116}$$

and obtained from a preliminary restricted Hartree–Fock (RHF) calculation. One should notice that the distortion of these two localized MOs a and b with respect to the free-atom RHF AOs $\tilde{a}'$ and $\tilde{b}'$ is very important. It involves, besides orthogonalization tails, contractions and distortions in the molecular field and significant hybridizations which minimize the electronic repulsion, avoiding for instance the neighbouring core:

$$a = \tilde{a}' + \sum_{\substack{p \in A \\ p \neq \tilde{a}}} C_{ap}p + \sum_{q \in B} C_{aq}q$$

These preliminary distortions are of major energetic importance.

Then one may introduce orthogonal atomic orbitals a_i' and b_i', which are excited atomic orbitals of A and B orthogonalized to a and b, and one will generate single and double excitations towards these virtual MOs from either neutral (ϕ_1, ϕ_2) or ionic (ϕ_3, ϕ_4) valence determinants.

Single excitations from ϕ_1 or ϕ_2 will generate

$$\langle \sigma\bar{\sigma}a_i'\bar{b}|H|\phi_1 \rangle = \langle a_i'| - K_b|a \rangle = (a_i'b, ab)$$

i.e. rather small corrections (due to the weakness of the interatomic distributions). K_b is the exchange operator associated with orbital b. Double excitations from ϕ_1 or ϕ_2 will generate

$$\langle \sigma\bar{\sigma}a_i'\bar{b}_i|H|\phi_1 \rangle = \langle a_i'b_i'|ab \rangle = (a\acute{a}_i, bb_i')$$

i.e. stabilizing interatomic dispersion forces, of moderate amplitude. Single excitations from ϕ_3 or ϕ_4 will lead to much larger corrections since the matrix element represents

$$\langle \sigma\bar{\sigma}a_i'\bar{a}|H|\phi_3 \rangle = \langle a_i'|J_a - J_b|a \rangle$$

i.e. the coupling of the $(a_i'a)$ dipolar distribution with the valence dipolar field of the A^-B^+ distribution, where J_a and J_b are the coulomb operators associated with orbitals a and b respectively. This represents both an atomic reorganization of the A^- electron pair (which tends to be more diffuse) and its molecular distortion towards the B^+ centre. Double excitations from ϕ_3

or ϕ_4) will lead to

$$\langle \sigma\bar{\sigma}a'_i\bar{a}'_j | H | \phi_3 \rangle = \langle a'_i a'_j | aa \rangle = (aa'_i, aa'_j)$$

corrections correlating the electron pair of the negative ion A^-, either radially if a'_i and a'_j have the same l value l_a, or angularly if a'_i and a'_j have a value of $l = l_a \pm 1$. Double excitations from the ionic valence components therefore introduce the radial and angular correlation of the monocentric electron pair.

The conclusion is thus that the extension of the basis set with respect to the minimal atomic basis set is important (i) to distort the AOs in order to minimize the interatomic repulsion in the *neutral* forms, and (ii) to lower the energies of the *ionic* components of the VB wavefunction by important instantaneous repolarization and correlation effects.[82-84]

If the atoms bear more than one electron, for instance for a carbon atom, the VB wavefunction of a C-containing molecule will introduce some components where the C atom is neutral, in several possible s^2p^2, sp^3, p^4 valence states (the latter being hybridized), and other components of $C^+, C^-, C^{2+}, C^{2-}, \ldots$ singly or multiply ionic character. The use of large basis sets in CI calculations essentially results in lowering of the effective energies of the various excited neutral, ground and excited ionic valence states appearing in the full valence CI wavefunction.

The influence of core excitations may be studied according to a similar analysis. One may first consider $\sigma \rightarrow \sigma^*$ single excitations. The effect on the neutral determinants ϕ_1, ϕ_2

$$\langle \phi_1 | H | a^+_{\sigma^*} a_\sigma \phi_1 \rangle = \langle \sigma | - K_b | \sigma^* \rangle$$

is weak, due to Brillouin's theorem. In contrast, single excitations acting on the ionic determinants ϕ_3, ϕ_4, lead to a very important coupling

$$\langle \phi_3 | H | a^+_{\sigma^*} a_\sigma \phi_3 \rangle = \langle \sigma | - J_a + J_b | \sigma^* \rangle$$

which represents the coupling between the $\sigma\sigma^*$ transition dipole and the A^-B^+ valence dipole.[85] The corresponding correction represents the repolarization of the σ core under the instantaneous valence field of the ionic valence determinants. This correction will be a major molecular correction.

Double excitations involving two core electrons introduce core correlation effects, almost independent of the valence electron distribution; they have a pure translational effect on the energies.

Double excitations involving both one core and one valence electron $a^+_{\sigma^*} a^+_{a_i} a_a a_\sigma$ lead to $(\sigma\sigma^*, aa'_i)$ matrix elements when they act on either the neutral or the ionic determinants. Their effect is essentially translational on the valence states. They stabilize the Rydberg states much less and become zero in the positive ion. They thus have an important spectroscopic effect,[86-90] but they are not crucial for the energy differences between *valence* states of the neutral molecule.

One should therefore remember that *the main effect of the core appears from single excitations on the valence ionic VB components.*

One thus sees that the core electrons will essentially act as a polarizable system, sensitive to the electric field instantaneously created by the valence electrons. This field is especially large in ionic VB components and the core–valence interaction should essentially result in a differential stabilization of the valence VB ionic structures with respect to the valence-only approach.

Therefore, both valence basis set enlargement and core correlation effects might be seen as going essentially through a coupling with the *ionic* components of the valence wavefunction. One may thus imagine immediately that their effect will be some specific energy lowering of these ionic VB component energies, with respect to those that one would predict from a frozen-core and/or from a minimal basis set calculation. In Eq. (115) the larger corrections will be $\delta E'$. In an effective Hamiltonian calculation the effect of these corrections will essentially be a *dressing of the ionic VB components,* i.e. a lowering of the self-repulsion, as well known in π-electron theories.

3. Simultaneous Reduction of the Number of Particles and Basis Set; the Valence Effective Hamiltonians of K. Freed

In a large series of papers, Freed and coworkers have tried to define valence-only minimal basis set effective Hamiltonians. This idea was first applied to the π system of conjugated hydrocarbons, eliminating the explicit treatment of the σ core and attempting to find rigorous foundations[91-94] for the π semi-empirical Hamiltonians of the Pariser–Parr–Pople family,[95,96] or to obtain non-empirical derivations of these simplified Hamiltonians. The CNDO valence-only Hamiltonians[97] excluded the $1s^2$ core electrons of the atoms and belonged to the same category. The same idea therefore received numerous other applications to atoms[98] (including transition metals[99,100]), hydrides.[101,102] alkali diatoms,[103,104] diatoms with more than two valence electrons (O_2),[105-107] etc.

The method[108-110] is frequently referred to as the partitioning (Feshbach–Löwdin) technique (see Section II.2). The model space is the valence complete active space, i.e. the whole set of determinants involving the same (unspecified) frozen core and all possible electronic distributions in the valence orbitals. This space is, of course, very large and spread on a huge range of energies. For O_2, for instance, it involves $O(2p^6) + O(2p^6)$ configurations, which are unbound. It also involves, for instance, $O^{2-}(2s^2p^6) + O^{2+}(2p^4)$ situations, which also generate unbound states at all distances. If the effective Hamiltonian was that produced by the partitioning technique, it would be energy-dependent. In other words, one would have a different valence effective

Hamiltonian for each state. The effective integrals would then be very difficult to transfer to larger problems. The first papers on π electron effective Hamiltonians were actually presented as energy-dependent.

In most papers, Freed et al.[108,110] (i) develop the energy denominator of Eq. (9) in a power expansion, thus going back to a Rayleigh–Schrödinger version of the quasi-degenerate perturbation theory (QDPT); (ii) use a monoelectronic definition of $H_0 = \sum_i \varepsilon_i a_i^+ a_i$ (Moller–Plesset definition of the unperturbed Hamiltonian); (iii) introduce a full degeneracy of the valence space, which, while this complete degeneracy results from the choice of H_0 in the case of a single band, further requires giving equal energies to s and p AOs in systems involving both s and p orbitals; (iv) define a set of rationally orthogonalized AOs by a proper combination of Schmidt and $S^{-1/2}$ transformations. Alternative definitions of localized equivalent orbitals may be obtained from the valence MOs of the upper multiplet. The QDPT expansion is performed in a basis of OVB determinants.

One should notice first that, from the very theory, the effective Hamiltonian of an n-active-electron problem spanned in a basis of N determinants is a series of n-electron operators

$$H^{\text{eff}} = \sum_{I,J=1}^{N} |\phi_I\rangle\langle\phi_I|H^{\text{eff}}|\phi_J\rangle\langle\phi_J| \tag{117}$$

which are quite difficult to handle if one wants to transfer them into a larger problem. Freed et al.[108–110] succeed in expressing the effective Hamiltonian in terms of one-, two-, three-,... particle operators. This is always possible in principle but quite arbitrary. If ϕ_I and ϕ_J differ by two spin orbitals only

$$\phi_J = a_j^+ a_l^+ a_k a_i \phi_I \tag{118}$$

one may be tempted to define the effective bielectronic integral as

$$(ij, kl)^{\text{eff}} = \langle\phi_I|H^{\text{eff}}|\phi_J\rangle$$

but if one considers another couple of determinants, also differing by the same spin orbitals

$$\phi_L = a_j^+ a_l^+ a_k a_i \phi_K \tag{119}$$

the effective interaction has no reason to be the same

$$\langle\phi \ |H^{\text{eff}}|\phi \ \rangle \neq \langle\phi_J|H^{\text{eff}}|\phi_I\rangle \tag{120}$$

since they include high-order processes which are different. If

$$\phi_K = a_m^+ a_n \phi_I \tag{121}$$

one may be tempted to write

$$\langle\phi_K|H^{\text{eff}}|\phi_L\rangle = (ij, kl)^{\text{eff}} + \delta_{ijkl,mn} \tag{122}$$

$\delta_{ijkl,mn}$ representing an increment of the (ij, kl) effective interaction when MO n is empty and the MO m is occupied.

In terms of operators this may be written as a six-body operator

$$\delta_{ijkl,mn} a_j^+ a_l^+ a_k a_i a_n a_n^+ a_m^+ a_m$$

This procedure may be generalized and is well defined, but it depends on the order of appearance of the ϕ_I. If one of the ϕ_I played a special role, as the Hartree–Fock (HF) determinant, it might be taken as the vacuum state, and a hierarchy of singly, doubly, ... excited determinants might be defined, but since one is looking for local (transferable) interactions involving orthogonal AOs, all VB determinants play the same role and any ordering would be arbitrary.

Freed et al.[108-110] therefore do not follow this way. They expand the effective Hamiltonian to second order, and since all the determinants of the model space are kept degenerate, the second-order corrections only introduce one-, two- and three- particle operators which only depend on the index of the connected propagation lines in the diagram. If one defines the reference description of the core as the vacuum, the model space states are defined by a certain number of upwards propagation lines, equal to the number of valence electrons. For a four-electron problem, the state ϕ_I is defined by the four indices

defining $\phi_I = a_i^+ a_j^+ a_k^+ a_l^+ \phi_C$.

Then a one-body operator may change ϕ_I into $\phi_J = a_p^+ a_l \phi_I$. A second-order correction may be viewed as

$$= \frac{(cc^*, \lambda l)(\lambda p, cc^*)}{E_c + E_l - E_\lambda - E_{c^*}} \qquad (123)$$

where the double arrows symbolize MOs which do not belong to the active space. The downward line necessarily concerns a core MO while

and

belong to the virtual space. Now it becomes clear from hypothesis (iii) that the process

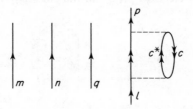

will result in an equal correction and one may say that

$$\left(\sum_c \sum_{c^*} \sum_\lambda \frac{(cc^*, \lambda l)(\lambda p, cc^*)}{E_c + E_0 - E_\lambda - E_{c^*}} \right) \tag{124}$$

is the amplitude of an effective second-order correction to the first-order interaction through the core Fock operator.

Typical two-body operators are

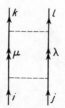

or

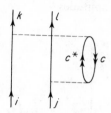

the second diagram involving a core excitation. The three-body terms are of the type

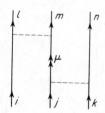

and may change there valence orbitals, but only involve one outer orbital.

One should notice, however, that the third-order corrections will destroy this apparent simplicity. The third-order corrections are given from Eq. (43) by

$$\langle \phi_I | H^{(3)} | \phi_J \rangle = \sum_{\alpha,\beta \notin S} \frac{\langle \phi_I | V | \alpha \rangle \langle \alpha | V | \beta \rangle \langle \beta | V | \phi_J \rangle}{(E_J - E_\alpha)(E_J - E_\beta)}$$
$$- \sum_{\alpha \notin S} \sum_{L \in S} \frac{\langle \phi_I | V | \alpha \rangle \langle \alpha | V | \phi_L \rangle \langle \phi_L | V | \phi_J \rangle}{(E_J - E_\alpha)^2} \qquad (125)$$

and since the model space involves some valence states of very high energy (for instance, highly hybridized VB structures, or multiply ionic components), some of the diagonal terms of the perturbation operator are actually *huge*. The second summation in Eq. (125) will involve some tremendous terms, which will make the perturbation expansion quite unreliable and will destroy the apparent simplicity of the above-mentioned one-, two-, and three-body operators. For instance the same one-body operator

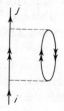

will be associated with contributions involving other valence MOs, which may have completely different amplitudes:

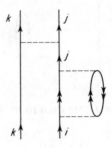

involves J_{jk} while

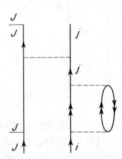

involves J_{jj}, which may be much larger. (These diagrams should be folded to enter the canonical diagrammatic QDPT expansion.)

The trick of introducing full degeneracy of the model space is only efficient at the second order, and the transformation of an effective Hamiltonian into one-, two-, three-,... body operators is questionable since from their very definition the effective Hamiltonians are actually N-body operators.

In practice, several questions arise:

1. When one uses a low-order perturbation expansion, resting on a very crude H_0 definition, are the lowest valence potential curves equally correct? The large series of papers by Freed and coworkers devoted to diatoms[101,102,105] actually seem to support a positive answer.

2. Is the transferability to positive and negative (di)atomic ions ensured, when the extraction has been done from neutral (di)atoms? A large series of papers have given satisfactory results for the positive ions of atoms[98] (even when multi-ionized) and of diatoms.[101,106] For negative ions Freed et al.[107] claimed that the use of valence effective interactions might solve the difficult resonance problem for negative ions, since it would directly give the potential curves of a dressed valence negative ion, which is supposed to generate the discrete state embedded in the continuum. Again, one should say that from the very principles the effective Hamiltonians for X_2^+ or X_2^-

are completely different from that of X_2. If $X = Li$ for instance the effective Hamiltonian for Li_2^+ cannot in principle be reduced to the one-particle effective operator of Li_2. The most recent attempts to calculate Li_2^+ from Li_2 have actually failed.[103,104]

3. Is the transferability to larger molecules ensured? The derivation of effective interactions from a diatomic problem is useless except for transferability to larger systems. An approach of this main goal has not been even attempted, except for the π systems of conjugated hydrocarbons. In that precise case[94] an effective Hamiltonian has been proposed for butadiene, and keeping the nearest-neighbour interactions from a similar calculation on ethylene, it was possible to estimate long-distance one- and two-body interactions and many-body terms. The results were very discouraging since the one-electron 1–4 hopping integral between the terminal atoms appeared to be as large as $-1.6\,eV$, i.e. of the same order of magnitude as the hopping integral between adjacent atoms ($\beta_{12} = -4.0\,eV$).

In view of these practical difficulties, and in order to separate the convergence problems of the perturbative approach from basic problems inherent to the effective Hamiltonians themselves, we would like to reanalyse the core–electron elimination and the reduction of the basis set in basic terms, i.e. referring to the definitions of the exact effective Hamiltonians.

4. Reduction of the Number of Particles

Let us define ϕ_c as an HF or internally correlated description of the core shell of an atom (or a molecule). Then one may choose as a valence model space the set of functions which are antisymmetrized products of ϕ_c by a valence function ϕ_I, in which the n_v valence electrons occupy valence, or Rydberg, or arbitrary orbitals orthogonal to the core orbitals (and not involved in its description)

$$P_0 = \sum_{I \in S} |\phi_c \cdot \phi_I\rangle\langle\phi_c \cdot \phi_I| \qquad (126)$$

For the lithium atom, for instance, ϕ_c may be a $1s^2$ product or a linear combination

$$\phi_c = \alpha(1s^2) + \beta(1s'^2) + \gamma(p^2)$$

including the largest part of the radial and angular correlation of the $1s^2$ pair. ϕ_I may include a set of n valence plus diffuse plus oscillating AOs, orthogonal to the core AOs. Then the target subspace, of dimension n, has to be defined. If one defines it as the n lowest roots of the Li problem treated in the same basis set

$$P = \sum_{k=1}^{n} |\psi_k\rangle E_k\langle\psi_k| \qquad (127)$$

with

$$H_P|\psi_k\rangle = E_k|\psi_k\rangle \tag{128}$$

H_P being the exact Hamiltonian restricted to the finite AO basis set, one should note two difficulties: (i) If the basis set involves spatially concentrated (oscillating) AOs, some of the roots may be unbound, i.e. embedded in the continuum; they are then strongly dependent on basis set. (ii) Some core excited solutions close, for instance, to $|1s2sns\rangle$ may be lower in energy than some roots keeping a frozen core $((1s^2ks)$ where ks would be a very concentrated and oscillating AO). This will result in an intruder state problem.

To avoid the arbitrariness noticed in (i), one may introduce a spectroscopic basis set of AOs spanning only the valence and lowest Rydberg states of the atoms. But it is known that the molecular construction implies some orbital contraction, which can hardly be mimicked from spectroscopic AOs. It is also very important when an atom A approaches an atom B that the basis set of the atom B involves same oscillating spatially contracted functions which are used by the outer electron orbitals of A to minimize their repulsion with the electrons of B. It seems almost compulsory for a correct molecular treatment that the atomic information concerns the energy of unbound spatially concentrated atomic distributions. Then the occurrence of the second problem (i.e. the occurrence of core excited intruder states) is not excluded.

The problem is worse when one considers atoms involving several valence electrons. If one wants to obtain the information relative to a double-zeta valence basis set of the C atom, it will be necessary, in principle, to find states of the type $1s^22p^4$ and also $1s^23p^4$; if $2p$ and $3p$ are the RHF AOs in the basis set, the latter $(1s^23p^4)$ determinant will certainly be higher in energy than core excited states of the type $(1s2s^22p^3)$, which will act as intruder states. The definition of the target space will then be quite difficult, since for some basis sets it will be difficult to discriminate between the eigenstates of the problems which keep a frozen core from those which are core excited.

Once the model and target spaces have been chosen, the atomic effective Hamiltonian is defined and its use in molecular calculations may be examined. For the Li atom, for instance,

$$H_A^{\text{eff}} = \sum_{k=1}^{n} |1s^2k\rangle E_k \langle 1s^2k| \tag{129}$$

H_A^{eff} is purely monoelectronic and monocentric. If one goes to Li_2 one may first assume a purely monoelectronic form of the Hamiltonian

$$H_{AB}^{\text{eff}} = H_A^{\text{eff}} + H_B^{\text{eff}} \tag{130}$$

while the basis set will involve $2n$ AOs on A and on B. Then the effective Hamiltonian will reduce to a valence Hückel-type problem

$$(H^{\text{eff}} - ES)C = 0$$

where for instance

$$\langle 2s_A | H^{eff} | 2s_a \rangle = E_{2s} + \sum_{k_b} \langle 2s_A | k_B \rangle^2 E_k$$

$$\langle 2s_A | H^{eff} | 2s_b \rangle = 2E_{2s} \langle 2s_a | 2s_b \rangle$$

It is clear that such a Hamiltonian can only rationalize a Hückel-type Hamiltonian (eventually in a non-minimal basis set) and that it essentially leads to a pure Mulliken-type approximation for the bicentric integrals (which play the key role in the construction of the bond through electron delocalization). It is known that such approximations usually fail; an empirical k parameter is used in the proportionality of the hopping integrals to the overlap and atomic energies (Wolfsberg–Helmholtz approximation).

An alternative strategy consists of considering that

$$E_k = \langle 1s^2 k | T - Z_A/r_A + v_A | 1s^2 k \rangle \tag{131}$$

i.e. that the effect of the core electrons goes through a monocentric correction to the integrals of the (kinetic + nuclear attraction) operator

$$v_A = \sum_k (E_k - \langle 1s^2 k | T - Z_A/r_A | 1s^2 k \rangle) | 1s^2 k \rangle \langle 1s^2 k |$$

The molecular calculation of Li_2 will then be performed through a two-electron Hamiltonian

$$H = T - \frac{Z_A Z_B}{r_A - r_B} + \frac{1}{r_{12}} + v_A + v_B \tag{132}$$

and developed in the basis of determinants $|1s_A^2 1s_B^2 kl\rangle$. The correction operator v_A, which may be considered as a core-potential correction written in a non-local finite expansion, will act on the atomic orbitals centred on atoms A and B.

This procedure offers in principle an alternative to the pseudopotential approaches for the treatment of electron cores. It faces two problems: (i) The non-local expansion of v_A is quite arbitrary. (ii) The process is difficult to generalize to atoms having several valence electrons. For boron atoms, for instance, the effective Hamiltonian should be written

$$H^{eff} = |1s^2 2s^2 2p\rangle \langle 1s^2 2s^2 2p | E(^2P)$$
$$+ |(^4D)1s^2 2s 2p^2\rangle \langle (^4D)1s^2 2s 2p^2 | E(^4D)$$
$$+ |(^2S)1s^2 2s 2p^2\rangle \langle (^2S)1s^2 2s 2p^2 | E(^2S) + \cdots \tag{133}$$

i.e. it is intrinsically trielectronic, without any rigorous reduction to a monoelectronic operator. Attempts along this way have never been practised, and should be performed carefully. The fact that multi-electronic operators are required does not condemn this approach, since these multi-electronic operators should receive a monocentric expansion and would only generate the calculation of overlap integrals in molecular calculations.

To summarize this section one should say that an effective Hamiltonian treatment of the core electron effect faces a contradiction between the necessity to use extended valence basis sets for the extraction and the risk of appearance of core excited intruder states. One should also recognize that this approach leads to p-electron operators for atoms involving p valence electrons and seems much more difficult to handle than the monoelectronic core pseudopotentials extracted by simulation techniques and discussed in Section IV of the present contribution. As a counterpart one should mention that this core effective Hamiltonian would be much superior, since it would include for instance the core–valence correlation effects which play such an important role in alkali- or alkaline-earth-containing molecules.

5. Reduction to Minimal Basis Sets

The problem consists of defining a valence minimal basis set effective Hamiltonian which would reproduce the valence part of the *exact* molecular spectrum. As already mentioned the problem might concern atoms involving core electrons, described in the same frozen wavefunctions. For the sake of simplicity we shall first consider hydrogen atoms, and a minimal basis set would be spanned by a single 1s AO per atom. For the H atom, the effective operator is $-0.5(\text{a.u.})|1s\rangle\langle 1s|$. For H_2 the model space is spanned by four determinants; calling a and b and $S^{-1/2}$ orthogonalized AOs $1s_A$ and $1s_B$ respectively, these four determinants are

$$\begin{aligned}\phi_1 &= |a\bar{b}| \\ \phi_2 &= |b\bar{a}|\end{aligned} \Bigg\} \text{neutral OVB components}$$

$$\begin{aligned}\phi_3 &= |a\bar{a}| \\ \phi_4 &= |b\bar{b}|\end{aligned} \Bigg\} \text{ionic OVB components}$$

$$P_0 = \sum_{i=1}^{4} |\phi_i\rangle\langle\phi_i|$$

as already discussed in Section III.C.2.

The first main problem concerns the choice of the target space. The (b) $^3\Sigma_u^+$ $||a\bar{b}| - |b\bar{a}||/\sqrt{2}$ combination will always correspond to the lowest (b) $^3\Sigma_u^+$ exact state, which dissociates into ground-state atoms. The ground state ($^1\Sigma_g^+$) has very large components in the model space, on neutral determinants only for large interatomic distances, and on neutral *and* ionic valence determinants at short interatomic distances. The two lowest eigenstates of H_2 should definitely belong to the target space.

For short interatomic distances the lowest (B)$^1\Sigma_u^+$ state is known to have large components on the antisymmetric combination of ionic determinants $||a\bar{a}| - |b\bar{b}||/\sqrt{2}$, and it should therefore definitely belong to the target space. The model space contains a second $^1\Sigma_g^+$ state, a linear combination of valence neutral and ionic components, and one may assume that the fourth eigenstate of the target space is the lowest excited E ($^1\Sigma_g^+$) state. Thus

$$P = |(X)^1\Sigma_g^+\rangle\langle(X)^1\Sigma_g^+| + |(E)^1\Sigma_g^+\rangle\langle(E)^1\Sigma_g^+|$$
$$+ |(b)^3\Sigma_u^+\rangle\langle(b)^3\Sigma_u^+| + |(B)^1\Sigma_u^+\rangle\langle(B)^1\Sigma_u^+| \qquad (134)$$

P_0, P and the exact energies being known, the effective Hamiltonian is defined in one of its usual versions.

As a first remark, one should mention that the projections of the two $^1\Sigma_g^+$ states into the model space have no reason to be orthogonal, since in

$$\psi(X)^1\Sigma_g^+ = \alpha(|a\bar{b}| + |b\bar{a}|) - \beta(|a\bar{a}| + |b\bar{b}|) + \cdots$$
$$\psi(E)^1\Sigma_g^+ = \beta'(|a\bar{b}| + |b\bar{a}|) + \alpha'(|a\bar{a}| + |b\bar{b}|) + \cdots$$
$$\alpha, \beta > 0 \qquad (135)$$

β' and β (resp. α' and α) are of the same order of magnitude but different. If $\beta'/a' = \beta/\alpha$ one might write

$$\begin{aligned}
H^{\text{eff}} = {} & E((b)^3\Sigma_u^+)[||a\bar{b}| - |b\bar{a}|\rangle\langle|a\bar{b}| - |b\bar{a}||+||ab|\rangle\langle|ab||\\
& + ||\overline{ab}|\rangle\langle|\overline{ab}||]\\
& + E((B)^1\Sigma_u^+)||a\bar{a}| - |b\bar{b}|\rangle\langle|a\bar{a}| - |b\bar{b}||\\
& + E((X)^1\Sigma_g^+)[|\alpha(|a\bar{b}| + |b\bar{a}|) - \beta(|a\bar{a}| + |b\bar{b}|)\rangle\langle\alpha(|a\bar{b}| + |b\bar{a}|)\\
& \qquad - \beta(|a\bar{a}| + |b\bar{b}|)|]\\
& + E((E)^1\Sigma_g^+)[|\beta(|a\bar{b}| + |b\bar{a}|) + \alpha(|a\bar{a}| + |b\bar{b}|)\rangle\langle\beta(|a\bar{b}| + |b\bar{a}|)\\
& \qquad + \alpha(|a\bar{a}| + |b\bar{b}|)|] \qquad (136)
\end{aligned}$$

This will be the general expression of the Schmidt-orthogonalized effective Hamiltonian. If $\beta/\alpha \neq \alpha'/\beta'$, the Bloch effective Hamiltonian will be non-Hermitian. The relative advantages of the Bloch, des Cloizeaux, Schmidt-orthogonalized effective Hamiltonians, or of the intermediate Hamiltonians has never been tested, especially in transfers to large systems.

A second remark concerns the definition of the target space of large interatomic distances, and the occurrence of intruder states. For large enough interatomic distances the lower $(B)^1\Sigma_u^+$ state ceases to be valence (and ionic) to become of $H(n = 2) + H(n = 1)$ character. It dissociates into the $H(2s) + H(1s)$ asymptote, i.e. an avoided curve crossing occurs between the ionic state and a Rydberg state. The same remark is valid for the second $^1\Sigma_g^+$ state, which is known to have two minima,[111] one of them being of Rydberg character. This state is only ionic for intermediate atomic distances, it is Rydberg on both short and large interatomic distances (cf. Fig. 10).

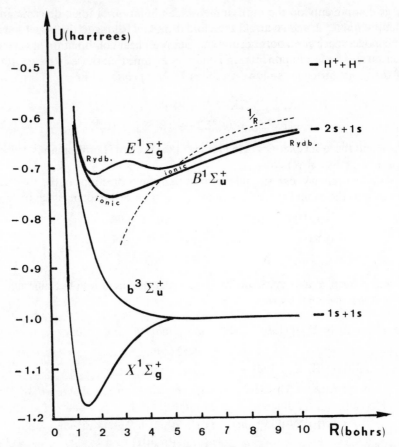

Fig. 10. Potential curves of the valence states of the H_2 molecule, showing the mixing with non-valence states.

Then one faces a dilemma:

1. One may either choose as target eigenvectors those which have the largest components onto the model space (maximum projection criterion) and in the $^1\Sigma_u^+$ symmetry, for instance, this eigenvector is the lowest $^1\Sigma_u^+$ root at short interatomic distances, then the second root besides a certain distance r_c, and may even go to the third root if a second Rydberg state appears below the ionic vector, etc. The problem is even worse with the $^1\Sigma_g^+$ symmetry. The consequence is that the effective Hamiltonian matrix elements are not continuous functions of the interatomic distances! The use of a discontinuous operator in larger systems (hydrogen clusters) seems to be very difficult.

2. One may forget this disappearance of the ionic components into higher eigenvectors and take an adiabatic definition of the target space assuming that the concerned roots are always the lowest of their symmetry, whatever their physical content. Then the effective operator matrix elements will be continuous functions of the interatomic distance, but the eigenvectors

$$\Phi_m = P_0 \psi_m$$

tend to have a vanishing amplitude when R_{AB} increases. This may result in numerical instability. It may also be dangerous to assume that in a cluster H_n a valence situation

$$H \cdots H^- \cdots H \cdots H \cdots H^+ \cdots H$$

will have the same energy as the situation

$$H \cdots H(2s) \cdots H \cdots H \cdots H(1s) \cdots H$$

6. Transferability; Appearance of New Situations

It one wants to treat an H_n problem using a bielectronic effective Hamiltonian derived from exact calculations on H_2, several questions arise concerning:

1. The choice of the basis set. One may perform a new $S^{-1/2}$ transform for the H_n conformation and identify the new orthogonalized AOs with that of the bicentric problem. This is the most direct solution but the orthogonaliz-ation tails will be different and the use in this new basis of the effective bielectronic Hamiltonian given in Eq. (136) for instance may result in uncontrolled effects. One may also express the bielectronic operator of Eq. (136) in the non-orthogonal basis set and calculate the Hamiltonian matrix of H_n in the basis of non-orthogonal determinants, antisymmetrized products of 1s AOs. The problem to solve is then of $(H\text{-}ES)$ type and it faces a typical non-orthogonality problem of VB methods, which has been a major drawback of these approaches.
2. The occurrence of new situations. In a linear H system, for instance, one must consider (a) neutral determinants, for instance $|a\bar{b}c\bar{d}|$ or $|\bar{a}bc\bar{d}|$, etc.; (b) singly ionic determinants, some of them having dipoles between adjacent atoms ($|ab\bar{b}d|$, i.e. AB^-C^+D), while others introduce long-distance electron jumps ($|a\bar{a}b\bar{c}|$, i.e. A^-BCD^+); and (c) doubly ionic determinants such as $A^-B^+C^-D^+$ or $AA^-B^-C^+D^+$

One may wonder whether a bielectronic Hamiltonian extracted from the H_2 problem is able to deal with some of these situations and to assign reasonable energies to them. The neutral determinants only involve neutral–neutral interatomic interactions which appeared in the H_2 problem. Assuming that

$$\langle a\bar{b}c\bar{d}|H^{\text{eff}}|a\bar{b}c\bar{d}\rangle = \langle a\bar{b}|H^{\text{eff}}|a\bar{b}\rangle + \langle ac|H^{\text{eff}}|ac\rangle + \langle a\bar{d}|H^{\text{eff}}|a\bar{d}\rangle + \text{etc.}$$

essentially neglects the possible three-body terms governed by the overlap expansion. But when one goes to singly ionic determinants in the energy of $|ab\bar{b'}\bar{d}|$ (B^-C^+), while information concerning the B^-C^+ interaction or the AD interaction is contained in the H_2 effective Hamiltonian, information concerning the AB^- or the C^+D interactions are lacking. They should be extracted from H_2^- and H_2^+ problem respectively, i.e. from the one- and three-electron diatomic problems. This might be done in principle although one may notice that H_2^- is unbound at short interatomic distances, i.e. that it is impossible to define *exact* valence states of H_2^-.

The A^-BC^+D determinants involve strong polarizations of the intermediate B atom, which are not given by numerical transfers from A^-B or BC^+, due to the non-additivity of polarization energies.

When one goes to doubly ionic structures, $A^-B^-C^+D^+$ for instance, the relevant information concerning the A^-B^- interaction should be extracted from the H_2^{2-} valence state, which of course cannot be defined.

The transferability of a valence effective Hamiltonian defined on H_2 to H_n clusters therefore faces a series of basic difficulties, which leaves little hope of success. The situation would be even worse of course if one dealt with boron or carbon atoms since for C_2 already one should introduce strongly hybridized (for instance $C(p^4) + C(p^4)$) or multiply ionic (for instance $C^{4+} + C^{4-}(s^2p^6)$) states which are unbound. The choice of the target space is already impossible on the diatom, and the definition of an exact (Bloch, des Cloizeaux,...) effective Hamiltonian from knowledge of the spectrum of the diatom is either impossible or perfectly arbitrary. Even if it were possible, the treatment of B_4 or C_4 would introduce some multiply ionic valence ($C_4^{4+}C^{4-}C^{4-}C^{4+}$) determinants for which the assessment of an effective energy would be impossible.

There is thus little hope, in our opinion, for a rigorous definition of valence minimal basis set effective Hamiltonians. To build them, the use of the diatomic effective Hamiltonian may be useful, but some supplementary assumptions should be made, along a physically grounded model, to define for instance three-body polarization energies and the energies of highly hybridized or multi-ionic VB structures. One should realize the physical origin of these numerous troubles; they essentially come from the inclusion of the *ionic* determinants in the model space. This inclusion first resulted in intruder state problems for the diatom; it also leads to the appearance of multiply ionic structures in the valence minimal basis set space of the cluster. It seems that, even for H, the definition of a full valence space is too ambitious.

Besides the effort to give theoretical grounds (and non-empirical versions) for the semi-empirical quantum chemistry or solid-state physics simplified Hamiltonians, which are valence minimal basis set Hamiltonians, Freed's attempt rested on a theoretical property, namely the size consistency (and the linked-cluster diagrammatic expansion) of the Rayleigh–Schrödinger QDPT development when the model space is a *complete* active space. The price to pay for the benefit of this theoretical guarantee is so large and so dramatic that one may wonder whether obtaining less ambitious effective Hamiltonians would not be preferable. (Notice that an effective Hamiltonian restricted to neutral non-hybridized plus neutral singly hybridized and singly ionic structures would be size-inconsistent, as is the double CI truncated treatment of the electronic correlation problem for the same fundamental reasons.)

This step towards simplification may proceed along two different ways:

1. One may resign oneself to treat the ionic states *exactly*, and use an intermediate Hamiltonian spanned by the *full* valence space but which concentrates on the neutral states and does not try to reproduce the ionic eigenstates. Work is in progress along this line; it shows that this approach solves many difficulties discussed above. The effective energies of the ionic determinants are not as critical, since they simply appear through their interaction with the neutral determinants to stabilize the lowest neutral states through electronic delocalization.
2. One may limit the model space to neutral situations. This is the philosophy of the Heisenberg Hamiltonians, which we discuss now.

D. Effective Hamiltonians Spanned by Neutral-only Valence-bond Determinants: Magnetic (Heisenberg) Hamiltonians and Their Possible Generalizations

The last class of effective Hamiltonians rests on the choice of a very limited model space, spanned by the neutral structures of an orthogonal VB development. This choice of the model space is grounded on the facts that (i) the neutral VB determinants are usually those of lowest energy, and (ii) the lowest eigenstates have large components on the neutral VB determinants. This model space is a part of the full valence space previously defined, and it is no longer a complete active space. The development of the corresponding effective Hamiltonians is especially simple for homogeneous systems involving only one type of active orbital (i.e. a single band), each atom having p electrons in p AOs (half-filled band). The simplest problem concerns systems where each atom brings one active electron in one AO. This is the case of clusters of hydrogen, alkali, or noble-metal atoms. The conjugated hydrocarbons may also be considered as belonging to that family if the active electrons are the π electrons, one per C atom in a $2p_z$ AO.

1. Half-filled Bands with One Active Electron, One Atomic Orbital per Centre

If a and b are two orthogonal localized orbitals of H (obtained for instance from the SCF calculation of the $(b)^3\Sigma_u^+$ state), the model space of neutral determinants is defined by

$$P_0 = ||ab|\rangle\langle|ab|| + ||\overline{ab}|\rangle + \langle|\overline{ab}|| + ||a\overline{b}|\rangle\langle|a\overline{b}|| + ||\overline{a}b|\rangle\langle|\overline{a}b| \quad (137)$$

Then, as previously discussed (see Section III.C.2) the most neutral eigenstates of 1s character are the ground $(X)^1\Sigma_g^+$ and lowest $(b)^3\Sigma_u^+$ states. Thus

$$P = |(b)^3\Sigma_u\rangle\langle(b)^3\Sigma_u| + |(X)^1\Sigma_g^+\rangle\langle(X)^1\Sigma_g^+| \quad (138)$$

where $(b)^3\Sigma_u^+$ involves its three $S_z = 0, \pm 1$ components.

Owing to space and spin symmetry, the eigenvectors projected into the model space are necessarily orthogonal

$$P_0|(b)^3\Sigma_u\rangle = \begin{cases} \|ab|\rangle & \text{for } S_z = 1 \\ \|\overline{ab}|\rangle & \text{for } S_z = -1 \\ (\|a\overline{b}|\rangle - \|b\overline{a}|\rangle)/\sqrt{2} & \text{for } S_z = 0 \end{cases}$$

$$P_0|(X)^1\Sigma_g^+ = (\|a\overline{b}|\rangle + \|b\overline{a}|\rangle)/\sqrt{2} \quad (139)$$

Then the effective Hamiltonian is entirely defined. It is Hermitian.

$$H^{\text{eff}} = E(^3\Sigma_u^+)[||ab|\rangle\langle|ab|| + ||\overline{ab}|\rangle\langle|\overline{ab}|| + ||a\overline{b}| - |b\overline{a}|\rangle\langle|a\overline{b}| - |b\overline{a}||]$$
$$+ E(^1\Sigma_g^+)||a\overline{b}| + |b\overline{a}|\rangle\langle|a\overline{b}| + |b\overline{a}|| \quad (140)$$

It may be written in second quantization form

$$H^{\text{eff}}(a, b) = E(^3\Sigma_u^+)_{ab} + [E(^1\Sigma_g^+)_{ab} - E(^3\Sigma_u^+)_{ab}]$$
$$\cdot(a_a^+ a_{\overline{b}}^+ + a_b^+ a_{\overline{a}}^+)(a_{\overline{b}} a_a + a_{\overline{a}} a_b) \quad (141)$$

There are only two parameters in the model, $E(^3\Sigma_u^+)$ and $E(^1\Sigma_g^+)$, which are distance-dependent, and may be expressed as

$$R(r_{ab}) = E(^3\Sigma_u^+)$$
$$g(r_{ab}) = \tfrac{1}{2}[E(^1\Sigma_g^+) - E(^3\Sigma_u^+)] \quad (142)$$

r_{ab} being the interatomic distance.

For the isoelectronic problem of the ethylene molecule, where the model space is reduced to two determinants having the same $\phi_c \sigma$ core

$$\phi_1 = |\phi_c \cdot a| \qquad \phi_2 = |\phi_c \cdot b| \quad (143)$$

a and b are now $2p_z$ AOs on atoms A and B, and the target eigenstates are the $X(^1A_1)$ ground state and the lowest $(^3B_2)$ $\pi\pi^*$ triplet state. The two basic parameters are then functions of both the r_{ab} distance and the torsional angle around the C–C bond θ_{ab}:

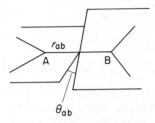

$$R(r_{ab}, \theta_{ab}) = E(^3B_2) \tag{144}$$

$$g(r_{ab}, \theta_{ab}) = \tfrac{1}{2}[E(^1A_1) - E(^3B_2)] \tag{145}$$

(Other parameters such as the pyradimalization angle of a C atom might be considered as well.)

Since the two determinants of the model space have the same space part and only differ by their spin, the effective Hamiltonian may be written as a spin-dependent Hamiltonian, the space part becoming implicit, but well defined (close to that of the separate atoms). A direct algebraic derivation shows that, if one defines E_0 as the barycentre of the configuration (which includes three triplets and one singlet),

$$E_0 = [3E(^3\Sigma_u^+) + E(^1\Sigma_g^+)]/4 \tag{146}$$

it becomes possible to write Eq. (141) as

$$H_{ab}^{\text{eff}} = E_0 + 2g\mathbf{S}_a \cdot \mathbf{S}_b \tag{147}$$

where $\mathbf{S}_a$ is the spin angular momentum on atom A, and g may be seen as an effective exchange integral.

One recognizes here the expression of a magnetic or Heisenberg Hamiltonian.[112] Heisenberg Hamiltonians were first proposed as phenomenological Hamiltonians and used to fit[113,114] the spectroscopic splittings between the multiplets of atoms or of molecular systems having several centres with unpaired electrons. Anderson[81] (see also Ref. 115) recognized that the Heisenberg Hamiltonians might be understood as effective Hamiltonians, deduced from the exact Hamiltonian by the choice of a model space reduced to the neutral VB components of the system of n electrons in n AOs. For a problem with two electrons, two AOs and two centres, the amplitude of the magnetic coupling $2g$ is directly reducible to the gap between the lowest singlet and triplet states (Eq. 145)); when working in an orthogonal basis set, this difference is known to come from the mixing between neutral and ionic components, which occurs in the singlet manifold, while it is impossible for the triplet stage. Turning back to the 4 × 4 matrix (Eq. (114)), it is clear[116,117] that after symmetry transformations and diagonalization of the 2 × 2 matrix concerning the $^1\Sigma_g^+$ symmetry

$$E(^3\Sigma_u^+) = -K_{ab} \qquad (148)$$

$$E(^1\Sigma_g^+) = K_{ab} + \tfrac{1}{2}[\Delta E - (\Delta E^2 + 16F^2)^{1/2}] \qquad (149)$$

i.e.

$$2g = 2K_{ab} - \Delta E - (\Delta E^2 + 16F^2)^{1/2} \qquad (150)$$

where K_{ab} is the direct (first-order) exchange integral between the a and b AOs, F the hopping integral between them, and ΔE the energy difference between neutral and ionic determinants. If $|F| \ll \Delta E$, one may write

$$g = K_{ab} - 2F^2/\Delta E \qquad (151)$$

K_{ab} is always positive. It lowers the triplet state energy and is said to be *ferromagnetic*. The second term lowers the singlet state and is called *antiferromagnetic*; and in most cases it predominates over the direct exchange, most systems therefore being antiferromagnetic.

Equation (151) may be obtained directly in a second-order (Q)DPT derivation of an effective Hamiltonian spanned by $|a\bar{b}|$ and $|b\bar{a}|$, and where the ionic determinants $|a\bar{a}|$ and $|b\bar{b}|$ span the outer space

$$\begin{aligned}
\langle |a\bar{b}\| H^{\text{eff}} \| a\bar{b}| \rangle &= K_{ab} - \langle |a\bar{b}\| H \| a\bar{a}| \rangle \langle |a\bar{a}\| H \| a\bar{b}| \rangle / \Delta E \\
&\quad - \langle |a\bar{b}\| H \| b\bar{b}| \rangle \langle |b\bar{b}\| H \| bb| \rangle / \Delta E \\
&= K_{ab} - 2F^2/\Delta E \qquad (152)
\end{aligned}$$

$$\begin{aligned}
\langle |a\bar{b}\| H^{\text{eff}} \| b\bar{a}| \rangle &= K_{ab} - \langle |a\bar{b}\| H \| a\bar{a}| \rangle \langle |a\bar{a}\| H \| b\bar{a}| \rangle / \Delta E \\
&\quad - \langle |a\bar{b}\| H \| a\bar{b}| \rangle \langle |b\bar{b}\| H \| b\bar{a}| \rangle / \Delta E \\
&= K_{ab} - 2F^2/\Delta E \qquad (153)
\end{aligned}$$

This perturbative approach is of course only valid when $|F|/\Delta E \ll 1$, i.e. when the electronic delocalization, governed by F, is smaller than the increase in the electronic repulsion when going from neutral to ionic determinants

$$\begin{aligned}
\Delta E &\simeq J_{aa} - J_{ab} \\
&= U \qquad \text{in the Hubbard Hamiltonian}^{116}
\end{aligned}$$

The outer space may be extended to any kind of determinant, involving angular and radial correlations of the ionic pairs $|a\bar{a}|$, or instantaneous repolarizations of the core in the ionic structures, as discussed in Section III.C.2. It becomes possible to treat these effects, which are additional to the second-order correction, as a dressing of the ionic structure energy, i.e. as a change of ΔE

$$\Delta E \to \overline{\Delta E} = \Delta E - \sum_\sigma \frac{\langle \sigma | J_a - J_b | \sigma^* \rangle^2}{\Delta E'} - \sum_{a_i' a_i''^*} \frac{(aa_i', aa_i'')^2}{\Delta E''} \qquad (154)$$

through infinite summations of diagrams.^{117}

We would like to stress the fact, however, that the convergence of the (Q)DPT perturbation expansion is by no means necessary to define a Heisenberg Hamiltonian, which is perfectly defined by the model space, the target space and the corresponding eigenenergies, all of which are known unambiguously in this precise problem. The possibility to define a Heisenberg Hamiltonian is *not* restricted to the case $|F| \ll \Delta E$, i.e. to problems which are weakly delocalized or strongly correlated, as usually believed.

a. Transfer

One may be tempted to transfer the Heisenberg Hamiltonian from the diatom to a cluster of atoms, or from ethylene to larger conjugated molecules,[118-128] i.e. to write

$$H^T = \sum_{i,j} H_{ij}^{eff} \tag{155}$$

This Hamiltonian will act on the neutral OVB determinants of the cluster, the space part of which are identical to the product of the valence AOs (i) centred on the various atoms (I), and supposed to be orthogonal. The various determinants differ by the spin distribution

$$\phi_K = \left(\prod_i i \right) \left(\prod_i \sigma_{iK} \right) \tag{156}$$

where σ_{iK} is the spin (α or β) borne by the atom I in the determinant ϕ_K. For the $S_z = 0$ (resp. $\frac{1}{2}$) manifold, which contains all possible multiplets of the problem, in a $2n$ (resp. $2n + 1$) centre problem, the size of the model space is C_{2n}^n (resp. C_{2n+1}^n), i.e. the number of ways to put n electrons on $2n$ (resp. $2n + 1$) centres.

This size is much smaller than that of the full VB matrix in the same minimal basis set, an of course fantastically smaller than the size of the CI in a large basis set. The treatment can of course only give *neutral* eigenstates, i.e. those which have *the largest projections onto* the neutral determinant subspace. Notice that this sentence has a very precise meaning; it does not mean that the corresponding eigenstates have a larger component into the model space than on the outer space; $\langle \psi_m | P_0 \psi_m \rangle$ may be smaller than $\langle \psi_m | Q_0 \psi_m \rangle$ as easily seen by considering nH_2 diatoms at infinite distances: if for H_2

$$\langle \psi({}^1\Sigma_g^+) | P_0 \psi({}^1\Sigma_g^+) \rangle = \alpha \tag{157}$$

for $(H_2)_n$

$$\langle \psi_0 | P_0 \psi_0 \rangle = \alpha^n \ll 1 - \alpha^n \qquad \text{for } n \text{ sufficiently large}$$

The Heisenberg Hamiltonian, of course, cannot deliver the energies of the states which are essentially ionic. It cannot give the energy of the lowest excited singlet state of H_2 (dipole-allowed ${}^1\Sigma_u^+$) nor the lowest ${}^1B_2$ ($\pi\pi^*$) singlet state of ethylene. The lowest dipole-allowed singlet states of conjugated hydrocarbons are essentially ionic and they cannot be reached by such an effective Hamiltonian.

The legitimacy of the transfer from the two-centre to the n-centre problem may be discussed along two lines:

1. As previously discussed in Section III.C, about the transferability of the valence minimal basis set effective Hamiltonians, the orthogonal valence AOs are not he same in the two-centre and n-centre problems, due to large orthogonalization tails if one uses $S^{-1/2}$ procedures, or to many-body distortions if one uses the localized MOs of the upper valence multiplet.[81] But it has been noticed[127] that the two-centre Heisenberg Hamiltonian is *entirely determined by symmetry, the amplitude of its operators being independent of the precise content of the two orthogonal valence orbitals a and b*. One may thus define truncated *ad hoc* AOs with weak (or even zero) spatial overlap. In that precise case the model space AOs are transferable. The second-order corrections inducing spin exchanges between atoms B and C in an ABCD cluster are identical to those concerning the BC spin exchange in the BC diatom. The same is true for higher-order terms involving only the two *ad hoc* valence AOs b and c.

2. Many-body (i.e. many-centre) terms would appear[112] if one derived directly the effective Hamiltonian spanned by the neutral OVB determinants of the H_n problem. Even the two-body terms between, say, two determinants differing by a spin exchange between atoms B and C in an ABC cluster will be different from the effective exchange in the BC diatom

$$\langle |a\bar{b}c\bar{d}\| H^{\text{eff}}_{\text{ABCD}} \| ab\bar{c}\bar{d}|\rangle \neq \langle |b\bar{c}\| H^{\text{eff}}_{\text{BC}} \| b\bar{c}|\rangle$$

due to perturbation orders larger than two, which imply for instance A^-C^+ singly ionic determinants between non-adjacent atoms, which do not appear in the BC problem. This question has been well studied by Maynau *et al.*[123,124,128] on model problems (i.e. starting from a PPP type Hamiltonian of π systems). They concluded that

a. two-body terms are quite transferable,
b. three-body terms are negligible,
c. four-body terms are large in compact (square) structures and negligible when the four atoms do not define a cyclic structure, and
d. six-body terms are lower than four-body terms but remain significant for cyclic structures.

The Heisenberg Hamiltonians should therefore include many-spin operators, which are essentially functions of the $|F|/\Delta E$ (i.e. β/U) ratio. The derivation of spin effective Hamiltonians is again always possible, whatever the $|F|/\Delta E$ ratio, but the negligibility of many-spin (> 2) operators is only valid for small $|F|/\Delta E$ values, i.e. for the highly correlated case, at least in compact structures. This restriction is not valid for systems involving only linear or branched chains or large cycles.

The most convincing applications of Heisenberg Hamiltonians in chemistry concern the X_3 problem (X being an H or alkali atom),[130] for which the two-body Heisenberg Hamiltonian is very efficient in predicting geometries and energies, and conjugated hydrocarbons. In a series of papers[123-126] it has been shown that most chemical concepts for these problems might be translated in terms of spin organization. Even the Woodward–Hoffmann rules may be demonstrated in this language where they concern the cyclic many-body operators. A careful extraction of the R and g parameters from accurate MO-CI calculations on ethylene as functions of bond distance and torsion angle gives a fantastically cheap and accurate tool[124,125] in the simplest two-body version of the Heisenberg Hamiltonian. Geometries are provided within 0.015 Å; the isomerization enthalpies and the rotational barriers are excellent; the transition energies towards neutral excited states are in very good agreement with the experimentally known values; and the model is able to treat the geometry reorganizations in neutral excited states, which are so difficult to calculate in MO-CI calculations. The open-shell problems are as easy to treat as are the closed-shell ones. The simplicity of the Heisenberg Hamiltonian has made possible research on asymptotic laws[125-127] for linear or cyclic polyenes concerning bond alternation, rotational barriers, solitonic deformation and excitation energies to the lowest triplet state (or lowest doublet excited state). The fantastic successes of these approaches—completely foreign to the conceptual background of chemistry—is in strong contrast with the difficulties encountered in research on valence-only effective Hamiltonians.

The model has even been used with success for the study of nitrogen-containing conjugated systems, $> C = N -$ bonds, i.e. weakly polar molecules[131,132] and is being extended to $> C = O$ containing systems.

The number of systems defining a half-filled band with one e^-, one AO per centre is however very limited. The alkali-metal and noble-metal clusters and solids belong to this category, but they are conducting metals and it seems risky to treat a conducting metal through a model that has been developed especially for a peculiar class of insulators (Mott insulators), which only treats explicitly neutral structures, in which each atom of the lattice keeps one electron. This challenge has been attempted, however, by Malrieu et al.,[129] who extracted accurate two-body Heisenberg Hamiltonians from the diatom lowest potential curves, and treated the solid by perturbing the spin wave presenting the largest number of spin alternancy between neighbouring atoms. This zeroth-order wavefunction is perturbed to fourth order by the spin exchanges. One may optimize the lattice parameters and compare the cohesive energy of various crystalline lattices. The results are surprisingly good, concerning the preferred crystallization mode (b.c.c. $\simeq$ compact ones for Na, Li, compact forms more stable than b.c.c. for Cu), the lattice parameters, the cohesive energy and even the bulk modulus (compressibility).

This highly unorthodox model of the conducting solid cannot explain the conductivity, although it is not incompatible with it since the wave operator Ω would build ionic components from the projected (neutral-only) wavefunction. From the principles, and as shown by the previously mentioned success of Heisenberg Hamiltonians on the most metallic chemical systems (aromatic molecules), almost any system of which the lowest VB determinants are neutral may be treated either by the independent-particle approach followed by a treatment of electronic correlation which reduces the ionic components[133,134] or by a Heisenberg-type effective Hamiltonian. This statement seems to be true whatever the β/U ratio. Malrieu et al.[129] also noticed that the many-body effects (for instance four-body cyclic contributions), which are so important on small molecules and clusters, play a much less important role in the solid.

One should point out, however, that the Heisenberg Hamiltonian fails to predict correctly the planar rhombus structure of Li_4 (and Na_4 or Cu_4) clusters. This failure has two origins: one is the involvement of the p band, which begins to be significantly populated in X_4 systems; the other is the highly ionic character of the ground state in its rhombus equilibrium geometry

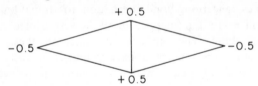

where the short diagonal has almost lost one electron. This means that the

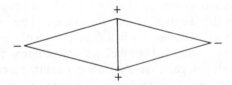

di-ionic component is quite low in energy, and the problem apparently cannot be treated by taking a model space of neutral VB structures. This opens two general questions concerning conducting solids: (i) Are the lowest-energy VB determinants the neutral ones? Might not multiply and regularly ionized determinants be lower in energy, if the + and − atoms regularly placed along a sublattice induce a stabilizing Madelung's field and polarize the neutral atoms? (ii) Is the Hubbard Hamiltonian, generally used to treat the correlation effects in solids, relevant since it cannot take into account this collective electrostatic stabilization?

These fundamental questions regarding the description of the simplest metals are actually fascinating and a major challenge on the borderline between quantum chemistry and solid-state physics.

2. Half-filled Band with p Electrons in p Atomic Orbitals per Centre

It might be thought that the foundation of Heisenberg Hamiltonians for half-filled bands where each atom brings p electrons in p AOs is evident. The Heisenberg Hamiltonian would distinguish intra-atomic (generally ferromagnetic, due to Hund's rule) and inter-atomic (generally antiferromagnetic) effective exchanges

$$
\begin{aligned}
H^{\text{eff}} = &\sum_{AB} R_{ab} + \sum_{A} \sum_{i_A j_A} K_{i_A j_A}(a_{j_A}^+ a_{\bar{j}_A}^+ + a_{j_A}^+ a_{\bar{i}_A}^+(a_{\bar{j}_A} a_{i_A} + a_{\bar{i}_A} a_{j_A}) \\
&+ \sum_{AB} \sum_{i_A} \sum_{j_A} g_{i_A j_B}(a_{i_A}^+ a_{\bar{j}_B}^+ + a_{j_B}^+ a_{\bar{i}_A}^+)(a_{j_B} a_{i_A} + a_{i_A} a_{j_B}) \\
&= E_0 + \sum_{ij} g_{ij} \mathbf{S}_i \cdot \mathbf{S}_j
\end{aligned}
\tag{158}
$$

In general the atomic orbitals i_A and j_A on the same centre are orthogonal by symmetry. As examples of half-filled bands, one might consider a cluster of nitrogen (N) atoms, keeping the s^2p^3 configuration with one electron in each p_x, p_y, p_z AO, as occurs in the free atom ground state. Then in N, the p_x, p_y and p_z AOs belong to three different symmetries (π_x, π_y and σ) and

$$
\langle i_A | F | j_A \rangle = \langle i_A | F | j_B \rangle = 0 \qquad \text{if } i \neq j
$$

The only hopping integrals occur between AOs of the same molecular symmetry ($\langle i_A | F | j_A \rangle \neq 0$) and Anderson's delocalization mechanism will take place within each symmetry subspace. The integrals $K_{i_A j_A}$ are essentially first-order terms, responsible for the atomic preference for high-spin order (the ground state of N is (^{4}S), i.e. $s^2 p_x p_y p_z$) while the interatomic effective exchange will again come from the second-order coupling between the neutral and singly ionic determinants; for instance for $S_z = 0$

$$
\begin{aligned}
&\langle s_A^2 x_A y_A z_A s_B^2 \bar{x}_B \bar{y}_B \bar{z}_B | H^{\text{eff}} | s_A^2 x_A \bar{y}_A z_A s_B^2 \bar{x}_B y_B \bar{z}_B \rangle = (y_A y_B, y_A y_B) \\
&+ \langle s_A^2 x_A y_A z_A s_B^2 \bar{x}_B \bar{y}_B \bar{z}_B | H | s_A^2 x_A y_A \bar{y}_A z_A a_B^2 \bar{x}_B \bar{z}_B \rangle (\Delta E_{y_A \leftarrow y_B})^{-1} \\
&\quad \times \langle s_A^2 x_A y_A \bar{y}_A z_A s_B^2 \bar{x}_B \bar{z}_B | H | s_A^2 x_A y_A z_A s_B^2 \bar{x}_B y_B \bar{z}_B \rangle \\
&+ \text{inverse } (y_A \rightarrow y_B) \text{ term} \\
&= (y_A y_B, y_A y_B) + 2 F_{y_A y_B}^2 / \Delta E_{y_A \rightarrow y_B}
\end{aligned}
\tag{159}
$$

where the interatomic direct exchange integral is very small, while the second term is due to the electronic delocalization in the π_y subsystem. Maynau

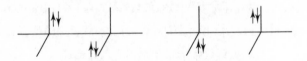

et al.[135] have tried to analyse the effect of higher orders on a problem involving two electrons and two orthogonal AOs per centre, namely the acetylene π problem when the σ core is taken frozen. Using accurate large basis set

MO-CI wave functions, they noticed that the Bloch effective hamiltonian was highly non-Hermitian, and that its diagonal energies did not follow the Heisenberg structure. The des Cloizeaux effective Hamiltonian also deviated strongly from the structure of a Heisenberg Hamiltonian. Both the Bloch and des Cloizeaux effective Hamiltonians when transferred into the upper homologue ($H-C'=C=C'-H$) were unable to predict correctly the ground-state nature. In contrast with these failures, the Schmidt-orthogonalized effective Hamiltonian appeared to keep the Heisenberg structure and predicted correctly the lower parts of the linear HC_3H molecule. The reasons for these unexpected results were analysed, and were shown to be due to the occurrence of low-lying neutral VB vectors in the outer space. For the acetylene problem these states are $x_A^2 y_B^2$ and $y_A^2 x_B^2$. They have a different

space part than $(\sigma)x_A y_A x_B y_B$ and they cannot therefore enter the model space generating a spin-only effective Hamiltonian. These determinants are neutral, the atoms are no longer in their ground state, but such VB structures are quite low in energy, and they will act as intruder states. They appear at second order in the wavefunction expansion through processes like

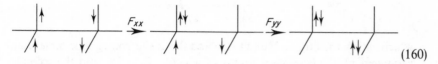

$$(160)$$

i.e. with a large coefficient. Note that the model space necessarily involves situations of the type

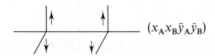

in which the overlapping AOs (x_A, x_B) and (y_A, x_B) have the same spin, and therefore do not permit any delocalization. These determinants, coupled with the lower ones through the *intra*-atomic spin exchange, are higher in energy than the neutral VB determinants belonging to the outer space. This generates trouble in the perturbation expansion and non-orthogonality between the projections of the eigenfunctions $\langle \Phi_m | \Phi_n \rangle \neq 0$ in the $^1\Sigma_g^+$ symmetry, due to strong components of the second and third eigenvectors on the (model space) $|x_A \bar{y}_A x_B \bar{y}_B|$ and (outer space) $|x_A \bar{x}_A y_B \bar{y}_B| VB$ components. This analysis shows that, for systems where each atom brings more than one active electron in one AO, (i) the perturbative (QDPT) generation of the Heisenberg Hamiltonian would diverge (the second-order result, which gives the expected Heisenberg structure, can only be considered as asymptotic), and (ii) the Heisenberg Hamiltonian is obtained from the Schmidt orthogonalization of the projections Φ_m of the eigenvectors into the model space (i.e. one accepts to lose more information on the upper eigenvectors than on the lowest ones, as seems reasonable). The same Heisenberg structure is obtained by using the intermediate Hamiltonian theory.[34]

3. Non-heisenberg Effective Hamiltonians Spanned by All Neutral Valence-bond Structures for Half-filled Bands

An attractive solution[136] would simply consist of enlarging the model space, leaving the constraint that all its determinants have the same $x_A y_A x_B y_B$ space part with one electron per active AO. If one includes the neutral VB determinants with zero or two electrons in the same AO, i.e. the determinants which acted previously as intruder states, one enlarges the model space, and one also introduce states of a different space part. For C_2H_2 again the Heisenberg Hamiltonian was spanned by six determinants (for $S_z = 0$) of three types, namely

1.

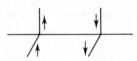

which satisfy the atomic Hund's rule and the antiferromagnetic molecular arrangement, the first-order energy of which is $E_0 - 2K$, and the second-order energy $E_0 - 2K - 2g$;

2.

which do not satisfy the atomic Hund's rule but keep the bond antiferromagnetic alignment, the first-order energy of which is E_0, and the second-order energy $E_0 - 2g$;

3.

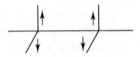

which do not satisfy neither the atomic Hund's rule nor the bond antiferromagnetism, and have an energy E_0 at both first and second orders.

Then one must add two types of determinants:

4.

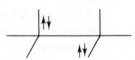

which have a first-order energy $E_0 + 2U$, due to the repulsion of the electrons in the same AO, and a second-order energy $E_0 + 2U - 2g$; and

5.

which do not permit electron delocalization, remaining very repulsive, with an energy $E_0 + 2U$.

Notice that in a minimal basis set $U = 2K$ and the coupling between the last two types of determinants is $K(\langle x_A^2 y_B^2 | H | x_A^2 y_A^2 \rangle = K_{xy})$, so that the number of parameters is not enlarged at this low order of perturbation. The coupling between the neutral (0 or 2 e^- per AO) determinants and those spanning the Heisenberg Hamiltonian (1e^- per AO) occurs at the second order, as already mentioned (cf. Eq. (160)), and its magnitude is $2F_{xx}F_{yy}/\Delta E$, i.e. $2F^2/\Delta E = g$ if the two subsystems π_x and π_y are equivalent (or $\sqrt{(g_x g_y)}$ if the two subsystems are not equivalent). This enlarged effective Hamiltonian spanned by *all* the neutral VB determinants of the band does not introduce severe complexity and a tremendous number of parameters. We do not give here a detailed second quantization formulation of the effective Hamiltonian, which belongs to a more general category given below, but one may say that it introduces some operators

$$\sum_i U_i(a_i^+ a_{\bar{i}}^+ a_{\bar{i}} a_i)$$

i.e. an operator counting the number of electron pairs in the same AO and terms involving *four* atomic orbitals, two of them belonging to atom A and two of them belonging to atom B, since the extradiagonal terms

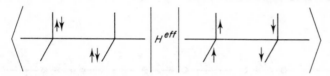

for instance, involve *four* AOS.

This effective Hamiltonian, which may be seen as a generalized Heisenberg Hamiltonian, since it is inspired by the same philosophy, solves some previously mentioned intruder states and thus convergence difficulties since it includes the intruder states of the strict Heisenberg Hamiltonian in its enlarged model space. One should note, however, that the price to pay is the inclusion of the strongly repulsive $|x_A^2 x_B^2|$ determinants in the model space, which may in turn introduce other intruder states (for instance $\sigma \to \sigma^*$ excited determinants).

4. Non-half-filled Bands: Example of Systems Having One Electron in p Atomic Orbitals per Centre (Magneto-angular Effective Hamiltonians)

Up till now we have only considered half-filled bands. One might wonder whether the basic idea which leads to Heisenberg Hamiltonians might not be generalized, again considering a model space spanned by the neutral VB determinants, even when these determinants do not have the same space part. The simplest problem concerns atoms having one valence electron which can occupy p equivalent AOs (i.e. degenerate in the atom and thus belonging to the same band). As a tentative example[137] one may consider for instance B atoms, keeping an $s^2 p$ character, where each atom has only one 2p electron which can occupy one of the three p_x, p_y, p_z AOs. There are thus two degrees of freedom, the space angular momentum of each atom, and its spin momentum. The 'radial' parts of the molecular wavefunctions of the model space are identical and one may immediately see that the effective Hamiltonian will be a *magneto-*

angular effective Hamiltonian. This has been proposed by Marinelli *et al.*,[137] who analysed the problem on the B_2 diatom, derived an 'exact' effective Hamiltonian from accurate large basis set MO-CI calculations and tested its transferability to the B_3 linear molecule. The effective interactions may concern four AOs, for instance the space part of atoms A and B are changed in the second-order coupling between the two determinants spanning the $^3\Sigma_u^-\,(S_z = 1)$ ground state

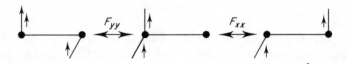

(where ● represents one $(2s)^2$ electron pair). The third-order corrections, for instance

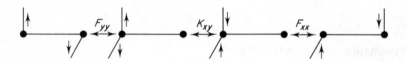

change both the space and spin parts of the atoms; they are however less important. The 'exact' effective Hamiltonian does not deviate strongly from the *structure* predicted by the low-order expansion although the terms appearing at order larger than 2 are damped with respect to their theoretical value.

The model space for B_2 does not span the state $^5\Sigma_g^-$ which is nearly degenerate with the $^3\Sigma_u^-$ ground state; the $^5\Sigma_g^-$ state is spanned by hybridized VB determinants involving the lowest excited state (sp^2) of one atom $s_A z_A x_A$ and a ground-state atom $s_B^2 y_B$

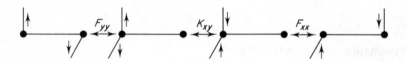

(where ○ represents the 2s AO). Despite this near-degeneracy, the effective Hamiltonian was able to predict correctly the ordering and spacing of the four lowest states of the B_3 linear molecule, independently determined by accurate MO-CI calculations. The model suggests that the linear B_n chain might be ferromagnetic.[137]

The remark concerning the low-lying hybridized state questions, however, the reliability of effective Hamiltonians spanned by the neutral non-hybridized OVB determinants, i.e. involving a single band.

5. General Effective Hamiltonians Spanned by All Neutral OVB Determinants, and Involving Several Bands[138]

In a further step to generalize the Heisenberg Hamiltonians, one may decide that they will be spanned by all the possible OVB valence neutral determinants, without any assumption concerning their hybridization state. The carbon atom will be either s^2p^2, sp^3 or p^4 for instance. In the language of solid-state physics, one would say that the two bands s and p are both involved. The various zeroth-order energies of the determinants belonging to the model space are no longer degenerate, since in C_2, for instance, the VB determinant $s_A^2 x_A y_A - s_B^2 \bar{x}_B \bar{y}_B$,

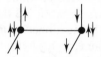

in which both atoms are in their atomic ground state, is much lower in energy than the determinant $p_{x_A} p_{y_A} p_{z_A}^2 \overline{p_{x_B} p_{y_B} p_{z_B}^2}$

in which both atoms are excited and which induces large interatomic repulsions in the σ symmetry.

One should therefore introduce mono- and bielectronic diagonal terms which take into account both the spectroscopic state of the atom and the interatomic repulsion energies, which depend on the orbital occupancies. The second-order Anderson-type corrections would then concern all back-and-forth processes from a neutral to a neutral OVB determinant through a singly ionic VB determinant. For instance in C_2, the second-order process

involving four AOs through an $A \leftarrow B$ electron jump followed by an $A \rightarrow B$ reverse electron jump, may be seen as a generalization of Anderson's process leading the antiferromagnetic effective exchange in a half-filled band.

The general back-and-forth operators may be written

$$(i_A j_B, k_A l_B)^{\text{eff}}(a_{i_A}^+ a_{j_B})(a_{l_B}^+ a_{k_A})$$

where

$$(i_A j_B, k_A l_B)^{eff} \simeq 2F_{i_A j_B} F_{k_A l_B}/\Delta E \qquad (161)$$

where i_A, j_B, k_A, l_B are now *spin orbitals*. Notice that one may have $j_B = l_B$, $i_A = k_A$, i.e. diagonal corrections. The problem concerns the definition of ΔE, which should be the energy of the two intermediate (singly ionic) outer determinants, with respect to the ket neutral determinant, in a Bloch-type expansion. The non-degeneracy of the model space, due to the hybridization, leads to the already discussed non-hermiticity problems. These problems may certainly be solved either through the hierarchic (i.e. Schmidt-type) ortho-gonalization procedure of the projected eigenvectors Φ_m or by the use of an intermediate Hamiltonian approach in which the main model space would only involve the non-hybridized neutral VB determinants.

Two other difficulties must be mentioned, namely (i) the possible occurrence of ionic intruder states since some highly hybridized neutral VB determinants may be high in energy, and (ii) the difficulty of calculating and selecting the set of exact wavefunctions ψ_m to define the target space, if one wants to use an *a posteriori* definition of the effective Hamiltonian, from knowledge of the target space.

Again the use of an intermediate Hamiltonian formalism may solve these difficulties.

The last problem concerns the size of the model space, which rapidly becomes very larger when the number of atoms increases and new approxim-ations would certainly be required.

One may nevertheless think that the above-mentioned strategy, which generalized Anderson's derivation of the Heisenberg Hamiltonian, is both conceptually interesting and practically useful especially if some simple parametrized formulae were used to obtain the effective integrals. Work is under progress[138] to define some Anderson–Hoffmann model where the diagonal energy differences (and monocentric exchanges) would be taken from the atomic spectra, the repulsive terms would be simple functions of interatomic overlaps and distances, and the effective bielectronic terms would be governed by simple AO overlap dependences.

6. Effective Hamiltonians for Non-neutral Systems; an Effective Hamiltonian for the Cation of a Half-filled Band[139]

So far the systems under study in this section have been neutral, and the effective VB was spanned by the neutral VB determinants. One might wonder whether it would not be possible to treat a non-neutral system in a similar way. To do this one must again consider among the OVB determinants those which have the lowest energy. While for neutral systems one had an energetic hierarchy according to the inonicity of the determinants

$$\text{neutral} < \text{singly ionic} < \text{doubly ionic} < \cdots$$

for the cations one may introduce the following hierarchy of determinants

one hole < two holes < three holes < $\cdots$

The holes are the number of atomic positive charges. A two-hole determinant necessarily involves two positively charged atoms or an atom bearing two positive charges, and another atom bearing one negative charge.

The simplest problem again concerns the cations of a half-filled band, for instance the cations of conjugated molecules, for which one may consider an $S_z = 0$ (or $\frac{1}{2}$) model space of dimension

$$n \times C_{n-1}^{(n-1)/2}$$

since each atom may be positively charged, the other atoms bearing one electron of α or β spin. This dimension is larger than that of the neutral problem. The effective Hamiltonian involves first-order hopping integrals between determinants differing only by the hole, for instance between $|a\bar{b}d|$ and $|a\bar{c}d|$ in $[ABCD]^+$

Second-order effective exchanges occur, as in neutral systems, between atoms which do not bear the hole,

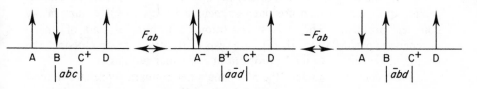

but one may also notice the occurrence of effective hopping integrals between non-adjacent atoms

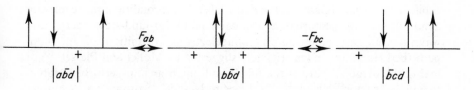

In this process the hole jumps to the second-neighbour atom. Once may notice a similar second-order process which simultaneously changes one hole and permutes the spins of two electrons

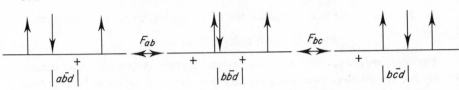

This effective Hamiltonian proposed by Gadéa et al.[139] for the cations of conjugated molecules is able to give many more eigenvectors of the positive ion than Koopman's theorem. It provides a direct estimate of the spectrum of the positive ion, involving the non-Koopmans states, which appear to occur at quite low energy and are described as two-hole one-particle states in the delocalized MO-CI language.

One may then establish that generalizations of the effective Hamiltonians to non-neutral systems are also possible. It is clear that in that case hole delocalization appears explicitly, while for neutral systems the effective Hamiltonian treated the electron delocalization through effective interatomic exchange operator only, or more generally through bielectronic bicentric effective integrals reflecting the back-and-forth electronic movement coupling neutral VB structures. These effects also appear for the cations, when they do not concern the hole, but first-order delocalization effects are also present.

A general scheme picturing the generalization of Heisenberg Hamiltonians is given in Table I.

E. Numerical Applications in the Search of Configuration-interaction Solutions

One should simply mention briefly the application of the effective Hamiltonian approaches which use them as technical tools to solve numerically complex problems. The uses of partitioning techniques and of quasi-degenerate perturbation theory are especially frequent in solving the configuration-interaction (CI) problem in molecular physics.

In most chemical systems, the SCF single determinant represents a good starting point for the description of the ground-state wavefunction. It has a large overlap with the exact wavefunction in the finite basis set, and is energetically well separated from the 'excited' determinants, obtained by single, double, triple,... substitutions of the ground-state occupied molecular orbitals by virtual MOs above the Fermi level. These conditions make relevant a non-degenerate perturbative approach of the ground-state energy and wavefunction, assuming purely monoelectronic definition of the non-perturbed Hamiltonian (as originally suggested by Moller and Plesset[140]) or taking the diagonal part of the full Hamiltonian as non-perturbed Hamiltonian (a version named[141-143] Epstein–Nesbet according to its former applications).

For excited states of molecules, it rapidly appeared that it was quite

TABLE I

Hierarchy of generalizations of Heisenberg Hamiltonians.

Type of problem	Model space	Neutral systems — Neutral VB determinants		Cationic systems — One-hole VB determinants	
		1 e⁻/AO	2, 1 or 0 e⁻/AO	0, 1 e⁻/AO	0, 1, 2 e⁻/AO
One half-filled band	1 e⁻, 1 AO/centre	Heisenberg Hamiltonian (alkali atoms, π system)		hopping integrals + effective exchange + mixed term (π systems)	
	p e⁻, p AO/centre	Heisenberg Hamiltonian ← Bloch, des Cloizeaux ← Schmidt orthog. ← interm. Hamilt.	non-Heisenberg Hamiltonian self-repulsion 4 AO operators	unexplored	
Non-half-filled band	1 e⁻, p AO/centre	magneto-angular effective Hamilt. (B_n)	unexplored		
	k e⁻, p AO/centre	unexplored			
Several bands		impossible	back-and-forth bicentric bielectronic effect. integrals, monocentric monoelectr. and bielectr. integrals tentative explorations	unexplored	

TABLE II

Comparison of multi-reference CI schemes; after a preliminary choice of a small zeroth-order space S.

1. Partitioning technique/S + simplification of the outer space matrix Shavitt[144] non-size-consistent[a]	2. Preliminary diagonalization of S + independent perturbations of the eigenvectors CIPSI[66,67] MRDCI[145] − artefacts in weakly avoided crossings − imperfect size consistency[a] $\varepsilon \propto 2N/3$ for CIPSI $\varepsilon \propto \sqrt{N}$ for MRDCI	3. QDPT construction of an effective Hamiltonian on S[147] → intruder states, divergence → size consistency problems $(\varepsilon \propto 2N$ for second order)[a] if S is not a complete active space	4. Preliminary diagonalization of S + definition of a smaller multi-configuration model space S[150−152] good convergence properties no intruder states approx. size consistency[a]	5. Use of generalized degenerate perturbation theory[34] ('shifted B_k approximation'[154]) with S as model space, avoids most of intruder state problems, good size consistency[a]

[a]Ref. 155; a correct size consistency would be $\varepsilon \propto N$.

impossible to find a satisfactory single-determinantal zeroth-order description. While the ground-state SCF determinant only interacts with doubly substituted determinants of much higher energy (due to the Brillouin's theorem), a singly excited (with two open shells) SCF configuration interacts with all other singly excited configurations, some of them being almost degenerate. For excited-state problems, it becomes necessary to use multi-configurational zeroth-order descriptions, a goal which is also desirable for some ground-state strongly correlated problems, for instance when some bonds are broken or when the molecule has large diradical character.

Methods have then been built to start multi-configuration (multi-reference) zeroth-order descriptions. The full CI space is first partitioned into a (small) main subspace involving the leading configurations for the descriptions of the states to be studied. Then

1. One may use the partitioning technique as suggested by Shavitt *et al.*[144] 20 years ago. If it was applied strictly the method would require the inversion of a very large matrix (cf. Eq. (8)) and would be as expensive as the direct diagonalization. The large matrix is then supposed to be diagonal and the process reduces to the multiplication of rectangular submatrices. The method has not been used frequently for CI calculations because it is no longer size-consistent at this level of approximation.

2. One may diagonalize first the small main subspace S (which involves all the nearly degenerate configurations, the number of which is usually taken from 20 to 200); if P_S is the projector on the main subspace

$$P_S = \sum_{K=1}^{N} |\phi_K\rangle\langle\phi_K|$$

$$P_S H P_S |\psi_{mS}^0\rangle = E_{mS}^0 |\psi_{mS}^0\rangle$$

$$|\psi_{mS}^0\rangle = \sum_{K\in S} C_{mK} |\phi_K\rangle \tag{162}$$

the multi-configurational wavefunctions $|\psi_{mS}^0\rangle$ should no longer be degenerate and might (in principle but *vide infra*) be treated independently. Some of these functions may be perturbed to the second order as in the original CIPSI algorithm[66] by configurations outside of S. Other algorithms, such as the MRDCI (multi-reference double CI) [145] of Buenker and Peyerimhoff or the improved CIPSI algorithm,[67] define a second class of determinants of mean importance and treat their effect variationally (i.e. to infinite order). These determinants are usually between 1000 and 10 000, while the others (in number larger than 10^5 or 10^6) are treated to second order only. These procedures do not lead to effective Hamiltonians and are simply mentioned for comparison and forthcoming discussion.

3. One may use directly the quasi-degenerate perturbation theory, or an iterative algorithm, to solve the Bloch equation, taking the main subspace S

as model space. But this procedure usually faces major intruder state problems, since some of the main configurations in the model space are high in energy (for instance, doubly excited configurations embedded in the continuum, without any spectral meaning, but strongly coupled to the ground or low excited configurations). Some more or less diffuse excited configurations which do not belong to S are lower in energy than these doubly excited configurations and happen to behave as intruder states, making the perturbation expansion divergent. This difficulty cannot be solved by including the intruder states in the model space for two reasons: (i) The enlargement of the model space results in a rapid increase of the computational cost. (ii) The newly introduced configurations generate their own set of intruder states and the series remain divergent. Despite this practical problem (see Robb and Hegarty[146]), Bartlett et al.[147] have implemented a second-order Rayleigh–Schrödinger QDPT expansion, which has not received many applications (see however Refs. 148 and 149).

4. Instead of using the QDPT expansion in a model space of single configurations, one may choose a model space of multi-configurational wavefunctions resulting from a preliminary diagonalization of a small subspace.[150–152] The preliminary diagonalization of 200 determinants gives a few physically relevant eigenvectors $|\psi^0_{mS}\rangle$ (the number of which will typically be lower than 10). Let us calls this new subspace S_0

$$P_{S_0} = \sum_{m=1}^{n \simeq 10} |\psi^0_{mS}\rangle\langle\psi^0_{mS}| \qquad (163)$$

which will be taken as a model space for the QDPT expansion. The resulting effective Hamiltonian has much smaller size than the effective Hamiltonian built on S (i.e. on N determinants). Owing to the multi-configurational nature of $|\psi^0_{mS}\rangle$, one must generate all the determinants from the N $|\phi_k\rangle$, and the cost of the perturbative expansion is equal to that of a direct expansion in the basis of configurations, but the convergence behaviour is better by far, since one only perturbs the lower eigenvectors of $P_S H P_S$, which are energetically far from the determinants outsides of S. The other eigenvectors of $P_S H P_S$ which do not belong to S_0 are closer in energy, but they are not coupled with the vectors spanning S_0 and do not contribute to the second order.

This procedure had been proposed by Daudey and Malrieu.[150] To second order the diagonal corrections of the effective Hamiltonian are identical to those calculated in the original CIPSI algorithm. The procedure simply consists of adding off-diagonal perturbative corrections between the zeroth-order multi-reference wavefunctions. These off-diagonal terms lead, after diagonalization of the effective Hamiltonian, to mixing of the zeroth-order wavefunctions, this mixing being of perturbative origin.

This mixing has proved to be very important in treating correctly the weakly avoided crossings in multi-reference CI procedures, as shown by Spiegelmann and Malrieu.[151] The construction of a small-sized effective Hamiltonian is necessary to avoid the artefacts noticed by Bonacic–Koutecky et al.[153] in the MRDCI scheme,[153] also present in the CIPSI algorithm.

The procedure now currently used in CIPSI calculations under the label 'CIPSI–Brandow' has also been suggested by Schneider et al.[152]

5. Another type of effective Hamiltonian has been proposed by Davidson and coworkers[154] for the practical treatment of CI problems in nearly degenerate cases, under the name 'shifted B_k' approximation (which refers to a version of the partitioning technique proposed by Gershgorn et al.[144] and already discussed). To second order the effective Hamiltonian

$$H^{\text{eff}} = P_{S_0} H P_{S_0} + P_{S_0} H Q_0 (E_0 - H_0)^{-1} Q_0 H P_{S_0} \qquad (164)$$

is identical to the usual Rayleigh–Schrödinger second-order QDPT effective Hamiltonian, except for the fact that all denominators are taken from a common zeroth-order energy E_0 (that of the lowest-energy configuration, for instance). This procedure, which may be shown to be size-consistent[155] on typical problems, might a priori be viewed as a level shift in the model space. It is more relevant to see it as the second-order expansion of the generalized degenerate perturbation theory recently proposed by Malrieu et al.[34] and briefly outfield in Section II.A. This procedure has the only defect of being explicitly energy-dependent. It might be used as well after a preliminary first-order diagonalization of the model space, and a redefinition of a main-model subspace on the lowest eigenvectors, as previously proposed in the CIPSI–Brandow scheme.

A general summary of the main CI techniques proposed for the treatment of excited states in nearly degenerate situations is pictured in Table II, mentioning their advantages and defects.

One should notice that these techniques may be used as well to search the solutions of Heisenberg Hamiltonians when the number of neutral VB determinants (i.e. of spin distributions) becomes too large. Sanchez-Marin et al. have recently treated[156] the molecular spectroscopy of large conjugated molecules through a Heisenberg Hamiltonian by truncating the basis of determinants and dressing the matrix through the generalized degenerate perturbation theory.[34] The results are very encouraging.

IV. APPLICATIONS OF SIMULATION TECHNIQUES

The applications presented in this section are limited to a selected choice of pseudo-Hamiltonians that will be determined by the non-empirical simulation method presented in Section II.B. The emphasis will be put more on the methodological aspects than on the results of these methods.

Note that our purpose of rigorous modelling cannot be completely separated from earlier research on semi-empirical or model Hamiltonians. On one side these Hamiltonians could be parametrized by theoretical simulation techniques and on the other some experimental data could also be introduced in the simulation techniques, for example in the characterization of truncated Hamiltonians. Finally it should be emphasized that research on pseudo-Hamiltonians and model Hamiltonians is always guided by some intuitive knowledge of the passive and active constituents of the system (atomic cores, atoms in molecules, functional group,...) and by the assumption of transferability of their potentials and interactions.

A. Pseudopotentials

A reasonable assumption for any chemist is that molecules are made up of fixed cores (the nucleus and the inner electrons) and of chemically active valence electrons moving in the field of these fixed cores (frozen-core approximation). Obviously for accurate investigations in spectroscopy this assumption would fail, for example for alkali or alkaline-rich elements on the left of the periodic table, which possess highly polarizable cores. In the following only fixed atomic cores will be considered.

Pseudopotentials describe the interaction of a valence electron with the core of the atoms. They are known in the literature under various names, such as model potentials, effective core potentials,.... Model potentials are generally parametrized from atomic spectroscopic data whereas effective core potentials and pseudopotentials are most often derived from *ab initio* calculations. There is a huge literature on the subject and several review articles.[157-160] The recent paper by Krauss and Stevens is recommended for an overall survey of the subject with applications and comparisons with all-electron calculations.[159] The recent review paper of Pelissier *et al.* is devoted to transition elements.[160] In the following we shall only review the main characteristics of the determination of atomic pseudopotentials by the *ab initio* simulation techniques of Section II.B.

The total Hamiltonian of an atom can be written in atomic units (a.u.) as

$$H = \sum_{i=1}^{N} (-\tfrac{1}{2}\Delta_i - Z/r_i) + \sum_{i<j} 1/r_{ij}. \tag{165}$$

For an even number of electrons, the Fock operator derived from H is

$$F = -\tfrac{1}{2}\Delta + \sum_c (2J_c - K_c) + \sum_v (2J_v - K_v)$$

$$= \sum_c \varepsilon_c |\varphi_c\rangle\langle\varphi_c| + \sum_v \varepsilon_v |\varphi_v\rangle\langle\varphi_v| + \sum_{i*} \varepsilon_{i*} |\varphi_{i*}\rangle\langle\varphi_{i*}| \tag{166}$$

where the indices c, v and $i*$ label the core, valence and the excited levels respectively. J and K are the usual Coulomb and exchange operators.

In a pseudopotential approach, one considers explicitly only the N_v valence electrons. Without considering the fixed core energy, the pseudo-Hamiltonian is assumed to be of the form

$$H^{ps} = \sum_{i=1}^{N_v} [-\tfrac{1}{2}\Delta_i + V^{ps}(i)] + \sum_{i<j}^{N_v} 1/r_{ij} \qquad (167)$$

In Eq. (167) the atomic *pseudopotential* is a one-electron operator which takes into account the interaction of a valence electron with the core. At large distance from the nuclei, V^{ps} tends to the Coulomb potential $-z/r$ where z is the net charge of the core of the atom. The valence pseudo-Fock operator takes the form

$$F^{ps} = -\tfrac{1}{2}\Delta + V^{ps} + \sum_v (2J'_v - K'_v)$$

$$= \sum_v \varepsilon'_v |\varphi'_v\rangle\langle\varphi'_v| + \sum_{i*} \varepsilon'_i |\varphi'_{i*}\rangle\langle\varphi'_*| \qquad (168)$$

The valence energies become the lowest. Here ε'_v and φ'_v resemble as much as possible the exact solutions ε_v, φ_v of (166). The determination of V^{ps} in (168) may be achieved by minimizing a reduced distance between F^{ps} and F:

$$\| F^{ps} - F \|_{\text{minimum}} \qquad (169)$$

Expressions (81)–(83) can be used by considering

$$P = \sum_v |\varphi_v\rangle\langle\varphi_v| \qquad \text{and} \qquad P' = \sum_v |\varphi'_v\rangle\langle\varphi'_v| \qquad (170)$$

The method provides valence orbitals in (168) with internal nodes, which closely resemble the original valence Hartree–Fock orbitals of (166). This method has been developed mainly by Huzinaga and colleagues, who determine atomic pseudopotentials (model potentials in their terminology) of the form[161]

$$V^{ps} = -\frac{z}{r}\left(1 + \sum_i C_i r^{n_i} e^{-\alpha_i r^2} \right) + 2\sum |\varepsilon_c| \cdot |\varphi_c\rangle\langle\varphi_c| \qquad (171)$$

The pseudopotential is the sum of a function of r(which depends on parameters C_i, n_i, α_i) and a non-local operator diagonal on the basis set of the core orbitals which shifts the core energies above the valence energies in F^{ps}. The parameters of V^{ps} are determined at best from a condition which is very similar to (169).

The advantage of Huzinaga's approach is that the model structure of the valence orbitals is preserved but the price one has to pay is the use of large basis sets of atomic orbitals which are not drastically reduced with respect to those used in standard all-electron calculations. Another approach to the problem is to start from *coreless* valence orbitals which can be expanded in a smaller atomic basis of Gaussian functions. For computational reasons, this last approach has been advocated by the majority of authors in the last 10 years. Note that these coreless orbitals, hereafter called pseudo-orbitals, are not defined in a unique way. Their definition is slightly different from one group to another. However, in all cases, they are norm-preserving (in contrast with the early pseudopotential methods in solid-state physics). Outside the core their amplitudes are as close as possible to the exact Hartree–Fock solutions and they also tend smoothly towards zero near the nucleus (Fig. 11). All the information used for determining the pseudopotential is contained in the truncated valence Fock-like operator

$$F^{\text{tr}} = \sum_v \varepsilon_v |\phi_v\rangle\langle\phi_v| \tag{172}$$

The ε_v in (172) are still exact energies of the original Fock operator and the ϕ_v are the *a priori* defined nodeless pseudo-orbitals. Instead of using (169) the pseudo-orbital is now determined by[162,163]

$$\| F^{\text{ps}} - F^{\text{tr}} \|_{\text{minimum}} \tag{173}$$

The distance is given by (85) and the projectors on the model spaces of F^{tr} and H^{ps} are

$$P = \sum_v |\phi_v\rangle\langle\phi_v| \qquad P' = \sum_v |\varphi_v'\rangle\langle\varphi_v'| \tag{174}$$

as where φ_v' is a valence pseudo-orbital obtained by the variational solution of the pseudo-Hamiltonian (168).

For most applications, atomic pseudopotentials are in local, semi-local or non-local form. The simplest local form

$$V^{\text{ps}} = V^{\text{ps}}(r) \tag{175}$$

consists of choosing a simple function of r; it is a too crude approximation for

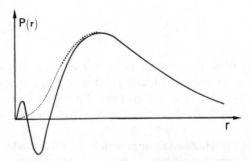

Fig. 11. Radial amplitude $P(r) = R(r)\cdot r$ of a valence Hartree–Fock orbital φ_v (full curve) and of a nodeless pseudo-orbital ϕ_v (dotted curve).

accurate molecular calculations, especially for atoms of the first row of the periodic table. The semi-local form is at the present time the most popular:[164]

$$V^{ps} = \sum_l V_l^{ps}(r)P_l \tag{176}$$

P_l is the projector on the l-space of spherical harmonics. $V_l^{ps}(r)$ is a function of r describing the potential acting on an electron locally of l symmetry. The matrix elements of semi-local pseudopotentials can be expressed in terms of almost analytical expressions, the computational time of which is negligible with respect to that of the two-electron matrix elements. The most general pseudopotentials can be written in the non-local form

$$V^{ps} = -z/r + \sum_{p,q} C_{pq}|f_p\rangle\langle f_q| \tag{177}$$

The functions f_p are generally Gaussian functions. This non-local form is very convenient for molecular calculations which involve the calculation of the overlap between the f_p and the atomic orbitals and for calculation of energy derivation (gradient techniques) for geometry optimizations.

If the Hartree–Fock equations associated with the valence pseudo-Hamiltonian (167) are solved with extended basis sets, then all the above V^{ps} are almost basis-set-independent. At the present time, and for practical reasons, most of the *ab initio* valence-only molecular calculations use coreless pseudo-orbitals. The reliability of this approach is still a matter of discussion. Obviously the nodal structure is important for computing observable quantities such as the diamagnetic susceptibility which implies an operator proportional to $1/r^3$.[165] From the computational point of view, it is always easy to recover the nodal structure of coreless valence pseudo-orbitals by orthogonalizing the valence molecular orbitals to the core orbitals. This procedure has led to very accurate results for several internal observables in comparison with all-electron results. The problem of the shape of the pseudo-orbitals in the core region is also important in relativity. For heavy atoms, the valence electrons possess high instantaneous velocities near the nuclei. Schwarz has recently investigated the compatibility between the internal structure of valence orbitals and the representations of operators such as the spin–orbit which vary as $1/r^3$ near the nucleus.[166] One should stress the fact that the simulation runs on the ground-state HF solution of the atom only. Despite this fantastic limitation, the method is able to provide reasonable estimates

1. of the correlation energy of this state (slightly overestimated by the nodeless structure of the pseudo-orbital),[167]
2. of the valence multiplet splittings (but *vide infra*),
3. of the Rydberg energies, and
4. of the ionization potentials and electron affinities of the atom,

and of the corresponding energies of molecules involving the pseudo-atom. This success is essentially due to the physically grounded form given to the pseudopotential. The essential defect of this method concerns the lack of core–valence correlation which may be treated later on.

Let us note however that the present methodology of pseudopotentials is based purely on one-electron methods. There is no guarantee with these techniques that the total energies of the various multiplets of an atom will be correctly reproduced from the valence Hamiltonian (167). The problem is particularly difficult for transition elements which involve nearly degenerate configurations, for understanding and computing their properties.[160] Some attempts have already been made to parametrize these pseudopotentials from total energies of the multiplets.[168] Much work remains to be done in this field.

One should also mention discussions concerning the relevant border between the core and the valence shells.[169-172] For copper atom, for instance, one may either consider a $(19e^-)$ $3s^23p^63d^{10}4s$, an $(11e^-)$ $3d^{10}4s$ or even a $(1e^-)$ $4s$ valence shell. In the latter case the core polarizability effects must be treated explicitly. As an extreme case, one might introduce $(0e^-)$ pseudo-potentials, for a rare gas for instance, which would be very useful to treat molecular interactions and matrix spectroscopy results. Attempts to define Ar $(0e^-)$ pseudopotentials from the virtual MO spectrum of Ar have failed to give relevant intermolecular energies and diatomic extraction appeared to be necessary.[173]

B. Groups and Fragments

Very often the chemical properties of a molecule or of a functional group are governed by a few electrons. For instance, the donor properties and proton affinity of ammonia can be understood from the character of its lone pair; bonding properties of an alkyl radical depend on its unpaired electron. As these molecules or functional groups are basic entities in chemistry, it can be conjectured that they could be described by fragment pseudopotentials quite similar to the atomic pseudopotentials associated with the cores of the atoms.[174]

Such a pseudopotential has been derived by Morokuma et al.[175] for ammonia. The 10-electron molecule is reduced to a fictitious system of two active electrons, those of the lone pair moving in the field of the three inactive electron pairs of the N–H bonds. The interaction between the lone pair and the other electrons and the nuclei of the molecule is described by means of a fragment potential (an effective fragment potential according to the termi-nology of the authors).[175] Their method is a straightforward extension of the techniques developed by Huzinaga. The interest in such an investigation is obviously to obtain transferable fragment potentials. Morokuma et al. have checked that their pseudopotential led to accurate results for predicting bond distances and bond energies when ammonia considered as a fictitious two-electron systems reacted with electron acceptors such as H^+ or BH_3. This is an encouraging result which suggests that determining fragment potentials for $CO, PH_3,\ldots$ would be very useful in the treatment of coordination chemistry problems. In the following the main steps of the derivation of a fragment potential by simulation techniques is applied to an open-shell one-electron pseudo-atom.

Let us consider a crystal of silicon and suppose that we are interested in a study of the local distortion associated with the creation of a vacancy in the crystal. For such local properties in the vicinity of the vacancy it may be reasonable to simulate the infinite crystal by a finite cluster of atoms. The silicon atoms on the surface of the cluster have to be replaced by one-electron pseudo-atoms denoted Si*. The saturation of a finite cluster by H atoms would be irrelevant since the Si–H bonds are polar and short, inducing artefactual polarization effects in the vacancy region. If the Si–H bonds are lengthened to the Si–Si bond length, the Si–H bond is too weak and introduces artefacts in the monoelectronic energy spectrum. A correct pseudo-atom Si* should bring one electron only, building a non-polar single bond with one Si atom. A convenient monoelectronic pseudopotential describing Si* can be chosen to be of the form

$$V_{\text{Si}*}^{\text{ps}} = -1/r + \sum_{p,q} C_{pq}|f_p\rangle\langle f_q| \tag{178}$$

The first term in (178) provides a correct asymptotic Coulomb dependence. The second term is a non-local operator of symmetry C_{3v} projected onto a finite basis of functions f_p. The C_{pq} are coefficients that will be best determined by simulation. The theoretical parametrization of (178) can be obtained from a full *ab initio* all-electron calculation on the disilane molecule and on the fictitious system SiH_3Si^*. These two systems contain the relevant information on the Si–Si bond (Fig. 12).

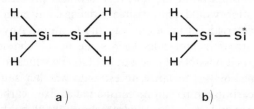

a)   b)

Fig. 12. (a) Disilane (14 valence electrons). (b) Fictitious
molecule with a one-electron pseudo-atom of silicon Si*
(eight valence electrons).

One can define a truncated Fock Hamiltonian as

$$F^{tr} = PFP = \sum_{p,q=1}^{4} \varepsilon_{pq} |\varphi_p^{loc}\rangle \langle \varphi_q^{loc}| \qquad (179)$$

where F is the original Fock operator of the disilane molecule and P is the projector corresponding to the six Si–H bonds and to the Si–Si bond. The localized orbitals φ_p^{loc} are determined by a proper orthogonal transformation of the delocalized Hartree–Fock orbitals. Other definitions of these bond orbitals are possible in close analogy with the definition of the coreless pseudo-orbitals reviewed in Section IV.A. Finally V_{Si*}^{ps} is obtained by minimizing a convenient distance between F^{ps} and F^{tr}. F^{ps} is the Fock pseudo-Hamiltonian arising from the solution of the pseudo-Hamiltonian of SiH_3Si^*.

The same technique could be used for deriving fragment potentials of the alkyl radical to be used for instance in the treatment of alkylated conjugated hydrocarbons. One should note that the pseudo-alkyl group may be a single-electron group, and one will only require σ type AOs in the subsequent molecular treatment if one is only interested in the inductive effect. If one is interested in the hyperconjugative effect too, one should introduce π type AOs on the pseudo-CH_3 and determine a π symmetry potential to mimic the acceptor ability of the σ^* MOs.

C. Atomic Operators for Pseudo-fock Molecular Calculations[45]

Despite their strong limitations, purely monoelectronic pictures, as the Hückel scheme (especially extended Hückel theory) or the tight-binding band model in solid-state physics, support a basic representation of the electronic population. These semi-empirical systematics are supposed to mimic the exact Fock monoelectronic operator, *without* calculating the integrals of the static electronic field and the exchange integrals. It might be tempting to avoid the semi-empirical fitting of parameters and to define a purely monoelectronic molecular operator which would simulate as closely as possible the exact Fock operator. This would ensure better reliability of Hückel-type models, which are so convenient.

From a conceptual point of view, this technical problem is related to a qualitative question: would it be possible to define a universal monoelectronic potential characteristic of an atom, say the C atom? Any chemist would consider that atoms in molecules keep some of the characteristics of the isolated atoms. Is it possible to give a quantitative justification of this idea? It would be interesting, for instance, to estimate whether singly, doubly and triply bonded carbon atoms can be considered as fixed carbon atoms in the series of saturated and conjugated hydrocarbons. It will be shown below that quantum chemistry is able to determine *purely monoelectronic* (pseudo)potentials of atoms in molecules and to check their transferability.

This will be done by the theoretical simulation techniques described in Section II.B by extracting information from molecular *ab initio* all-electron calculations.

Let us consider a molecule made up of atoms A, B, C,.... It is assumed that the Fock operator can be written as the sum of the kinetic energy and of the various atomic potentials in the molecule

$$F^{ps} = -\tfrac{1}{2}\Delta + \sum_A V_A^{ps} \tag{180}$$

V_A is the monoelectronic pseudopotential of atom A. (Notice the disappearance of any r^{-1}, J or K operator.) For computational simplicity it may be chosen in the form of a non-local operator:

$$V_A^{ps} = \sum_{p,q} C_{pq} |f_p\rangle\langle f_q| \tag{181}$$

The theoretical determination of the coefficients C_{pq} proceeds as follows. In a first step one determines the valence Fock operator of a molecule containing A. This can be done by solving the valence pseudo-Hamiltonian which leads to

$$F = \sum_v \varepsilon_v |\varphi_v\rangle\langle\varphi_v| \tag{182}$$

For economy we intend to perform further molecular calculations within a minimal basis set of atomic orbitals. Then the parametrization of F has to be performed not from F but from the truncated valence Fock operator

$$F^{tr} = \sum_v \varepsilon_v |\varphi_v^{appr}\rangle\langle\varphi_v^{appr}| \tag{183}$$

The φ_v are exact Hartree–Fock energies of (182) and the φ_v^{appr} are approximate valence molecular orbitals computed within a minimal basis of Slater orbitals (2s and 2p orbitals for a carbon atom). Finally the determination of the C_{pq} in (181) is obtained by minimizing a reduced distance between F^{ps} and F^{tr}, the trace being kept in the subspace of the φ_v^{appr}. Most often the information contained in one molecule is insufficient to obtain *transferable* potentials. For example, the extraction from CH_4 gives information which is only valid for C–H bonds. To determine universal atomic potentials able to reproduce the formation of single, double and triple carbon–carbon bonds, one needs to extract information from molecules containing these bonds. This has been done by minimizing a distance involving ethane, butadiene and acetylene:

$$(\| F^{ps} - F^{tr} \|_{butane} + \| F^{ps} - F^{tr} \|_{ethylene} + \| F^{ps} - F^{tr} \|_{acetylene})_{minimum} \tag{184}$$

This technique has provided a potential for the carbon atom which is *universal* in the sense that it can be transferred with good accuracy in the whole series of hydrocarbon compounds.[45,176] The main advantage of simulating the Fock operator by the kinetic operator plus the sum of atomic potentials is to produce approximate valence Fock Hamiltonians which can be solved with computational times similar to those of the Hückel method. Under the (confusing) name of valence effective Hamiltonian (VEH) the method has been widely applied by Brédas *et al.* to band calculations of organic polymers especially for polyacetylene, polyethylene, polythiophene and polypyrole. Useful estimates have been found from these methods concerning ionization potentials, band widths, band gaps and electrical conductivity properties of doped polymers.[177-182]

D. Simulation of Bielectronic Operators

To our knowledge, a rigorous simulation of bielectronic operators has never been attempted. One might use this technique for instance to fit bielectronic integrals. Turning again to the H_2 problem, one might wish to work in a minimal basis set to use exact monoelectronic operators and to reduce the bielectronic integrals to (aa, aa) and (aa, bb), as occurs in CNDO approximations, determining their values to give the best simulation of the lowest states, both the energy *and* the wavefunction. As previously discussed (Section III.C) the valence states which one may reproduce in a minimal basis set are the $(X)(^1\Sigma_g^+)$, $(b)^3\Sigma_u^+$, $(B)^1\Sigma_u^+$ and $(E, F)^1\Sigma_g^+$ states. The wavefunctions of these states should be reduced to their valence components according to Eqs. (135) and renormalized. Since one searches Hermitian operator, the two $^1\Sigma_g^+$ states should be made orthogonal, either by a symmetrical or a Schmidt orthogonalization, the second solution appearing to be preferable to save information. Then the most relevant truncated Hamiltonian would be defined from

$$^1\psi_g = [(\cos\varphi)(|a\bar{b}| + |b\bar{a}|) + (\sin\varphi)(|a\bar{a}| + |b\bar{b}|)]/\sqrt{2}$$
$$^3\psi_u = (|a\bar{b}| - |b\bar{a}|)/\sqrt{2}$$
$$^1\psi_u = (|a\bar{a}| - |b\bar{b}|)/\sqrt{2}$$
$$^1\psi_g' = [(\sin\varphi)(|a\bar{b}| + |b\bar{a}|) - (\cos\varphi)(|a\bar{a}| + |b\bar{b}|)]/\sqrt{2}$$
$$H^{tr} = E((X)^1\Sigma_g)|^1\psi_g\rangle\langle^1\psi_g| + E((b)^3\Sigma_u^+)|^3\psi_u\rangle\langle^3\psi_u|$$
$$+ E((B)^1\Sigma_u^+)|^3\psi_u\rangle\langle^3\psi_u| + E((E, F)^1\Sigma_g^+)|^1\psi_g'\rangle\langle^1\psi_g'|$$

The simulation determining the effective integrals (aa, aa) and (aa, bb) may then concern

1. the ground state only, which involves two degrees of freedom, namely the energy and the angle φ, and implies both (aa, aa) and (aa, bb),
2. the ground state and the triplet state, the latter implying (aa, bb) only,
3. the ground state, the triplet state and the purely ionic singlet state, the latter implying (aa, aa) only, or
4. the four states.

It would be interesting to compare the values of the integrals so obtained, the reduced distances between H^{ps} and H^{tr}, and the errors on the valence states which were not involved in the simulation process. (This exercise should preferably be performed simultaneously on the mono- and bielectronic pseudo-integrals, in order to avoid a simulation of the hybridization effect (largely monoelectronic) through the bielectronic integrals.)

One should note that some of the difficulties mentioned in Section III.C may be met here, if one wants to simulate too many states. If the simulation concerns only the two neutral states (X) and (b) (solutions 1. and 2.), there is no ambiguity and no continuity problem, but the ionic states may be quite erroneous. If the simulation also concerns the lowest ionic state (solution 3.) there will be a problem in assigning its energy at large interatomic distances, due to the change of the $(B)^1\Sigma_u^+$ state into a Rydberg state. The same trouble occurs even at short interatomic distances for the upper $(E, F)^1\Sigma_g^+$ state, which is no longer valence for $r \simeq r_e$.

A preliminary diabatization, defining valence ionic $^1\Sigma_u^+$ and $^1\Sigma_g^+$ diabatic states, would be necessary before defining H^{tr}, in order to avoid these intruder state problems.

It is interesting to note that the intruder state problem, which appeared explicitly in the effective Hamiltonian approach, is also present in pseudo-Hamiltonian formalisms when the simulation is too ambitious and claims to concern some states strongly mixed with other states out of the model.

V. CONCLUSIONS

Both the methodological part and the review of applications have shown the similarities and differences between effective Hamiltonians and pseudo-Hamiltonians. The similarities sometimes concern the purpose of the modelling (for instance the reduction to a minimal basis set) which may be attained in one way or another. They also concern the use of some *reduction of information* to a definite part (in general the lowest one) of the spectrum. This reduction is explicit in the effective Hamiltonian theory, through the choice of a *model space*, while in the pseudo-Hamiltonian approach it goes through the choice of a *reduced distance* between the exact Hamiltonian and the pseudo-Hamiltonian.

But it must be clear that this reduction of information and this focus on some low part of the spectrum proceed differently and lead to completely different tools. The effective Hamiltonians appear as *N-electron operators* acting in well defined finite bases of N-electron functions. The effective Hamiltonians obtained from the exact bielectronic Hamiltonian introduce three- and four-body interactions. They may essentially be expressed as *numbers* multiplied by products of creation and annihilation operators. In contrast, the pseudo-Hamiltonians keep an *a priori* defined *analytic* form, sometimes simpler than the exact Hamiltonian to mimic. For instance, the core pseudopotentials are monoelectronic while they simulate complex one- and two-body effects.

As another important difference between the two philosophies one may stress the fact that the effective Hamiltonians acting out of the model space simply give zero, while the pseudo-Hamiltonians give some flexible answer, the reliability of which essentially depends on the physical realism of the pseudo-operators assumed at the beginning of the simulation. The core pseudopotentials are an excellent example of this flexibility; extraction proceeds on the ground (valence) state of the atom, and the resulting pseudopotential is apparently able to give with a good accuracy the various valence multiplets, the Rydberg spectrum, and the positive and negative ion energies, provided that the pseudopotential has a reasonable shape with a short-range repulsive potential and a Z^{eff}/r long-range tail. This flexibility, this ability to follow physical situations which were not involved in the simulation process (since not concerned in the reduced distance), explain the greater success of pseudo-Hamiltonians over the more rigorous but too rigid effective Hamiltonian approaches.

The two methodologies should not be considered as contradictory; they may be used in conjunction, as has been mentioned. It may be useful for instance to use the projection approach, defining a valence effective Hamiltonian, which will be later mimicked (as H^{tr}) by simulation techniques. The diabatization potential energy surfaces might be an important step to define valence states, in regions where non-valence intruder states appear, before simulating them by pseudo-Hamiltonian techniques.

One of the contributions of the present review concerns the question of the intruder state. This question is frequently seen as convergence trouble in a perturbation expansion. Moving back to the basic equations defining the wave operator and the effective Hamiltonian, it appears that the intruder state problem is a problem concerning the *definition of the target space*, when the correspondence between the model space and the target space becomes ambiguous. Two choices have been proposed, based on either a projection criterion or an adiabatic energy following, but both solutions lead to major difficulties if transferability of effective interactions is desired.

One should finally stress the fact that more flexible algorithms should be sought in the field of effective Hamiltonian theory. For a given model space (of dimension $N_m + N_i$) the lowest part of the spectrum involving only N_m roots, for instance, must satisfy the basic properties of effective Hamiltonians (i.e. the roots must be exact energies and projections of exact eigenvectors) while the upper part of the spectrum must simply be realistic, continuously varying with internuclear distance. In simulation techniques this flexibility may be obtained by the weighting coefficients appearing in the reduced distance. In effective Hamiltonian techniques, Gram–Schmidt or hierarchized hermitization of the effective Hamiltonians, which proved to be a fruitful technique, is a step along this direction, since it favours the lowest part of the spectrum, the eigenvectors of the upper roots deviating significantly from the projections of the exact eigenvectors in the model space. The introduction of intermediate Hamiltonians as a new class of more flexible effective Hamiltonians is inspired by the same philosophy. It consists of accepting, for consistency reasons, to work in a rather large model space, but to lose some accuracy in both the energies and eigenvectors of the upper roots. These techniques, which directly realize some approximate diabatization when intruder states appear, may solve many of the intruder state problems.

Both effective Hamiltonians and pseudo-operators achieve rigorous simplifications of the *ab initio* schemes. They offer a way to recombine the two fundamental tasks of quantum chemistry, namely the desire for numerical accuracy and efficiency and the desire for understanding the forces governing the electronic population.

References

1. Kvasnička, V., in (Eds. I. Prigogine and S. A. Rice), *Adv. Chem. Phys.*, Vol. XXXVI, p. 345, Wiley, New York, 1977.
2. Lindgren, I., and Morrison J., *Atomic Many-Body Theory*, p. 200, Springer-Verlag, Berlin and Heidelberg, 1982.
3. Brandow, B. H., in *Effective Interactions and Operators in Nuclei* (Ed. B. R. Barrett), Lecture Notes in Physics, Vol. 40, p. 1, Springer, Berlin, 1975.
4. Feschbach, H., *Ann. Phys. N. Y.*, **5**, 357 (1958); *ibid.*, **19**, 287 (1962).
5. Feschbach, H., *Annu. Rev. Nucl. Sci.*, **B**, 49 (1958).
6. Löwdin, P. O., *J. Math. Phys.*, **3**, 969 (1962).
7. Bloch, C., *Nucl. Phys.*, **6**, 329 (1958).
8. des Cloizeaux, J., *Nucl. Phys.*, **20**, 321 (1960).
9. Kvasnička, V., *Int. J. Quantum Chem.*, **24**, 335 (1983).
10. Ôkubo, S., *Prog. Theor. Phys.*, **12**, 603 (1954).
11. Jorgensen, F., *Mol. Phys.*, **29**, 1137 (1975).
12. Brandow, B., *Int. J. Quantum Chem.*, **15**, 207 (1979).
13. Durand, Ph., *Phys. Rev. A*, **28**, 3184 (1983).
14. Lindgren, I., *J. Phys. B: At. Mol. Phys.*, **7**, 2441 (1974).
15. Kato, T., *Prog. Theor. Phys.*, **4**, 514 (1949): *ibid.*, **5**, 95, 207 (1950).
16. Messiah, A., *Mecanique Quantique*, p. 614, Dunod, Paris, 1960.
17. Suzuki, K., and Okamoto, R., *Prog. Theor. Phys.*, **70**, 439 (1983).
18. Suzuki, K., and Okamoto, R., *Prog. Theor. Phys.*, **71**, 1221 (1984).
19. Durand, E., *Solutions Numériques des Equations Algébriques*, Vol. 1, p. 31, Masson, Paris, 1960.
20. Suzuki, K., and Lee, S. Y., *Prog. Theor. Phys.*, **64**, 2091 (1980).
21. Durand, Ph., *J. Phys. Lett.*, **43**, L-461 (1980).
22. Suzuki, K., and Okamoto, R., *Prog. Theor. Phys.*, **72**, 534 (1984).
23. Van Vleck, J. H., *Phys. Rev.*, **33**, 467 (1929).
24. Primas, H., *Helv. Phys. Acta*, **34**, 331 (1961).
25. Primas, H., *Rev. Mod. Phys.*, **35**, 710 (1963).
26. Klein, D. J., *J. Chem. Phys.*, **61**, 786 (1974).
27. Shavitt, I., and Redmon, L. T., *J. Chem. Phys.*, **73**, 5711 (1980).

28. Soliverez, C. E., *Phys. Rev. A*, **24**, 4 (1981).
29. Kutzelnigg, W., *Chem. Phys. Lett.*, **83**, 156 (1981).
30. Kutzelnigg, W., *J. Chem. Phys.*, **77**, 3081 (1982).
31. Kutzelnigg, W., and Koch, J., *J. Chem. Phys.*, **79**, 4315 (1983).
32. Kutzelnigg, W., *J. Chem. Phys.*, **80**, 822 (1984).
33. Kutzelnigg, W., *J. Chem. Phys.*, **82**, 4166 (1985).
34. Malrieu, J. P., Durand, Ph., and Daudey, J. P., *J. Phys. A: Math. Gen.*, **18**, 809 (1985).
35. Brueckner, K. A., *Phys. Rev.*, **100**, 36 (1955).
36. Hugenholtz, N. M., *Physica*, **23**, 481 (1957).
37. Brueckner, K. A., *The Many-Body Problem, Les Houches 1958*, (Ed. C. De Witt), Dunod, Paris, 1959.
38. Goldstone, J., *Prog. Phys. Soc. A*, **239**, 267 (1957).
39. Brandow, B. H., *Rev. Mod. Phys.*, **39**, 771 (1967).
40. Lindgren, I., *Int. J. Quantum Chem. Symp.*, **12**, 33 (1978).
41. Hose, G., and Kaldor, U., *J. Phys. B: At. Mol. Phys.*, **12**, 3827 (1979).
42. Hose, G., and Kaldor, U., *Phys. Scr.*, **21**, 357 (1980).
43. Hose, G., and Kaldor, U., *J. Phys. Chem.*, **86**, 2133 (1982).
44. Sheppard, M. G., *J. Chem. Phys.*, **80**, 1225 (1984).
45. Nicolas, G., and Durand, Ph., *J. Chem. Phys.*, **72**, 453 (1980).
46. Moss, R. E., *Advanced Molecular Quantum Mechanics*, Chapman and Hall, London, 1973.
47. Foldy, L. L., and Wouthuysen, S. A., *Phys. Rev.*, **78**, 29 (1958).
48. Blinder, S. M., *J. Mol. Spectrosc.*, **5**, 17 (1960).
49. Löwdin, P. O., *J. Mol. Spectrosc.*, **14**, 131 (1964).
50. Morrison, J. D., and Moss, R. E., *Mol. Phys.*, **41**, 491 (1980).
51. Ketley, I. J., and Moss, R., *Mol. Phys.*, **49**, 1289 (1983).
52. Moss, R. E., *Mol. Phys.*, **53**, 269 (1984).
53. Durand, Ph., *C.R. Acad. Sci. Paris*, II, **303**, 119 (1986).
54. Valance, A., and Nguyen Tuan, Q., *J. Phys. B: At. Mol. Phys.*, **15**, 17 (1982).
    Jeung, G. H., *J. Phys. B: At. Mol. Phys.*, **16**, 4289 (1983).
55. Verges, J., Effantin, C., d'Incan, J., Cooper, D. L., and Barrow, R. F., *Phys. Rev. Lett.*, **53**, 46 (1984).
56. Landau, L. D., *Phys. Z. Sowjetunion*, **2**, 46 (1932).
    Zener, C., *Proc. R. Soc. A*, **137**, 696 (1932).
57. Sidis, V., and Lefebvre-Brion, H., *J. Chem. Phys.*, **73**, 5711 (1980).
    Tully, J. C., *J. Chem. Phys.*, **50**, 5122 (1973).
    O'Malley, T. F., *Adv. At. Mol. Phys.*, **7**, 223 (1971).
58. Hirsh, G., Bruna, P. J., Buenker, R. J., and Peyerimhoff, S. D., *Chem. Phys.*, **45**, 335 (1980).
59. Mead, C. A., and Trulhar, D. G., *J. Chem. Phys.*, **77**, 6090 (1982).
60. Petrongolo, C., Buenker, R. J., and Peyerimhoff, S. D., *J. Chem. Phys.*, **78**, 7284 (1983).
61. Cimiraglia, R., Malrieu, J. P., Persico, M., and Spiegelmann, F., *J. Phys. B: At. Mol. Phys.*, **18**, 3073 (1985).
62. Hendekovic, J., *Chem. Phys. Lett.*, **90**, 153 (1982).
63. Levy, B., *Current Aspects of Quantum Chemistry* (Ed. R. Carbo), p. 127, Elsevier, Amsterdam, 1982.
64. Bacchus-Montabonel, M. C., Cimiraglia, R., and Persico, M., *J. Phys. B: At. Mol. Phys.*, **17**, 1931 (1984).
65. Spiegelmann, F., and Malrieu, J. P. *J. Phys. B: At. Mol. Phys.*, **17**, 1259 (1984).

66. Huron, B., Rancurel, P., and Malrieu, J. P., *J. Chem. Phys.*, **58**, 5745 (1973).
67. Evangelisti, E., Daudey, J. P., and Malrieu, J. P., *Chem. Phys.*, **75**, 91 (1983).
68. Gadéa, F. X., Spiegelmann, F., Pelissier, M., and Malrieu, J. P., *J. Chem. Phys.*, **84**, 4873 (1986).
69. Connor, J. N. L., *Comput. Phys. Commun.*, **17**, 117 (1979).
    Waterland, R. L., and Delos, J. B., *J. Chem. Phys.*, **80**, 2034 (1984).
70. Kuntz, P. J., *Chem. Phys. Lett.*, **16**, 581 (1972).
    Tully, J. C., in *Modern Theoretical Chemistry* (Ed. G. A. Segal), Vol. 7, p. 173, Plenum, New York, 1977.
71. Komiha, N., Daudey, J. P., and Malrieu, J. P., to be published.
72. Rajzman, M., and Spiegelmann, F., to be published.
73. Hazi, A. U., and Taylor, H. S., *Phys. Rev. A*, **1**, 1109 (1970).
    Eliezer, I., Taylor, H. S., and William, J. K., *J. Chem. Phys.*, **47**, 2165 (1967).
74. Lefebvre-Brion, H., *Chem. Phys. Lett.*, **19**, 456 (1973).
    Pearson, P. K., and Lefebvre-Brion, H., *Phys. Rev. A*, **31**, 2106 (1976).
75. Bettendorf, M., Buenker, R. J., and Peyerimhoff, S. D., *Mol. Phys.*, **50**, 1363 (1983).
76. Bardsley, J. N., Herzenberg, A., and Mandl, F., *Proc. Phys. Soc.*, **85**, 305 (1966).
77. Fano, U., *Phys. Rev.*, **124**, 1866 (1961).
78. Freed, K. F., *Acc. Chem. Res.*, **16**, 137 (1983).
79. Westhaus, P., *Int. J. Quantum Chem. Symp.*, **7**, 463 (1973).
    Westhaus, P., Bradford, E. G., and Hall, D., *J. Chem. Phys.*, **62**, 1607 (1975).
    Westhaus, P., and Bradford, E. G., *J. Chem. Phys.*, **63**, 5416 (1975).
    Bradford, E. G., and Westhaus, P., *J. Chem. Phys.*, **64**, 4276 (1976).
    Westhaus, P., *J. Chem. Phys.*, **73**, 5197 (1980).
80. Mukherjee, D., Moitra, R. K., and Mukhopadhyay, A., *Pramana*, **4**, 247 (1975); *Mol. Phys.*, **30**, 1861 (1975); *Mol. Phys.*, **33**, 955 (1977).
    Mukhopadhyay, A., Moitra, R. K., Mukherjee, D., *J. Phys. B: At. Mol. Phys.*, **12**, 1 (1975).
    Pal, S., Prasad, M. D., and Mukherjee, D., *Theor. Chim. Acta*, **66**, 311 (1984).
81. Anderson, P. W., *Solid State Phys.*, **14**, 99 (1963).
82. Goodgame, M. M., and Goddard, W. A., *Phys. Rev. Lett.*, **54**, 661 (1985).
83. Spiegelmann, F., Malrieu, J. P., Maynau, D., and Zurru, F., *J. Chim. Phys.*, **83**, 69 (1986).
84. Denis, A., and Malrieu, J. P., *J. Chem. Phys.*, **52**, 6076 (1970).
85. Denis, A., and Malrieu, J. P., *J. Chem. Phys.*, **52**, 4769 (1970).
86. Bottcher, C., and Dalgarno, A., *Proc. R. Soc. A*, **340**, 187 (1974).
87. Bardsley, J. N., *Case Stud. At. Phys.*, **4**, 299 (1974).
88. Rosmus, P., and Meyer, W., *J. Chem. Phys.*, **65**, 492 (1976).
89. Jeung, G. H., Malrieu, J. P., and Daudey, J. P., *J. Chem. Phys.*, **77**, 3571 (1977).
90. Jeung, G. H., Daudey, J. P., and Malrieu, J. P., *J. Phys. B: At. Mol. Phys.*, **16**, 699 (1983).
91. Freed, K. F., in *Modern Theoretical Chemistry* (Ed. G. A. Segal), Vol. 7, p. 201, Plenum, New York, 1977.
92. Freed, K. F., *J. Chem. Phys.*, **60**, 1765 (1974).
93. Iwata, S., and Freed, K. F., *J. Chem. Phys.*, **61**, 1500 (1974).
94. Lee, Y. S., Freed, K. F., Sun, H., and Yeager, D. L., *J. Chem. Phys.*, **79**, 3862 (1983).
95. Pariser, R., and Parr, R. G., *J. Chem. Phys.*, **21**, 466 (1953); *ibid.*, **21**, 767 (1953).
96. Pople, J. A., *Trans. Faraday Soc.*, **49**, 1375 (1953).
97. Pople, J. A., Santry, D. P., and Segal, G. A., *J. Chem. Phys.*, **43**, S129 (1965).
    Pople, J. A., and Segal, G. A., *J. Chem. Phys.*, **43**, S136 (1965).

98. Sun, H., Freed, K. F., Herman, M. F., and Yeager, D. L., *J. Chem. Phys.*, **72**, 4158 (1980).
99. Lee, Y. S., Sun, H., Sheppard, M. G., and Freed, K. F., *J. Chem. Phys.*, **73**, 1472 (1980).
100. Lee, Y. S., and Freed, K. F., *J. Chem. Phys.*, **79**, 839 (1983).
101. Sun, H., and Freed, K. F., *Chem. Phys. Lett.*, **78**, 531 (1981).
102. Sun, H., Sheppard, M. G., and Freed, K. F., *J. Chem. Phys.*, **74**, 6842 (1981).
103. Takada, T., Sheppard, M. G., and Freed, K. F., *J. Chem. Phys.*, **79**, 325 (1983).
104. Takada, T., and Freed, K. F., *J. Chem. Phys.*, **80**, 3253 (1984).
105. Takada, T., and Freed, K. F., *J. Chem. Phys.*, **80**, 3696 (1984).
106. Olesik, J. J., Takada, T., and Freed, K. F., *Chem. Phys. Lett.*, **113**, 249 (1985).
107. Sun, H., and Freed, K. F., *J. Chem. Phys.*, **76**, 5051 (1982).
108. Iwata, S., and Freed, K. F., *J. Chem. Phys.*, **65**, 1071 (1976).
109. Sheppard, M. G., Freed, K. F., Herman, M. F., and Yeager, D. L., *Chem. Phys. Lett.*, **61**, 577 (1979).
110. Freed, K. F., and Sheppard, M. G., *J. Phys. Chem.*, **86**, 2130 (1982).
111. Kolos, W., and Wolniewicz, L., *J. Chem. Phys.*, **50**, 3228 (1969).
112. Herring, C., *Magnetism*, **2B**, 1 (1962).
113. Dirac, P. A. M., *Proc. R. Soc. A*, **123**, 714 (1929).
114. Van Vleck, J. H., *Phys. Rev.*, **45**, 405 (1934).
115. Brandow, B. H., *Adv. Quantum Chem.*, **10**, 187 (1977).
116. Hubbard, J., *Proc. R. Soc. A*, **276**, 283 (1963).
117. De Loth, Ph., Cassoux, Daudey, J. P., and Malrieu, J. P., *J. Am. Chem. Soc.*, **103**, 4007 (1981).
118. Ovchinnikov, A. O., *Theor. Chim. Acta*, **47**, 297 (1978).
119. Bulaewski, L. N., *Zh. Eksp. Teor. Fiz.*, **51**, 230 (1966).
120. Klein, D. J., and Garcia-Bach, M. A., *Phys. Rev. B*, **19**, 877 (1979).
121. Garcia-Bach, M. A., and Klein, D. J., *Int. J. Quantum Chem.*, **12**, 273 (1977).
     Klein, D. J., *J. Chem. Phys.*, **77**, 3098 (1982).
122. Kuwajima, S., *J. Chem. Phys.*, **77**, 1930 (1982).
     Soos, Z. G., Kuwajima, S., and Mihalick, J. E., *Phys. Rev. B*, **32**, 3124 (1985).
123. Malrieu, J. P., and Maynau, D., *J. Am. Chem. Soc.*, **104**, 3021 (1982).
124. Maynau, D., and Malrieu, J. P., *J. Am. Chem. Soc.*, **104**, 3029 (1982).
125. Said, M., Maynau, D., Malrieu, J. P., and Garcia-Bach, M. A., *J. Am. Chem. Soc.*, **106**, 571 (1984).
126. Said, M., Maynau, D., and Malrieu, J. P., *J. Am. Chem. Soc.*, **106**, 580 (1984).
127. Mujica, V., Correia, N., and Goscinski, O., *Phys. Rev. B*, **32**, 4178 (1985).
128. Maynau, D., Durand, Ph., Daudey, J. P., and Malrieu, J. P., *Phys. Rev. A*, **28**, 3193 (1983).
129. Malrieu, J. P., Maynau, D., and Daudey, J. P., *Phys. Rev. B*, **30**, 1817 (1984).
130. Malrieu, J. P., *Nouv. J. Chim.*, **10**, 61 (1986).
131. Sanchez-Marin, J., and Malrieu, J. P., *J. Phys. Chem.*, **89**, 978 (1985).
132. Sanchez-Marin, J., and Malrieu, J. P., *J. Am. Chem. Soc.*, **107**, 1985 (1985).
133. Gutzwiller, M. C., *Phys. Rev. A*, **134**, 993 (1964).
134. Stolhoff, G., and Fulde, P., *Z. Phys. B*, **26**, 257 (1977); *ibid.*, **29**, 231 (1978); *J. Chem. Phys.*, **73**, 4548 (1980).
135. Maynau, D., Garcia-Bach, M. A., and Malrieu J. P., *J. Physique*, **47**, 207 (1986).
136. Maynau, D., Garcia-Bach, M. A., and Malrieu, J. P., to be published.
137. Pellegati, A., Marinelli, F., Roche, M., Maynau, D., and Malrieu, J. P., *Jole Physique*, in press.
138. Sanchez-Marin, J., Maynau, D., Trinquier, G., and Malrieu, J. P., to be published.

139. Gadéa, F. X., Maynau, D., and Malrieu, J. P., *Int. J. Quantum Chem.*, **26**, 1 (1984).
140. Moller, C., and Plesset, M. S., *Phys. Rev.*, **46**, 618 (1934).
141. Claverie, P., Diner, S., and Malrieu, J. P., *Int. J. Quantum Chem.*, **1**, 751 (1967).
142. Epstein, P. S., *Phys. Rev.*, **28**, 695 (1926).
143. Nesbet, R. K., *Proc. R. Soc. A*, **230**, 312 (1955).
144. Gershgorn, Z., and Shavitt, I., *Int. J. Quantum Chem.*, **2**, 751 (1968).
145. Buenker, R. J., and Peyerimhoff, S. D., *Theor. Chim. Acta*, **12**, 183 (1968); *ibid.*, **35**, 33 (1974); *ibid.*, **39**, 217 (1975).
    Buenker, R. J., Peyerimhoff, S. D., and Butsher, W., *Mol. Phys.*, **35**, 771 (1978).
146. Hegarty, D., and Robb, M. A., *Mol. Phys.*, **37**, 1445 (1979).
147. Redmon, L. T., and Bartlett, R. J., *J. Chem. Phys.*, **76**, 1983 (1982).
148. Salomonson, S., Lindgren, I., and Martenson, A. M., *Phys. Scr.*, **21**, 351 (1980).
149. Kaldor, U., *J. Chem. Phys.*, **81**, 2406 (1984).
150. Daudey, J. P., and Malrieu, J. P., in *Current Aspects of Quantum Chemistry* (Ed. R. Carbo), p. 35, Elsevier, Amsterdam, 1982.
151. Spiegelmann, F., and Malrieu, J. P., *J. Phys. B: At. Mol. Phys.*, **17**, 1235 (1984).
152. Sheppard, M. G., Schneider, B. I., and Martin, R. L., *J. Chem. Phys.*, **79**, 1364 (1983).
153. Bonacic-Koutecky, V., Persico, M., Donhert, B., and Sevin, A., *J. Am. Chem. Soc.*, **104**, 6900 (1982).
154. Nitzsche, L. E., and Davidson, E. R., *J. Chem. Phys.*, **68**, 3103 (1978).
    Nitzsche, L. E., and Davidson, E. R., *J. Am. Chem. Soc.*, **100**, 1201 (1978).
    Davidson, E. R., McMurchie, L. E., and Day, S. J., *J. Chem. Phys.*, **74**, 5451 (1981).
155. Malrieu, J. P., *Theor. Chim. Acta*, **62**, 163 (1982).
156. Sanchez-Marin, J., and Malrieu, J. P., *Int. J. Quant. Chem.*, in press.
157. Dixon, R. N., and Robertson, I. L., in *Theoretical Chemistry*, Vol. 3, Specialist Periodical Reports, The Chemical Society, London, 1978.
158. Durand, Ph., and Barthelat, J. C., *Gazz. Chim. Ital.*, **108**, 225 (1978).
159. Krauss, M., and Stevens, W. J., in *Annu. Rev. Phys. Chem.*, **35**, 357 (1984).
160. Pelissier, M., Daudey, J. P., Malrieu, J. P., and Jeung, G. H., *Quantum Chemistry: The Challenge of Transition Metals and Coordination Chemistry* (Ed. A. Veillard), Reidel, Dordrecht, in press.
161. Sakai, Y., and Huzinaga, S., *J. Chem. Phys.*, **76**, 2537 (1982).
162. Durand, Ph., and Barthelat, J. C., *Chem. Phys. Lett.*, **27**, 191 (1974).
163. Durand, Ph., and Barthelat, J. C., *Theor. Chim. Acta*, **38**, 283 (1975).
164. Hay, P. J., and Wadt, W. R., *J. Chem. Phys.*, **82**, 270 (1985).
165. Barthelat, J. C., Durand, Ph., and Serafini, A., *Mol. Phys.*, **33**, 159 (1977).
166. Collignon, G., and Schwarz, W. H. E., *Private Communication*.
167. Teichteil, Ch., Malrieu, J. P., and Barthelat, J. C., *Mol. Phys.*, **33**, 181 (1977).
168. Pelissier, M., *J. Chem. Phys.*, **75**, 775 (1981).
169. Jeung, G. H., and Barthelat, J. C., *J. Chem. Phys.*, **78**, 2097 (1983).
170. Illas, F., and Rubio, J., *Chem. Phys. Lett.*, **119**, 397 (1983).
171. Stoll, H., Fuentealba, P., Dolg, M., Flad, J., V., Szentpály, L., and Preuss, H., *J. Chem. Phys.*, **79**, 5532 (1983).
172. Stoll, H., Fuentealba, P., Schwerdtfeger, P., Flad, J., V., Szentpály, L., and Preuss, H., *J. Chem. Phys.*, **81**, 2732 (1984).
173. Hliwa, M., and Daudey, J. P., to be published.
174. Durand, Ph., and Barthelat, J. C., in *Localization and Delocalization in Quantum Chemistry* (Eds. O. Chalvet *et al.*), Vol. II, p. 91, Reidel, Dordrecht, 1976.
175. Ohta, K., Yoshioka, Y., Morokumá, K., and Kitaura, K., *Chem. Phys. Lett.*, **101**, 12 (1983).

176. Nicolas, G., and Durand, Ph., *J. Chem. Phys.*, **70**, 2020 (1979).
177. Brédas, J. L., Chance, R. R., Silbey, R., Nicolas, G., and Durand, Ph., *J. Chem. Phys.*, **75**, 255 (1981).
178. Brédas, J. L., Chance, R. R., Baughman, R. H., and Silbey, R., *J. Chem. Phys.*, **76**, 3673 (1982).
179. Brédas, J. L., Chance, R. R., Silbey, R., Nicolas, G., and Durand, Ph., *J. Chem. Phys.*, **77**, 361 (1982).
180. Brédas, J. L., Silbey, R., Elsenbaumer, R., and Chance, R. R., *J. Chem. Phys.*, **78**, 5656 (1983).
181. Brédas, J. L., Silbey, R., Boudreaux, D. S., and Chance, R. R., *J. Am. Chem. Soc.*, **105**, 6559 (1983).
182. Brédas, J. L., Thémans, B., and André, J. M., *J. Chem. Phys.*, **78**, 6137 (1983).

Ab Initio Methods in Quantum Chemistry—I
Edited by K. P. Lawley
© 1987 John Wiley & Sons Ltd.

MOLECULAR CALCULATIONS WITH THE DENSITY FUNCTIONAL FORMALISM

R. O. JONES

Institut für Festkörperforschung der Kernforschungsanlage Jülich, D-5170 Jülich, Federal Republic of Germany

CONTENTS

I. INTRODUCTION

The microscopic understanding of the bonding between atoms is an important goal in a variety of fields of physics and chemistry. This is self-evident in theoretical chemistry, where a basic problem is the calculation of the total energy of different molecular states as a function of geometry. Total energy calculations, however, also provide an essential ingredient in the solution of an important class of problems in solid-state physics.

This is illustrated in Fig. 1, where we show two atoms moving in the neighbourhood of an ideal, unreconstructed surface, as well as a defect atom in the bulk. A calculation of the total energy as a function of the positions of the constituent atoms would yield the equilibrium structure of the surface and the distortions arising from the additional atoms. Furthermore, binding energies and the activation energies for diffusion of adsorbate and defect atoms would

413

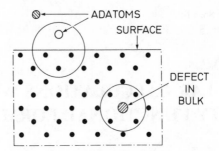

Fig. 1. Two atoms near the ideal, unreconstructed surface as well as a defect atom in a
perfect crystal.

also be accessible, as well as the paths and heats of reactions at the surface. Such problems cover a wide range of current activity in solid-state physics.

An important simplification results if we can consider the bonding between atoms to be a local phenomenon. In this event, we would need to consider only the immediate neighbours of the adsorbate or defect atoms, and we arrive at the 'cluster' models circled in Fig. 1. Of course, some properties of the system will depend on its extended nature. Others, including the variation in total energy with small displacements of atoms, should be described satisfactorily by a cluster calculation. In such cases, the problem has been reduced to one of molecular dimensions, so that the methods of molecular physics or theoretical chemistry could be used. For many systems of interest to the solid-state physicist, where a typical problem might be the chemisorption of a carbon monoxide molecule on the surface of a ferromagnetic metal surface such as nickel, the methods discussed in much of the rest of the present volume are inappropriate. It is necessary to seek alternatives, and this chapter is concerned with one of them, the density functional (DF) formalism. While the motivation of the solid-state physicist is perhaps different from that of the chemist, the above discussion shows that some of the goals are very similar. Indeed, it is my view that the density functional formalism, which owes much of its development and most of its applications to solid-state physicists, can make a useful contribution to theoretical chemistry.

It is not my purpose to describe here all the details of the density functional formalism, or to document all the molecular systems which have been studied. I shall also say little about numerical methods. Several review volumes have appeared recently,[1-4] and the reader is referred to them for more detailed bibliographies. However, I shall address the interested—and often sceptical— chemist, and try to explain just what the DF formalism is, and also to correct some prevalent misconceptions. With the help of selected examples, I shall point out the advantages of the approach, without trying the hide those aspects of the field which are presently less than satisfactory. I hope that the reader will obtain a clearer view of the relationship of the method to other schemes for electronic structure calculations, and be in a better position to judge future work in the field.

In Section II, we survey some of the methods which have been used for molecular structure calculations, particularly the Thomas–Fermi (TF), Hartree–Fock (HF) and configuration-interaction (CI) methods. In Section III we describe the essential features of the DF formalism. The applications of the local spin density (LSD) approximation described in Section IV illustrate the strengths of the method and indicate areas where more work is essential. In the final section, I shall give an assessment of the future prospects.

II. CONVENTIONAL TREATMENTS OF MANY-ELECTRON SYSTEMS

The class of problems addressed above requires the determination of the total energy of a system of N interacting electrons in a given external field v_{ext}—in the present case the Coulomb potential of the nuclei. It may be noted here, however, that the total energy of such a system is not generally measurable, and our focus will be on energy *differences*. A further quantity which plays a central role in our discussion is the electron density distribution, $n(\mathbf{r})$. One of the earliest and most familiar approaches to this problem is that of Thomas and Fermi.[5] In this model, the electron motions are assumed to be uncorrelated, and the kinetic energy is determined locally from $n(\mathbf{r})$ using the corresponding result for an electron gas of uniform density, i.e. the contribution from point $\mathbf{r}$ is proportional to $[n(\mathbf{r})]^{5/3}$. The potential energy of the system follows from the solution of Poisson's equation, and the requirement of constant chemical potential leads to the Thomas–Fermi equation for $n(\mathbf{r})$.

This approach gives a reasonable description of the density distribution in heavy atoms. In fact, it can be shown to be exact for atoms, molecules and solids in the limit as the atomic number $Z \rightarrow \infty$.[6] However, one should not be surprised that it has serious flaws. In the present context, it is noteworthy that molecular bonding is impossible, i.e. the lowest energy of an aggregate of atoms is always reached when the nuclei are infinitely far apart.[7] Furthermore, the atoms do not show the shell structure familiar from the periodic table and, in fact, *shrink* with increasing Z (as $Z^{-1/3}$).[6] The electron density is infinite at the nucleus, so that the description of the kinetic energy is seriously in error. Nevertheless, the approach continues to be of interest as a simple, well defined model, and its mathematical properties have been subject to considerable scrutiny. The focus on the charge density is also shared by the density functional formalism, as we discuss below.

The Thomas–Fermi method incorporates quantum effects only to the extent that the electrons obey the exclusion principle. A natural goal of a quantum-mechanical description of molecular bonding would be the wave-function of the N-electron system, $\Psi(\mathbf{r}_1,\ldots,\mathbf{r}_n,\ldots)$, where $\mathbf{r}_n$ denotes the coordinates and spin of the nth electron. A classic approximation for Ψ is that due to Hartree,

$$\Psi(\mathbf{r}_1,\ldots,\mathbf{r}_n,\ldots) = \psi_1(\mathbf{r}_1)\cdots\psi_n(\mathbf{r}_n)\cdots \tag{1}$$

where the functions ψ satisfy the single-particle Hartree equations (in atomic units),

$$[-\tfrac{1}{2}\nabla^2 + v_{\text{ext}} + \Phi_i(\mathbf{r})]\psi_i(r_i) = \varepsilon_i\psi_i(r_i) \tag{2}$$

where v_{ext} is the external potential of the nuclei and the Coulomb potential, Φ_i, is given by the solution of Poisson's equation

$$\nabla^2\Phi_i = 4\pi \sum_{\substack{j=1 \\ j \neq i}}^{N} |\psi_j(r_j)|^2 \tag{3}$$

In other words, the potential term in the equation determining the single-particle function ψ_i is to be found self-consistently from the mean field of the other electrons. The Hartree–Fock (HF) approximation incorporates Fermi statistics into this picture by replacing the product wavefunction (Eq. (1)) by a single determinantal function (Slater determinant) constructed from the ψ_i. In this case, the single-particle functions satisfy an equation similar to (2) but with an additional non-local potential term.

A great advantage of these approximations is the single-particle picture which results. The wavefunction is described in terms of molecular orbitals, ψ_i, which have occupation numbers appropriate to the state in question. Furthermore, the single-particle eigenvalues in the HF approximation can be related by Koopmans' theorem to excitation energies of the system. For over 50 years, the HF approximation has been a cornerstone of molecular structure calculations. The ongoing development of both numerical methods and computers means that such calculations can now be performed routinely for relatively large molecules. The results obtained for equilibrium geometries, for example, often reproduce the experimental values well. However, it is now generally accepted that a single determinant cannot lead to a satisfactory result for the total energy. There are usually many Slater determinants with comparable HF energies, and an improvement results by expressing the wavefunction as a linear combination of these determinants or 'configurations' (configuration interaction, CI). In principle, this approach enables one to determine the exact wavefunction of the interacting N-particle system and all quantities, such as the total energy, which can be found from it.

The Hartree–Fock approximation is free of experimentally determined or adjustable parameters, allows a simple, single-particle interpretation, is often satisfactory for ground-state geometries, and can be improved systematically if needed. It is perhaps not surprising that many chemists view the development of alternative approaches as unnecessary. However, there are many cases where the corrections to the HF energy ('correlation' effects) are large, or differ significantly from one state or geometry of a molecule to another. A well known example is the inability of the HF approximation to describe adequately the dissociation of diatomic molecules. In the case of F_2, the error is so large that the minimum energy of the molecule is substantially *above* that of two fluorine atoms. In some cases, such qualitative errors can be overcome by the inclusion of a few additional configurations. In general, however, many are required, and the number increases explosively with increasing electron number. To make calculations tractable, workers in this field must find ways to limit the numbers of configurations considered, and the size of the atomic-orbital basis sets used to describe them.

III. DENSITY FUNCTIONAL FORMALISM

A. Exact Single-particle Description of a Many-particle System

The density functional formalism of Hohenberg and Kohn[8] may be viewed, in principle, as the exact quantum-mechanical generalization of the Thomas–Fermi theory. Hohenberg and Kohn showed that the ground-state energy of the system considered above—N interacting electrons in a given external field, v_{ext}—is determined uniquely by $n(\mathbf{r})$. In other words, the ground-state energy is a 'functional' of the density, $E = E[n]$. Furthermore, the minimum value of this functional is equal to the exact ground-state energy of the system, E_{gs}, and is found for the exact ground-state density, n_{gs}.

The original proof of these results has been simplified and generalized by Levy,[9] who gave an explicit expression for the functional form. Consider a density which is N-representable, i.e. can be derived from some N-particle antisymmetric wavefunction Ψ. We define the functional

$$Q[n] \equiv \min \langle \Psi_n | T + V_{ee} | \Psi_n \rangle \tag{4}$$

where T and V_{ee} are the kinetic and electron–electron repulsion energies, and the minimum is sought over all antisymmetric functions, Ψ_n, which give the density n. If Ψ_n^{min} is the wavefunction corresponding to the minimum of $Q[n]$, the Rayleigh–Ritz variational principle for the system gives

$$\langle \Psi_n^{min} | T + V_{ee} + v_{ext} | \Psi_n^{min} \rangle \geqslant E_{gs} \tag{5}$$

It follows immediately that

$$Q[n] + \int d\mathbf{r} n(\mathbf{r}) v_{ext}(\mathbf{r}) \geqslant E_{gs} \tag{6}$$

For a given external field, the ground-state energy can then be determined variationally from a universal functional of the density. The second result of DF theory, that the minimum is found for the exact ground-state density, n_{gs}, follows in a straightforward fashion.[9] In principle, this scheme determines the exact ground-state wavefunction and all quantities derivable from it.

Levy's elegant proof of the existence of $Q[n]$ is important. It involves, however, a search through *all* antisymmetric wavefunctions, and is obviously impracticable. The minimization scheme can, however, be performed using the scheme introduced by Kohn and Sham,[10] who suggested a particularly useful separation of the energy functional,

$$E[n] = T_0[n] + \int dr n(\mathbf{r})[v_{ext}(\mathbf{r}) + \tfrac{1}{2}\Phi(\mathbf{r})] + E_{xc}[n] \tag{7}$$

$T_0[n]$ is the kinetic energy which a system with density n would have if there were no electron–electron interactions, and E_{xc} may be viewed as a definition of the exchange–correlation energy. In principle, the other terms may be evaluated with arbitrary numerical accuracy, so that E_{xc} plays a central role in the following discussion.

The variational principle then yields

$$\frac{\delta E[n]}{\delta n(r)} = \frac{\delta T_0[n]}{\delta n(r)} + v_{ext}(r) + \Phi(r) + \frac{\delta E_{xc}[n]}{\delta n(r)} = \mu \tag{8}$$

where μ is the Lagrange multiplier associated with the requirement of constant particle number. If we compare this with the corresponding equation for a system with an effective potential $V(r)$ but *without* electron–electron interactions,

$$\frac{\delta E[n]}{\delta n(r)} = \frac{\delta T_0[n]}{\delta n(r)} + V(r) = \mu \tag{9}$$

we see that the mathematical problems are identical, provided that

$$V(r) = v_{ext} + \Phi(r) + \frac{\delta E_{xc}[n]}{\delta n(r)} \tag{10}$$

The solution of (9) can be found simply by solving the Schrödinger equation for non-interacting particles,

$$[-\tfrac{1}{2}\nabla^2 + V(r)]\psi_i(r_i) = \varepsilon_i \psi_i(r_i) \tag{11}$$

yielding

$$n(\mathbf{r}) = \sum_{i=1}^{N} |\psi_i|^2 \tag{12}$$

It is necessary to satisfy the condition (10), and this can be achieved in a self-consistent procedure.

The solution of this system of equations leads then to the energy and density of the ground state, or the lowest state with a given set of quantum numbers,[11] and all quantities, such as those discussed in Section I, which may be derived from them. Instead of seeking these quantities by determining the wavefunction of the system of interacting electrons, the density functional method reduces the problem *exactly* to the solution of a single-particle equation of Hartree form (Eq. (2)). It shares the main advantage of the HF method, namely the single-particle interpretation of the results. In addition, we note that the Thomas–Fermi energy functional can be recovered from Eq. (7) by writing

$$T_0^{TF} = \tfrac{3}{10}(3\pi^2)^{2/3} \int d\mathbf{r}[n(\mathbf{r})]^{5/3} \tag{13}$$

However, the present method suffers from none of the drawbacks of the TF approach listed above.

B. Exchange–Correlation Energy, E_{xc}

The numerical advantages of the approach described are obvious. Since the pioneering work of Slater,[12] efficient methods have been developed for solving single-particle Schrödinger equations with a local effective potential, and there is no restriction to small systems. We have noted, however, that the exchange–correlation energy, E_{xc}, is defined as the difference between the exact energy and other contributions which may be evaluated numerically exactly. In practice, it is necessary to make approximations for this term. To provide a perspective on such approximations, we now examine E_{xc} in some detail.

The crucial simplification in the density functional scheme is the relationship between the interacting system, whose energy and density we seek, and the fictitious, non-interacting system for which we solve Eqs. (9) and (10). This can be studied by considering the interaction $\lambda/|\mathbf{r} - \mathbf{r}'|$ and varying λ from 0 (non-interacting system) to 1 (physical system). This is done in the presence of an external potential, V_λ, such that the ground state of the Hamiltonian,

$$H_\lambda = -\tfrac{1}{2}\nabla^2 + v_{ext}(r) + V_\lambda + \lambda V_{ee} \tag{14}$$

has density $n(\mathbf{r})$ for all λ. The exchange–correlation energy of the interacting system can then be expressed *exactly* in terms of an integral over the coupling constant λ,[11,13]

$$E_{xc} = \tfrac{1}{2} \int d\mathbf{r}n(\mathbf{r}) \int d\mathbf{r}' \frac{1}{|\mathbf{r} - \mathbf{r}'|} n_{xc}(\mathbf{r}, \mathbf{r}' - \mathbf{r}) \tag{15}$$

with

$$n_{xc}(\mathbf{r}, \mathbf{r}' - \mathbf{r}) \equiv n(\mathbf{r}') \int_0^1 d\lambda[g(\mathbf{r}, \mathbf{r}', \lambda) - 1] \tag{16}$$

The function $g(\mathbf{r}, \mathbf{r}', \lambda)$ is the pair-correlation function of the system with density $n(\mathbf{r})$ and Coulomb interaction λV_{ee}. The exchange–correlation hole, n_{xc}, describes the effect of the interelectronic repulsions, i.e. the fact that an electron present at the point $\mathbf{r}$ reduces the probability of finding one at $\mathbf{r}'$. The exchange–correlation (XC) energy may then be viewed as the energy resulting from the interaction between an electron and its exchange–correlation hole.

The isotropic nature of the Coulomb interaction, V_{ee}, has an important consequence. A variable substitution $\mathbf{R} \equiv \mathbf{r}' - \mathbf{r}$ in (15) yields

$$E_{xc} = \tfrac{1}{2} \int d\mathbf{r} n(\mathbf{r}) \int_0^\infty dR R^2 \frac{1}{R} \int d\Omega n_{xc}(\mathbf{r}, \mathbf{R}) \qquad (17)$$

Equation (17) shows that the XC energy depends only on the spherical average of $n_{xc}(\mathbf{r}, \mathbf{R})$. It is an important result, showing that approximations for E_{xc} can give an *exact* value, even if the description of the non-spherical parts of n_{xc} is quite inaccurate. Furthermore, a sum rule requires that the XC hole contains one electron, i.e. for all $\mathbf{r}$,

$$\int d\mathbf{r}' n_{xc}(\mathbf{r}, \mathbf{r}' - \mathbf{r}) = -1 \qquad (18)$$

This means that we can consider $- n_{xc}(\mathbf{r}, \mathbf{r}' - \mathbf{r})$ as a normalized weight factor, and define locally the radius of the XC hole,

$$\langle 1/R \rangle_r = - \int dR \frac{n_{xc}(\mathbf{r}, R)}{|\mathbf{R}|} \qquad (19)$$

This leads to

$$E_{xc} = - \tfrac{1}{2} \int d\mathbf{r} n(\mathbf{r}) \langle 1/R \rangle_r \qquad (20)$$

showing that, provided the sum rule (18) is satisfied, the exchange–correlation energy depends only weakly on the details of n_{xc}.

C. Local Spin Density Approximation

It is natural that the early work on the density functional formalism examined approximations which could be used for E_{xc}. Hohenberg and Kohn[8] studied two situations: (i) an electron gas of almost constant density, and (ii) the case of slowly varying density. In these cases, one can show that

$$E_{xc}^{LD} = \int d\mathbf{r} n(\mathbf{r}) \varepsilon_{xc}[n(\mathbf{r})] \qquad (21)$$

where $\varepsilon_{xc}(n)$ is the exchange and correlation energy per electron of a uniform gas of density n. For systems with a net spin, this local density (LD) approximation may be generalized[14] to the case of a spin-polarized electron gas, with spin-up and spin-down densities, $n_\uparrow$ and $n_\downarrow$, respectively. This refinement, known as the local spin density (LSD) approximation, forms the basis of the results discussed in Section IV. While Kohn and Sham[10] thought that this approximation should give a good representation of exchange and correlation effects in metals, alloys and small-gap semiconductors, they noted that 'we do not expect an accurate description of chemical binding'. In fact, it was a decade before self-consistent density functional calculations using an accurate description of $V(r)$ were attempted on molecules.[15] The results for the binding energy curves, particularly for H_2, were astonishingly good. It should be noted, however, that the electron density is far from homogeneous in atoms, molecules *and* solids, so that arguments based on small departures from homogeneity cannot be justified. However, the LSD approximation satisfies the sum rule (18), so that the inaccurate description of the non-spherical parts of the XC hole[16] does not affect the calculation of E_{xc}. The extent to which we obtain a satisfactory description of total energy differences will be discussed in the following section.

In common with other methods of electronic structure calculations,[17] it is often possible to obtain insight into bonding mechanisms by studying the variation of the single-particle eigenvalues with changing geometry (Walsh diagrams). If we assume that the atomic core densities are frozen and do not overlap, the total energy E can be separated rigorously into core (E_c) and valence (E_v) contributions.[18] E_c is geometry-independent, and E_v is a functional of n_v alone. It can be written in terms of the valence eigenvalues of Eqs. (9) and (10) as

$$E_v[n_v] = \sum_n^{val} f_n \varepsilon_n + \int d\mathbf{r} n(\mathbf{r})\{\varepsilon_{xc}[n(\mathbf{r})] - \varepsilon_{xc}[n_c(\mathbf{r})]\}$$

$$+ \int d\mathbf{r} n_v(\mathbf{r})[\tfrac{1}{2}\phi_v(\mathbf{r}) + \phi_c(\mathbf{r}) + v_{ext}(\mathbf{r}) - V(\mathbf{r})]$$

$$+ \tfrac{1}{2}\sum_{i \neq j} \frac{Z_{ci}Z_{cj}}{|\mathbf{r}_i - \mathbf{r}_j|} \tag{22}$$

Here $Z_{ci} \equiv Z_i - \int d\mathbf{r} n_{ci}(\mathbf{r})$ is the net charge on the core of the ith atom (with atomic number Z_i), and ϕ_v and ϕ_c are the Coulomb potentials corresponding to n_v and n_c, respectively. Apart from the first, all contributions to E_v (Eq. (22)) increase as the atoms approach each other.[18] The eigenvalue sum is then the term which determines the strength of the bond.

IV. MOLECULAR CALCULATIONS WITH LOCAL SPIN DENSITY APPROXIMATION

A. Local Spin Density and Numerical Approximations

The last decade has seen a large number of density functional calculations of molecules. Most have been based on local density approximations, not only using electron gas parametrizations of $\varepsilon_{xc}(n)$, but also on the so-called $X\alpha$ approximation,

$$E_x^{X\alpha} = \tfrac{3}{2}\alpha \int dr\, n(\mathbf{r})\varepsilon_x[n_\uparrow(\mathbf{r}), n_\downarrow(\mathbf{r})]$$

$$= -\tfrac{3}{2}\alpha C \int d\mathbf{r}\{[n_\uparrow(\mathbf{r})]^{4/3} + [n_\downarrow(\mathbf{r})]^{4/3}\} \qquad (23)$$

where $\varepsilon_x(n_\uparrow, n_\downarrow)$ is the exchange energy per particle of a spin-polarized electron gas, and $C = 3(3/4\pi)^{1/3}$. The parameter α has historical origins,[19] but energy differences for a given atom or molecule depend only weakly on α for values near 2/3, the exchange energy value. Energy differences calculated using different electron gas parametrizations are usually very similar. Most of the calculations discussed below have used the results of electron gas calculations of Ceperley and Alder,[20] as parametrized by Vosko et al.[21] Use of the $X\alpha$ approximation (23) leads to overestimates in the relative stability of states with larger spin densities. If molecular formation is accompanied by spin flips, as in the case of N_2 or CO, there are then substantial differences between dissociation energies calculated using electron gas and $X\alpha$ functionals.[18,22] In most cases, however, these approximations to E_{xc} give remarkably similar equilibrium geometries and vibration frequencies, as well as energy differences in cases where there is little change in the spin density. In LSD calculations, the density and spin density are usually constructed from the orbitals, ψ_i, using occupation and spin occupation numbers corresponding to the state in question.

In discussing the results of calculations using local density approximations, it is essential to separate the consequences of the LD approximation from those of a numerical nature. An important example is provided by the many calculations of molecular and cluster properties which use the $X\alpha$ approximation and the scattered-wave[19] (or Korringa–Kohn–Rostoker, KKR) method to solve equations of the form (9) and (10). In a number of cases, such calculations gave spectacularly incorrect predictions of ground-state geometries, examples being a repulsive energy curve in C_2 (Ref. 23) (experimental well depth 6.3 eV) and a linear geometry for H_2O (Ref. 24) (experimental bond angle 104.5°). These results, and some of the exaggerated claims by protagonists of the method, have not helped density functional methods gain acceptance among chemists. However, scattered-wave calculations use an additional assumption in evaluating the energy, namely the muffin-tin approximation for the potential $V(r)$. This assumes that the potential is spherically symmetric in spheres centred on the nuclei, and constant elsewhere. It is a poor approximation in open systems such as molecules, and recent work shows that this is the cause of the qualitatively incorrect behaviour of the energy surfaces. It is encouraging that calculations which include the full potential (i.e. include non-muffin-tin components) but use different basis sets (Slater-type orbitals,[25] Gauss-type orbitals,[22,26] and linearized muffin-tin orbitals[27]) give very similar results for the first-row dimers, so that we may confidently ascribe remaining discrepancies from experiment to the local density approximation.

The examples discussed below comprise only a selection of the many which have been carried out, and they serve to illustrate the level of accuracy to be expected from LSD calculations. Of particular interest is the question of whether the single-particle DF method can give satisfactory results in cases where the single-particle HF approximation leads to qualitatively incorrect results.

B. First-row Dimers

Diatomic molecules of first-row atoms have provided a testing ground for numerous methods of electronic structure calculations, including density functional methods. In Table I, we compare measured well depths[28] for first-row dimers with values calculated using Hartree–Fock,[29] LSD and $X\alpha$[22] approximations. The HF approximation leads to substantial underestimates of the binding energies, particularly for singlet ground states. The LSD values consistently overestimate the stability of these molecules, although the deviations from experiment are small for H_2 and Li_2. As noted above, the LSD and $X\alpha$ approximations give similar energy differences in cases where the change in spin density on bonding is small.

The equilibrium separations calculated using the LSD and $X\alpha$ approximations are generally in good agreement with experiment, with an overestimate of 1–2% being common.[22,25-27] There is a corresponding underestimate in the ground-state vibration frequencies. In cases where it leads to an energy minimum, the HF approximation usually leads to a small underestimate of r_e.

TABLE I

Experimental and calculated well depths for the experimental ground states of first-row dimers. HF calculations for Be_2 give a purely repulsive energy curve.

	Experiment[28]	LSD[22]	$X\alpha$[22]	HF[29]
H_2	4.75	4.91	3.59	3.64
Li_2	1.07	1.01	0.21	0.17
Be_2	0.10	0.50	0.43	–
B_2	3.09	3.93	3.79	0.89
C_2	6.32	7.19	6.00	0.79
N_2	9.91	11.34	9.09	5.20
O_2	5.22	7.54	7.01	1.28
F_2	1.66	3.32	3.04	−1.37

We note here that the dipole moment of CO and its variation with internuclear separation are given significantly better by the LSD approximation than by Hartree–Fock.[18]

C. Alkaline-earth Dimers

While the LSD and HF approximations generally lead to similar bond lengths in first-row molecules, there is a striking difference in the case of Be_2. In this case, as in the other group IIA dimers (Mg_2, Ca_2, Sr_2,...), the lowest-lying state is $^1\Sigma_g^+$ ($\sigma_g(\uparrow\downarrow)\sigma_u(\uparrow\downarrow)$), i.e. there is an equal occupancy of bonding and antibonding orbitals. Hartree–Fock calculations[30] lead to repulsive curves, and it has been the general view for many years that binding, if present, must be due to polarization (van der Waals) forces.[31] The bond strengths should then increase in the order $Be_2 \rightarrow Mg_2 \rightarrow Ca_2 \ldots$, due to the corresponding increase in atomic polarizabilities.[32] The earliest CI calculations supported this picture, with Be_2 having a very weak bond with a minimum at about 9 a.u.[33]

The local density approximation gives qualitatively different bounding trends in this family of molecules.[34] The energy curves show minima in all cases, and the equilibrium separations in Mg_2 and Ca_2 agree very well with measured values (7.351 a.u. and 8.083 a.u., respectively). The bond length in

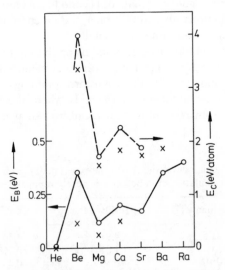

Fig. 2. Well depths calculated for $^1\Sigma_g^+$ state of group IIA dimers (————, left scale, Ref. 34) and cohesive energies of bulk materials (-----, right scale, Ref. 35). Corresponding experimental values (×) are given where known (Refs. 28 and 35).

TABLE II
Spectroscopic constants of $Be_2\,({}^1\Sigma_g^+)$.

	r_e(a.u)	ω_e(cm^{-1})	D_e(eV)
Experiment[a]	4.658	223.4	~0.11
CI[b]	4.73 ± 0.03	–	0.09 ± 0.01
CI[c]	4.78	–	0.10 (est)
CI[d]	4.9	–	0.04
LD-LMTO[e]	4.67	360	0.48
LD-LACO[f]	4.63	362	0.50

[a] Bondybey and English.[28]
[b] Lengsfield et al.[36]
[c] Harrison and Handy[37]: full CI, 361 468 configurations.
[d] Blomberg et al.[30]: CI.
[e] Jones, unpublished.[34]
[f] Painter and Averill.[22]

Be_2 (4.86 a.u.) was found to be much shorter than in previous calculations. Most striking, however, is the variation in well depth shown in Fig. 2. The energy minimum in Be_2 is *deeper* than in Mg_2, and the variation with atomic number parallels the irregular behaviour observed in the bulk cohesive energies (Fig. 2).[35] The binding energies in the molecules and solids are overestimated by the local density approximation in those cases where experimental values are known. The error is significant in Be_2, as we show in Table II, where we include the results of subsequent LD[22,34] and CI[36,37] calculations, as well as the gas-phase experimental results of Bondybey and English.[28] The experimental value of the equilibrium separation is in remarkably good agreement with the density functional result, and the extensive CI calculations reproduce both the well depth and equilibrium separation satisfactorily. The Be_2 molecule has proved to be a very severe test of CI methods, since essentially *all* the correlation energy must be found in order to obtain a satisfactory energy curve.

The density functional picture of bonding in this series is very different from that of weakly interacting closed-shell systems bound by polarization forces. The plots of the valence orbitals (Fig. 3) show that, except in He_2, there is a substantial overlap between the densities on the two atoms. This is particularly true in Be_2. It is also interesting that the radial extent of the orbitals does not increase smoothly with increasing core size, but shows a secondary periodicity. In Mg, for example, the 3s orbital is relatively compact. The introduction of 2p functions into the core has no repulsive (orthogonalization) effect on the 3s function, and is sufficiently extended that the increased core charge is incompletely screened. A similar effect is evident in Sr and Ra, where 3d and 4f functions are present in the core for the first time. The unoccupied p orbitals are generally more extended than the s valence functions. Figure 3 shows that

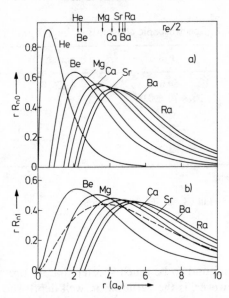

Fig. 3. Tails of valence wavefunctions for group IIA atoms: (a) s functions for the ^1S (ns^2) states; (b) p functions for the ^3P (ns^1np^1) state. The broken curve is the 2p function of He (1s2p). Also shown (arrows) are distances corresponding to half the calculated equilibrium separation in each dimer (Ref. 34).

Be is an exception. There are no p states in the core, and the effective potential experienced by the 2p state is more attractive. The similar extent of the 2s and 2p orbitals suggests that sp mixing will be favoured.

The significance of sp hybridization effects can also be seen in Fig. 4, where we show the self-consistent valence eigenvalues, ε_i, for molecules in this series. In He_2, the σ_g and σ_u eigenvalues are remarkably symmetrical about the atomic eigenvalue. The wavefunction overlap is very small, so that the weak minimum in the energy curve may be considered a van der Waals minimum. The Be_2 eigenvalues show a pronounced symmetry, i.e. the eigenvalue sum in Eq. (22) gives a bonding contribution. The Mg_2 curves are more symmetric, but there is a net bounding in the heavier molecules associated with the increasing importance of sd mixing. This picture of bonding in these molecules is very much in accord with the pseudopotential theory of cohesion in the bulk, namely a stronger sp hybridization in Be than in Mg, and the increasing importance of sd hybridization with increasing atomic number. In my experience, few solid-state theorists are surprised by the parallel between cohesion and binding energies shown in Fig. 2.

It is encouraging that the bounding trends predicted by the DF calculations,

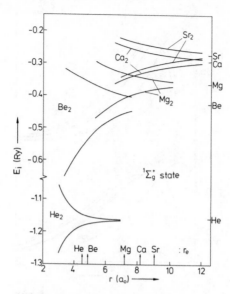

Fig. 4. Self-consistent valence eigenvalues for the $^1\Sigma_g^+$ state He$_2$ to Sr$_2$. The ns eigenvalue for the ground state of the corresponding atom is also shown. [34]

particularly the relative strength of the bond in Be$_2$, have been confirmed by both exhaustive CI calculations and by experiment. The picture provided of the bond in these molecules is also simple and plausible. However, there is a substantial overestimate in the calculations of all bond strengths discussed here, as well as in most of the first-row dimers. This is an important point, and we return to it below.

D. C$_2$, Si$_2$, C$_3$ and Si$_3$

The absence of core p states and the relatively compact 2p valence functions apply to all first-row atoms, not just beryllium. In fact, these atoms show qualitative differences in their bonding properties from atoms in the remainder of the periodic table. An important example is given by the group IVA dimers—C$_2$, Si$_2$, Ge$_2$ Sn$_2$ and Pb$_2$. It is striking that the last four have the same ground state ($^3\Sigma_g^-$, $2\sigma_g^2 1\pi_u^2$), with an excitation energy to the $^1\Sigma_g^+$ state ($1\pi_u^4$) of 1.0–1.5eV.[38,39] This is strikingly different from the situation in C$_2$, where the experimental ground state is $^1\Sigma_g^+$ with excitation energies (T_e) to the $^3\Pi_u$ ($2\sigma_g\pi_u^3$) and $^3\Sigma_g^-$ states of 0.09 and 0.80 eV, respectively.[28] The C–C bond then shows a remarkable ease of $\sigma_g \to \pi_u$ transfer, and substantially stronger π bonds than in the other atoms in this series.

The reason for the relative strength of C–C π bonds is apparent from Fig. 5,

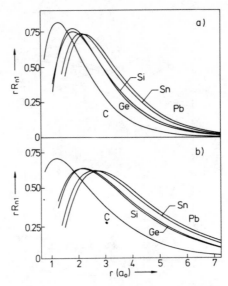

Fig. 5. Radial valence functions for group IVA
atoms: (a) s functions; (b) p functions.

where we show the tails of the valence orbitals in the group IVA atoms. The secondary periodicity noted above in group IIA atoms is even more apparent here. The s and p functions are remarkably similar in Si and Ge, for example, because the relatively diffuse 3d core density in Ge imperfectly screens the additional nuclear charge. This is reflected in the very similar properties, such as ground-state geometries, shown by these two elements. The special status of the first-row atom (C) is again apparent, in that the maxima of the radial s and p functions occur at almost the same radius.

The behaviour of the self-consistent eigenvalues of the $^3\Sigma_g^-$ state in C_2, Si_2 and Ge_2 are shown in Fig. 6. As expected from the above discussion, the Si_2 and Ge_2 eigenvalues are very similar. The C_2 values are fully consistent with the relative stability of the π_u occupancy in the C–C bond.

Further insight into the differences between the bonds in carbon and the other group IVA atoms can be found by studying the bonds in C_3 and Si_3.[40,41] The linear C–C–C bond is remarkably flexible, with a bending vibration frequency of only about 63 cm^{-1}, and this feature is reproduced very well by LSD calculations.[40] The vibration of total energy with bond angle in Si_3 is quite different. The minimum energy of the $^1A_1$ (C_{2v}) state is found for a bond angle of about 85° with an energy approximately 0.5 eV below that of the linear geometry ($^1\Sigma_g^+$). This is in satisfactory agreement with calculations performed independently by Diercksen *et al.* (CI).[42] Grev and Schaefer (CI)[43] and Raghavachari (HF with fourth-order Moller–Plesset perturbation correc-

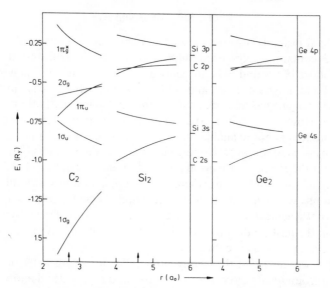

Fig. 6. Self-consistent eigenvalues for $^3\Sigma_g^-$ state of C_2, Si_2 and Ge_2. Also shown are the valence eigenvalues for the isolated atoms. The arrows denote the calculated equilibrium separations in each case.

tions).[44] It is of particular interest that the LSD calculations and two of the others[43,44] predict a near-degeneracy between the minimum of the $^1A_1$ and that of the lowest $^3A_2'$ (D_{3h}, bond angle $\alpha = 60°$). There is little experimental information available on Si_3, but the calculated vibration frequencies of the $^1A_1$ state (HF values: $206\,cm^{-1}$, $560\,cm^1$ and $582\,cm^{-1}$)[44] are much higher than the C_3 bending vibration, in spite of the increased mass in Si.

The calculations discussed in this section then provide a very plausible picture of the differences between C–C bonds and other bonds involving group IVA atoms. The C_2 bond is stronger, and is characterized by an ease of $\sigma_g \rightarrow \pi_u$ transfer. For the remaining dimers, the p orbitals are more extended than the valence s orbitals, and occupancy of the π orbitals is less favourable. A simple extension of this argument[38] explains why the C–C bond is stronger in H—C≡C—H than in C_2, while the H—Si≡Si—H bond is weaker than in Si_2. Carbon is unique in forming strong and flexible bonds in the trimer, as well as showing such ease of σ–π transfer in the dimer. This is quite consistent with the special role of the C–C bond in nature. It is again very satisfying that the energy surfaces obtained using density functional calculations are in good agreement with traditional *ab initio* results where these are available. However, it must not be overlooked that the bond strengths discussed in this section are overestimated by the LSD approximation by 0.7–1.0 eV.

E. O_3, SO_2, SOS and S_3

As a final example, we study the energy surfaces of a family of triatomic molecules containing oxygen and sulphur, each with 18 valence electrons. The ground states of O_3 and SO_2 are known to have $^1A_1$ (C_{2v}) symmetry, with similar bond angles (116.8° and 119.4°, respectively).[45] The absorption spectrum of ozone has also been studied in detail, particularly for ultraviolet radiation. However, detailed experimental information of the excited-state energy surfaces is not available for any of these molecules, and theory has played an essential role in developing our understanding of them.[46] It is worth noting that the HF approximation gives a qualitatively incorrect ordering of the low-lying states of O_3, since there are two configurations with substantial contributions to the ground-state wavefunction. These systems then provide ideal tests of schemes for incorporating correlation effects. Attention has been focused in the past on the presence of *two* minima in the $^1A_1$ energy surface, and the energy difference between them.

In Fig. 7 we show that the energy surfaces for low-lying states of O_3 and SO_2. The ground-state geometries are reproduced very well and the excitation energy between the two $^1A_1$ minima (1.4 eV) is in reasonable agreement with the most recent CI calculations, which give results between 1.0 and 1.4 eV.[46] In both molecules, the bond lengths increase in the order $X^1A_1 \rightarrow 1^3B_1 \rightarrow$

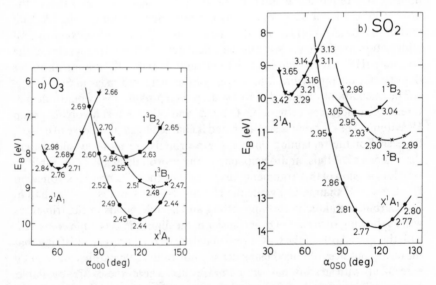

Fig. 7. Energy surfaces for X^1A_1 (●), 1^3B_1 (×), 1^3B_2 (■) and 2^1A_1 (▼) of (a) O_3 and (b) SO_2. For each bond angle α, we show the bond length which optimizes the energy.

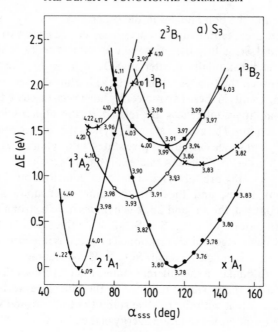

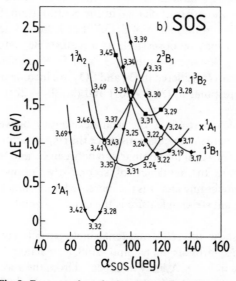

Fig. 8. Energy surfaces for low-lying states of (a) S_3 and (b) SOS. The optimum bond length for each bond angle α is also shown.

$1^3B_2 \to 2^1A_1$, and the bond angles in the order $2^1A_1 \to 1^3B_2 \to x^1A_1 \to 1^3B_1$. However, there are qualitative differences between the *energy* ordering of the states in O_3 and SO_2. In particular, the excitation energy of the low-lying triplet states is substantially greater in SO_2 then in O_3, and the energy separation between the two $^1A_1$ states (4.1 eV) is almost three times as large.

The bonding trends in these two molecules have been discussed elsewhere,[46] where the qualitative differences are related to the higher-lying valence p eigenvalue in sulphur. The ground state (X^1A_1) has the valence configuration[47] $1a_2^2 3b_2^2 4a_1^2 2b_1^0$, and single excitations give states with symmetries 1^3B_1 ($4a_1 \to 2b_1$) and 1^3B_2 ($1a_2 \to 2b_1$). The 2^1A_1 state corresponds to the excitation ($3b_2^2 \to 2b_1^2$). The $1a_2$ and $3b_2$ orbitals have significant contributions only from the outer (oxygen) atoms, and the corresponding eigenvalues show similar trends.[46] By contrast, the $4a_1$ and $2b_1$ eigenvalues have a strong contribution from the central atom (O in O_3, S in SO_2). Since these are the highest-lying amongst the valence eigenvalues mentioned above, it is not surprising that the eigenvalue spread is much larger in SO_2.[46] If we assume that the density distributions are similar in the different states, Eq. (22) shows that the energy difference between them will be reflected in the difference in the eigenvalue sum. This is consistent with the relative stability of the ground state in SO_2, and also explains the large excitation energy of the 2^1A_1 state in SO_2, since this state corresponds to a double excitation to the high-lying $2b_1$ orbital.

This is a very simple picture, and can be checked by comparing the S_3 (Ref. 48) and SOS (Ref. 49) molecules, where the change from trimer to mixed molecule reverses the order of change in the central atom ($S \to O$). The SOS molecule is not the ground state of S_2O,[50] but is bound according to LSD calculations[49] and may be observable as a metastable state. The calculated energy surfaces for low-lying states of S_3 and SOS are shown in Fig. 8. In addition to the states considered in O_3 and SO_2, we include the 1^3A_2 and 2^3B_1 states, which correspond to ($3b_2 \to 2b_1$) and ($1a_2 3b_2 \to 2b_1^2$) excitations from the ground state.

The first interesting result is the near-degeneracy between the two $^1A_1$ minima in S_3,[51] where the calculations place the ring structure less than 0.1 eV below the open form. As in the comparison between C_3 and Si_3, the O_3 and S_3 results demonstrate the tendency of second-row atoms to favour bent geometries. A contributing factor is the increased importance of d functions in this row. The second striking feature is the closed ground state found for the SOS molecule.

In Fig. 9, we show the self-consistent single-particle eigenvalues for the uppermost valence orbitals in S_3 and SOS, calculated for the optimum geometries shown in Fig. 8. As may be expected from the above discussion, the $4a_1$ and $2b_1$ orbitals, which have an important contribution from the central atom, lie *lower* in SOS than in S_3, so that the triplet states and the ring $^1A_1$

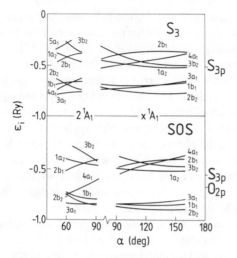

Fig. 9. Self-consistent eigenvalues for the upper-most valence orbitals in S_3 and SOS. For each bond angle α, we show the results for the optimum bond length.

state will be stabilized relative to the open $^1A_1$ state. This is fully consistent with the calculated energy surfaces.

The single-particle picture means that excitation energies between states can often be associated with simple changes in the occupation numbers. This is useful if we calculate the energy surface for one state (e.g. the ground state) and note the geometries where the single-particle eigenvalues are degenerate. The energy surfaces for the ground state and the state corresponding to the change in occupation numbers should then interest near this geometry. Such arguments must be treated with caution, particularly if electron transfer results in a change in spin degeneracy, but the results are remarkably consistent with the calculated excited-state surfaces in all molecules discussed in this section.

V. CONCLUDING REMARKS

As noted at the outset, I have not attempted to catalogue all the molecular applications of density functional theory. The examples given above illustrate, however, the usefulness of density functional calculations in molecular physics and chemistry. Even in Be_2 and ozone, where the Hartree–Fock approximation provides qualitatively incorrect descriptions and where CI calculations have only recently been successful, DF calculations provide a very satisfactory description of the geometries of low-lying states. The interpretation of the results is facilitated by the single-particle nature of the equations to be solved.

Misconceptions concerning the density functional formalism abound among chemists. Although the Thomas–Fermi theory can be derived as a special case, DF theory is *not* a minor modification of the former and it suffers from none of its qualitative inadequacies. Although it is related to the $X\alpha$ scattered-wave scheme, it is *not* restricted to the muffin-tin-potential construction which is the cause of the errors in total energy calculations using this scheme. Furthermore, the use of a local spin density approximation based on exchange–correlation energies for a homogeneous electron gas by no means invalidates results for systems where the density is far from homogeneous. The discussion in Section III shows that *exact* values of the exchange–correlation energy can result, even if some details of the electron–electron interaction, namely the non-spherical parts of the exchange–correlation hole, are poorly described. This means that it is *not* in principle necessary to know the exact wavefunction of the interacting system in order to obtain an exact ground-state energy. As be seen from the numerous applications in solid-state physics, DF calculations are not restricted to systems with few electrons. The results for the families of molecules discussed above show, in fact, that first-row atoms have features which are not typical for the periodic system as whole. This is an important observation, since chemical intuition and rules (such as Valsh's rules) are based largely on results for systems with atomic numbers $Z \leqslant 10$, i.e. the least typical atoms.

There remain important problems with the DF formalism and its application. Particular interest has focused on the properties to be satisfied by the exact form of E_{xc},[4] and the consequences for the calculation of excitation energies in extended systems, particularly bulk semiconductors.[52] There has also been much discussion of possible improvements to the LSD approximation to E_{xc}. Recently, Gunnarsson and Jones[53] performed an extensive series of calculations on atoms and small molecules, and compared the local density description of exchange energy differences to the exact values. In cases where the states in question had the same nodal structures, e.g. bonding in H_2 and Li_2, the LSD approximation gives a satisfactory description of exchange energy differences and the LSD results are is very good agreement with experiment. In cases such as Be_2, however, where there is a qualitative change in angular nodal structure bonding (two valence s functions give rise to a bonding *and* an antibonding orbital), the local density approximation overestimates the stability of the states with additional nodal planes. This is consistent with the overestimates found in the binding energies of all sp bonded molecules studied to date. These errors are also greater in molecules comprising atoms in the second half of a row of the periodic table (e.g. O, F, S), where antibonding π orbitals are occupied, than in the first (C, B, Si). There have been numerous suggestions for modifying the LSD approximation,[54] but no practicable scheme for doing this systematically.

This appears to me to be the fundamental problem facing the application of DF methods in chemistry. Although the CI approach cannot be applied to systems with more than a few atoms, and even in very small systems such as Be_2 and ozone can lead to unreliable results unless performed with great care, it is, in principle, and exact method for improving on Hartree–Fock calculations. Practical calculations with the DF formalism require an approximation to E_{xc}, and the results can be viewed as tests of this approximation, irrespective of the accuracy of the numerical work. These tests have given results of a quality which no one could have anticipated a decade ago. The search for a systematic method for improving on them as an important challenge for solid-state physicists *and* chemists.

References

1. Lundqvist, S., and March, N. H. (Eds.), *Theory of the Inhomogeneous Electron Gas*, Plenum, New York, 1983.
2. Dahl, J. P., and Avery, J. (Eds.), *Local Density Approximations in Quantum Chemistry and Solid State Physics*, Plenum, New York, 1984.
3. Phariseau, P., and Temmerman, W. M., (Eds.), *The Electronic Structure of Complex Systems*, Plenum, New York, 1984.
4. Dreizler, R. M., and da Providencia, J. (Eds.), *Density Functional Methods in Physics*, Plenum, New York, 1985.
5. For a survey of Thomas–Fermi and related theories, see Lieb, E. H., *Rev. Mod. Phys.*, **53**, 603 (1981).
6. Lieb, E. H., and Simon, B., *Phys. Rev. Lett.*, **31**, 681 (1973).
7. Teller, E., *Rev. Mod. Phys.*, **34**, 627 (1962).
8. Hohenberg, P., and Kohn, W., *Phys. Rev.*, **136**, B864 (1964).
9. Levy, M., *Proc. Nat. Acad. Sci. USA*, **76**, 6062 (1979). See also Percus, J. K., *Int. J. Quantum Chem.*, **13**, 89 (1978).
10. Kohn, W., and Sham, L. J., *Phys. Rev.*, **104**, A1133 (1965).
11. Gunnarsson, O., and Lundqvist, B. I., *Phys. Rev. B*, **13**, 4274 (1976).
12. Slater, J. C., *Phys. Rev.*, **81**, 385 (1951); *ibid.*, **82**, 538 (1951).
13. Harris, J., and Jones, R. O., *J. Phys. F: Metal Phys.*, **4**, 1170 (1974); Perdew, J. P., and Langreth, D. C., *Solid State Commun.*, **17**, 1425 (1975). For a further discussion of the connection between physical and non-interacting systems, see Harris, J., *Phys. Rev. A*, **29**, 1648 (1984).
14. von Barth, U., and Hedin, L., *J. Phys. C: Solid State Phys.*, **5**, 2064 (1972).
15. Gunnarsson, O., and Johansson, P., *Int. J. Quantum Chem.*, **10**, 307 (1976).
16. See, for example, Gunnarsson, O., and Jones, R. O., *Phys. Scr.*, **21**, 394 (1980).
17. Walsh, A. D., *J. Chem. Soc.*, **1953**, 2260 (1953) and following articles; Mulliken, R. S., *Rev. Mod. Phys.*, **14**, 204 (1942). For a review, see Buenker, R. J., and Peyerimhoff, S. D., *Chem. Rev.*, **74**, 127 (1974).
18. Gunnarsson, O., Harris, J., and Jones, R. O., *Phys. Rev. B*, **15**, 3027 (1977); *J. Chem. Phys.*, **67**, 3970 (1977).
19. See, for example, Slater, J. C., *Quantum Theory of Molecules and Solids*, Vol. IV, McGraw-Hill, New York, 1974.
20. Ceperley, D. M., and Alder, B. J., *Phys. Rev. Lett.*, **45**, 566 (1980).
21. Vosko, S. H., Wilk, L., and Nusair, M., *Can. J. Phys.*, **58**, 1200 (1980).
22. Painter, G. S., and Averill, F., *Phys. Rev. B*, **26**, 1781 (1982).
23. Danese, J. B., *J. Chem. Phys.*, **61**, 3071 (1974).
24. Connolly, J. W. D., and Sabin, J. R., *J. Chem. Phys.*, **56**, 5529 (1972).

25. Baerends, E. J., and Ros, P., *Int. J. Quantum Chem. Symp.*, **12**, 169 (1978).
26. Dunlap, B. I., Connolly, J. W. D., and Sabin, J. R., *J. Chem. Phys.*, **71**, 4993 (1979).
27. Jones, R. O., *J. Chem. Phys.*, **76**, 3098 (1982).
28. Huber, K. P., and Herzberg, G., *Molecular Structure and Molecular Spectra*, Vol. IV *Constants of Diatomic Molecules*, Van Nostrand Reinhold, New York, 1979. The value for Be_2 is from Bondybey, V. E., and English, J. H., *J. Chem. Phys.*, **80**, 568 (1984).
29. The HF molecular energies (for the experimental equilibrium internuclear separation, r_e) are taken from Cade, P. E., and Wahl, A. C., *At. Data Nucl. Data Tables*, **13**, 339 (1974). The atomic energies are taken from Clementi, E., and Roetti, C., *At. Data Nucl. Data Tables*, **14**, 177 (1974). The HF dissociation energies (for the HF value of r_e) will be somewhat greater.
30. Blomberg, M. R. A., Siegbahn, P. E. M., and Roos, B. O., *Int. J. Quantum Chem. Symp.*, **14**, 229 (1980) (HF and CI calculations).
31. The following is a selection from the theoretical and experimental literature on Be_2. 'Our calculations show that the resulting molecular state is repulsive': Bartlett, J. H., Jr, and Furry, W. H., *Phys. Rev.*, **38**, 1615 (1931). 'There is no evidence from this calculation that the ground state of Be–Be is bound': Bender, C. F., and Davidson, E. R., *J. Chem. Phys.*, **47**, 4972 (1967). 'Our second order perturbation theory calculations ... indicate that the ground $^1\Sigma_g^+$ state potential curve of Be_2 is indeed repulsive. Of course the ground states of the analogous diatomics (Mg_2, Hg_2, Cd_2) are bound due to the much larger dispersion interactions which occur in these more polarizable species': Jordan, K. D., and Simons, J., *J. Chem. Phys.*, **65**, 1601 (1976). 'The van der Waals bound Be_2 molecule should have a dissociation energy somewhat less than 1.2 kcal, the experimental D_0^0 for Mg_2': Dykstra, C. E., Schaefer, H. F., III, and Meyer, W., *J. Chem. Phys.*, **65**, 5141 (1976). 'The ground state well depth ... surely increases in the series Be_2, Mg_2, Ca_2, Sr_2': Miller, J. C., Ault, B. S., and Andrews, L., *J. Chem. Phys.*, **67**, 2478 (1977).
32. Maeder, F., and Kutzelnigg, W., *Chem. Phys. Lett.*, **37**, 285 (1976). These authors calculated the coefficients of the van der Waals potential coefficients C_6, C_8 and C_{10} for He_2, Be_2, Mg_2 and Ca_2. These coefficients increase dramatically with increasing atomic number.
33. See, for example, Blomberg, M. R. A., and Siegbahn, P. E. M., *Int. J. Quantum Chem.*, **14**, 583 (1978).
34. Jones, R. O., *J. Chem. Phys.*, **71**, 1300 (1979) and unpublished.
35. The measured cohesive energies are from Gschneidner, K. A., Jr, *Solid State Phys.*, **16**, 276 (1964). LD values are from Moruzzi, V. L., Janak, J. F., and Williams, A. R., *Calculated Electronic Properties of Metals*, Pergamon, New York, 1978.
36. Lengsfield, B. H., III, McLean, A. D., Yoshimine, M., and Liu, B., *J. Chem. Phys.*, **79**, 1891 (1983).
37. Harrison, R. J., and Handy, N. C., *Chem. Phys. Lett.*, **98**, 97 (1983).
38. Harris, J., and Jones, R. O., *Phys. Rev. A*, **18**, 2159 (1978); *Phys. Rev. A*, **19**, 1813 (1979).
39. Northrup, J. E., Yin, M. T., and Cohen, M. L., *Phys. Rev. A*, **28**, 1945 (1983).
40. Jones, R. O., *J. Chem. Phys.*, **82**, 5078 (1985).
41. Jones, R. O., *Phys. Rev. A.*, **32**, 2589 (1985).
42. Diercksen, G. H. F., Grüner, N. E., Oddershede, J., and Sabin, J. R., *Chem. Phys. Lett.*, **117**, 29 (1985).
43. Grev, R. S., and Schaefer, H. F., III, *Chem. Phys. Lett.*, **119**, 111 (1985).
44. Raghavachari, K., *J. Chem. Phys.*, **83**, 3520 (1985).
45. Herzberg, G., *Electronic Structure and Electronic Spectra of Polyatomic Molecules*, Van Nostrand, Princeton, 1969.

46. Jones, R. O., *J. Chem. Phys.*, **82**, 325 (1985). This paper contains a survey of the literature on O_3 and SO_2.
47. To simplify the comparison between different molecules in this family, we label only the valence orbitals.
48. Jones, R. O., *J. Chem. Phys.*, **84**, 318 (1986).
49. Jones, R. O., *Chem. Phys. Lett.*, **125**, 221 (1986).
50. The experimental ground state has the asymmetric bent structure S–S–O with $\alpha_{sso} = 118.26°$. See Tiemann, E., Hoeft, J., Lovas, F. J., and Johnson, D. R., *J. Chem. Phys.*, **60**, 5000 (1974) and references therein.
51. Ahlrichs, R., (private communication) has performed CI calculations in S_3 which also lead to a near-degeneracy.
52. See, for example, Perdew, J. P., and Levy, M., *Phys. Rev. Lett.*, **51**, 1884 (1983); Sham, L. J., and Schlüter, M., *Phys. Rev. Lett.*, **51**, 1888 (1983).
53. Jones, R. O., and Gunnarsson, O., *Phys. Rev. Lett.*, **55**, 107 (1985); Gunnarsson, O., and Jones, R. O., *Phys. Rev. B*, **31**, 7588 (1985).
54. Perdew, J. P., *Phys. Rev. Lett.*, **55**, 1665 (1985) and references therein.

Ab Initio Methods in Quantum Chemistry—I
Edited by K. P. Lawley
© 1987 John Wiley & Sons Ltd.

BASIS SETS

STEPHEN WILSON*

Theoretical Chemistry Department, Oxford OX1 3TG, UK

CONTENTS

*SERC Advanced Fellow.

I. INTRODUCTION

Almost all applications of quantum mechanics to the problem of quantitatively describing molecular electronic structure begin with the choice of a suitable basis set in terms of which the electronic wavefunction is then parametrized. This choice of basis set is crucial, since it ultimately determines the accuracy of the calculation, whether it be a matrix Hartree–Fock calculation, a configuration-interaction study or a many-body perturbation theory expansion, whether it is a calculation of the total energy of a system, the energy of interaction between two subsystems, or the determination of some electric or magnetic property.

The history of computational quantum chemistry can be traced in terms of the use of basis sets of increasing size—a development which can be seen as a result of, on the one hand, the availability of increasingly powerful computers and, on the other hand, the growing awareness of the fact that the truncation of the basis set is an important, and frequently dominant, source of error in molecular electronic structure calculations. In the early years, much of the 'art' in performing quantum-chemical calculations lay in the construction of basis sets affording a reasonably accurate description of the system being investigated. The success of such calculations frequently depended on a fortuitous cancellation of basis set truncation errors and other effects, such as truncation of the expansion of the correlation energy. More recent work has recognized the need for a more systematic approach to the problem of constructing basis sets with the aim of reducing the basis set truncation error in molecular electronic structure calculations.

The purpose of this chapter is to review the progress which has been made over the past few years on the problem of reducing the error associated with basis set truncation in molecular calculations and in atomic calculations using the algebraic approximation or basis set expansion technique. It is clearly not possible, within the space available, to give a completely comprehensive account of all developments which have recently been made in the field of basis set construction, a field that forms the foundation upon which the vast majority of contemporary atomic and molecular electronic structure studies are based. The review is, therefore, necessarily selective but, nevertheless, should provide an up-to-date account of the most important aspects of current thinking on the basis set expansion method or algebraic approximation.

Basis set construction has been reviewed over the past 10 years by a number of authors. Dunning and Hay[1] surveyed methods for constructing basis sets of Gaussian-type functions and contracted Gaussian-type functions. Ahlrichs and Taylor[2] also discussed the choice of Gaussian-type basis sets. Čársky and Urban's book,[3] *Ab Initio Calculations*, provides a particularly useful survey of both exponential-type and Gaussian-type basis sets. Szabo and Ostlund[4] provide some discussion of the basis set problem in their book, *Modern Quantum Chemistry*. Huzinaga and his coworkers have recently published a monograph[5] dealing with the construction of Gaussian-type basis sets. My own previous review[6] has some degree of overlap with the present work. The problem of constructing basis sets for electron correlation studies is discussed in the monograph[7] *Electron Correlation in Molecules*. Very recently, Huzinaga has described methods for systematically constructing basis sets and also basis set superposition effects.[8] The basis set problem will also be addressed in a forthcoming book[9] by the author.

This review is divided into 11 main parts. In Section II, the nature of the algebraic approximation is considered in somewhat more detail than is usual in atomic and molecular physics textbooks. The third section is devoted to an overview of the various types of basis functions which are employed in atomic and molecular electronic structure studies. Exponential-type functions, Gaussian-type functions and contracted Gaussian-type functions are considered together with less frequently employed functions, elliptical-type functions and piecewise polynomials. Section IV is devoted to the problem of constructing basis sets from the various types of basis functions described in Section III. One-centre and multi-centre basis sets, minimum basis sets, double-zeta basis sets and polarization functions are discussed. Even-tempered basis sets and universal basis sets are considered together with systematic sequences of basis sets. Attention is turned to the calculation of electron correlation effects in Section V. The basis set problem in both the configuration-interaction expansion and the many-body perturbation theory is addressed. A simple model problem is employed to highlight some of the problems that arise. The use of universal basis sets in correlation energy calculations is described and the magnitude of the basis set truncation errors in some typical calculations is assessed. The design of basis sets for calculations of atomic and molecular properties, other than the energy, is discussed in Section VI. The discussion is subdivided into consideration of electric properties, magnetic properties and relativistic properties. Section VII is devoted to the treatment of basis set superposition effects. These effects can be crucially important in calculations of small interaction energies such as van der Waals interactions. The function counterpoise technique is discussed and the use of systemic sequences of even-tempered basis sets in calculations of van der Waals interaction potentials is described together with the problems which arise in calculations of many-body van der Waals interactions. In Section VIII to the special problems which arise in performing relativistic electronic structure calculations within the algebraic approximation are reviewed. The so-called 'finite basis set disease' is described and the importance of using matched basis sets for the parametrization of the large and small components of the relativistic wavefunction is underlined. Prototype relativistic calculations using basis sets are described. The design of basis sets for the quantum-chemical study of extend systems is addressed in Section IX. The computational aspects of the basis set expansion method are overviewed in Section X, paying particular attention to the problem of practical linear dependence in large basis sets and also to the efficiency with which matrix operations, which are fundamental elements of molecular electronic structure calculations performed within the algebraic approximation, can be executed on parallel processing computers. In the final part, Section XI, the current status of research into the basis set problem is overviewed and possible future directions are indicated.

II. THE ALGEBRAIC APPROXIMATION

By employing the basis set expansion technique, quantum-mechanical differential equations and integro-differential equations are converted into matrix equations, algebraic equations for the expansion coefficients of the unknown wavefunction. For example, Hall[10] and, independently, Roothaan[11] showed that the integro-differential Hartree–Fock equations become a set of matrix pseudo-eigenvalue problems upon invoking the algebraic approximation. In this section, the nature of the algebraic approximation is considered in somewhat more detail than is usual in textbooks dealing with atomic and molecular physics. The matrix representation of operators is described and then the use of the algebraic approximation in independent-electron models and in methods which take account of electron correlation is discussed.

A. Matrix Representation of Operators

There has, in fact, been very little systematic study of the properties of finite-dimensional matrix approximations of quantum-mechanical operators generated by employing basis sets. This is not a trivial question, since unbounded operators, such as, for example, position and momentum, cannot be represented by finite matrices. A well known example of this fact is the canonical relation $[q, p] = i\hbar$, which cannot be satisfied using finite-dimensional representations of the operators q and p.[12] In spite of this problem, many calculations are performed within the algebraic approximation and in the case of molecular electronic structure studies this approach is almost universal.

In the discussion that follows, it will be assumed that the reader is familiar with the theory of linear operators on Hilbert spaces, as presented in standard textbooks on quantum theory (for example, Messiah).[13] Whilst this is adequate for most purposes, the somewhat more mathematical presentation to be found in books such as those by Richtmyer,[14] Kato[15] or Reed and Simon[12] will be followed here. The reader may wish to consult these books for background information. The presentation in this section will be informal and reference to these works avoided.

Let H be a Hilbert space and

$$A: \mathsf{H} \to \mathsf{H} \tag{1}$$

a linear operator. A is said to be linear if its domain $D(A)$, that is the subset of H on which its actions are defined, maps onto another linear subset $R(A)$, the range of A, which is also in H. A is a bounded operator if we can find a positive number k such that

$$\|Au\| \leqslant k\|u\| \qquad \forall u \in D(A) \tag{2}$$

The smallest such k is said to be the norm of the operator A, $\|A\|$, and we have the well known inequality

$$\|Au\| \leqslant \|A\|\|u\| \qquad \forall u \in D(A) \tag{3}$$

If no finite k can be found, then A is said to be unbounded. An example in the case of $\mathsf{H} = L^2(-\infty, \infty)$, the Hilbert space of functions $\phi(x)$ on the real line such that

$$\|\phi\|^2 = \int_{-\infty}^{\infty} |\phi(x)|^2 \, dx \leqslant \infty \tag{4}$$

(in the Lesbesgue sense), are the operators q and p, where

$$q\phi(x) := x\phi(x) \tag{5}$$

and

$$p\phi(x) := -i\hbar \frac{d\phi(x)}{dx} \tag{6}$$

The Hellinger–Toeplitz theorem states that such operators cannot be defined on the whole of H; for example, $D(q)$ must be a proper subset of H, consisting of all $\phi(x)$ for which

$$\| q\phi \|^2 = \int_{-\infty}^{\infty} |x\phi(x)|^2 \, dx \tag{7}$$

is bounded. *Many quantum mechanics texts fail to stress the necessity of defining the domain and the range of an operator as a part of its specification.*

Further problems appear when, as is done in molecular electronic structure calculations in which the algebraic approximation is almost invariably invoked, $\phi(x)$ is represented by a finite linear superposition of suitably chosen linearly independent basis functions, say

$$\mathbb{M} = \{u_i | i = 1, 2, \ldots, M\} \subset \mathbb{H} \tag{8}$$

If the element Au_i is to have any meaning, it is clearly essential that

$$\mathbb{M} \subseteq D(A) \tag{9}$$

though, of course, this may not be the case if the set $\mathbb{M}$ has been chosen in some arbitrary fashion. In general, the basis functions will not be orthogonal. If we define for real functions u and v in $L^2(-\infty, \infty)$ the inner product

$$(u, v) = \int_{-\infty}^{\infty} u(x)v(x) \, dx \tag{10}$$

then the matrix representation of the operator A on $\mathbb{M}$ can be introduced as the matrix with elements

$${}^m A^m{}_{ij} = (u_i, Au_j) \qquad i, j = 1, 2, \ldots, M \tag{11}$$

the superscripts m being intended to emphasize that both the 'bra' and the 'ket' vectors are drawn from the same finite set $\mathbb{M}$. The matrix representation of the identity operator is

$${}^m S^m{}_{ij} = S_{ij} = (u_i, v_j) \qquad i, j = 1, 2, \ldots, M \tag{12}$$

which reduces to the $M \times M$ unit matrix for an orthonormal set.

We note, for completeness, that an arbitrary vector, f, in the linear span of $\mathbb{M}$, can be expanded in the form

$$f = \sum_{i=1}^{M} u_i(u_i', f) \tag{13}$$

where u_i' denotes a reciprocal basis vector, which is defined by the relation

$$(u_i', u_j) = \delta_{ij} = \begin{cases} 1 & \text{if } i = j \\ 0 & \text{if } i \neq j \end{cases} \qquad i, j = 1, 2, \ldots, M \tag{14}$$

Thus, it follows that

$$u_i = \sum_{j=1}^{M} S_{ij} u_j' \qquad i = 1, 2, \ldots, M \tag{15}$$

and

$$u'_j = \sum_{k=1}^{M} S'_{jk} u_k \qquad j = 1, 2, \dots, M \qquad (16)$$

where

$$S'_{jk} = (u'_j, u'_k) \qquad (17)$$

so that

$$\sum_{j=1}^{M} S_{ij} S'_{jk} = \delta_{ik} \qquad (18)$$

The M reciprocal basis vectors also span $\mathbb{M}$ and become identical with the original basis set when it is orthonormal.

B. Independent-electron Models

Most tractable schemes employed in the study of atomic and molecular electronic structure use an independent-electron model as an initial approximation.[16] The wavefunction for an N-electron system is then written as an antisymmetrized product of one-electron functions, orbitals or spin–orbitals. These one-electron functions are usually taken to be the eigenfunctions of some model Hamiltonian, h, say,

$$h\phi_i = \varepsilon_i \phi_i \qquad (19)$$

Once the algebraic approximation has been invoked, the operator h is replaced by its matrix representation and the one-electron function ϕ_i is approximated by a linear combination of basis functions

$$\phi_i = \mathbf{u} \mathbf{c}_i \qquad (20)$$

The operator eigenvalue problem (19) is thus replaced by the matrix eigenvalue problem

$$\mathbf{h} \mathbf{c}_i = \mathbf{s} \mathbf{c}_i \mathbf{e} \qquad (21)$$

where the metric matrix $\mathbf{s}$ has been introduced to allow for the fact that the basis functions may be non-orthogonal,

$$^m h^m_{\ jk} = (u_j, h u_k) \qquad j, k = 1, 2, \dots, M \qquad (22)$$

$$^m S^m_{\ jk} = (u_j, u_k) \qquad j, k = 1, 2, \dots, M \qquad (23)$$

and $\mathbf{e}$ is a diagonal $M \times M$ matrix.

The simplest independent-electron model results from a complete neglect of electron–electron interactions. This is the bare-nucleus model for which the Hamiltonian h is merely a sum of kinetic energy and nucleus–electron attraction terms. The bare-nucleus model has a number of unique features which make its use in atomic and molecular studies attractive (see, for example, Refs. 7 and 17–22). The most widely used independent-electron model is the Hartree–Fock model. In this model, as is well known, the averaged electron–electron interactions are taken into account. The eigenvalue equation (19) is now an integro-differential equation, the operator h being the sum of the bare-nucleus operator and certain integral operators which depend on the functions ϕ_i. The corresponding matrix equation (Eq. (21)) is a pseudo-matrix eigenvalue problem, the matrix $\mathbf{h}$ depending on the expansion coefficient vectors $\mathbf{c}_i$.

We shall not discuss independent-electron models further here; the reader is referred elsewhere for more details.[7,23,24]

C. Electron Correlation

Just as the use of a finite basis set in independent-electron models restricts the domain of the relevant one-electron operator, h, so the algebraic approximation results in the restriction of the domain of the total Hamiltonian to a finite-dimensional subspace of the Hilbert space. In most applications of quantum mechanics to atoms and molecules which go beyond the independent-electron models, the N-electron wavefunction is expressed in terms of the Nth-rank direct product space $\mathbb{M}^N$ generated by a finite-dimensional single-particle space $\mathbb{M}^1$, that is

$$\mathbb{M}^N = \mathbb{M}^1 \otimes \mathbb{M}^1 \otimes \cdots \otimes \mathbb{M}^1 \qquad (24)$$

The algebraic approximation may be implemented by defining a suitable orthonormal basis set of $M \, (> N)$ one-electron spin–orbitals and constructing all unique N-electron determinants using the M one-electron functions. The number of unique determinants which can be formed in this way is

$$\eta = \binom{M}{N} \qquad (25)$$

and η is the dimension of the subspace spanned by the set of determinants. In the algebraic approximation, the domain of the total Hamiltonian is restricted to this η-dimensional subspace.

If account is taken of the fact that the wavefunction should be an eigenfunction of the operators associated with the square of the total spin and the component of spin in the z direction, the dimension of the subspace which should be considered is reduced from that given in Eq. (25) to

$$W = W(N, S, M) = \frac{2S+1}{M+1} \binom{M+1}{\frac{1}{2}N - S} \binom{M+1}{M - \frac{1}{2}N - S} \qquad (26)$$

which is the well known Weyl dimension formula,[25] in which N is the number of electrons, S is the total spin and M is the number of basis functions employed. In Table I the dependence of Weyl's number on the number of basis functions is illustrated by considering a five-electron system with a total spin of $\frac{1}{2}$. It can be seen to increase rapidly with the number of basis functions.

TABLE I

The dependence of Weyl's number on the number of basis functions, M. A five-electron system with a total spin of $\frac{1}{2}$ is considered as an example.

Number of basis functions, M	Weyl's number, W
5	1 200
6	4 200
7	11 760
8	28 224
9	60 480
10	118 800
11	217 800
12	377 520
13	624 624
14	993 720
16	2 284 800
18	4 744 224
20	9 097 200
30	109 498 200
40	631 924 800
60	7 389 466 800
80	42 059 347 200
100	161 683 830 000

III. BASIS FUNCTIONS

There is a considerable degree of freedom in the choice of basis functions used in atomic and molecular electronic structure calculations. Any set of functions which form a complete set can be employed. In practice, the choice of functional form is governed by two major factors: the rate of convergence of the orbitals when parametrized in terms of a particular basis set, and the ease with which the integrals over the basis functions can be evaluated. These two factors are, of course, not unrelated in that, if a particular type of basis function provides a rapidly convergent representation of an orbital, then the number of integrals required is smaller than it would be if the basis functions formed a poorly convergent expansion. In this section, we discuss the most commonly used types of basis function together with some of the less conventional choices which have been shown to yield high accuracy.

A. Exponential-type Functions[3,6,7,9]

Exponential-type basis functions are widely used in calculations for atomic systems, diatomic systems and linear polyatomic systems. They are most

TABLE II
Spherical harmonics, $Y_{lm}(\theta, \phi)$.

l	m	$Y_{lm}(\theta, \phi)$
0	0	$+\frac{1}{2}\pi^{-1/2}$
1	-1	$+\frac{1}{2}\left(\frac{3}{2\pi}\right)^{1/2} \sin\theta \exp(-i\phi)$
1	0	$+\frac{1}{2}\left(\frac{3}{\pi}\right)^{1/2} \cos\theta$
1	$+1$	$-\frac{1}{2}\left(\frac{3}{2\pi}\right)^{1/2} \sin\theta \exp(+i\phi)$
2	-2	$+\frac{1}{4}\left(\frac{15}{2\pi}\right)^{1/2} \sin^2\theta \exp(-2i\phi)$
2	-1	$+\frac{1}{2}\left(\frac{15}{2\pi}\right)^{1/2} \sin\theta\cos\theta \exp(-i\phi)$
2	0	$+\frac{1}{4}\left(\frac{5}{\pi}\right)^{1/2} (3\cos^2\theta - 1)$
2	$+1$	$-\frac{1}{2}\left(\frac{15}{2\pi}\right)^{1/2} \sin\theta\cos\theta \exp(+i\phi)$
2	$+2$	$+\frac{1}{4}\left(\frac{15}{2\pi}\right)^{1/2} \sin^2\theta \exp(+2i\phi)$
3	-3	$+\frac{1}{8}\left(\frac{35}{\pi}\right)^{1/2} \sin^3\theta \exp(-3i\phi)$
3	-2	$+\frac{1}{4}\left(\frac{105}{2\pi}\right)^{1/2} \sin^2\theta\cos\theta \exp(-2i\phi)$
3	-1	$+\frac{1}{8}\left(\frac{21}{\pi}\right)^{1/2} (5\cos^2\theta - 1)\sin\theta \exp(-i\phi)$
3	0	$+\frac{1}{4}\left(\frac{7}{\pi}\right)^{1/2} (5\cos^2\theta - 1)\cos\theta$
3	1	$-\frac{1}{8}\left(\frac{21}{\pi}\right)^{1/2} (5\cos^2\theta - 1)\sin\theta \exp(+i\phi)$
3	2	$+\frac{1}{4}\left(\frac{105}{2\pi}\right)^{1/2} \sin^2\theta\cos\theta \exp(+2i\phi)$
3	3	$-\frac{1}{8}\left(\frac{35}{\pi}\right)^{1/2} \sin^3\theta \exp(+3i\phi)$

frequently taken to have the form

$$\chi_{nlm}(r, \theta, \phi; \rho) = \frac{(2\zeta)^{n+1/2}}{(2n)!} r^{n-1} \exp(-\zeta r) Y_{lm}(\theta, \phi) \tag{27}$$

where ρ is the orbital exponent or screening constant and Y_{lm} is a spherical harmonic, which is defined in Table II.

Exponential-type basis functions usually provide a rapidly convergent expansion for atomic and molecular orbitals. They afford an accurate representation of the wavefunction in the region of space close to the nucleus and in the long-range region. Unfortunately, multi-centre integrals over exponential-type basis functions are difficult to compute. Not only is the evaluation of integrals, and two-electron integrals in particular, time-consuming but also the accuracy to which they are usually computed for non-linear polyatomic molecules is not as high as it is when, for example, Gaussian-type basis functions are used. This can be particularly important when computing small interaction energies such as van der Waals interactions.

B. Gaussian-type Functions[1-9]

Gaussian-type functions are the most widely used basis functions in molecular electronic structure calculations. Two forms of Gaussian-type functions are in common usage. Cartesian Gaussian-type functions have the form

$$\chi_{pqr}(x, y, z; \zeta) = (\pi/2\zeta)^{3/2} \frac{(2p-1)!!(2q-1)!!(2r-1)!!}{2^2(p+q+r)\zeta(p+q+r)}$$

$$\times x^p y^q z^r \exp(-\zeta r^2) \tag{28}$$

Use of these Cartesian Gaussian-type functions leads, for example, to six components of d symmetry instead of the true five components. This can be shown to be equivalent to the addition of a 3s function to the basis set and can lead to numerical problems associated with near-linear dependence if the s basis set is sufficiently large. The use of spherical Gaussian-type functions, which are often defined as

$$\chi_{nlm}(r, \theta, \phi; \zeta) = \left[\frac{2^{2n}(n-1)!}{(2n-1)!} \left(\frac{(2\zeta)^{2n+1}}{\pi} \right)^{1/2} \right]^{1/2}$$

$$\times r^{n-1} \exp(-\zeta r^2) Y_{lm}(\theta, \phi) \tag{29}$$

where the Y_{lm} are the spherical harmonics defined in Table II, avoids this problem.

The major strength of the Gaussian-type functions lies in the fact that multi-centre integrals, especially those involving the operator r_{12}^{-1}, can be easily evaluated since the product of two Gaussian-type functions on centres A and B, say, is also a Gaussian-type function, with appropriate normalization constant, with centre P on the line AB. Explicitly, for a function of s symmetry, it can be shown that

$$\exp(-\zeta_A r_A^2)\exp(-\zeta_B r_B^2) = K \exp(-\zeta_P r_P^2) \tag{30}$$

in which the constant K is given by

$$K = \exp\{[-\zeta_A\zeta_B/(\zeta_A + \zeta_B)]R_{AB}^2\} \tag{31}$$

and

$$R_{AB}^2 = (x_A - x_B)^2 + (y_A - y_B)^2 + (z_A - z_B)^2 \tag{32}$$

$$\zeta_P = \zeta_A + \zeta_B \tag{33}$$

$$\mu_P = (\zeta_A\mu_A + \zeta_B\mu_B)/(\zeta_A + \zeta_B) \qquad \mu = x, y, z \tag{34}$$

The integrals over Gaussian-type functions can, therefore, be evaluated efficiently and accurately, a property which makes them by far the most popular choice of basis functions for studies of polyatomic molecules.

The major disadvantage of Gaussian-type functions as basis functions in atomic and molecular electronic structure calculations is that they have an inappropriate form both in regions of space close to the nucleus upon which they are centred (they have no cusp) and in the long-range region. More Gaussian-type functions than, say, exponential-type functions are required to approximate a given molecular orbital to a certain accuracy, particularly in the case of core orbitals. However, although Gaussian-type functions are not well suited for the description of eigenfunctions of Hamiltonians involving Coulomb potentials and, therefore, relatively large numbers of basis functions are required, this disadvantage is usually more than offset by the ease with which the integrals can be computed.

C. Contracted Gaussian-type Functions[1-9]

The efficiency of sets of Gaussian-type functions can be increased, particularly in the parametrization of core molecular orbitals, by employing contracted Gaussian-type functions instead of the primitive Gaussian-type functions discussed in the preceding subsection. Contracted Gaussian-type functions are defined as

$$\phi = N\sum_i \chi_i c_i \tag{35}$$

where $\{\chi_i\}$ denotes a set of primitive functions of the same symmetry type and centred on the same nucleus, the c_i are a set of fixed contraction coefficients and N is a normalization constant.

The contraction coefficients may be determined in a number of ways. They can be chosen to provide a best, in the least-squares sense, representation of an exponential-type function. They may be contracted on the basis of atomic matrix Hartree–Fock calculations. Usually, it is found that the innermost Gaussian-type functions can be contracted with little loss in accuracy; however, the outer Gaussian-type functions should not be contracted, the additional flexibility obtained by using primitive functions more than offsetting the additional computation involved.

D. Elliptical-type Functions

Elliptical-type basis functions have been shown to be useful in calculations for diatomic molecules by a number of authors.[26-36] They are most often defined in terms of the elliptical coordinates λ, μ and ϕ as follows

$$\chi_{pq\nu}(\lambda, \mu, \phi; \alpha, \beta) = (2\pi R^3)^{-1/2} \exp(-\alpha\lambda)\exp(-\beta\mu)\lambda^p \mu^q$$
$$\times [(\lambda^2 - 1)(1 - \mu^2)]^{|\nu|/2} \exp(i\nu\phi) \tag{36}$$

where α and β are screening constants, ν determines the orbital symmetry and R is the nuclear separation. Some typical elliptical-type basis functions are

TABLE III

Comparison of self-consistent field calculations using elliptical-type basis functions with fully numerical studies of diatomic molecules.[a]

System	Property[b]	Value	Reference
H_2	E_{total}	$-1.133\,629\,57$	Wells and Wilson[36]
		$-1.133\,629\,57$	Laaksonen et al.[38]
	$\varepsilon_{orbital}$	$-0.594\,658\,57$	Wells and Wilson[36]
		$-0.594\,658\,57$	Laaksonen et al.[38]
	Q_2	$+0.243\,288\,9$	Wells and Wilson[36]
		$+0.243\,288\,8$	Laaksonen et al.[38]
	$\langle r^2 \rangle$	$+2.573\,929\,9$	Wells and Wilson[36]
		$+2.573\,930$	Laaksonen et al.[38]
HeH$^+$	E_{total}	$-2.933\,103\,22$	Wells and Wilson[36]
		$-2.933\,103\,25$	Laaksonen et al.[38]
	$\varepsilon_{orbital}$	$-1.637\,450\,63$	Wells and Wilson[36]
		$-1.627\,450\,62$	Laaksonen et al.[38]
	$\langle z \rangle$	$-0.494\,460\,15$	Wells and Wilson[36]
		$-0.494\,459\,96$	Laaksonen et al.[38]
	Q_2	$+0.373\,728\,4$	Wells and Wilson[36]
		$+0.373\,726\,9$	Laaksonen et al.[38]
	$\langle r^2 \rangle$	$+1.340\,833\,4$	Wells and Wilson[36]
		$+1.340\,832\,3$	Laaksonen et al.[38]

Taken from Refs. 36 and 38.
[a]All results are given in atomic units.
[b]$\langle z \rangle$ and Q_2 are defined with respect to the molecular midpoint.

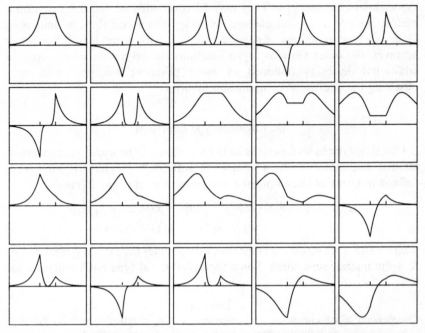

Fig. 1. Some examples of elliptical basis functions. In each sketch the positions of the nuclei are marked on the internuclear axis. From left to right, top to bottom, the functions shown are: $(\alpha, \beta, p, q, v) = (1,0,0,0,0), (1,0,0,1,0), (1,0,0,2,0), (1,0,0,3,0), (1,0,0,4,0), (1,0,0,5,0), (1,0,0,6,0), (1,0,1,0,0), (1,0,2,0,0), (1,0,3,0,0), (1,0.6,0,0,0), (1,0.6,1,0,0), (1,0.6,2,0,0), (1,0.6,3,0,0), (1,0.6,0,1,0), (1,0.6,0,2,0), (1,0.6,0,3,0), (1,0.6,0,4,0), (1,0.6,1,1,0), (1,0.6,2,1,0).$

displayed in Fig. 1 for $v = 0$ with various values of the screening constants and the integers p and q.

Elliptical-type basis functions can provide a very accurate representation of orbitals in diatomic molecules. In Table III, we compare the results of calculations employing basis sets of elliptical-type functions[36] with fully numerical studies.[37,38] It can be seen that the two approaches yield results of comparable accuracy. It should be emphasized that the basis set approach, in contrast to the fully numerical technique, affords a compact representation of the wavefunction.

E. Piecewise Polynomial Functions

Silverstone and his collaborators[39-41] have proposed the use of piecewise polynomial basis functions and have demonstrated their use for atomic systems. By using such functions, they aim to overcome the computational linear dependence problems which may be associated with large basis sets of exponential-type functions or Gaussian-type functions.

Piecewise polynomial basis functions are defined as

$$\chi(r, \theta, \phi) = R(r)r^l Y_{lm}(\theta, \phi) \tag{37}$$

where $Y_{lm}(\theta, \phi)$ is the usual spherical harmonic and $R(r)$ is a piecewise polynomial. If we subdivide the r axis by mesh points

$$r_0 = 0 < r_1 < r_2 < \cdots < r_N < \infty \tag{38}$$

$R(r)$ is a polynomial on each subinterval (r_i, r_{i+1}) but the polynomials of different subintervals are not identical. Different classes of piecewise polynomial are determined by (i) the degree of the polynomial, (ii) the number of derivatives that remain continuous at each mesh point, and (iii) the choice of the mesh points.

The integrals over piecewise polynomial basis functions are elementary. However, it is found that the number of terms in the final integral formulae is much larger for piecewise polynomial functions than for exponential-type functions, for example.

IV. BASIS SETS

Having discussed in Section III the various types of basis functions which can be employed in atomic and molecular calculations, we turn now to the problem of constructing a basis set.

A. One-centre Basis Sets

In principle, any molecular wavefunction can be expanded in terms of a complete set of functions centred at any convenient point in space. This approach, which is referred to as the one-centre expansion method, the central-field approximation or the united-atom method, goes back to the earliest days of atomic and molecular physics.[7,42,43] Many more terms have to be included in a one-centre basis set than in the more usual multi-centre basis set in order to obtain an adequate representation of the wavefunction close to off-centre nuclei. However, all of the integrals over the basis functions which arise in the one-centre method can be evaluated analytically. They can be calculated accurately and efficiently and it may be more convenient to recompute the integrals involving the two-electron operator r_{12}^{-1} each time they are required and thus avoid the necessity of handling large files of integrals.

The basis functions in a one-centre calculation belong to representations of the spherical symmetry group. There is, therefore, a higher degree of symmetry in the integrals over the basis functions than in the integrals over the molecular orbitals. This can be exploited in self-consistent field calculations and in orbital transformations.

For a one-centre basis set, in contrast to multi-centre basis sets, problems arising from overcompleteness can usually be controlled if not avoided. The one-centre approach provides control over the convergence of a calculation with respect to the size of the basis set and control over computational linear dependence. Furthermore, because of the ease with which integrals involving one-centre functions can be handled, the method can be used to explore the use of alternative types of basis function. The one-centre method is ideally suited to the calculation of energy derivatives with respect to the nuclear coordinates.

The one-centre method does, however, usually provide a poor representation of off-centre nuclei. Most applications have been to hydrides with a single heavy atom which is used as an expansion centre. However, the one-centre method holds some promise in the treatment of extended systems and we shall return to this aspect in Section IX.B.

B. Multi-centre Basis Sets

The vast majority of contemporary molecular calculations overcome the convergence problems of the one-centre expansion method, which are associated with the description of off-centre nuclei, by employing basis functions located on a number of centres usually coinciding with the nuclei in the molecule. Rather accurate calculations for molecules containing more than one non-hydrogenic nucleus can be performed by using a multi-centre basis set. The use of such basis sets does, however, give rise to new problems.

Firstly, the evaluation of integrals, and in particular two-electron integrals involving the interelectronic distance, over basis functions on different centres is usually more complicated than in the one-centre case. For example, integrals involving exponential-type basis functions and three or four different centres are particularly difficult to evaluate efficiently and accurately.

Secondly, the use of a multi-centre basis set often leads to overestimates of interaction energies between atoms and/or molecules. This is mainly attributable to the basis set superposition effects and will be discussed further in Section VII.

C. Minimum Basis Sets[1,3,8,9]

In matrix Hartree–Fock calculations for atomic systems, there must be at least one basis function for each occupied atomic orbital. Such a basis set is termed a minimum basis set. A minimum basis set for the He atom consists of a single 1s function, whilst a minimum basis set for C, N and O consists of five functions: 1s, 2s, $2p_x$, $2p_y$ and $2p_z$.

Minimum basis sets were widely used in quantum-chemical calculations in the late 1950s and early 1960s. The accuracy which can be achieved in calculations of total energies using such basis sets is illustrated in Table IV

TABLE IV

Total matrix Hartree–Fock energies for some closed-shell atoms obtained by using basis sets of exponential-type functions.[a]

Atom	Minimum basis set	Double-zeta basis set	Hartree–Fock limit
He(^{1}S)	−2.847 656 3	−2.861 672 6	−2.861 679 9
Be(^{1}S)	−14.556 740	−14.572 369	−14.573 021
Ne(^{1}S)	−127.812 18	−128.535 11	−128.547 05
Mg(^{1}S)	−198.857 79	−199.607 01	−199.614 61
Ar(^{1}S)	−525.765 25	−526.815 11	−526.817 39
Ca(^{1}S)	−675.633 90	−676.755 94	−676.758 02
Zn(^{1}S)	−1771.150 9	−1777.669 9	−1777.847 7
Kr(^{1}S)	−2744.519 7	−2751.961 3	−2752.054 6

[a]Based on the work of Roetti and Clementi.[44] All energies are in hartree.

where results for a number of closed-shell systems are presented. Today, minimum basis sets are employed in qualitative discussions of the form of the molecular orbitals, in calculations on large molecules (see Section IX for further discussion) and sometimes as a method for generating an initial guess for self-consistent-field calculations using extended basis sets.

D. Double-zeta Basis Sets[1,3,8,9]

'Double-zeta' basis sets were introduced by Roetti and Clementi[44] to provide greater flexibility in the orbital expansion and to avoid the need to reoptimize the orbital exponents when the basis set is used in a molecular calculation. Double-zeta basis sets contain two functions for every function in a minimum basis set. The accuracy which can be achieved in calculations of total energies using such basis sets is illustrated in Table IV where calculations using double-zeta basis sets are compared with those using minimum basis sets and the Hartree–Fock limit.

E. Polarization Functions[1,3,8,9]

In calculations on molecules within the matrix Hartree–Fock approximation, it is found to be important to add polarization functions to double-zeta basis sets. Such basis functions do not improve the energies of the isolated component atomic species but contribute significantly to calculated bond energies and to the accuracy of calculated equilibrium bond angles. 'Double-zeta plus polarization' basis sets (usually designated DZP or DZ + P) became widespread in quantum chemistry in the 1970s. In such a basis set the hydrogen atom is described by two s functions and one set of p functions; the carbon, nitrogen and oxygen atoms are described by four s functions, two sets of p functions and one set of d functions.

F. Even-tempered Basis Sets

Large basis sets can be efficiently generated by utilizing the concept of an even-tempered basis set. Such a basis set consists of pure exponential or pure Gaussian functions multiplied by a solid spherical harmonic, that is a spherical harmonic multiplied by r^l. Thus an even-tempered basis set consists of 1s, 2p, 3d, 4f,... functions. A set of even-tempered basis functions is thus defined by

$$\chi_{klm}(r, \theta, \phi) = N \exp(-\zeta_k r^p) r^l Y_{lm}(\theta, \phi) \tag{39}$$

where $p = 1$ (2) for exponential (Gaussian) functions. Even-tempered atomic orbitals for a given $Y_{lm}(\theta, \phi)$ do not differ in the power of r and thus in linear combinations of primitive functions the solid harmonic can be factored. The orbital exponents, ζ_k, are taken to form a geometric sequence

$$\zeta_k = \alpha \beta^k \qquad k = 1, 2, \dots, N \tag{40}$$

The use of such a series is based on the observation that independent optimization of the exponents with respect to the energy in self-consistent field calculations yields an almost linear plot of $\ln(\zeta_k)$ against k. The use of orbital exponents which form a geometric series was originally advocated by Reeves[45] and the idea was revived and extensively employed by Ruedenberg and his collaborators.[46]

A number of advantages accrue to the use of even-tempered basis sets:

1. They have only two parameters, α and β, which have to be determined for each group of atomic functions belonging to the same symmetry species, as opposed to one optimizable orbital exponent per basis function. The determination of orbital exponents by energy minimization is a non-linear optimization problem and there is little possibility of performing a full optimization for polyatomic molecules if all orbital exponents are independent.
2. The further restriction of using the same exponents for all values of l, so that there are only two non-linear parameters, α and β, per atom, does not produce a very large difference in the calculated energies. Clementi and his collaborators[55,56] have recently reinvestigated this idea.
3. The proper mixing of basis functions, in terms of principal quantum number, is superfluous since no mixing is employed.
4. It is evident that an even-tempered basis set approaches a complete set in the limit $\alpha \to 0$, $\beta \to 1$, $\beta^{k_{max}} \to \infty$ as $k_{max} \to \infty$.
5. An even-tempered basis set cannot become linear-dependent if $\beta > 1$.
6. The parameter β provides control over practical linear dependence in the

TABLE V
Structure of the overlap matrix for an even-tempered basis set of
(a) exponential-type 1s basis functions and (b) Gaussian-type 1s
basis functions.

(a) $\rho_k = \alpha\beta^k (k = 1, 2, \ldots, 5)$, $\alpha = 0.5$, $\beta = 1.5$

1.000	0.941	0.787	0.592	0.409
0.941	1.000	0.941	0.787	0.592
0.787	0.941	1.000	0.941	0.787
0.592	0.787	0.941	1.000	0.941
0.409	0.592	0.787	0.941	1.000

(b) $\rho_k = \alpha\beta^k (k = 1, 2, \ldots, 5)$, $\alpha = 0.25$, $\beta = 2.25$

1.000	0.887	0.639	0.402	0.235
0.887	1.000	0.887	0.639	0.402
0.639	0.887	1.000	0.887	0.639
0.402	0.639	0.887	1.000	0.887
0.235	0.402	0.639	0.887	1.000

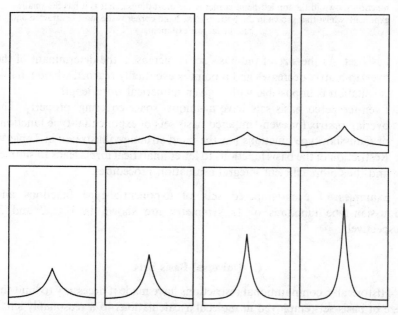

Fig. 2. Orbital amplitude plots for a set of even-tempered 1s exponential-type functions.
The function shown in the top left-hand corner is the most diffuse, that is it has the smallest
exponent, whilst that shown in the bottom right-hand corner is the most contracted and
has the largest exponent.

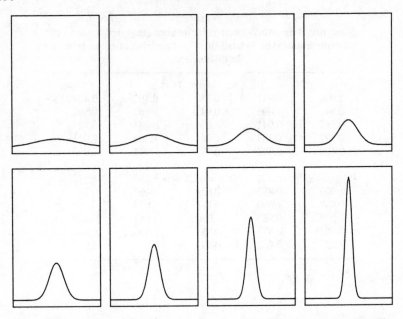

Fig. 3. Orbital amplitude plots for a set of even-tempered 1s Gaussian-type functions. The function shown in the top left-hand corner is the most diffuse, that is it has the smallest exponent, whilst that shown in the bottom right-hand corner is the most contracted and has the largest exponent.

basis set. As the size of the basis set is increased, the determinant of the overlap matrix decreases and a point is eventually reached where reliable calculation is impossible with a given numerical word length.

7. Even-tempered basis sets have a unique 'space-covering' property. The overlap matrix for even-tempered basis sets of exponential-type functions or Gaussian-type functions has the band structure illustrated in Table V.

8. Restriction of the basis functions to fewer analytical forms leads to simpler and thus more efficient integral evaluation procedures.

Examples of even-tempered sets of exponential-type functions and Gaussian-type functions of 1s symmetry are shown in Figs. 2 and 3, respectively.

G. Universal Basis Sets

Historically, computational restrictions have made it necessary to limit the size of basis sets employed in molecular calculations to a reasonably small number of functions in order to keep the computations tractable. However, with the advent over the past 10 years or so of powerful new computers with vector processing and parallel processing capabilities and with projected rates

TABLE VI

Matrix Hartree–Fock ground-state energies for first-row atoms obtained by using optimized basis sets and a universal even-tempered basis set of exponential-type functions.[a]

Atom	Basis set[b]						Universal
	B	C	N	O	F	Ne	
B(^{2}P)	−24.529 06	−24.528 52	−24.525 06	−24.503 21	−24.476 27	−24.428 09	−24.528 92
δ[c]		0.000 54	0.004 00	0.025 85	0.052 79	0.100 97	0.000 14
C(^{3}P)	−37.686 72	−37.688 61	−37.687 95	−37.684 62	−37.677 93	−37.659 60	−37.688 54
δ	0.001 89		0.000 66	0.003 99	0.010 68	0.029 01	0.000 07
N(^{4}S)	−54.397 43	−54.398 90	−54.400 92	−54.400 25	−54.399 03	−54.394 97	−54.400 84
δ	0.003 49	0.002 00		0.000 67	0.001 89	0.005 95	0.000 08
O(^{3}P)	−74.798 98	−74.801 09	−74.807 15	−74.809 37	−74.808 67	−74.807 48	−74.809 33
δ	0.010 39	0.008 28	0.002 22		0.000 70	0.001 89	0.000 04
F(^{2}P)	−99.316 09	−99.391 14	−99.399 38	−99.407 13	−99.409 30	−99.408 61	−99.409 15
δ	0.093 21	0.018 16	0.009 92	0.002 17		0.000 69	0.000 15
Ne(^{1}S)	−128.120 75	−128.491 79	−128.527 52	−128.536 24	−128.544 97	−128.547 05	−128.546 81
δ	0.426 30	0.055 26	0.019 53	0.010 81	0.002 08		0.000 24

[a] Taken from the work of Silver, Wilson and Nieuwpoort.[47] All energies are in hartree.

[b] The optimized basis sets are taken from the work of Clementi and Roetti.[44]

[c] δ is the difference between the energy obtained by using the universal basis set and that given by Clementi and Roetti.[44]

TABLE VII

Diagrammatic many-body perturbation theory calculations of the correlation energy of various diatomic molecules in their ground states using universal basis sets of even-tempered exponential-type functions. Comparison with other approaches.*

Method	E_{total}	Percentage of e_{exp}
LiH (empirical correlation energy, $e_{exp} = -0.083$)		
Bender and Davidson[a] CI	-8.0606	88.3
Meyer and Rosmus[b] PNO-CI	-8.0647	93.3
CEPA	-8.0660	94.8
Wilson and Silver[c] DPT, E[2/1]	-8.0642	92.8
Ẽ[2/1]	-8.0652	93.9
Diagrammatic perturbation E[2/1]	-8.0653	94.0
theory/Universal basis set[d] Ẽ[2/1]	-8.0661	94.9
Li$_2$ (empirical correlation energy, $e_{exp} = -0.126$)		
Werner and Reinsch[e] MCSCF-CI	-14.9649	74.3
Diagrammatic perturbation E[2/1]	-14.9842	89.6
theory/Universal basis set[f] Ẽ[2/1]	-14.9845	89.9
FH (empirical correlation energy, $e_{exp} = -0.381$)		
Bender and Davidson[a] CI	-100.3564	75.1
Meyer and Rosmus[b] PNO-CI	-100.3274	67.5
CEPA	-100.3392	70.6
Wilson and Silver[c] E[2/1]	-100.3727	
Ẽ[2/1]	-100.3707	78.9
Diagrammatic perturbation E[2/1]	-100.3837	82.3
theory/Universal basis set[d] Ẽ[2/1]	-100.3770	80.6
N$_2$ (empirical correlation energy, $e_{exp} = -0.538$)		
Langhoff and Davidson[g] CI	-109.2832	58.4
Wilson and Silver[h] DPT, E[2/1]	-109.4180	79.5
Diagrammatic perturbation		
theory/Universal basis set[i] E[2/1]	-109.443	83.7
CO (empirical correlation energy, $e_{exp} = -0.525$)		
Siu and Davidson[j] CI	-113.1456	69.4
Bartlett et al.[k] DPT, E [2/1]	-113.1952	77.4
Diagrammatic perturbation		
theory/Universal basis set[i] E[2/1]	-113.2286	82.7
BF (empirical correlation energy, $e_{exp} = -0.531$)		
Bender and Davidson[a] MCSCF	-124.235	
Wilson et al.[l] DPT, E[2/1]	-124.5028	65.2
Diagrammatic perturbation		
theory/Universal basis set[i] E[2/1]	-124.5782	77.1

*All energies are in atomic units. The following abbreviations are used: CI, configuration interaction, PNO-CI, pair natural orbital–configuration interaction; CEPA, coupled electron-pair approximation; DPT, diagrammatic perturbation theory; MCSCF, multi-configuration self-consistent field; E[2/1] denotes the [2/1] Padé approximant to the perturbation series based on the Hartree–Fock model zero-order Hamiltonian and Ẽ[2/1] denotes the [2/1] Padé approximant to the shifted denominator expansion.

of computation of several gigaflops in the near future, the situation is changing radically. Furthermore, in order to achieve high accuracy in molecular studies, particularly in studies of electron correlation effects, moderately large basis sets are ultimately required. Since the flexibility of a basis set generally increases with its size, the need to optimize orbital exponents becomes less important. It is now well established that it is almost always more profitable to add extra functions to a given basis set than to optimize the orbital exponents exhaustively. These considerations have led to the concept of a universal basis set.[47-49] Such a basis set is moderately large and thus has a considerable degree of flexibility. It is, therefore, transferable from system to system with little loss of accuracy even though the orbital exponents are not changed as the nuclear charges vary.

Several advantages accrue to the use of a universal basis set:

1. Molecular electronic structure calculations begin with the evaluation of one- and two-electron integrals over the basis functions. For a given set of nuclear positions, the integrals for a universal basis set can be evaluated one and then used in all subsequent studies without regard to the identity of the constituent atoms. This transferability extends to all integrals arising in the evaluation of the energy and molecular properties.
2. A universal basis set is, almost by definition, capable of providing a rather uniform description of a series of atoms and molecules. This uniformity is illustrated in Table VI where matrix Hartree–Fock energies for some first-row atoms obtained by using a universal basis set of even-tempered exponential-type functions are given.
3. Since universal basis sets are not optimized with respect to the total energy or any other property, it is expected that they will afford a uniform description of a range of properties. Modifications of a universal basis set may, of course, be necessary in order to evaluate properties which are particularly sensitive to the quality of the basis set in one region of space.
4. In order to be flexible, a universal basis set is necessarily moderately large and, therefore, it is capable of yielding high accuracy. The high accuracy

[a] Bender, C.F., and Davidson, E.R., *J. Phys. Chem.* **70**, 2675 (1966); *Phys. Rev.*, **183**, 23 (1969.
[b] Meyer, W., and Rosmus, P., *J. Chem. Phys.*, **63**, 2356 (1975).
[c] Wilson, S., and Silver, D.M., *J. Chem. Phys.*, **66**, 5400 (1977).
[d] Wilson, S., and Silver, D.M., *J. Chem. Phys.*, **77**, 3674 (1982).
[e] Werner, H.-J., and Reinsch, E.-A., *Proc. Fifth Seminar on Computational Methods in Quantum Chemistry* (eds. P.Th. van Duijnen and W. C. Nieuwpoort), Groningen, 1981.
[f] Wilson, S., *Theoretical Chemistry*, Vol. 4, p. 1, Specialist Periodical Reports, The Royal Society of Chemistry, London, 1981.
[g] Langhoff, S., and Davidson, E.R., *Int. J. Quantum Chem.*, **8**, 61 (1974).
[h] Wilson, S., and Silver, D.M., *J. Chem. Phys.*, **67**, 1689 (1977).
[i] Wilson, S., and Silver, D.M., *J. Chem. Phys.*, **72**, 2159 (1980).
[j] Siu, A.K.Q., and Davidson, E.R., *Int. J. Quantum Chem.* **4**, 223 (1970).
[k] Bartlett, R.J., Wilson, S., and Silver, D.M., *Int. J. Quantum Chem.* **13**, 737 (1977).
[l] Wilson, S., Silver, D.M., and Bartlett, R.J., *Mol. Phys.*, **33**, 1177 (1977).

which can be attained by using universal basis sets of exponential-type functions in electron correlation energy calculations is illustrated in Table VII.

5. A universal basis set can have a higher degree of symmetry than the particular molecule under investigation.

A universal basis set need not necessarily be an even-tempered set; however, the concept of a universal even-tempered basis set has been shown to be useful and enables large bases to be generated easily and efficiently. The parameters defining the even-tempered set (see Eq. (40)) should be chosen according to the following guidelines: (i) α should be small enough to ensure a wide range of orbital exponents; (ii) β should be large enough to avoid near-linear dependence in the basis set; (iii) N should be large enough to generate a 'near-complete' basis set. The following values of α, β and N have been shown to be useful in studies of first-row atoms and molecules containing them using basis sets of exponential-type functions.[50-54]

$$
\begin{array}{llll}
\text{1s} & \alpha = 0.5, & \beta = 1.55, & N = 9 \\
\text{2p} & \alpha = 1.0, & \beta = 1.60, & N = 6 \\
\text{3d} & \alpha = 1.5, & \beta = 1.65, & N = 3
\end{array}
\tag{41}
$$

This universal even-tempered basis set was employed in all of the correlation energy calculations summarized in Table VII. Clementi and his collaborators[55,56] have investigated the use of this type of basis set within the Hartree–Fock approximation. They impose the further restriction of using the same exponents for all values of l and term such basis sets 'geometric basis sets'. Huzinaga[8] comments that 'these remarkable universal geometric basis sets ... will certainly usher us to a new plateau of computational chemistry.'

For a given configuration of the nuclei, the integrals over universal basis sets of exponential-type functions or Gaussian-type functions can be evaluated once and stored for use in all subsequent calculations for that particular geometry without regard to the identity of the constituent atoms. For diatomic molecules, elliptical basis functions not only offer the possibility of high accuracy but also the transferability of integrals from calculations for one nuclear separation to another.[33] For elliptical basis functions as defined in Eq. (36), the overlap integral can be shown to be

$$
\begin{aligned}
\mathbb{S}_{jk} &= \langle \chi_{p_j q_j \nu_j}^{(j)} | \chi_{p_k q_k \nu_k}^{(k)} \rangle \\
&= \tfrac{1}{8}\delta(\nu_j, \nu_k)[A_{p_j + p_k + 2}^{\nu_j}(\alpha_j + \alpha_k)B_{q_j + q_k}^{\nu_j}(\beta_j + \beta_k) \\
&\quad - A_{p_j + p_k}^{\nu_j}(\alpha_j + \alpha_k)B_{q_j + q_k + 2}^{\nu_j}(\beta_j + \beta_k)]
\end{aligned}
\tag{42}
$$

where

$$
A_p^\nu(\alpha) = \int_1^\infty x^p \exp(-\alpha x)(x^2 - 1)^{|\nu|}\,dx
\tag{43}
$$

and

$$B_q^v(\beta) = \int_{-1}^{+1} x^q \exp(-\beta x)(1 - x^2)^{|v|}\, dx \qquad (44)$$

Thus the overlap integral is independent of the nuclear separation. The kinetic energy integrals can also be written in terms of the intermediates (43) and (44) and can be shown to satisfy the relation

$$T_{jk}(R) = R^{-2}\mathbb{T}_{jk} \qquad (45)$$

where $\mathbb{T}_{jk}$ is an integral which is independent of R, the nuclear separation. Similarly, the nuclear–electron attraction integrals satisfy the relation

$$V_{jk}(R) = R^{-1}\mathbb{V}_{jk} \qquad (46)$$

and the electron–electron repulsion integrals the relation

$$I_{ijkl}(R) = R^{-1}\mathbb{I}_{ijkl} \qquad (47)$$

where $\mathbb{V}_{jk}$ and $\mathbb{I}_{ijkl}$ are independent of R. Thus once the integrals over a

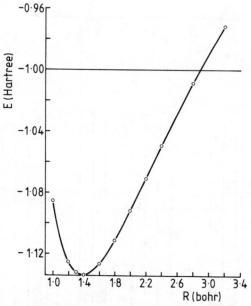

Fig. 4. Potential energy curve for the ground state of the hydrogen molecule obtained by Kolos and Roothaan (*Rev. Mod. Phys.*, **32**, 169) (1960) using the Hartree–Fock approximation together with some energy values obtained by performing matrix Hartree–Fock calculations with a universal basis set of elliptical functions.

universal basis set of elliptical functions have been evaluated, they can be stored and used for any diatomic molecule, provided that it is sufficiently large and flexible, and at any finite internuclear distance. Furthermore, the possibility of performing calculations on several points of a potential energy curve simultaneously is opened up. To illustrate the use of the same set of integrals over elliptical functions in calculations at different values of R, we display in Fig. 4 the potential energy curve for the ground state of the hydrogen molecule obtained by Kolos and Roothaan using the Hartree–Fock approximation together with some energy values obtained by performing matrix Hartree–Fock calculations with a universal basis set of elliptical functions.[34] More extensive studies of this type have been reported elsewhere.[35,36]

H. Systematic Sequences of Basis Sets

Atomic and molecular electronic structure calculations are often performed using a single basis set which is constructed in an *ad hoc* fashion using experience gained in previous studies of similar systems and little effort is made

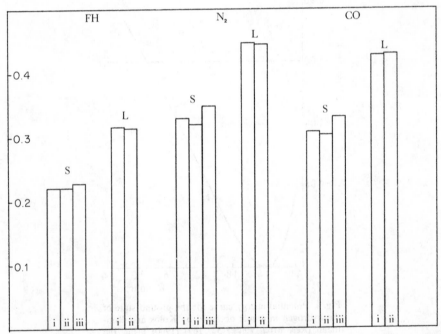

Fig. 5. Magnitude of the basis set truncation error in calculations of electron correlation energies for some closed-shell diatomic molecules. 'S' indicates the calculations performed using smaller basis sets, and 'L' designates calculations with larger basis sets. (i), (ii) and (iii) denote many-body perturbation theory calculations of the correlation energy through second, third and fourth order, respectively.

to examine the dependence of calculated properties on the basis set. Ruedenberg and his coworkers[57,58] and the present author[59,60] have reiterated the view first put forward by Schwarz[61,62] that the convergence of calculations with respect to the size of the basis set is a very important problem. Figure 5 illustrates the magnitude of the basis set truncation error in calculations of electron correlation energies for diatomic molecules.

Ruedenberg and his coworkers[57,58] devised schemes for systematically extending even-tempered basis sets. They noted that, in order for an even-tempered basis set to tend to a complete set as the number of functions, N, tends to infinity, α and β must be functions of N such that

$$\alpha \to 0 \qquad \beta \to 1 \qquad \beta^N \to \infty \qquad (48)$$

On the basis of atomic matrix Hartree–Fock calculations using even-tempered sets of Gaussian-type functions of various sizes in which α and β were optimized with respect to the energy, Ruedenberg and his coworkers proposed empirical functional forms for the dependence of α and β on N:

$$\ln \ln \beta = b \ln N + b' \qquad (49)$$

and

$$\ln \alpha = a \ln(\beta - 1) + a' \qquad (50)$$

with

$$-1 < b < 0 \qquad (51)$$

and

$$a > 0 \qquad (52)$$

Equations (49) and (50) are equivalent to the recursions

$$\alpha(M) = \left(\frac{\beta(M) - 1}{\beta(M-1) - 1} \right)^a \alpha(M-1) \qquad (53)$$

and

$$\ln \beta(M) = \left(\frac{N}{N-1} \right)^b \ln \beta(M-1) \qquad (54)$$

respectively.

By employing a sequence of basis sets which are constructed in a systematic fashion, reliable extrapolation procedures can be used to determine the basis set limit of the energy. Ruedenberg and his coworkers[57,58] showed that the Hartree extrapolation technique[63,64] led to an empirical upper bound to the basis set limit. The extrapolated energy value is given by

$$E_\infty[N_3] = (E[N_1]E[N_3] - E[N_2]^2)/(E[N_3] - 2E[N_2] + E[N_1]) \quad (55)$$

where N_1, N_2 and N_3 are three basis sets in the systematic sequence. Schmidt and Ruedenberg[57] demonstrated that an empirical lower bound to the basis

TABLE VIII

Matrix Hartree–Fock energies for neutral atoms obtained by using a universal systematic sequence of even-tempered basis sets of exponential-type functions.[a]

Basis set	B	C	N	O	F
3s/2p	−23.216 495	−36.137 816	−53.070 927	−73.477 881	−97.354 435
6s/4p	−24.516 657	−37.685 322	−54.399 953	−74.802 404	−99.400 352
9s/6p	−24.528 915	−37.688 542	−54.400 841	−74.809 335	−99.408 088
12s/8p	−24.529 060	−37.688 611	−54.400 933	−74.809 399	−99.409 345
15s/10p	−24.529 059	−37.688 618	−54.009 934	−74.809 400	−99.409 350

[a]Based on the work of Cooper and Wilson.[66] All energies are in hartree.

set limit is given by

$$\hat{E}_\infty[N] = E_\infty[N] - (E[N] - E_\infty[N]) \tag{56}$$

Combining (55) and (56), we obtain a 'best' estimate of the basis set limit

$$E_{av}[N] = \tfrac{1}{2}(E_\infty[N] + \hat{E}_\infty[N]) \tag{57}$$

and an estimate of its accuracy

$$D[N] = \tfrac{1}{2}(E_\infty[N] - \hat{E}_\infty[N]) \tag{58}$$

The use of a universal systematic sequence of even-tempered basis sets, that is a single sequence of basis sets which can be used for any atom irrespective of its charge or its environment, has been investigated by the present author and his coworkers[60,65] using both Gaussian-type basis functions and exponential-type basis functions. Some typical results are given in Table VIII where the convergence of the energies of the ground states of neutral first-row atoms with

TABLE IX

Comparison of self-consistent field calculations using exponential-type basis functions with fully numerical studies for atomic systems.[a]

Atom	'Basis set' calculation	'Numerical' calculation
B(^{2}P)	−24.529 059	−24.529 061
C(^{3}P)	−37.688 619	−37.688 619
N(^{4}S)	−54.400 934	−54.400 934
O(^{3}P)	−74.809 401	−74.809 398
F(^{2}P)	−99.409 350	−99.409 349

[a]Based on the work of Cooper and Wilson[66] and Fischer.[69] All energies are in hartree.

TABLE X

Comparison of self-consistent field calculations
using Gaussian-type basis functions with fully
numerical studies for atomic systems.[a]

Atom	'Basis set' calculation	'Numerical' calculation
$B(^2P)$	$-24.529\,061$	$-24.529\,061$
$C(^3P)$	$-37.688\,619$	$-37.688\,619$
$N(^4S)$	$-54.400\,935$	$-54.400\,934$
$O(^3P)$	$-74.809\,400$	$-74.809\,398$
$F(^2P)$	$-99.409\,353$	$-99.409\,349$

[a]Based on the work of Schmidt and Ruedenberg[57] and
Fischer.[69] All energies are in hartree.

increasing size of basis sets of exponential-type functions is demonstrated.
That the calculated energies obtained in studies using basis sets of
exponential-type functions converge to the basis set limit can be seen from
Table IX where the 'best' estimates of the basis set limit energy values are
compared with the results of numerical calculations. That a similar level of
accuracy can be achieved in calculations employing basis sets of Gaussian-
type functions is demonstrated in Table X. Universal systematic sequences of
even-tempered basis sets of exponential-type functions have also been
developed for negative ions.[67,68]

V. BASIS SETS FOR ELECTRON CORRELATION ENERGY CALCULATIONS

A. The Basis Set Problem in the Method of Configuration Mixing

The method of configuration mixing or configuration interaction is formally
the most straightforward approach to the electron correlation problem in
atoms and molecules. Practical applications of the method usually employ
some limited expansion for the wavefunction which is then optimized
according to the Rayleigh–Ritz variational principle. However, limited
configuration mixing is, in general, not size-consistent; that is it does not scale
linearly with the number of electrons in the system. Furthermore, this problem
becomes increasingly problematic as the number of electrons in the system
under investigation is increased.[7]

Full configuration interaction avoids the size-consistency problem but can
only be performed routinely[70,71] for small atomic systems, which are
simplified by the high degree of symmetry they possess, if basis sets of an
adequate size are to be employed. For molecular systems there is no possibility

of performing full configuration mixing calculations using basis sets of an adequate size within the foreseeable future. Such calculations would involve the diagonalization of a matrix whose dimensions is given by Weyl's number, typical values of which are given in Table I. For small basis sets it is possible to perform full configuration mixing studies which can often provide useful benchmarks.[72,73] However, the basis set truncation errors almost invariably overshadow other errors which may be present in calculations with which these full configuration mixing studies are compared. This point will be discussed further in the following sections.

B. The Basis Set Problem in Many-body Perturbation Theory

The many-body perturbation theory[7,74] affords a size-consistent description of electron correlation effects in atoms and molecules. In this section, we wish to draw attention to a number of properties of the many-body perturbation theory when formulated within the algebraic approximation.

First, we draw attention to the fact that the convergence properties of the perturbation expansion may be significantly affected by the basis set employed.[75] When the algebraic approximation is invoked, both the domain and the range of the operators are modified. It is, therefore, not surprising that the convergence properties of the perturbation series can be critically dependent on the particular basis set employed. A number of comparisons have been made of finite-order many-body perturbation theory and full configuration interaction for atoms[76] and for molecules.[77-79] A common feature of the molecular studies is that it has been necessary to employ basis sets of modest size in order to keep the computations tractable. Any conclusions concerning the convergence properties of the perturbation series may, therefore, have a strong basis set dependence.

The results of a simple model[75] calculation will be presented here which demonstrate quite dramatically the dependence of the convergence properties of the perturbation series on the basis set employed. Consider the model problem of a hydrogenic atom with nuclear charge Z perturbed by the potential $-Z'/r$, i.e. the problem with Hamiltonian

$$\mathcal{H} = -\tfrac{1}{2}\nabla^2 - \frac{Z}{r} - \lambda\frac{Z'}{r} \tag{59}$$

in which λ is the perturbation parameter. The exact energy is given by the second order energy sum

$$E = -\tfrac{1}{2}Z^2 - ZZ'\lambda - \tfrac{1}{2}Z'^2\lambda^2 \tag{60}$$

with third- and higher-order terms in the energy series being individually zero. Consider now a calculation for this system within the algebraic approximation using a basis set of just two exponential-type functions, one with orbital exponent Z, which solves the ground-state zero-order problem exactly, and one with exponent $Z + Z'$, which solves the perturbed problem exactly. By invoking the variation theorem, both the zero-order and the perturbed problems can be solved exactly (i.e. by solving the 2×2 secular equation with $\lambda = 0$ and $\lambda = 1$, respectively). However, solution of the perturbed, two-basis-function problem by means of perturbation theory necessitates going beyond second order in the expansion for the energy. In particular, for the case $Z = 1$ and $Z' = 1$ the perturbation series will diverge as is clearly shown in

TABLE XI

Energy coefficients in the minimum basis set model perturbation expansion with $Z = 1$ and $Z' = 1$. The exact energy in this case is -2.0 hartree.

Order of perturbation	Energy coefficient[a]
0	−0.5000
1	−1.5000
2	−1.8512
3	−2.0602
4	−2.1113
5	−2.0543
6	−1.9729
7	−1.9357
8	−1.9601
9	−2.0140
10	−2.0481
11	−2.0365
12	−1.9940
13	−1.9599
14	−1.9633
15	1.9995
20	−1.9585
30	−2.0392
40	−1.9780
50	−1.9613
60	−2.1904
70	−1.5050
80	−2.9796
90	−0.5227
100	−3.2448
110	−3.7603
120	+9.8607
140	+78.4708
160	+136.5224

[a] In hartree.

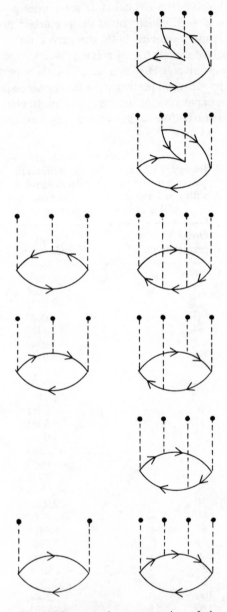

Fig. 6. Diagrammatic representation of the perturbation series for the model problem of a hydrogenic atom with nuclear charge Z perturbed by the potential $-Z'/r$. The perturbation is represented by a broken line terminated by a heavy dot.

TABLE XII

Components of the perturbation series for the hydrogenic system with $Z = 4$ and $Z' = 1$ using a universal systematic sequence of even-tempered basis sets of exponential-type functions.[a]

N	E_0	E_1	E_2	$E_3(p)$	$E_3(h)$
2	$-6.592\,299(+0)$	$-2.380\,827(+0)$	$-1.233\,837(-2)$	$-1.430\,613(-3)$	$5.981\,195(-3)$
4	$-7.988\,631(+0)$	$-4.016\,263(+0)$	$-5.203\,589(-1)$	$-1.958\,593(-1)$	$2.134\,141(-1)$
6	$-7.999\,366(+0)$	$-4.000\,274(+0)$	$-5.014\,825(-1)$	$-1.882\,146(-1)$	$1.885\,384(-1)$
8	$-7.999\,962(+0)$	$-3.999\,929(+0)$	$-5.000\,585(-1)$	$-1.875\,268(-1)$	$1.874\,090(-1)$
10	$-8.000\,000(+0)$	$-3.999\,997(+0)$	$-4.999\,880(-1)$	$-1.874\,916(-1)$	$1.874\,886(-1)$
12	$-8.000\,000(+0)$	$-4.000\,001(+0)$	$-5.000\,006(-1)$	$-1.875\,002(-1)$	$1.875\,020(-1)$
14	$-8.000\,000(+0)$	$-4.000\,000(+0)$	$-5.000\,000(-1)$	$-1.875\,000(-1)$	$1.874\,997(-1)$
16	$-8.000\,000$	$-4.000\,000(+0)$	$-5.000\,000(-1)$	$-1.875\,000(-1)$	$1.875\,000(-1)$
18	$-8.000\,000(+0)$	$-4.000\,000(+0)$	$-5.000\,000(-1)$	$-1.875\,000(-1)$	$1.875\,000(-1)$
20	$-8.000\,000(+0)$	$-4.000\,000(+0)$	$-5.000\,000(-1)$	$-1.875\,000(-1)$	$1.875\,000(-1)$
Exact	-8	-4	-0.5	$-1.187\,5$	$+0.187\,5$

[a]Based on the work of Quiney, Grant and Wilson.[81] All energies are in hartree. Powers of 10 are given in parentheses.

Table XI where the energy coefficients in the perturbation series are displayed. This simple model calculation serves to emphasize the dramatic effects that the algebraic approximation may have on the convergence properties of a perturbation series in that the two-basis-function calculation diverges and the large-basis-set calculation (i.e. exact) converges to the exact answer by second order.

The second point concerning the formulation of many-body-perturbation theory within the algebraic approximation again relates to the hydrogenic model problem described above. For this model, the diagrammatic represent-ation of the perturbation series is shown in Fig. 6, the broken line terminated by a heavy dot representing the perturbing potential. It is possible to evaluate each of these diagrams analytically using a technique due to Dalgarno and Lewis.[80] Although the total energy coefficient in each order beyond second order is identically zero, the components corresponding to each of the diagrams are non-zero. For example, in third order it can be shown that

$$E_3(p) = -\frac{3}{4}\frac{Z'^3}{Z} \qquad E_3(h) = \frac{3}{4}\frac{Z'^3}{Z} \qquad (61)$$

Hence, for this model it is possible to investigate the dependence of each diagrammatic component of the perturbation expansion on the basis set employed. The results of such an investigation[81,82] using a universal systematic sequence of even-tempered basis sets of exponential-type functions is presented in Table XII, where it can be seen that each component does indeed converge to the corresponding exact value. It is also of interest to compare the accuracy obtained in the basis set study with the achieved in numerical methods,[83] and this is done in Table XIII, where it can be seen that

TABLE XIII

Comparison of diagrammatic perturbation theory energy components obtained by using numerical methods with those obtained by employing basis set expansions of exponential-type functions.[a]

Energy component	Exact	Numerical method	Basis set expansion
(i) $Z = 1, Z' = 1$			
E_2	-0.5	-0.4996	-0.5000
$E_3(p)$	-0.75	-0.7492	-0.7500
$E_4(pp)$	-1.125	-1.1236	-1.1250
(ii) $Z = 4, Z' = 1$			
E_2	-0.5	-0.4987	-0.5000
$E_3(p)$	-0.1875	-0.1866	-0.1875
$E_4(pp)$	-0.0703125	-0.0698	-0.0703125

[a]Based on the work of Rossky and Karplus[83] and Quiney, Grant and Wilson.[81] All energies are in hartree.

the former approach is in fact more accurate. The reader is referred to Refs. 81 and 83 for a discussion of the problems associated with the numerical approach.

Many-body perturbation theory provides an efficient scheme for calculating the energy associated with triple[84-86] and quadruple[86,87] excitations, such terms arising for the first time in fourth order.[88] As a third point, attention is drawn to a study of the effects of basis set completeness on the relative importance of triply and quadruply excited configurations in correlation energy calculations. Calculations[89] for the ground state of the water molecule, which are shown in Table XIV, demonstrate that[89] 'the triple-excitation component increases in importance relative to the quadruple-excitation energy as the quality of basis set is improved. This suggests that

TABLE XIV

Relative importance of the fourth-order triple-excitation and quadruple-excitation components of the correlation energy as a function of basis set quality for the ground state of the water molecule.[a]

Basis set							
Primitive	Contracted	E_2	E_{4T}	E_{4Q}	$E_2\Delta_{11}$	$\lambda^3 E_2 \Delta_{11}$	$\mathscr{R}(\mu)$
[9s5p/4s]	$(3s2p/2s)^b$	-111.836	-1.286	-0.489	-4.222	-4.055	0.27
	$(3s2p/2s)^c$	-123.976	-1.274	-0.778	-4.963	-4.561	0.22
	$(4s2p/2s)$	-125.082	-1.285	-0.814	-5.043	-4.612	0.22
	$(4s3p/2s)$	-139.850	-3.218	-0.732	-5.578	-4.871	0.50
	$(5s3p/2s)$	-140.049	-3.232	-0.734	-5.589	-4.876	0.51
	$(5s3p/3s)$	-141.182	-3.271	-0.684	-5.678	-4.954	0.51
	$(9s5p/4s)$	-145.888	-3.673	-0.627	-5.873	-5.094	0.57
[10s6p/5s]	$(5s3p/3s)$	-144.324	-3.379	-0.740	-5.978	-5.160	0.50
	$(5s4p/3s)$	-147.882	-3.953	-0.664	-6.138	-5.268	0.58
[10s6p/5s]	$(5s4p/3s)$						
	$+1d(O)$	-211.140	-4.860	$+1.425$	-10.691	-9.603	0.52
	$+1p(H)$	-170.604	-5.320	$+0.157$	-7.985	-6.848	0.68
	$+1d(O)1p(H)$	-226.090	-6.148	$+2.158$	-12.451	-11.025	0.60
	$+2d(O)2p(H)$	-244.691	-7.934	$+3.059$	-14.578	-12.692	0.69

[a]Taken from the work of Wilson and Guest.[89] In millihartree.
[b]Contraction coefficients taken from the work of Dunning.[90]
[c]Contraction coefficients taken from the work of Dunning and Hay.[91]

E_2 denotes the second-order energy. E_{4T} and E_{4Q} denote the fourth-order energy component associated with linked diagrams containing triply excited and quadruply excited intermediate states respectively. $\Delta_{11} = \langle \phi_1 | \phi_1 \rangle$, where ϕ_1 is the first-order wavefunction. $E_2\Delta_{11}$ is, therefore, the unlinked diagram fourth-order component, $\lambda^3 E_2 \Delta_{11}$ contains a variational parameter, λ. $\mathscr{R}(\mu)$ is the ratio of the fourth-order triple excitation energy to the total fourth-order quadruple excitation energy:

$$\mathscr{R}(\mu) = \frac{E_{4T}}{E_{4Q} + E_2\Delta_{11}}$$

configuration interaction calculations which use small basis sets lead to an underestimate of the contribution made by triply substituted states when reasonably large basis sets are employed.'

C. Basis Set Truncation Effects

It has been clearly demonstrated in Fig. 5 that the basis set truncation is often a more significant source of error in electron correlation energy calculations than the neglect of higher-order terms in the perturbation series. In Fig. 5, 'S' labels calculations made with 'smaller' basis sets and 'L' denotes those made with 'larger' basis sets. There (i), (ii), and (iii) are used to denote the correlation energy taken through second, third and fourth order, respectively.

In Fig. 7, calculated correlation energies for first- and second-row closed-shell diatomic hydrides,[92] which are represented by the first bar in each of the histograms, are compared with empirical estimates of the total correlation energy, represented by the second bar, and empirical estimates of the relativistic component of the energy, represented by the third bar. It can be seen that, except for the molecules containing very light atoms, i.e. LiH and

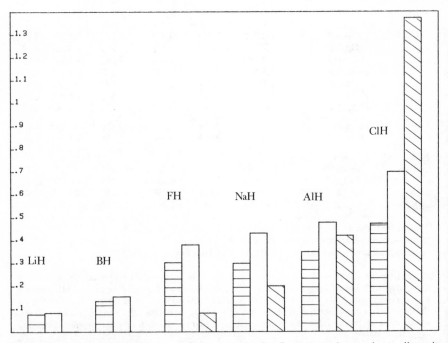

Fig. 7. Comparison of calculated correlation energies for first-row and second-row diatomic hydrides (bars with horizontal hatching) with empirical estimates of the total correlation energy (bars with no hatching) and the empirical relativistic component of the total energy (bars with diagonal hatching).

BH, the relativistic component of the energy is comparable to, and in fact usually larger than, the remaining error in the correlation energy. (It should, of course, be remembered that the correlation energy and the relativistic component of the energy are not strictly additive as has been assumed here.) The problem of calculating relativistic wavefunctions for atoms and molecules within the algebraic approximation will be considered further in section VIII.

VI. BASIS SETS FOR ATOMIC AND MOLECULAR PROPERTIES

A. Electric Properties

The basis set truncation error is often particularly significant in calculations of atomic and molecular properties other than the energy. We begin by considering electric properties.

In the presence of an external field, the perturbed Hamiltonian can be written in terms of the multiple moment tensors. For an uncharged molecule

$$\mathscr{H} = \mathscr{H}_0 - \mu_\alpha F_\alpha - \tfrac{1}{3}\Theta_{\alpha\beta}F_{\alpha\beta} - \tfrac{1}{15}\Omega_{\alpha\beta\gamma}F_{\alpha\beta\gamma} - \tfrac{1}{105}\Phi_{\alpha\beta\gamma\delta}F_{\alpha\beta\gamma\delta}\cdots \quad (62)$$

and the corresponding energy is

$$\begin{aligned}
E = E_0 &- \mu_\alpha^0 F_\alpha - \tfrac{1}{3}\Theta_{\alpha\beta}^0 F_{\alpha\beta} - \tfrac{1}{15}\Omega_{\alpha\beta\gamma}^0 F_{\alpha\beta\gamma} - \tfrac{1}{105}\Phi_{\alpha\beta\gamma\delta}^0 F_{\alpha\beta\gamma\delta} + \cdots \\
&- \tfrac{1}{2}\alpha_{\alpha\beta}F_\alpha F_\beta - \tfrac{1}{3}A_{\alpha,\beta\gamma}F_\alpha F_{\beta\gamma} - \tfrac{1}{6}C_{\alpha\beta,\gamma\delta}F_{\alpha\beta}F_{\gamma\delta} - \tfrac{1}{15}E_{\alpha,\beta\gamma\delta}F_\alpha F_{\beta\gamma\delta} \\
&+ \cdots - \tfrac{1}{6}\beta_{\alpha\beta\gamma}F_\alpha F_\beta F_\gamma - \tfrac{1}{6}B_{\alpha\beta,\gamma\delta}F_\alpha F_\beta F_{\gamma\delta} + \cdots - \tfrac{1}{24}\gamma_{\alpha\beta\gamma\delta}F_\alpha F_\beta F_\gamma F_\delta \\
&+ \cdots
\end{aligned} \quad (63)$$

where F_α, $F_{\alpha\beta}$, ... are the field, field gradient, etc., at the origin; F_0, μ_α^0, $\Theta_{\alpha\beta}^0$, $\Omega_{\alpha\beta\gamma}^0$, $\Phi_{\alpha\beta\gamma\delta}^0$ the energy and permanent moments of the free species; and $\alpha_{\alpha\beta}$, $\beta_{\alpha\beta\gamma}$, $\gamma_{\alpha\beta\gamma\delta}$, $A_{\alpha,\beta\gamma}$, $C_{\alpha\beta,\gamma\delta}$, $E_{\alpha,\beta\gamma\delta}$ and $B_{\alpha\beta,\gamma\delta}$ the polarizabilities. Greek suffixes denote Cartesian components and summation over repeated suffixes is implied.

Bounds and Wilson[93] have recently investigated the calculation of molecular multipole moments using universal systematic sequences of even-tempered basis sets Gaussian-type functions. Their results for the FH molecule are displayed in Table XV. They are compared with the results of Bishop and Maroulis,[94] who employed basis sets of Gaussian-type functions specifically designed for the FH molecule, and with the fully numerical calculations of Sundholm, Pyykko and Laaksonen.[95] Bounds and Wilson[93] conclude that the higher-order multipoles are increasingly sensitive to the quality of the basis set and 'that basis set truncation errors are much more important in calculations of higher multipole moments than in calculations of dipole moments and quadrupole moments. Whereas electron correlation effects are the most important remaining source of error in the dipole moments and quadrupole moments which we obtain with our largest basis set, this is not like to be the case for the higher moments.' Bishop and Maroulis[94] also

TABLE XV
Calculations of multipole moments of the FH molecule.[a]

| | Universal basis sets[b] | | | | | |
	6s3p1d	12s6p2d	18s9p2d	24s12s4d	B & M[c]	Numerical[d]
Q_1	0.7543	0.7718	0.7634	0.7586	0.7571	0.7561
Q_2	1.7577	1.7507	1.7389	1.7397	1.7407	1.7321
Q_3	2.4121	2.5710	2.6086	2.5887	2.6265	2.5924
Q_4	4.1614	4.9922	5.0573	5.0472	5.0496	5.0188
Q_5	6.8869	8.3407	9.0205	8.7779		8.7298
Q_6	9.5598	13.4058	15.5591	14.9293		14.9888
Q_7	9.2280	20.7394	26.4297	24.9108		25.407
Q_8	−3.8503	30.0947	44.0550	40.9537		42.77
Q_9	−57.7270	37.9782	71.8062	66.6105		71.6
Q_{10}	−227.157	30.9475	113.775	107.407		120.

[a]Multipole moments are given relative to the centre of mass. The unit of the ith moment Q_i, is ea_0^i.
[b]From the work of Bounds and Wilson.[93]
[c]From the work of Bishop and Maroulis.[94]
[d]From the work of Sundholm, Pyykko and Laaksonen.[95]

investigated static polarizabilities and hyperpolarizabilities and concluded that 'The choice of the basis set is still the main problem for SCF calculations and its quality will undoubtably influence correlation treatments as well. Most calculations have been carried out with conventionally constructed basis sets, which can hardly meet the requirements for accurate molecular polarizability predictions.'

B. Magnetic Properties

An interesting point arises in the calculation of magnetic properties of molecules. The magnetic vector potential **A** can be defined in an infinite number of ways since it can be subjected to a gauge transformation

$$\mathbf{A}' = \mathbf{A} + \nabla \phi \qquad (64)$$

where ϕ is any scalar function which can be differentiated twice. A change of origin is a gauge transformation and thus **A** is origin-dependent. Physical observables are clearly not dependent on the choice of origin; they are gauge-invariant. For a calculation which employed a complete basis set, calculated magnetic constants would not depend on either the choice of the magnetic vector potential **A** or the choice of origin. However, in practice, a finite and, therefore, incomplete basis set has to be employed and this yields calculated magnetic properties which are not gauge-invariant. The extent to which calculated magnetic properties are gauge-invariant is an indication of the quality of the calculation and, in particular, the quality of the basis set. There appears to have been little systematic study of the effects of basis set truncation on the accuracy with which magnetic properties can be computed.

C. Relativistic Properties

For atoms from the first row of the periodic table and molecules containing them, the Breit–Pauli approximation appears to afford an accurate description of relativistic effects. The most important of these effects is usually the spin–orbit coupling

$$\langle SLJM | \mathscr{H}_{so} | SLJM \rangle = \tfrac{1}{2} A [J(J+1) - L(L+1) - S(S+1)] \tag{65}$$

Cooper and Wilson[96,97] have investigated the calculation of spin–orbit coupling constants within the matrix Hartree–Fock approximation using systematic sequences of even-tempered basis sets of exponential-type functions. Some typical results are displayed in Table XVI.

A more unified description of relativistic effects, which can be used for systems containing atoms from the second row of the periodic table and beyond, is described in Section VIII where the Breit–Pauli approximation is avoided.

TABLE XVI

Spin–orbit coupling constants calculated within the matrix Hartree–Fock approximation using a universal systematic sequence of even-tempered basis sets of exponential-type functions. Comparison with results obtained using basis sets specifically designed for each of the molecules studied.

Molecule	One-electron component		Two-electron component		Total	
	Universal basis set	Specific basis set	Universal basis set	Specific basis set	Universal basis set	Specific basis set
OH^a	-215.772	-216.748	-74.941	-76.105	-140.832	-140.643
CH^a	$+56.471$	$+56.129$	-27.007	-27.301	$+29.364$	$+28.828$
$HeNe^{+b}$	-731.955	-733.568	$+213.267$	$+213.246$	-518.688	-520.321
BH^{+c}	$+29.450$	$+30.041$	-15.581	-15.953	$+13.869$	$+14.087$
LiO^d	-199.194	-193.590	$+70.103$	$+70.195$	-123.092	-123.394
CO^e	-195.991	-196.514	$+70.014$	$+70.277$	-125.977	-126.237

[a]Cade, P. E., and Huo, W. H., J. Chem. Phys., **47**, 617 (1967).
[b]Cooper, D. L., and Wilson, S., J. Chem. Phys., **76**, 6088 (1983).
[c]Cooper, D. L., and Richards, W. G., J. Phys. B: At. Mol. Phys., **15**, 491 (1982).
[d]Cooper, D. L., and Richards, W. G., J. Chem. Phys., **73**, 3515 (1980).
[e]McLean, A. D., and Yoshimine, M., IBM J. Res. Dev., **12**, 20 (1968)

VII. BASIS SET SUPERPOSITION EFFECTS

A. Basis Set Superposition in the Supermolecular Method

The use of multi-centre basis set often leads to overestimates of interaction energies between atoms and/or molecules. Such effects are particularly problematic in calculations small interaction energies such as van der Waals interaction energies, where they can completely mask the true interaction.[98-106] This error is mainly attributable to basis set superposition effects.

Consider two systems, X and Y, separated by a distance R. The interaction energy is defined by

$$\Delta E_{XY}(R) = E_{XY}(R) - E_X - E_Y \tag{66}$$

where $E_{XY}(R)$ is the total energy of the supersystem and E_X and E_Y are the energies of the isolated systems. Now let us assume that the system X is described by a basis set which we denote by S_X and the subsystem Y by a basis set S_Y. The interaction energy calculated using these basis sets is

$$\Delta E_{XY}(R/S_X \cup S_Y) = E_{XY}(R/S_X \cup S_Y) - E_X(S_X) - E_Y(S_Y) \tag{67}$$

The basis set truncation errors associated with the three energies on the right-hand side of this equation are

$$\varepsilon_{XY}(R) = E_{XY}(R) - E_{XY}(R/S_X \cup S_Y) \tag{68}$$

$$\varepsilon_X = E_X - E_X(S_X) \tag{69}$$

$$\varepsilon_Y = E_Y - E_Y(S_Y) \tag{70}$$

In general,

$$\varepsilon_{XY}(R) - \varepsilon_X - \varepsilon_Y \neq 0 \tag{71}$$

except in the limit of large and flexible basis sets. This is the basis set superposition effect. Even if the basis sets S_X and S_Y are so large that $\varepsilon_X \approx 0$ and $\varepsilon_Y \approx 0$, $\varepsilon_{XY}(R)$ may differ significantly from zero. When a calculation is performed on the supersystem XY, the description of the system X (Y) is improved by the additional flexibility afforded by the basis set S_Y (S_X), resulting in a non-physical lowering of the energy of the supersystem which may totally obliterate the true interaction.

It should be emphasized that basis set superposition effects exist in all theoretical and computational methods which involve the algebraic approximation, in matrix Hartree–Fock calculations, in configuration interaction and in many-body perturbation theory. It is frequently found that basis set superposition effects are larger in calculations which take account of electron correlation effects than they are in orbital theories such as the matrix Hartree–Fock method. We shall discuss this aspect further in Section VII.C. It should be pointed out that it is particularly important that methods employed in the calculation of van der Waals interaction potential are size-consistent. The linked-diagram theorem ensures that the energy resulting from a many-body perturbation theory calculation on the supersystem X...Y, with X and Y at infinite separation, is equal to the sum of the energies of the system X and Y calculated separately. The accurate treatment of van der Waals interaction potentials requires a careful consideration of basis set superposition effects and electron correlation effects.[107]

B. The Function Counterpoise Method

Boys and Bernardi[99] proposed the function counterpoise method to correct for the effects of basis set superposition in calculations of small interaction energies, such as van der Waals interaction potential. For a supersystem X...Y, the function counterpoise method is employed in the following manner. The energies E_X and E_Y of the subsystems X and Y are calculated using the full basis set employed in the calculation on the supersystem X...Y rather than just the basis sets for X or Y alone. The interaction energy is then given by

$$\Delta E_{XY}(R/S_X \cup S_Y) = E_{XY}(R/S_X \cup S_Y) - E_X(R/S_X \cup S_Y) - E_Y(R/S_X \cup S_Y) \quad (72)$$

Note that E_X and E_Y in this definition depend on the separation R between the subsystems X and Y. Computationally, the calculation of $E_X(R/S_X \cup S_Y)$ involves putting the nuclear charge on centre Y to zero and changing the number of electrons in the program employed in the study of the supersystem X...Y. The set of basis functions S_Y centred on nucleus Y of zero charge is referred to as a set of ghost functions.

It is clear, however, that the function counterpoise method will overestimate the basis set superposition error. In the supersystem X...Y, the Pauli principle will prevent subsystem X from fully utilizing the basis set of subsystem Y, while in the ghost system the calculation of $E_X(R/S_X \cup S_Y)$ does not involve such a restriction. This had led some workers to propose modified function counterpoise correction procedures, none of which can be rigorously justified.[108]

For extensive basis sets, an optimal description of the subsystems X and Y and the supersystem X...Y will be obtained. The basis set superposition error will then be very small. In recent work, Wells and Wilson[107] did not use the function counterpoise correction in the usual fashion described above. They pointed out not only that the Boys–Bernardi procedure overcorrects for basis set superposition effects but also that it cannot be uniquely generalized for the calculation of a many-body interaction. Wells and Wilson argue that the function counterpoise correction should be used as a test for basis set superposition errors.

C. Systematic Sequences of Basis Sets for van der Waals Interactions

When developing the large basis sets which are required for the reliable calculation of van der Waals interactions, it is important to ensure that the basis set be constructed and extended in a systematic fashion. In recent work, Wells and Wilson[106,107] have employed systematic sequences of even-tempered basis sets of Gaussian-type functions in conjunction with many-body perturbation theory calculations to study van der Waals interaction potentials, thereby ensuring basis set superposition errors and size-inconsistency problems are controlled. They used the Boys–Bernardi procedure as a test for the magnitude of the basis set superposition error rather than as a correction.

TABLE XVII

Function counterpoise corrections for the ground state of the neon atom calculated within the matrix Hartree–Fock approximation for a systematic sequence of even-tempered basis sets of Gaussian-type functions. In this table G represents a set of ghost orbitals. The internuclear separation in the NeG system is 5.0 bour.[a]

Basis set	$E(\text{Ne})$	$E(\text{NeG})$	ΔE
6s/3p	$-128\,079\,772.753$	$-128\,079\,891.221$	-118.468
8s/4p	$-128\,439\,171.661$	$-128\,439\,492.595$	-320.984
10s/5p	$-128\,519\,330.088$	$-128\,519\,496.915$	-166.827
12s/6p	$-128\,539\,358.980$	$-128\,539\,415.400$	-56.420
14s/p	$-128\,544\,744.842$	$-128\,544\,762.955$	-18.113
16s/8p	$-128\,546\,301.834$	$-128\,546\,308.899$	-7.065
18s/9p	$-128\,546\,816.546$	$-128\,546\,818.828$	-2.972

[a]Based on the work of Wells and Wilson.[106] All energies are in μ hartree.

TABLE XVIII

Function counterpoise correction for the ground state of the neon atom calculated using diagrammatic many-body perturbation theory and employing a systematic sequence of even-tempered basis sets of Gaussian-type functions. In this table G represents a set of ghost orbitals. The NeG internuclear separation is 5.0 bohr.[a]

Basis set	$E(\text{Ne})$	$E(\text{NeG})$	ΔE
8s/4p/1d	$-128\,597\,725.57$	$-128\,598\,212.12$	-486.55
	$(-128\,439\,171.61)^b$	$(-128\,439\,572.61)$	(-401.00)
12s/6p/2d	$-128\,737\,186.19$	$-128\,737\,357.65$	$(-171.47$
	$(-128\,539\,358.98)$	$(-128\,539\,432.67)$	(-73.69)
[12s/6p]3d	$-128\,777\,283.17$	$-128\,777\,333.64$	-50.47
	$(-128\,546\,301.83)$	$(-128\,546\,312.88)$	(-11.05)

[a]Based on the work of Wells and Wilson.[106] All energies are in μhartree.
[b]Matrix Hartree–Fock energies and energy differences are given in parentheses.

In Table XVII the function counterpoise correlations for the ground state of the neon atom are shown for calculations made within the matrix Hartree–Fock approximation with a systematic sequence of even-tempered basis sets of Gaussian-type functions. In these calculations, a ghost nucleus, designated G, was placed at a distance of 5 bohr from the neon nucleus. Calculations were also performed using many-body perturbation theory to take account of electron correlation. Some typical results are displayed in Table XVIII. It can be seen by comparing Tables XVII and XVIII that the basis set superposition error in this system is larger in the calculations which take account of electron correlation effects than in those performed at the matrix Hartree–Fock level. In both cases the basis set superposition error can be systematically reduced by employing a sequence of even-tempered basis sets.

D. Many-body Effects

The importance of the non-pairwise additive components of the interaction energy between atoms and molecules is widely recognized[109] and *ab initio* electronic structure calculations offer a route to important information about such effects. Attention has recently been drawn to the fact that there is no unique generalization of the Boys–Bernardi function counterpoise technique to clusters of molecules.[110-113] Two possible generalizations have been introduced, as follows.

1. Pairwise Additive Function Counterpoise

In this approach it is assumed that the basis set superposition error in the many-body cluster can be approximated by the sum of the Boys–Bernardi function counterpoise corrections for pairs of bodies. Hence the total interactions for an N-body cluster using the pairwise additive function counterpoise correction is given by

$$\Delta E(ij\ldots) = E(ij\ldots) - \sum_i E(i) - \sum_{ij} [E(iG_j) - E(i)] \tag{73}$$

where $R(ij\ldots)$ is the total energy calculated for the cluster $ij\ldots$ and $E(iG_j)$ is the energy obtained by considering the ith body in the presence of a ghost body at the jth position, a ghost body being obtained by putting all nuclear charges in that body to zero whilst retaining all of the basis functions which are used to describe that body.

2. Site–site Function Counterpoise

If it is assumed that the non-physical part of the energy lowering left at each body in the cluster is due to the mixing in of all other orbitals in the cluster, irrespective of whether these orbitals are occupied or not, then one obtains the site–site function counterpoise correction. The total interaction energy is then written.

$$\Delta E(ij\ldots) = E(ij\ldots) - \sum_i E(i) - \sum_i [E(iG_{jk\ldots}) - E(i)] \tag{74}$$

where $E(iG_{jk\ldots})$ is the energy obtained for the ith body in the presence of ghost bodies corresponding to all other bodies in the cluster. Unlike the pairwise additive function counterpoise method, the site–site function counterpoise approach affords a basis for a decomposition of the total interaction energy into its n-body components.

The total interaction energy in the absence of basis set superposition effects may be written in the form

$$\Delta E(ij\ldots) = \frac{1}{2!}\sum_{ij}' \varepsilon(ij) + \frac{1}{3!}\sum_{ijk}' \varepsilon(ijk) + \cdots \tag{75}$$

with the two-body term given by

$$\varepsilon(ij) = E(ij) - E(i) - E(j) \tag{76}$$

and the three-body terms given by

$$\varepsilon(ijk) = E(ijk) - \varepsilon(ij) - \varepsilon(jk) - \varepsilon(ik) - E(i) - E(j) - E(k) \tag{77}$$

If the basis set employed in a particular study is so large that basis set superposition effects can be neglected, then these equations serve to define the many-body potential. However, if the site–site function counterpoise approximation is used to take account of superposition effects in calculations performed with basis sets which are not so large, then equation (76) is replaced by

$$\varepsilon(ij) = E(ijG_{kl\ldots}) - E(iG_{jkl\ldots}) - E(jG_{ikl\ldots}) \tag{78}$$

Equation (77) is replaced by

$$\begin{aligned}\varepsilon(ijk) = E(ijkG_{lm\ldots}) &- \varepsilon(ij) - \varepsilon(jk) - \varepsilon(ik) \\ &- E(iG_{jkl\ldots}) - E(jG_{ikl\ldots}) - E(kG_{ijl\ldots})\end{aligned} \tag{79}$$

where $\varepsilon(ij)$ is now given by Eq. (78). Similar modifications must be made for the higher-order terms.

VIII. BASIS SETS FOR RELATIVISTIC ELECTRONIC STRUCTURE CALCULATIONS

A. The Dirac Equation in the Algebraic Approximation

Over the past few years has been increasing interest[114,115] in the use of the algebraic approximation in estimating the solutions of the Dirac equation and the Dirac–Hartree–Fock equations for atoms and, more particularly, for molecules. Such studies were often found to suffer from a problem which has been termed 'variational collapse' or, more correctly, 'finite basis set disease'.[116] When employing the algebraic approximation, spurious unphysical solutions of the Dirac equation or Dirac–Hartree–Fock equations were obtained with too small a kinetic energy, leading in turn to an overestimation of the binding energy. Furthermore, there was found to be no systematic variation of some eigenvalues with increasing size of basis set used. The eigenvalues of the relativistic problem did not pass to the corresponding non-relativistic limit if the speed of light was treated as a parameter and increased to infinity. To trace the origin of this problem, we follow Dyall, Grant and Wilson[117] and consider the Dirac equation for hydrogenic systems formulated within the algebraic approximation.

The radial Dirac equation for a hydrogen-like atom with nuclear charge Z may be written, in atomic units, in the form

$$H\Phi_\mu = \varepsilon_\mu \Phi_\mu \tag{80}$$

with

$$H = \begin{pmatrix} -Z/r & c\pi \\ c\pi^+ & -2c^2 - Z/r \end{pmatrix} \tag{81}$$

where r is the nucleus–electron distance, c is the speed of light and

$$\pi = \frac{\kappa}{r} - \frac{d}{dr} \tag{82}$$

$$\pi^+ = \frac{\kappa}{r} + \frac{d}{dr} \tag{83}$$

κ is the angular quantum number. The notation π and π^+ follows from the relation

$$\langle u|\pi|v \rangle = \langle v|\pi^+|u \rangle + (v(r)u(r))_0^\infty \tag{84}$$

which is merely integration by parts, so that π and π^+ are adjoints provided that u and v are finite and continuous on $(0, \infty)$ and their product vanishes at 0 and ∞.

The wavefunction has the form

$$\Phi_\mu = \begin{pmatrix} g_\mu \\ f_\mu \end{pmatrix} \tag{85}$$

where g_μ and f_μ are the large and small components of the relativistic wavefunction, respectively, and satisfy the normalization condition

$$\int_0^\infty [g_\mu(r)^2 + f_\mu(r)^2]\, dr = 1 \tag{86}$$

In the algebraic approximation, it is assumed that the large and small components of the wavefunction can be parametrized in terms of some finite set of basis functions. Thus for the large component, we put

$$g_\mu = \sum_{i=1}^{M} \chi_i a_{\mu_i} = \chi^+ \cdot \mathbf{a}_\mu \qquad (87)$$

where χ and $\mathbf{a}_\mu$ are column vectors of basis functions and expansion coefficients respectively. Similarly, for the small component, we put

$$f_\mu = \sum_{i=1}^{N} \omega_i b_{\mu_i} = \omega^+ \cdot \mathbf{b}_\mu \qquad (88)$$

ω and $\mathbf{b}_\mu$ being column vectors of basis functions and expansion coefficients, respectively.

Substituting expressions (87) and (88) into the Dirac equation for the hydrogen-like atom, expression (80), and assuming for the moment that the basis sets χ and ω consist of orthonormal elements, we obtain the Dirac equation in the algebraic approximation

$$\begin{pmatrix} -\mathbf{V} - \varepsilon_\mu \mathbf{I} & c\pi \\ c\pi^+ & -2c^2\mathbf{I}' - \mathbf{V} - \varepsilon_\mu \mathbf{I}' \end{pmatrix} \begin{pmatrix} \mathbf{a}_\mu \\ \mathbf{b}_\mu \end{pmatrix} = 0 \qquad (89)$$

in which $\mathbf{V}$ is an $M \times M$ matrix with elements

$$V_{ij} = \langle \chi_i | Z/r | \chi_j \rangle \qquad i,j = 1,2,\dots,M \qquad (90)$$

and $\mathbf{V}'$ is an $N \times N$ matrix with elements

$$V'_{ij} = \langle \omega_i | Z/r | \omega_j \rangle \qquad i,j = 1,2,\dots,N \qquad (91)$$

π is an $M \times N$ matrix with elements

$$\pi_{ij} = \langle \chi_i | \pi | \omega_j \rangle \qquad i = 1,2,\dots,M, \quad j = 1,2,\dots,N \qquad (92)$$

and π^+ is its transpose. $\mathbf{I}$ and $\mathbf{I}'$ are unit matrices of dimensions M and N, respectively.

If, as is more usually the case, the basis functions are non-orthogonal, then the Dirac equation for a hydrogen-like atom assumes the following form in the algebraic approximation

$$\begin{pmatrix} -\mathbf{V} - \varepsilon_\mu \mathbf{S} & c\pi \\ c\pi^+ & -2c^2\mathbf{S}' - \mathbf{V}' - \varepsilon_\mu \mathbf{S}' \end{pmatrix} \begin{pmatrix} \mathbf{a}_\mu \\ \mathbf{b}_\mu \end{pmatrix} = 0 \qquad (93)$$

where $\mathbf{S}$ and $\mathbf{S}'$ are the following matrices of overlap integrals

$$S_{ij} = \langle \chi_i | \chi_j \rangle \qquad i,j = 1,2,\dots,M \qquad (94)$$

$$S'_{ij} = \langle \omega_i | \omega_j \rangle \qquad i,j = 1,2,\dots,N \qquad (95)$$

B. The Finite Basis Set Problem

Early work on the finite basis set problem in relativistic calculations has been reviewed by Kutzelnigg.[116] Spurious unphysical solutions of the Dirac equation or the Dirac–Hartree–Fock equations are observed with too small a kinetic energy, leading to an overestimation of the binding energy. Furthermore, these solutions are found neither to tend to the solutions of the Schrödinger equation in the limit $c \to \infty$ nor to vary systematically with increasing size of basis set.

To trace the origin of the finite basis set problem, let us[117] observe that the Dirac equation for a hydrogenic system within the algebraic approximation may be written as two simultaneous equations. Thus Eq. (93) gives

$$- \mathbf{V} \mathbf{a}_\mu + c \pi \mathbf{b}_\mu = \varepsilon_\mu = \varepsilon_\mu \mathbf{S} \mathbf{a}_\mu \tag{96}$$

and

$$c \pi^+ \mathbf{a}_\mu - (2c^2 \mathbf{S}' + \mathbf{V}') \mathbf{b}_\mu = \varepsilon_\mu \mathbf{S}' \mathbf{b}_\mu \tag{97}$$

From the second of these two equations an expression can be obtained for the coefficient vector of the small component

$$\mathbf{b}_\mu = [\varepsilon_\mu \mathbf{S}' + (2c^2 \mathbf{S}' + \mathbf{V}')]^{-1} c \pi^+ \mathbf{a}_\mu \tag{98}$$

which can be substituted in the first equation to give

$$\{ - \mathbf{V} + c^2 \pi [\varepsilon_\mu \mathbf{S}' + (2c^2 \mathbf{S}' + \mathbf{V}')]^{-1} \pi^+ \} \mathbf{a}_\mu = \varepsilon_\mu \mathbf{S} \mathbf{a}_\mu \tag{99}$$

By using the matrix identity

$$(\mathbf{A} + \mathbf{B})^{-1} = \mathbf{A}^{-1} - \mathbf{A}^{-1} \mathbf{B} (\mathbf{A} + \mathbf{B})^{-1} \tag{100}$$

with

$$\mathbf{A} = 2c^2 \mathbf{S}' \tag{101}$$

and

$$\mathbf{B} = \varepsilon_\mu \mathbf{S}' + \mathbf{V}' \tag{102}$$

Eq. (99) may be put in the form

$$\{ - \mathbf{V} + \tfrac{1}{2} \pi (\mathbf{S}')^{-1} \pi^+ + \pi \tfrac{1}{2} (\mathbf{S}')^{-1} (\varepsilon_\mu \mathbf{S}' + \mathbf{V}')$$
$$\times (\varepsilon_\mu \mathbf{S}' + 2c^2 \mathbf{S}' + \mathbf{V}')^{-1} \pi^+ \} \mathbf{a}_\mu = \varepsilon_\mu \mathbf{S} \mathbf{a}_\mu \tag{103}$$

Now the third term on the left-hand side of this equation will tend to zero in the non-relativistic limit. The first two terms will tend to the Schrödinger limit for hydrogenic atoms provided that

$$\mathbf{T} = \tfrac{1}{2} \pi (\mathbf{S}')^{-1} \pi^+ \tag{104}$$

where $\mathbf{T}$ is the matrix of the non-relativistic kinetic energy operator $-\tfrac{1}{2} \nabla^2$. In general, of course, Eq. (104) will not be satisfied and the matrix product on the right-hand side of Eq. (104) will underestimate the kinetic energy unless special precautions are taken. We consider the general problem of constructing matrix representations of operator products in the next section and then return to the problem of formulating the Dirac equation within the algebraic approximation in the following section.

C. Matrix Representation of Operator Products

Using the notation which was introduced in II.A Section, we now consider the arbitrary composite operator AB.[118] This product only has meaning if the range of B, $R(B)$, lines in the domain of A

$$R(B) \subseteq D(A) \tag{105}$$

and then

$$D(AB) \subseteq D(A) \tag{106}$$

The matrix of AB on $\mathbb{M} = \{u_i | i = 1, 2, \ldots, N\} \subset \mathbb{H}$ is ${}^m(AB)^m$. How is this related to the product of the matrices ${}^mA^m$ of A and ${}^mB^m$ of B? If $\mathbb{M} = D(A) = D(B) = R(B)$ then

$$Bu_j = \sum_{i=1}^{N} u_i(u_i', Bu_j) \tag{107a}$$

$$= \sum_{i=1}^{N} u_i \sum_{k=1}^{N} S_{ik}'(u_k, Bu_j) \tag{107b}$$

$$= \sum_{i=1}^{N} u_i \sum_{k=1}^{N} S_{ik}' B_{kj} \tag{107c}$$

where for simplicity the left and right superscripts m have been omitted. Again

$$ABu_j = \sum_{k=1}^{N} (Au_i) \sum_{k=1}^{N} S_{ik} B_{kj} \tag{108}$$

from which we find

$$(AB)_{ij} = \sum_{k=1}^{N} \sum_{l=1}^{N} A_{ik} S_{kl}' B_{lj} \tag{109}$$

which can be written formally as

$${}^m(AB)^m = {}^mA^m \, {}^m(S^{-1})^m \, {}^mB^m \tag{110}$$

The weakness of the formal manipulation is that it fails if $R(B)$ is not completely contained in $\mathbb{M}$. This can be understood by consideration of the operator $PABP$. Suppose that $\mathbb{H}$ is decomposed not with reference to $\mathbb{M}$ (and its orthogonal complement $\mathbb{M}^\perp$ on $\mathbb{H}$) but with respect to some other set $\mathbb{N}$. Let

P' be the orthogonal projector onto $\mathbb{N}$

$$P'v = \begin{cases} v & \text{if} \quad v \in \mathbb{N} \\ 0 & \text{if} \quad v \in \mathbb{N}^{\perp} \end{cases} \tag{111}$$

$$\mathbb{H} = \mathbb{N} \cup \mathbb{N}^{\perp} \tag{112}$$

Thus Q' is the orthogonal projector onto $\mathbb{N}^{\perp}$; its orthogonal complement is such that

$$P' + Q' = 1 \tag{113}$$

and

$$P'^2 = P' \tag{114}$$

$$P'Q' = 0 \tag{115}$$

$$Q'^2 = Q' \tag{116}$$

on the whole of $\mathbb{H}$. So

$$\begin{aligned} PABP &= PA(P' + Q')BP \\ &= PAP'BP + PAQ'BP \end{aligned} \tag{117}$$

Now $R(BP)$, the range of B acting on $\mathbb{M}$, is the span of the elements Bu_i. If we make the identification,

$$\mathbb{N} = \{Bu_i | i = 1, 2, \ldots, N\} \tag{118}$$

then

$$Q'Bu_i = 0 \qquad i = 1, 2, \ldots, N \tag{119}$$

and the second term of Eq. (117) vanishes. It is easy to see then, when we choose $\mathbb{N}$ in this way only, Eq. (110) generalizes to

$$^m(AB)^m = {}^mA^n\,{}^n(S^{-1})^n\,{}^nB^m \tag{120}$$

where

$$^mA^n = (u_i, Av_j) \qquad i,j = 1, 2, \ldots, N \tag{121}$$

$\{v_j | j = 1, 2, \ldots, N\}$ being some set spanning $\mathbb{N}$.

From the above discussion, it is now clear why the expectation of the kinetic energy $T = +\frac{1}{2}p^2$ is always underestimated if we attempt to use the matrix product $^mp^m\,{}^mp^m$ to replace $^m(p^2)^m$ without taking special precautions. We have

$$Pp^2P = PpP'pP + PpQ'pP \tag{122}$$

Let

$$P\phi = \sum_{i=1}^{N} c_i u_i$$

then

$$(P\phi, TP\phi) = \sum_{i,j=1}^{N} c_i c_j (^m(p^2)^m)_{ij} \tag{123}$$

Similarly

$$\tfrac{1}{2}(P\phi, PpP' \cdot P'pP \cdot P\phi) = \tfrac{1}{2} \sum_{i,j} c_i c_j (^mp^n\,{}^n(S^{-1})^n\,{}^np^m)_{ij} \tag{124a}$$

TABLE XIX
Matrix representation of the operator product d/dz. d/dz.

(a) Matrix representation of $\left(v_i, \dfrac{d^2}{dz^2} v_j \right)$

$\{v_i\} = \{1s; \rho_k = \alpha\beta^k(k = 1, 2, 3, 4), \alpha = 0.5, \beta = 1.5\}$

-0.1875	-0.2645	-0.3318	-0.3748
-0.2645	-0.4219	-0.5952	-0.7466
-0.3318	-0.5952	-0.9492	-1.3393
-0.3748	-0.7466	-1.3393	-2.1357

(b) Matrix product defined in Eq. (125) with $\{v_i\}$ defined as in (a) and
$\{u_i\} = \{1p; \rho_k = \alpha\beta^k(k = 1, 2, 3, 4), \alpha = 0.5, \beta = 1.5\}$

-0.1875	-0.2645	-0.3318	-0.3748
-0.2645	-0.4219	-0.5952	-0.7466
-0.3318	-0.5952	-0.9492	-1.3393
-0.3748	-0.7466	-1.3393	-2.1357

(c) Matrix product defined in Eq. (125) with $\{v_i\}$ defined as in (a) and
$\{u_i\} = \{2p; \rho_k = \alpha\beta^k(k = 1, 2, 3, 4), \alpha = 0.5, \beta = 1.5\}$

-0.1870	-0.2631	-0.3280	-0.3642
-0.2631	-0.4179	-0.5844	-0.7163
-0.3280	-0.5844	-0.9198	-1.2570
-0.3642	-0.7163	-1.2570	-1.9058

---

and

$$(P\phi, PpQ' \cdot Q'pP \cdot P\phi) = \| Q'pP\phi \|^2 > 0 \tag{124b}$$

Thus (124a) will also underestimate (123) unless we choose the intermediate decomposition in terms of the set $\mathbb{N}$ defined as in (118).

We illustrate these ideas and concepts for the matrix representation of the operator product $d/dz \cdot d/dz = d^2/dz^2$ in Table XIX. The matrix of the operator product is compared with the product of the matrix representation of d/dz with itself, that is

$$K_{il} = \sum_{j,k}^{N'} \left(v_i, \frac{d}{dz} u_j \right) (S'^{-1})_{jk} \left(u_k, \frac{d}{dz} v_l \right)$$

$$i, l = 1, 2, \ldots, N \tag{125}$$

where

$$u_i \in \mathbb{M} \qquad \text{and} \qquad v_i \in \mathbb{N} \tag{126}$$

is compared with the matrix

$$\left(v_i, \frac{d^2}{dz^2} v_l \right) \qquad i, l = 1, 2, \ldots, N \qquad (127)$$

It can be seen from Table XIX that the matrix product is equal to the matrix of the operator product provided that the sets $\{u_i\}$ and $\{v_i\}$ are chosen so that

$$u_i = \frac{d}{dz} v_i \qquad i = 1, 2, \ldots, N \qquad (128)$$

D. One-electron Systems

We now return to the problem of formulating the Dirac equation in the algebraic approximation and avoiding the so-called finite basis set disease by considering applications to one-electron systems. Applications to many-electron systems are discussed in Section VIII.E. We shall first of all consider the minimum basis set description of one-electron systems[117] and then describe extended basis set calculations.[119]

For a hydrogenic atom with a point nucleus, the ground state ($\kappa = -1$) energy is given by the Sommerfeld eigenvalue

$$\varepsilon_{\text{Som}} = -c^2 \left[1 - \left(1 - \frac{Z^2}{c^2} \right)^{1/2} \right] = -\frac{1}{2} Z^2 - \frac{1}{8} \frac{Z^4}{c^2} - \frac{1}{16} \frac{Z^6}{c^4} - \cdots \qquad (129)$$

For a minimum basis set calculation, the large and small components of the relativistic wavefunction, P and Q, respectively, are each approximated by a single function. Writing

$$\begin{aligned} P(r) &= au(r) \\ Q(r) &= bv(r) \end{aligned} \qquad (130)$$

we determine a and b variationally, which leads to the secular equation

$$\begin{vmatrix} V_L - \varepsilon & c\pi \\ c\pi^+ & V_S - 2c^2 - \varepsilon \end{vmatrix} = 0 \qquad (131)$$

where $V_L = \langle u| - Z/r|u \rangle$, $V_S = \langle v| - Z/r|v \rangle$ and $\pi = \langle u|\pi|v \rangle$. We disregard the lowest eigenvalue, which corresponds to a positron solution with energy below $-2c^2$.

In Figs. 8 and 9 we show the energies obtained by Dyall, Grant and Wilson[117] using various minimum basis sets. In Fig. 8, the difference $\Delta\varepsilon = \varepsilon - \varepsilon_{\text{Som}}$ is shown as a function of the nuclear charge together with the exact relativistic correction to the energy, i.e. $-\frac{1}{2}Z^2 - \varepsilon_{\text{Som}}$. The low-$Z$ behaviour is emphasized in Fig. 9, where the eigenvalue difference is divided by Z^2. The

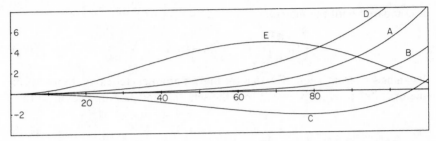

Fig. 8. Minimum basis set calculations on hydrogenic atoms (see text for details).

curves presented in Figs. 8 and 9 are as follows:

A. The relativistic correction, $-\frac{1}{2}Z^2 - \varepsilon_{Som}$.

B. A 1s exponential-type function is used for the large component, a 1p exponential-type function for the small component, both functions with exponents equal to Z. This choice ensures that the Schrödinger kinetic energy is represented satisfactorily and leads to an eigenvalue

$$\varepsilon_{1s,1p}^{ETF} = -c^2\left\{1 + \left(\frac{Z}{c}\right)^2 - \left[1 + \left(\frac{Z}{c}\right)^2\right]^{1/2}\right\}$$

$$= -\frac{1}{2}Z^2 - \frac{1}{8}\frac{Z^4}{c^2} + \frac{1}{16}\frac{Z^6}{c^4} + \cdots \tag{132}$$

which gives an upper bound to the exact value and the correct non-relativistic limit as $c \to \infty$.

C. If the small component is now represented by a 2p exponential-type function then the Schrödinger kinetic energy is underestimated and the

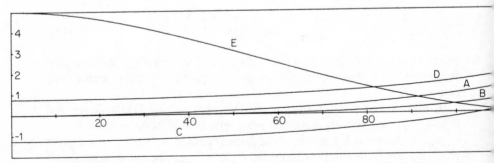

Fig. 9. Minimum basis set calculations on hydrogenic atoms (see text for details).

energy eigenvalue

$$\varepsilon_{1s,2p}^{ETF} = -c^2\left[1 + \frac{3}{4}\left(\frac{Z^2}{c^2}\right) - \left(1 + \frac{1}{4}\frac{Z^2}{c^2} + \frac{1}{16}\frac{Z^4}{c^4}\right)^{1/2}\right]$$

$$= -\frac{5}{8}Z^2 + \frac{3}{128}\frac{Z^4}{c^2} - \frac{3}{1024}\frac{Z^6}{c^4} + \cdots \qquad (133)$$

lies below the exact value and does not yield the correct non-relativistic limit.

D. A 1s Gaussian-type function is used for the large component and a 2p function for the small component. The exponent in both functions is taken to be $8Z^2/9\pi$, which is the optimal value for the corresponding non-relativistic problem. The eigenvalue is

$$\varepsilon_{1s,2p}^{GTF} = -c^2\left[1 + \frac{20}{9\pi}\frac{Z^2}{c^2} - \left(1 + \frac{16}{9\pi}\frac{Z^2}{c^2} + \frac{16}{81\pi^2}\frac{Z^4}{c^4}\right)^{1/2}\right]$$

$$= -\frac{4}{3\pi}Z^2 - \frac{8}{27\pi^2}\frac{Z^4}{c^2} + \frac{64}{273\pi^2}\frac{Z^6}{c^4} + \cdots \qquad (134)$$

Although a minimum basis set of Gaussian-type functions cannot be expected to yield results of any accuracy, this choice of basis sets leads to a correct representation of the Schrödinger kinetic energy, leads to an upper bound to the exact value and yields the corresponding non-relativistic limit as $c \to \infty$.

TABLE XX

Calculated relativistic total energies for the ground state of the Ne^{9+} ion obtained by using a systematic sequence of even-tempered basis sets of exponential-type functions.[a,b]

N	(1s1p) basis set		(1s2p) basis set	
	E_0	Δ_0	E_0	Δ_0
2	−21.015 420	+2.9(+)	−21.823 452	+2.8(+1)
4	−41.743 422	+8.3(+0)	−47.651 360	+2.4(+0)
6	−49.922 994	+1.4(−1)	−65.455 535	−1.5(+1)
8	−50.065 382	+1.4(−3)	−57.909 300	−7.8(+0)
10	−50.066 615	+1.3(−4)	−51.928 592	−1.9(+0)
12	−50.066 738	+2.4(−6)	−50.626 218	−5.6(−1)
14	−50.066 739	+1.9(−6)	−50.255 382	−1.9(−1)

[a]Based on the work of Dyall, Grant and Wilson.[119] All energies are given inhartree.
[b](1snp) denotes a basis set in which the large component of the relativistic wavefunction is parametrized in terms of 1s functions and the small component is parametrized in terms of np functions. N is the number of basis functions of each symmetry type. Δ_0 is the difference between the calculated energy and the exact value. Powers of 10 are given in parentheses.

E. This curve corresponds to the use of $\frac{1}{2}$s and $\frac{1}{2}$p exponential-type functions for the large and small components, respectively. Further details will be found in the work of Dyall et al.[117]

The minimum basis set calculations summarized in Figs. 8 and 9 clearly demonstrate the importance of using appropriate basis sets for the parametrization of the large and small components of the relativistic wavefunction in avoiding the so-called 'finite basis set' disease. That these ideas continue to be valid in calculations employing extended basis sets is illustrated in Table XX where calculations for the ground state of the Ne^{9+} ion using systematic sequences of even-tempered basis sets of exponential-type functions are shown. These calculations used 1s functions for the large component and either 1p or 2p functions for the small component. It can be seen that when 1p functions are employed for the small component the calculated energy remains above the exact value whereas when 2p functions are used it often falls below.

E. Many-electron Systems

The ideas and concepts concerning the use of basis sets in relativistic calculations which have been described in the previous subsections allow the Dirac–Fock equations for many-electron systems to be formulated within the algebraic approximation. A discussion of these equations lies outside the scope of the present chapter.

One feature of relativistic self-consistent field calculations to which attention should be drawn is the fact that the Breit interaction can be easily included in the self-consistent field iterations once the algebraic approximation has been invoked.[120] This should be contrasted with the situation in atomic calculations using numerical methods in which the Breit interaction is treated by first-order perturbation theory.

F. Relativistic Many-body Perturbation Theory

The relativistic many-body perturbation theory of atomic and molecular electronic structure can be formulated within the algebraic approximation in a manner analogous to the non-relativistic formulation. A detailed discussion of the method, which is still under development, lies outside the scope of this chapter but the technique's potential will be illustrated by displaying some results for the relativistic version of the model problem considered in Section V.B, a hydrogenic atom with nuclear charge Z perturbed by the potential $-Z'/r$.[121] The exact energy of the perturbed problem in its ground state is

$$\varepsilon = -c^2 \left\{ 1 - \left[1 - \left(\frac{Z + Z'}{c} \right)^2 \right]^{1/2} \right\} \qquad (135)$$

TABLE XXI
Relativistic perturbation theory for hydrogenic systems with $Z = 50$ and $Z' = 1, n = 1$, $\kappa = 1$. The systematic sequence of even-tempered basis sets of exponential-type functions of Quiney et al. was employed. M denotes the number of basis functions.[a]

M	ε_0	ε_1	ε_2	ε_3	ε_4
2	$-1.284\,600(3)$	$-5.180\,321(1)$	$-5.974\,942(-1)$	$-9.532\,053(-3)$	$-2.005\,620(-5)$
3	$-1.285\,051(3)$	$-5.323\,477(1)$	$-6.665\,710(-1)$	$-2.503\,827(-3)$	$9.840\,676(-5)$
4	$-1.293\,983(3)$	$-5.386\,028(1)$	$-6.184\,670(-1)$	$-5.588\,265(-4)$	$-2.289\,774(-5)$
5	$-1.294\,440(3)$	$-5.374\,605(1)$	$-6.165\,983(-1)$	$-1.805\,950(-3)$	$-3.474\,051(-5)$
6	$-1.294\,567(3)$	$-5.367\,568(1)$	$-6.168\,740(-1)$	$-2.011\,084(-3)$	$-2.020\,084(-5)$
7	$-1.294\,625(3)$	$-5.370\,194(1)$	$-6.195\,748(-1)$	$-1.921\,139(-3)$	$-1.312\,666(-5)$
8	$-1.294\,585(3)$	$-5.369\,816(1)$	$-6.196\,833(-1)$	$-1.904\,844(-3)$	$-1.521\,957(-5)$
9	$-1.294\,625(3)$	$-5.370\,195(1)$	$-6.193\,532(-1)$	$-1.884\,940(-3)$	$-1.665\,757(-5)$
10	$-1.294\,618(3)$	$-5.370\,171(1)$	$-6.194\,841(-1)$	$-1.897\,635(-3)$	$-1.684\,397(-5)$
11	$-1.294\,624(3)$	$-5.370\,148(1)$	$-6.194\,090(-1)$	$(-1.901\,249(-3)$	$-1.693\,645(-5)$
12	$-1.294\,624(3)$	$-5.370\,210(1)$	$-6.194\,712(-1)$	$-1.901\,208(-3)$	$-1.674\,402(-5)$
13	$-1.294\,624(3)$	$-5.370\,186(1)$	$-6.194\,750(-1)$	$-1.902\,630(-3)$	$-1.676\,917(-5)$
14	$-1.294\,625(3)$	$-5.370\,209(1)$	$-6.194\,781(-1)$	$-1.901\,686(-3)$	$-1.677\,660(-5)$
Exact	$-1.294\,636(3)$	$-5.370\,219(1)$	$-6.194\,923(-1)$	$-1.902\,709(-3)$	$-1.681\,852(-5)$

[a]Based on the work of Quiney, Grant and Wilson.[121] All energies are given in atomic units. Powers of 10 are given in parentheses.

from which it can be shown that the energy coefficients in the perturbation series are

$$\varepsilon_1 = -Z(1 - Z^2/c^2)^{-1/2}Z' \tag{136}$$

$$\varepsilon_2 = -\tfrac{1}{2}(1 - Z^2/c^2)^{-3/2}Z'^2 \tag{137}$$

$$\varepsilon_3 = -\tfrac{1}{2}(Z/c^2)(1 - Z^2/c^2)^{-5/2}Z'^3 \tag{138}$$

$$\varepsilon_4 = -\frac{1}{8}\frac{1}{c^2}\left(1 - \frac{Z^2}{c^2}\right)^{-7/2}\left[1 + 2\frac{Z^2}{c^2}\left(2 + 3\frac{Z^2}{c^2}\right)\right]Z'^4 \tag{139}$$

Energies calculated using these exact expressions can be compared with results obtained within the algebraic approximation. Such a comparison is made in Table XXI. It can be seen that, as the basis set is systematically extended, the various energy components converge to their exact values.

Applications of the relativistic perturbation theory to many-electron systems are presented elsewhere.[122]

IX. BASIS SETS FOR EXTENDED MOLECULE STUDIES

A. Multi-centre Basis Sets

The number of two-electron integrals that arise in calculations formulated within the algebraic approximation increases as

$$\tfrac{1}{2}m(m + 1) \tag{140}$$

where

$$m = \tfrac{1}{2}n(n + 1) \tag{141}$$

and n is the number of basis functions. This approximate n^4 dependence would appear to render calculations for large molecular systems prohibitive computationally. If advantage is taken of the fact that a large number of integrals for extended molecules will be negligibly small, and therefore need not be evaluated, then very significant reductions in computer times may be achieved. For example, in a study of the cytosine–guanine complex, Clementi[123] showed that the computer time requirements could be reduced by a factor of about 100 by not evaluating small integrals. Ahlrichs[124] has demonstrated that the approximate n^4 dependence of the number of integrals is reduced to an n^2 dependence for large molecules.

B. One-centre Basis Sets

The one-centre expansion method has already been discussed in Section IV.A. Here the use of the one-centre expansion method in calculations for large molecules which take account of electron correlation effects will be briefly discussed.[125,126]

In independent-electron models of atomic and molecular electronic structure, such as the Hartree–Fock approximation, only functions corresponding to the first few values of the angular momentum quantum number l contribute significantly to the energy, or to other expectation values, when multi-centre basis sets are employed. However, in treatments which take account of electron correlation effects, the higher harmonics are known to be important. A considerable amount of data is available on the convergence properties of the harmonic expansion for atoms[41,61,62,127] and the importance of higher-order terms in the harmonic expansion for molecular systems has also been demonstrated.[128]

It has been suggested[126] that, since basis functions with high l values are ultimately required to obtain accurate correlation energies and since such functions are also required to obtain accurate one-centre expansions, a universal basis set can be constructed for polyatomic molecules by making a one-centre expansion and including functions with high l values in the basis set. Such a universal basis set would allow any polyatomic molecule to be handled at any nuclear geometry employing the same set of two-electron integrals over the basis functions. Only the nucleus–electron attraction integrals would have to be evaluated for calculations at different geometries and these integrals can be evaluated efficiently, increasing as $\tfrac{1}{2}n(n + 1)$ with the number of basis functions n. Furthermore, the basis functions used in a one-centre calculation transform according to the representations of the spherical symmetry group. There is, therefore, a higher degree of symmetry associated with the integrals over the basis functions than with the integrals over the molecular orbitals. This can be exploited both in self-consistent field calculations and in the determination of electron correlation effects. For one-centre basis sets, unlike multi-centre basis sets, problems arising from overcompleteness can usually be controlled, if not avoided altogether. The one-centre approach provides control over the convergence of a calculation with respect to the size of the basis set and control over practical linear dependence. Moreover, because of the ease with which integrals involving one-centre basis functions can be handled, the method can be used in conjunction with alternative types of functions, for example, the piecewise polynomial functions suggested by Silverstone and his coworkers.[39-41]

X. COMPUTATIONAL ASPECTS OF THE ALGEBRAIC APPROXIMATION

A. Practical Linear Dependence

In the introduction to this chapter, it was noted that the history of computational quantum chemistry can be traced in terms of the use of basis sets of increasing size and that, to some extent, this has been due to the availability of increasingly powerful computers. However, as basis sets of larger and larger size are devised, numerical problems become increasingly acute on machines which have a finite word length. In particular, practical linear dependence problems have to be avoided. The author would suggest that it is the need to avoid computational linear dependence which is the key issue in constructing the large basis sets required for accurate electronic structure studies within the algebraic approximation and not the computing power required actually to carry out the calculation. Little work appears to have been done on this problem.[129]

B. Matrix Multiplication

Once the algebraic approximation has been invoked, the vast majority of the computations involved in the study of atomic and molecular electronic structure from first principles can be cast in the form of simple matrix operators. We note that matrix multiplication can be performed particularly efficiently on vector processing computers such as the Cray 1 or the Cyber 205.[130]

Matrix multiplication is generally stated to be an n^3 process, n being the dimension of the matrix. However, using the techniques of algebraic complexity theory, matrix multiplication can in fact be shown to be an $n^{2.495\cdots}$ process on a serial computer.[131] It can be shown that on a parallel processor (single instruction stream, multiple data stream), the maximum number of processors which can be usefully employed is[132]

$$p \leqslant n^{2.495\cdots}/\log_2 n \tag{142}$$

and that the matrix multiplication will increase as[132]

$$n^{2.495}/p \tag{143}$$

Provided that the large basis sets, which are obviously going to be required to perform very accurate electronic structure calculations, can be designed so as to avoid the numerical problems mentioned in the preceding section, it is likely that, with the use of increasingly parallel architectures in modern computers, we are poised to obtain a significant improvement in the accuracy of atomic and, particularly, molecular electronic structure calculations.

XI. SUMMARY AND CONCLUDING REMARKS

It has not been possible within the space available to give a completely comprehensive account of all of the developments which have recently been made in the field of basis set construction, a field that forms the foundation upon which the vast majority of contemporary atomic and molecular electronic structure studies are based. This review has necessarily been selective but should provide the reader with an up-to-date account of the most important aspects of current thinking on the algebraic approximation. It has concentrated on the construction of basis sets for electron correlation energy calculations, for calculations of atomic and molecular properties (other than the energy), for the study of small interaction energies such as van der Waals interactions, for the determination of relativistic effects and for studies of extended molecular systems.

In this concluding section, methods which completely avoid the use of basis sets will be briefly mentioned. Fully numerical[37,38] and semi-numerical[133-135] integration techniques have been employed in studies of diatomic molecules for which one angular variable may be separated and treated analytically. Such calculations have provided very useful benchmark results within the Hartree–Fock (and Hartree–Fock–Slater) approximations which can be employed in assessing the accuracy of calculations performed by basis set expansion methods. The numerical solutions of the molecular Schrödinger equation for arbitrary polyatomic molecules would involve a three-dimensional numerical integration and would be very demanding computationally. Quantum Monte Carlo methods[136] have recently been used in molecular structure studies and show some promise.[137-139] The variance associated with such calculations is often too large for chemical applications but techniques such as the differential Monte Carlo method avoid this problem to some extent.[140] In spite of these developments, the algebraic approximation remains competitive in terms of accuracy and flexibility; it certainly provides a more compact representation of the wavefunction.

With the use of increasingly parallel architectures in modern computers and the growing awareness that basis set truncation errors are frequently the dominant source of error in contemporary electronic structure studies, significant improvements in the accuracy of such studies can be expected in the new few years. The key issue in making this progress is the construction of large basis sets which will afford an accurate representation of atomic and molecular wavefunctions whilst avoiding computational linear dependence. This review was completed in September, 1985.

References

1. Dunning, T. H., Jr, and Hay, P. J., in *Methods of Electronic Structure Theory* (Ed. H. F. Schaefer III), Plenum, New York, 1977.
2. Ahlrichs, R., and Taylor, P. R., *J. Chim. Phys.*, **78**, 315 (1981).
3. Čársky, P., and Urban, M., *Ab Initio Calculations. Methods and Applications in Chemistry*, Springer-Verlag, Berlin, 1980.
4. Szabo, A., and Ostlund, N. S., *Modern Quantum Chemistry*, Macmillan, London, 1982.
5. Huzinaga, S., Andzelm, J., Klobukowski, M., Radzio-Andzelm, E., Sakai, Y., and Tatewaki, H. (Ed.), *Gaussian Basis Sets for Molecular Calculations*, Elsevier, Amsterdam, 1984.
6. Wilson, S., in *Methods in Computational Molecular Physics* (Eds. G. H. F. Diercksen and S. Wilson), Reidel, Dordrecht, 1983.
7. Wilson, S., *Electron Correlation in Molecules*, Clarendon Press, Oxford, 1984.
8. Huzinaga, S., *Comput. Phys. Rep.*, **2**, 281 (1985).
9. Wilson, S., *Computational Quantum Chemistry*, Clarendon Press, Oxford, 1987.
10. Hall, G. G., *Proc. R. Soc. A*, **205**, 541; **208**, 328 (1951).
11. Roothaan, C. C. J., *Rev. Mod. Phys.*, **23**, 69 (1951).
12. Reed, M., and Simon, B, *Methods of Modern Mathematical Physics*, Vol. 1, Academic Press, New York, 1972.
13. Messiah, A., *Quantum Mechanics*, North-Holland, Amsterdam, 1961.
14. Richtmyer, R. D., *Principles of Advanced Mathematical Physics*, Vol. 1, Springer-Verlag, New York, 1978.
15. Kato, T., *Perturbation Theory for Linear Operators*, 2nd Edn, Springer-Verlag, New York, 1976.
16. See, for example, Ref. 7, chapter 2.
17. Sinanoglu, O., *Phys. Rev.*, **122**, 493 (1961).
18. Riley, M., and Dalgarno, A., *Chem. Phys. Lett.*, **9**, 382 (1971).
19. Musher, J. I., and Schulman, J. M., *Phys. Rev.*, **173**, 93 (1968).
20. Goodisman, J., *J. Chem. Phys.*, **48**, 2981 (1968).
21. Goodisman, J., *J. Chem. Phys.*, **50**, 903 (1969).
22. Wilson, S., *J. Phys. B: At. Mol. Phys.*, **17**, 505 (1984).
23. McWeeny, R., and Sutcliffe, B. T., *Methods of Molecular Quantum Mechanics*, Academic Press, New York, 1976.
24. McWeeny, R., and Pickup, B. T., *Rep. Prog. Phys.*, **43**, 1065 (1980).
25. Paldus, J., *Theor. Chem. Adv. Perspect.*, **2**, 131 (1976).
26. Harris, F. E., *J. Chem. Phys.*, **32**, 3 (1960).
27. Davidson, E. R., *J. Chem. Phys.*, **33**, 1577 (1960).

28. Harris, F. E., and Taylor, H. S., *J. Chem. Phys.*, **38**, 2591 (1963).
29. Taylor, H. S., and Harris, F. E., *Mol. Phys.*, **6**, 183 (1963).
30. Taylor, H. S., *J. Chem. Phys.*, **39**, 3357 (1963).
31. Ebbing, D. D., *J. Chem. Phys.*, **36**, 1361 (1962).
32. Bishop, D. M., and Cheung, L. M., *J. Chem. Phys.*, **78**, 1396 (1983).
33. Wilson, S., *J. Phys. B: At. Mol. Phys.*, **16**, L429 (1983).
34. Wells, B. H., and Wilson, S., *Proc. Sixth Seminar on Computational Methods in Quantum Chemistry*, Schloss Ringberg, 1984.
35. Wells, B. H., and Wilson, S., *J. Phys. B: At. Mol. Phys.*, **19**, 17 (1986).
36. Wells, B. H., and Wilson, S., *J. Phys. B: At. Mol. Phys.*, **18**, 2731 (1985).
37. Laaksonen, L., Pyykko, P., and Sundholm, D., *Int. J. Quantum Chem.*, **23**, 309 (1983).
38. Laaksonen, L., Pyykko, P., and Sundholm, D., *Int. J. Quantum Chem.*, **23**, 319 (1983).
39. Gazquez, J. L., and Silverstone, H. J., *J. Chem. Phys.*, **67**, 1887 (1977).
40. Silverstone, H. J., Carroll, D. P., and Silver, D. M., *J. Chem. Phys.*, **68**, 616 (1978).
41. Carroll, D. P., Silverstone, H. J., and Metzger, R. M., *J. Chem. Phys.*, **71**, 4142 (1979).
42. Coolidge, A. S., *Phys. Rev.*, **42**, 189 (1932).
43. Bishop, D. M., *Adv. Quantum Chem.*, **3**, 25 (1967).
44. Roetti, C., and Clementi, E., *J. Chem. Phys.*, **60**, 4725 (1974).
45. Reeves, C. M., *J. Chem. Phys.*, **39**, 1 (1963).
46. Ruedenberg, K., Raffenetti, R. C., and Bardo, R., in *Energy Structure and Reactivity*, Proc. 1972 Boulder Res. Conf. Theor. Chem., Wiley, New York, 1973.
47. Silver, D. M., Wilson, S., and Nieuwpoort, W. C., *Int. J. Quantum Chem.*, **14**, 635 (1978).
48. Silver, D. M., and Nieuwpoort, W. C., *Chem. Phys. Lett.*, **57**, 421 (1978).
49. Silver, D. M., and Wilson, S., *J. Chem. Phys.*, **69**, 3787 (1978).
50. Wilson, S., and Silver, D. M., *Chem. Phys. Lett.*, **63**, 367 (1979).
51. Wilson, S., and Silver, D. M., *J. Chem. Phys.*, **72**, 2159 (1980).
52. Wilson, S., in *Theoretical Chemistry*, Vol. 4, p. 1, Specialist Periodical Reports, The Royal Society of Chemistry, London, 1981.
53. Wilson, S., and Silver, D. M., *J. Chem. Phys.*, **77**, 3674 (1982).
54. Wilson, S., *Mol. Phys.*, **49**, 1489 (1983).
55. Clementi, E., and Corongiu, G., *Geometrical Basis Sets for Molecular Computations*, IBM Research Report, Poughkeepsie, 1982.
56. Clementi, E., Corongiu, G., Gratarola, M., Habitz, P., Lupo, C., Otto, P., and Vercauteren, D., *Int. J. Quantum Chem. Symp.*, **16**, 409 (1982).
57. Schmidt, M. W., and Ruedenberg, K., *J. Chem. Phys.*, **71**, 3951 (1979).
58. Feller, D. F., and Ruedenberg, K., *Theor. Chim. Acta*, **52**, 231 (1979).
59. Wilson, S., *Theor. Chim. Acta*, **57**, 53 (1980).
60. Wilson, S., *Theor. Chim. Acta*, **58**, 31 (1980).
61. Schwarz, C. M., *Phys. Rev.*, **126**, 1015 (1962).
62. Schwarz, C. M., *Meth. Comput. Phys.*, **2**, 241 (1963).
63. Hartree, D. R., *Proc. Camb. Phil. Soc.*, **45**, 230 (1948).
64. Roothaan, C. C. J., and Bagus, P. S., *Meth. Comput. Phys.*, **2**, 47 (1963).
65. Bounds, D. G., and Wilson, S., *Mol. Phys.*, **54**, 445 (1985).
66. Cooper, D. L., and Wilson, S., *J. Chem. Phys.*, **77**, 5053 (1982).
67. Cooper, D. L., and Wilson, S., *J. Chem. Phys.*, **77**, 4551 (1982).
68. Cooper, D. L., and Wilson, S., *J. Chem. Phys.*, **78**, 2456 (1983).
69. Fischer, C. F., *The Hartree–Fock Method for Atoms*, Wiley, New York, 1977.

70. Bunge, C. F., *Phys. Rev. A*, **14**, 1965 (1976).
71. Bunge, C. F., *Phys. Scr.*, **21**, 328 (1980).
72. Handy, N. C., *Chem. Phys. Lett.*, **74**, 280 (1980).
73. Saxe, P., Schaefer, H. F., and Handy, N. C., *Chem. Phys. Lett.*, **79**, 202 (1981).
74. Wilson, S., *Comput. Phys. Rep.*, **2**, 389 (1985).
75. Wilson, S., *Proc. Sixth Seminar on Computational Methods in Quantum Chemistry*, Schloss Ringberg, 1984.
76. Silver, D. M., Wilson, S., and Bunge, C. F., *Phys. Rev. A*, **19**, 1375 (1979).
77. Wilson, S., Jankowski, K., and Paldus, J., *Int. J. Quantum Chem.*, **23**, 1781 (1983).
78. Knowles, P. J., Samasundrum, K., Handy, N. C., and Hirao, K., *Chem. Phys. Lett.*, **113**, 8 (1985).
79. Handy, N. C., Samasundrum, K., and Knowles, P. J., *Theor. Chim. Acta.*, **68**, 123 (1985).
80. Dalgarno, A., and Lewis, J. T., *Proc. R. Soc. A*, **233**, 70 (1955).
81. Quiney, H. M., Grant, I. P., and Wilson, S., *J. Phys. B: At. Mol. Phys.*, **18**, 577 (1985).
82. Quiney, H. M., Grant, I. P., and Wilson, S., *Proc. Sixth Seminar on Computational Methods in Quantum Chemistry*, Schloss Ringberg, 1984.
83. Rossky, P. J., and Karplus, M., *J. Chem. Phys.*, **67**, 5419 (1977).
84. Wilson, S., and Saunders, V. R., *J. Phys. B: At. Mol. Phys.*, **12**, L403 (1979); **13**, 1505 (1980).
85. Wilson, S., *J. Phys. B: At. Mol. Phys.*, **12**, L657 (1979); **13**, 1505 (1980).
86. Wilson, S., and Guest, M. F., *Mol. Phys.*, **43**, 1331 (1981).
87. Wilson, S., and Silver, D. M., *Theor. Chim. Acta*, **54**, 83 (1979).
88. Wilson, S., in *Correlated Wavefunctions* (Ed. V. R. Saunders), Science Research Council, Daresbury, 1978.
89. Wilson S., and Guest, M. F., *Chem. Phys. Lett.*, **73**, 607 (1980).
90. Dunning, T. H., Jr, *Chem. Phys*, **55**, 116 (1975).
91. Dunning, T. H., and Hay, P. J., in Methods in *Electronic Structure Theory* (Ed. H. F. Schaefer III), Plenum, New York, 1977.
92. Wilson, S., and Silver, D. M., *J. Chem. Phys.*, **66**, 5400 (1977).
93. Bounds, D. G., and Wilson, S., *Mol. Phys.*, **54**, 445 (1985).
94. Bishop, D. M., and Maroulis, G., *J. Chem. Phys.*, **82**, 2380 (1985).
95. Sundholm, D., Pyykko, P., and Laaksonen, L., *Mol. Phys.*, **56**, 1411 (1985).
96. Cooper, D. L., and Wilson, S., *J. Phys. B: At. Mol. Phys.*, **15**, 493 (1982).
97. Cooper, D. L., and Wilson, S., *J. Chem. Phys.*, **76**, 6088 (1982).
98. Clementi, E., *J. Chem. Phys.*, **46**, 3851 (1967).
99. Boys, S. F., and Bernardi, F., *Mol. Phys.*, **19**, 553 (1970).
100. Johansson, A., Kollman, P., and Rothenberg, S., *Theor. Chim. Acta*, **29**, 167 (1973).
101. Liu, B., and McLean, A. D., *J. Chem. Phys.*, **59**, 4557 (1973).
102. Urban, M., and Hobza, P., *Theor. Chim. Acta*, **36**, 215 (1975).
103. Ostlund, N. S., and Merrifield, D. L., *Chem. Phys. Lett.*, **39**, 612 (1976).
104. Dacre, P. D., *Chem. Phys. Lett.*, **50**, 147 (1977).
105. Price, S. L., and Stone, A. J. S., *Chem. Phys. Lett.*, **65**, 127 (1979).
106. Wells, B. H., and Wilson, S., *Mol. Phys.*, **50**, 1295 (1983).
107. Wells, B. H., and Wilson, S., *Mol. Phys.*, **54**, 787 (1985).
108. Daudey, J. P., Claverie, P., and Malrieu, J. P., *Int. J. Quantum Chem.*, **8**, 1 (1974).
109. Klein, M. L., and Venables, J. A. (Eds.), *Rare Gas Solids*, Vol. 1, Academic Press, New York, 1976.
110. Wells, B. H., and Wilson, S., *Chem. Phys. Lett.*, **101**, 429 (1983).
111. Wells, B. H., and Wilson, S., *Mol. Phys.*, **55**, 199 (1985).

112. Wells, B. H., and Wilson, S., *Mol. Phys.*, **57**, 21 (1985).
113. Wells, B. H., and Wilson, S., *Mol. Phys.*, **57**, 421 (1985).
114. Malli, G. L. (Ed.), *Relativistic Effects in Atoms, Molecules and Solids*, Plenum, New York, 1983.
115. Pyykko, P. (Ed.), *Symposium on Relativistic Effects in Quantum Chemistry*, Abo Akademie, Finland, 1982; *Int. J. Quantum Chem.*, **25**, 1 (1984).
116. Kutzelnigg, W., *Int. J. Quantum Chem.*, **25**, 107 (1984).
117. Dyall, K. G., Grant, I. P., and Wilson, S., *J. Phys. B: At. Mol. Phys.*, **17**, L45 (1984).
118. Dyall, K. G., Grant, I. P., and Wilson, S., *J. Phys. B: At. Mol. Phys.*, **17**, 493 (1984).
119. Dyall, K. G., Grant, I. P., and Wilson, S., *J. Phys. B: At. Mol. Phys.*, **17**, 1201 (1984).
120. Quiney, H. M., Grant, I. P., and Wilson, S., *J. Phys. B: At. Mol. Phys.*, in press.
121. Quiney, H. M., Grant, I. P., and Wilson, S., *J. Phys. B: At. Mol. Phys.*, **18**, 2805 (1985).
122. Quiney, H. M., Grant, I. P., and Wilson, S., *J. Phys. B: At. Mol. Phys.*, in preparation.
123. Clementi, E., *Proc. Nat. Acad. Sci.* (USA), **69**, 2942 (1972).
124. Ahlrichs, R., *Theor. Chim. Acta*, **33**, 157 (1974).
125. Ladik, J., and Čížek, J., *J. Chem. Phys.*, **73**, 2357 (1980).
126. Wilson, S., *Mol. Phys.*, in preparation.
127. Jankowski, K., and Malinowski, P., *Phys. Rev. A*, **22**, 51 (1980).
128. Wilson, S., *J. Phys. B: At. Mol. Phys.*, **15**, L191 (1982).
129. See Klahn, B., *Adv. Quantum Chem.*, **13**, 155 (1981).
130. Guest, M. F., and Wilson, S., in *Supercomputers in Chemistry* (Eds. P. Lykos and I. Shavitt), American Chemical Society, Washington DC, 1981.
131. Pippenger, N., *IBM J. Res. Dev.*, **25**, 825 (1981).
132. Chandra, A. K., IBM Research Report RC 6193 (26615), 1976.
133. McCullough, Jr, E. A., *Chem. Phys. Lett.*, **24**, 55 (1974).
134. McCullough, Jr, E. A., *J. Chem. Phys.*, **62**, 3991 (1975).
135. Adamowitz, L., and McCullough, Jr, E. A., *J. Chem. Phys.*, **75**, 2475 (1981).
136. Metropolis, N., and Ulam, S. M., *J. Am. Stat. Assoc.*, **44**, 335 (1949).
137. Anderson, J. B., *J. Chem. Phys.*, **63**, 1499 (1975).
138. Reynolds, P. J., Ceperley, D. M., Alder, B. J., and Lester, W. A., *J. Chem. Phys.*, **77**, 5593 (1982).
139. Moskowitz, J. W., Schmidt, K. E., Lee, M. A., and Kalos, M. H., *J. Chem. Phys.*, **77**, 349 (1982).
140. Wells, B. H., *Chem. Phys. Lett.*, **115**, 89 (1985).

Ab Initio Methods in Quantum Chemistry—I
Edited by K. P. Lawley
© 1987 John Wiley & Sons Ltd.

THE COUPLED PAIR APPROXIMATION

REINHART AHLRICHS AND PETER SCHARF

*Institut für Physikalische Chemie und Elektrochemie, Lehrstuhl für
Theoretische Chemie, Universität Karlsruhe, Kaiserstrasse 12, 7500
Karlsruhe, West Germany*

CONTENTS

NOTATION AND CONVENTIONS

The following notation is used throughout, unless stated otherwise:

ψ_0 zeroth-order wavefunction, usually of self-consistent field type

$E_0 = \langle \psi_0 | H | \psi_0 \rangle$ reference energy

φ_i spin–orbital, alternatively denoted i

$\phi_i^a, \phi_{ij}^{ab}, \ldots$ singly, doubly,... substituted (with respect to ψ_0) functions with occupied spin–orbitals $i, j, \ldots$ replaced by virtual spin–orbitals $a, b, \ldots$

C_i^a, C_{ij}^{ab} corresponding configuration expansion coefficients of the wavefunction

$$\psi_i = \sum_a C_i^a \phi_i^a, \quad \psi_{ij} = \sum_{a \leqslant b} C_{ij}^{ab} \phi_{ij}^{ab}$$

$$\psi_S = \sum_i \psi_i$$ sum of all single substitutions

ψ_D, ψ_T, ψ_Q analogous sums for double, triple and quadruple excitations

$P, Q, \ldots$ compound index, specifying the internal part of excited functions, e.g. $P = (ij)$ for doubles, $P = (i)$ for singles on the spin–orbit level, or as defined in text

ψ_c correlation function

$$\psi_c = \psi_S + \psi_D + \cdots = \sum_P \psi_P$$

$\psi = \psi_0 + \psi_c$ intermediate normalization

E_c correlation energy

$e_{ij} = \langle \psi_{ij} | H | \psi_0 \rangle$ pair energy, e_P used analogously

$$\langle ij \| ab \rangle = \int i(1) j(2) \frac{1}{r_{12}} (1 - P_{12}) a(1) b(2) \, d\tau$$

We will frequently rely on Moller–Plesset perturbation theory (MP-PT) as a means to order and classify terms occurring in the derivations. MP-PT uses Hartree–Fock as its zeroth approximation:

$$H_0(1, \ldots, n) = \sum_{v=1}^n F(v)$$

and further

$$V = H - H_0$$

I. INTRODUCTION AND HISTORICAL REVIEW

A. Antisymmetrized Product of Strongly Orthogonal Geminals

The first pair theory was proposed as long ago as 1953: the antisymmetrized product of strongly orthogonal geminals (APSG) of Hurley et al.[1]

$$\psi = \mathscr{A} \left[\prod_{k=1}^{n/2} w_k(2k - 1, 2k) \right] \tag{1}$$

The pair functions w_k are assumed to be strongly orthogonal.[1] The APSG method aimed at a concise description—including effects of electron correlation—of the electronic structure of atoms and especially molecules. The idea behind the APSG ansatz was clearly attractive. Two-electron systems were well understood and the accurate quantum-mechanical treatments for He, by Hylleraas,[2] and for H_2, by James and Coolidge,[3] could be hoped to be extendable to more general electron-pair systems. The APSG approach has an important feature: a system of separated pairs, such as He_n or $(H_2)_n$ at sufficiently large intersystem distances, is treated correctly, i.e. the method is size-consistent or size-extensive as it would now be stated.[4] Since then, the aim for size consistency has always been central to pair theories.

The APSG ansatz cannot be rigorous for interacting pairs, e.g. the bond electron pairs in CH_4. There was a strong hope, however, that the approximations involved would not lead to major losses in accuracy. Even in 1963 Clementi[5] estimated the bond correlation energy by simply dividing the valence-shell correlation energy by 4, the number of bond pairs. Unfortunately, matters were not so simple, as was clearly indicated by the first *ab initio* treatment of valence correlation effects of CH_4—which attributed only 56% of the valence correlation to intra-bond contributions[6]—and especially by applications of the APSG approach to BH.[7] As a consequence, the very first pair theory is too inaccurate for chemical applications and we will not discuss its attractive structural features and its properties any further (for details, see Kutzelnigg's review[8]).

There is clearly no extension of (1) aiming at the description of correlation effects among *all* possible pairs of electrons—or better (spin) orbitals—within a product ansatz for the total wavefunction. As a consequence, pair theories have developed in various directions and were not a really uniform undertaking. Their development was, of course, intimately tied to other techniques of electronic structure calculations, such as the configuration-interaction (CI) or perturbation theory methods.

Pair theories generally share the following two features:

1. The treatment of an n-electron system is reduced to (effective) two-electron systems, which may be coupled. (The scope of this fact is elucidated by the properties of two-electron functions—pair functions—which will be discussed below.)
2. The expression for the (total, correlation) energy scales properly with the size of the system (size extensivity of the energy).

A variety of pair approaches have been proposed and tested in applications which take advantage of the special properties of pair functions. All these methods are based on a reference or zeroth-order wavefunction ψ_0 and implicitly or explicitly assume the correlation effects to be relatively small. Until very recently ψ_0 was usually assumed to be a self-consistent field (SCF) wavefunction (the following discussion is done on the spin–orbit level unless stated otherwise):

$$\psi_0 = |\varphi_1 \varphi_2 \cdots \varphi_n| \tag{2}$$

Let us now briefly review some pair approaches in increasing order of complexity.

B. Independent Electron-pair Approach

The simplest way to take correlation effects into account is to treat explicitly only two electrons at a time and keep the others in their SCF orbitals.

To be more specific, one constructs two-hole functions

$$\phi_{ij} = a_i a_j \psi_0 \tag{3}$$

by means of the annihilation operators a_i and a_j referring to the occupied SCF orbitals φ_i and φ_j. The n-electron function

$$\psi_{ij} = \psi_0 + \mathscr{A}[\phi_{ij} u_{ij}] \tag{4}$$

is then obtained by coupling the two-electron function u_{ij} and the two-hole function ϕ_{ij}. The function ψ_{ij} includes correlation effects experienced by the pair (ij).

One then determines u_{ij} such that E_{ij} is minimized:

$$E_{ij} = E_0 + e_{ij} = \langle \psi_{ij}|H|\psi_{ij}\rangle / \langle \psi_{ij}|\psi_{ij}\rangle \tag{5}$$

where e_{ij} denotes the pair correlation energy of pair (ij), and the total correlation energy is approximated as

$$E_c = \sum_{i<j} e_{ij} = \left\langle \psi_0 \left| H \right| \sum_{i<j} \psi_{ij} \right\rangle \tag{6}$$

This approach is called the independent electron-pair approach (IEPA), since electron pairs are treated completely independently of each other. The IEPA has been developed and discussed extensively by Sinanoglu and by Nesbet,[9] although in different contexts and in variants which differ slightly from the one sketched above.

The IEPA is obviously size-extensive. A simple justification of the IEPA may be given by using a transition energy formula for E_c. Let

$$\psi = \psi_0 + \psi_S + \psi_D + \psi_T + \cdots \tag{7}$$

denote the decomposition of the full CI wavefunction; then

$$E_c = \langle \psi_0|H|\psi_D\rangle \tag{8}$$

The IEPA expression for E_c, Eq. (6), is in agreement with (8) as long as the ψ_{ij}

are sufficiently close to the double replacement contributions obtained from a full CI.

In other words, the IEPA is accurate up to second order of perturbation theory based on the Hartree–Fock (HF) model as zeroth-order approximation. The exact third-order contribution involves—besides the terms $\langle \psi_{ij} | H | \psi_{ij} \rangle$, which are properly included—the interaction between different pair correlation functions

$$E^{(3)}_{ij,kl} = \langle \psi_{ij} | H | \psi_{kl} \rangle \tag{9}$$

These terms are nowhere included within the IEPA, which is the major drawback of this procedure. As a consequence, one finds that E_c evaluated according to Eqs. (5) and (6) depends crucially on the actual choice of occupied molecular orbitals (MOs), i.e. E_c is not invariant with respect to a unitary transformation of occupied MOs. IEPA correlation energies may either overshoot or undershoot the exact E_c depending on the φ_i chosen. This deficiency may be rectified by an inclusion of (9) in E_c in the lowest order of perturbation theory,[10] but this approach will not been pursued further.

C. The Coupled Pair Approaches

Let us now turn to the coupled pair methods. The conventional discussion of these approaches starts from the well known $\exp(T)$ ansatz for the wave operator. Since this ansatz is reviewed elsewhere,[11] we will choose an alternative route, which is in fact more straightforward. From a practical (or technical) point of view, pair techniques aim at a size-extensive treatment at the computational expense of a CI calculation with single and double excitations (CI(SD)), which requires including effects of higher excitations in an approximate manner. This goal has been approached in basically three different ways:

1. The coupled electron-pair approximation (CEPA).[12] This starts from the hierarchy of the full CI equations. A truncation on the SD level is then achieved by approximating the Hamiltonian matrix elements which couple the singles and doubles with higher terms (triples, quadruples).
2. The coupled pair functional (CPF).[13] An alternative route is to begin with the convenient expression for the CI(SD) energy, to analyse its drawbacks (violation of size extensivity) and to remedy these shortcomings by means of appropriate modifications of the energy functional.
3. The variational CEPA (CEPA-V).[14,15] In this variant one constructs an energy functional which upon variation yields CEPA equations as closely as possible.

The first of these approaches, the CEPA method of W. Meyer,[12] leads to coupled pair equations of the form (if we concentrate on the doubles in this outline and neglect the singles)

$$\left\langle \phi_{ij}^{ab} \middle| H - E_0 - A_{ij} \middle| \psi_0 + \sum_{kl} \psi_{kl} \right\rangle = 0 \tag{10}$$

from which the coefficients C_{ij}^{ab}, which determine ψ_{ij}, are obtained. The total correlation energy is again determined according to Eq. (6). We have written the CEPA equations (10) in a way which comprises a variety of methods depending on the choice made for the shifts A_{ij}. Since a more detailed discussion is given below, we just mention some important aspects:

1. The CEPA equations (10) constitute a coupled system of equations for the pair correlation functions $\psi_{ij}(\psi_{ij}$ is, of course, an n-electron function but $(n-2)$ electrons are frozen in their HF orbitals and ψ_{ij} is essentially the pair correlation function for spin–orbitals i and j).
2. The choice $A_{ij} = E_c$ leads (strictly) to the CI(D) method, which is not size-extensive. The general implications of the choice for A_{ij} can be seen in the following way. In a first approximation one gets from Eq. (10)

$$C_{ij}^{ab} \approx - \langle \phi_{ij}^{ab} | H | \psi_0 \rangle / (\langle \phi_{ij}^{ab} | H | \phi_{ij}^{ab} \rangle - E_0 - A_{ij})$$

Size consistency requires C_{ij}^{ab} to be (roughly) independent of the size of the system under consideration. This condition is only met if A_{ij} does not increase with size—but is clearly violated if one puts $A_{ij} = E_c$. In order to verify this reasoning, consider a system of molecules B, C, D,... at sufficiently large intermolecular distances. Let i, j, a, b all refer to the same molecule, e.g. B, then C_{ij}^{ab} should not depend on the presence of the remaining subsystems C, D,..., as would be the case for $A_{ij} = E_c$.
3. A rather drastic remedy for this shortcoming is to put $A_{ij} = 0$. This leads to the linear version of Cizek's (coupled pair-many electron theory) CP-MET,[16,17] also called CEPA-0,[18] which is size-extensive.
4. An intermediate route is to put

$$A_{ij} = e_{ij} \tag{11}$$

(Compare Eq. (5).) This version is now usually called CEPA-2. It differs from IEPA in the important aspect that the ψ_{kl}, $(kl) \neq (ij)$, are included in (10), such that the coupling between ψ_{ij} and ψ_{kl} is now correctly described in the lowest order of perturbation theory.
5. There appears to be an agreement now that the best choice for A_{ij} is the one of the CEPA-1 version

$$A_{ij} = \tfrac{1}{2} \sum_k (e_{ik} + e_{jk}) \tag{12}$$

This choice has been justified by Meyer by means of a detailed consideration of separated electron pairs, e.g. He_n, where CEPA-1 gives the correct correlation energy.

The CEPA techniques can be derived in various ways. Size consistency can only be achieved by an approximate inclusion of higher excitations (than doubles). From this point of view, one starts best from the CP-MET method[17] and considers CEPA (versions 0, 1, 2) as simplifications of CP-MET. However, one may also start from a detailed consideration of the deficiencies of a CI(SD) and then remedy its shortcomings by appropriate modifications of the CI(SD) equations.

The CEPA versions do not obey the variational principle, i.e. E_c so obtained is not an upper bound to the eigenvalue of the Hamiltonian. Furthermore, CEPA-1 and CEPA-2 cannot be derived from variation of an energy functional, whereas this is possible for IEPA (compare Eq. (5)) and CEPA-0.[18] Recent efforts have concentrated on the development of a variational formulation of coupled pair techniques in the following sense: to define a correlation energy functional which leads to coupled pair equations. Such a functional (variational) formulation has several advantages, which are discussed in Section III.[14] Ahlrichs[18] has pointed out that Eq. (10) can be derived from variation of

$$F_c = \langle \psi | H - E_0 | \psi \rangle - \sum_{ij} A_{ij} \langle \psi_{ij} | \psi_{ij} \rangle \qquad (13)$$

$$\psi = \psi_0 + \psi_c$$

with respect to ψ_{ij} if A_{ij} are considered as external parameters not to be varied. It is in fact obvious that such a variation leads to the CEPA equations (10). Pulay et al.[14,15] have extended this idea and proposed to perform a complete variation of F_c (Eq. (13)) with A_{ij} from Eqs. (5) and (11), which implies a CEPA-2 type approximation. First applications have shown that results for energies, bond distances, etc., are virtually identical to the original CEPA-2, which has proved very useful in applications.

A different route has been taken by the present authors in the development of the CPF method, where one starts from the very beginning with a functional $F_c[\psi_c]$ which represents E_c in terms of ψ_c. The form of F_c is derived from the CI(SD) expression (compare the nomenclature for the definition of ψ_P, ψ_c)

CI(SD):

$$E_c = \frac{\langle \psi_0 + \psi_c | H - E_0 | \psi_0 + \psi_c \rangle}{1 + \langle \psi_c | \psi_c \rangle}$$

$$= \left[2 \sum_P \langle \psi_0 | H - E_0 | \psi_P \rangle + \sum_{P,Q} \langle \psi_P | H - E_0 | \psi_Q \rangle \right] \Big/ [1 + \langle \psi_c | \psi_c \rangle]$$

$$(14)$$

by approximately including effects of higher substitutions in order to achieve size consistency. Since higher excitations are dominantly of 'unlinked type' (i.e. their coefficients are essentially products of individual coefficients of singles and doubles, as will be discussed in Section III.A), their effect on E_c is largely to cancel $\langle \psi_c | \psi_c \rangle$ in the denominator of (14). This cancellation is accounted for in the CPF method by introducing individual partial normalization denominators N_P which replace the global denominator in (14)

CPF:

$$E_c = F_c[\psi_c] = 2 \sum_P \frac{\langle \psi_0 | H - E_0 | \psi_P \rangle}{N_P} + \sum_{P,Q} \frac{\langle \psi_P | H - E_0 | \psi_Q \rangle}{(N_P N_Q)^{1/2}} \qquad (15)$$

F_c is then minimized with respect to the ψ_P, which leads to variational equations closely related to the CEPA equations (10).

It is a simple exercise to verify that (15) is exact for the case of separated electron pairs provided the N_P are appropriately chosen, as discussed in Section III.C. This fact is actually not surprising: since separated pairs can be described accurately on the SD level—if the constituent subsystems are treated separately—there must be a way to carry this treatment over to the simultaneous treatment of pairs.

The accurate description of simple cases together with numerous applications strongly indicate that the CPF procedure gives a reliable account of cluster corrections arising from singles and doubles.

II. ELECTRON PAIR FUNCTIONS

Wavefunctions for two-electron systems have a number of special properties which result in drastic simplifications in their CI treatment. It is one of the basic ideas of pair theories to take advantage of these features when dealing with n-electron systems since this leads to considerable technical advantages.

First of all, two-electron functions factorize into a space part, ϕ, and a spin part, σ:

$$\psi(1, 2) = \phi(1, 2)\sigma(1, 2) \qquad (16)$$

σ is either a singlet, $S = 0$, or a triplet, $S = 1$, spin function. As a consequence of the Pauli principle, ϕ is symmetric or antisymmetric

$$\phi(1, 2) = p\phi(2, 1) \qquad (17)$$
$$p = 1 - 2S(= \pm 1) \qquad (18)$$

It has already been pointed out by Löwdin and Shull[19] that ϕ can be expanded in terms of simple products of orbitals a, b:

$$\phi = \sum_{a,b} C_{ab} \tilde{\phi}_{ab} \qquad (19)$$

$$\tilde{\phi}_{ab} = a(1)b(2) \qquad (20)$$
$$\mathbf{C}^+ = p\mathbf{C} \qquad (21)$$

The great advantage of Eqs. (19–21) becomes apparent if we consider linear transformations of orbitals from basis functions $\{a\}$ to $\{\alpha\}$

$$a = \sum_{\alpha} \alpha T_{\alpha a} \tag{22}$$

where $\mathbf{T}$ is only required to be regular, i.e. $\mathbf{T}^{-1}$ has to exist. This implies that the sets $\{\alpha\}$ or $\{a\}$ are linearly independent but need not be orthonormal, e.g.

$$\langle \alpha | \beta \rangle = S_{\alpha\beta} \neq \delta_{\alpha\beta} \tag{23}$$

The $\tilde{\phi}_{ab}$ are then given as

$$\tilde{\phi}_{ab} = \sum_{\alpha\beta} \tilde{\phi}_{\alpha\beta} T_{\alpha a} T_{\beta b} \tag{24}$$

i.e. the basis functions used in (19) transform as tensors of rank 2. In order that ϕ be invariant with respect to the basis set chosen

$$\phi = \sum_{\alpha\beta} D_{\alpha\beta} \tilde{\phi}_{\alpha\beta} = \sum_{ab} C_{ab} \tilde{\phi}_{ab} \tag{25}$$

one gets immediately

$$\sum_{ab} T_{\alpha a} C_{ab} T_{\beta b} = D_{\alpha\beta} \tag{26a}$$

$$\mathbf{D} = \mathbf{TCT}^{+} \tag{26b}$$

The coefficient matrix specifying ϕ according to Eq. (19) transforms as a tensor of rank 2.

$\mathbf{T}$ can especially be chosen to diagonalize the coefficient matrix

$S = 0$:

$$\phi = \sum_{\alpha} D_{\alpha} \alpha\alpha \tag{27}$$

$S = 1$:

$$\phi = \sum_{\alpha} D_{\alpha} [\alpha\tilde{\alpha} - \tilde{\alpha}\alpha] \tag{28}$$

where the orbitals involved can be required to be orthonormal. The representations (27) and (28) are the famous natural-orbital (NO) expansions of pair functions. They represent the most compact CI expansion of a two-electron wavefunction. The optimal convergence of the NO expansion is exploited in pair natural-orbital (PNO) methods (to treat n-electron systems) where either pair function is expanded in its NOs, the corresponding PNOs.[6,12,20,21]

The construction of NOs (PNOs) requires either knowledge of the pair function ϕ—the NOs are then obtained by diagonalizing the corresponding $\mathbf{C}$ matrix—or the solution of a multi-configuration SCF (MCSCF) problem, i.e. it requires the simultaneous optimizing of D_{α} and α in (27) or (28). However, very efficient methods have been developed to determine good approximations to NOs (PNOs) which result in a loss of 1–2% of the correlation energy only as compared to exact NOs.[12,20]

The PNO expansion of correlating pair functions considerably facilitates the processing of two-electron integrals since no complete integral transformation is required.[12,21] This has opened the way to employing—and virtually exhausting—relatively large basis sets at times when four-index transformations of two-electron integrals were still hard to perform routinely. Since the (single-reference) PNO CI technique has already been reviewed[22] we only mention the recent extension to the multiple-reference case by P. Taylor.[23] PNO methods have recently lost ground, mainly for reasons which will now be explained.

Consider the CI equations for an electron-pair system in spinless form, i.e. with ψ represented by ϕ as given in (19):

$$\langle \tilde{\phi}_{cd} | H - E | \phi \rangle = 0 \qquad (29)$$

which yields, for $H(1, 2) = h(1) + h(2) + g(1, 2)$,

$$\mathbf{R}[\mathbf{C}] := \mathbf{hCS} + \mathbf{SCh} + \mathbf{K}[\mathbf{C}] - E\mathbf{SCS} = 0 \qquad (30)$$

$\mathbf{R}[\mathbf{C}]$ denotes here the residual vector, $\mathbf{h}$ and $\mathbf{S}$ are the usual matrix representations, $h_{ab} = \langle a | h | b \rangle$, $S_{ab} = \langle a | b \rangle$, and

$$\mathbf{K}[\mathbf{C}]_{ab} = \sum_{cd} \langle \tilde{\phi}_{ab} | g | \tilde{\phi}_{cd} \rangle C_{cd} \qquad (31)$$

$\mathbf{K}[\mathbf{C}]$ is simply a generalization of the exchange operator familiar from the SCF theory

$$\mathbf{K}[\mathbf{C}]_{ab} = \sum_{cd} \langle a(1)c(1) | g(1, 2) | b(2)d(2) \rangle C_{cd} \qquad (32)$$

The present formulation of the CI equations is well suited for an integral driven procedure and constitutes a special version of the direct CI method of Roos and Siegbahn.[24,25]

The solution of the CI equations (30) is conveniently done in an iterative way. For a given $\mathbf{C}$ one evaluates the corresponding residuum $\mathbf{R}$ and then updates $\mathbf{C}$, until convergence is reached. Since (30) holds for any chosen orbital basis, it is clearly preferable to work in the original AO basis. This renders a transformation of two-electron integrals superfluous and the most time-consuming step in (30) concerns the evaluation of $\mathbf{K}[\mathbf{C}]$, Eq. (31).

There are various reasons in favour of working in an orthonormal basis, especially since the update procedure for $\mathbf{C}$ (with the aid of $\mathbf{R}[\mathbf{C}]$) is then much simpler. This is in fact easily achieved if one exploits the transformation properties (26), and the corresponding one for the exchange-type operator $\mathbf{K}$. For an orthonormal basis

$$\langle a | b \rangle = \delta_{ab}$$

the CI equations simplify to

$$\mathbf{R}[\mathbf{C}] = \mathbf{h}\mathbf{C} + \mathbf{C}\mathbf{h} + \mathbf{K}[\mathbf{C}] - E\mathbf{C} = 0 \tag{33}$$

where $\mathbf{K}[\mathbf{C}]$ is obtained as

$$\mathbf{K}[\mathbf{C}] = \mathbf{T}^+ \mathbf{K}^{\mathrm{AO}}[\mathbf{D}]\mathbf{T} \tag{34}$$

The superscript 'AO' for $\mathbf{K}$ indicates that the operator is constructed in the AO basis with $\mathbf{D}$ from Eq. (26), which is the CI vector (in matrix form) with respect to the AO basis.

The evaluation of $\mathbf{R}[\mathbf{C}]$, Eq. (32), requires the same computational work as a single SCF iteration, and the iterative solution of the CI equations (complete in the given basis) requires the same work as an SCF treatment since no integral transformation is required. The present formulation of the two-electron problem is not only formally simple, but it is also ideally suited for applications. These advantages are basically a consequence of the transformation properties (24)–(26), which result from the special ansatz (19) for the wavefunction.

The possibility of a very efficient CI treatment of two-electron wavefunctions was realized first by Ahlrichs and Driessler.[20] Meyer then showed with the development of the self-consistent electron-pair (SCEP) method[26] that the great structural simplicity, e.g. of Eq. (33), can basically be carried over to the treatment of n-electron systems, e.g. within the framework of a single-reference CI(D) treatment. Since then, various improvements and extensions have been proposed,[18,27] of which we mention especially the generalization to the MR-CI(SD) case.[28,29]

These techniques are clearly related to the direct CI method of Roos[24,25] and of Siegbahn[30,31] and are probably best called 'matrix oriented direct CI procedures'. The matrix formulation—essentially derived from the transformation properties (24)–(26)[32,33]—reduces logic in computer codes to a minimum and makes these methods ideally suited for vector computers.[34,35] The matrix oriented formulations do not require a complete integral transformation. This fact is an advantage mainly for large basis sets and small numbers of correlated electron pairs since the integral transformation is relatively unimportant otherwise.[35]

It could be considered as a drawback of the representation (19) for ϕ that the ϕ_{ab}, Eq. (20), do not possess proper permutation symmetry, and that the Pauli principle is expressed by (21). This deficiency is easily remedied if one defines ϕ_{ab} as

$$\phi_{ab} = [a(1)b(2) + pb(1)a(2)]/2 \tag{35}$$

which possesses the proper permutation symmetry. One can then replace $\tilde{\phi}_{ab}$ by ϕ_{ab} in all equations occurring above, especially in the important transformation properties (25) and (26b). For further details, the reader is referred to the literature,[33] where this approach is discussed in connection with the general MR-CI(SD) case.

No effort is made in this chapter to demonstrate how the gratifying properties of two-electron systems can be exploited in the treatment of n-electron systems and we refer the reader to the article of Werner[36] for a detailed account.

III. THEORY OF PAIR APPROACHES

A. The Original Coupled Electron-pair Approach

The CEPA procedures are most conveniently derived by starting from the hierarchy of the full CI equations, which are then truncated at the SD level in a way that maintains size extensivity. We will work at the spin–orbit level. The full CI equations then read, within the intermediate normalization,

$$E_c = \langle \psi_0 | H | \psi_S + \psi_D \rangle \tag{36}$$

$$\langle \phi_i^a | H - E_0 - E_c | \psi_0 + \psi_S + \psi_D \rangle = -\langle \phi_i^a | H | \psi_T \rangle \tag{37}$$

$$\langle \phi_{ij}^{ab} | H - E_0 - E_c | \psi_0 + \psi_S + \psi_D \rangle = -\langle \phi_{ij}^{ab} | H | \psi_T + \psi_Q \rangle \tag{38}$$

and so on for ϕ_{ijk}^{abc}, etc. The terms on the right-hand side of Eq. (37) and especially (38) introduce the coupling of singles and doubles to higher excitations. These terms are neglected in the CI(SD), which is the reason for the violation of size extensivity of this approach.

This shortcoming is rectified in CEPA by including the right-hand side in an approximate way. We will now demonstrate how this is done for the couplings between doubles and quadruples, $\langle \phi_{ij}^{ab} | H | \psi_Q \rangle$, occurring in (38). Unfortunately, a similar reasoning is not available for the terms involving the triples ψ_T, which also enter the wavefunction in second order and the energy in fourth order, as do the quadruples ψ_Q. Therefore, the influence of triples is usually neglected in pair approaches. This is justified to some extent since they contribute much less to the energy—but their neglect is certainly not very satisfying.

For an evaluation of the term $\langle \phi_{ij}^{ab} | H | \psi_Q \rangle$ we first of all need an approximation of ψ_Q, i.e. of the coefficients C_{ijkl}^{abcd}. For this purpose we resort to Moller–Plesset perturbation theory (MP-PT). A straightforward analysis yields (superscripts denote orders of perturbation theory (PT)), see e.g. Ref. 32,

$$C_{ijkl}^{(2)abcd} = C_{ij}^{(1)ab} C_{kl}^{(1)cd} - C_{ik}^{(1)ab} C_{jl}^{(1)cd} + \cdots \tag{39}$$

$$C_{ijkl}^{(2)abcd} = \tfrac{1}{32} \sum_{P,P'} \mathrm{sgn}(P)\,\mathrm{sgn}(P')\, C_{ij}^{(1)ab} C_{kl}^{(2)cd} \tag{40}$$

where P and P' denote permutations of $(ijkl)$ and $(abcd)$, respectively. This is an important result: the quadruple excitations are simply 'antisymmetrized products' of double substitutions in the lowest order of PT. An analogous result holds for higher terms: the $2n$-fold substitutions enter in nth order of PT and are just products of $C^{(1)ab}_{ij}$, properly antisymmetrized, of course.

The usual convergence problems of PT render the direct application of (40) inadvisable, but it appears reasonable to exploit the structure displayed by (40) in a self-consistency procedure:

$$C^{abcd}_{ijkl} \approx U^{abcd}_{ijkl} = \tfrac{1}{32} \sum_{P,P'} \text{sgn}(P)\,\text{sgn}(P') C^{ab}_{ij} C^{cd}_{kl}$$

$$= C^{ab}_{ij} C^{cd}_{kl} - C^{ab}_{ik} C^{cd}_{jl} + C^{ab}_{il} C^{cd}_{jk} \cdots \qquad (41)$$

It should be noted that the entity U displays the permutational symmetries required by the Pauli principle:

$$U^{abcd}_{ijkl} = \text{sgn}(P)\,\text{sgn}(P') U^{P'abcd}_{Pijkl}$$

If one inserts (41) into (38) and neglects singles and triples, one gets the famous CP-MET of Cizek[16]

$$\langle \phi^{ab}_{ij} | H - E_0 - E_c | \psi_0 + \psi_D \rangle = -\tfrac{1}{4} \sum_{\substack{kl \\ cd}} \langle kl \| cd \rangle U^{abcd}_{ijkl} \qquad (42)$$

which has to be used together with (36) and (41). The extension of this approach, which includes singles in addition to doubles, is now usually called CCSD (coupled-cluster singles and doubles).[11]

The CP-MET equations represent a non-linear coupled system of equations which for a long time has resisted direct solution. The first *ab initio* model calculation using CP-MET was published by Paldus *et al.*[37] in 1972, a more realistic application by Taylor *et al.*[38] in 1976. Although considerable progress has been made recently in this field,[11,39,40] it is probably fair to say that the CP-MET is too involved to call it a minor modification of the CI(SD).

Let us now analyse the structure of the term on the right-hand side of Eq. (42), which will open the way to further considerable simplifications leading to CEPA:

$$W^{ab}_{ij} = \sum_{klcd} \langle kl \| cd \rangle U^{abcd}_{ijkl}$$

$$= \sum_{klcd} \langle kl \| cd \rangle (C^{cd}_{kl} C^{ab}_{ij} - C^{cd}_{jl} C^{ab}_{ik} + \cdots) \qquad (43)$$

The first term can be summed immediately and yields $E_c C^{ab}_{ij}$. The remaining terms are more involved. However, since the indices of $\langle kl \| cd \rangle$ do not fit those of the doubles coefficients, one has—roughly speaking—of sum over products of more or less independent terms which can be expected to cancel to a large extent.

This reasoning can be supported in another way. The CP-MET equations are invariant with respect to a unitary transformation among the occupied and/or the virtual MOs, and one can work in a localized description. The correction (43) is non-negligible only if $\langle kl \| cd \rangle$ and at least one of the coefficient products are relatively large. The integral $\langle kl \| cd \rangle$ can only be relatively large if the MOs k, l, c and d are localized in the same region of space. The coefficient products in (43) beyond the first one are then relatively small if at least one of the MOs i, j, a and b is localized in a different region than that of MOs k, l, c and d, which constitutes the vast majority of cases.

Our conclusion is, therefore, that

$$W_{ij}^{ab} = \sum_{klcd} \langle kl \| cd \rangle U_{ijkl}^{abcd} \approx E_c C_{ij}^{ab} \tag{44}$$

If one simply uses, in Eq. (42),

$$W_{ij}^{ab} = E_c C_{ij}^{ab} \tag{45}$$

one gets the linear version of Cizek's CP-MET, which is now usually called CEPA-0

$$\langle \phi_{ij}^{ab} | H - E_0 | \psi_0 + \psi_D \rangle = 0 \tag{46}$$

The important difference between (46) and the CI(D) equations is that the right-hand side of (42) cancels (approximately) the correlation energy E_c on the left-hand side of (42).

The equations (46) are just the variational equations of the functional[18]

$$F_c = \langle \psi_0 + \psi_D | H - E_0 | \psi_0 + \psi_D \rangle \tag{47}$$

with respect to ψ_D. It is instructive to compare F_c with the expression of the correlation energy of the CI(D)

$$E_c = \frac{\langle \psi_0 + \psi_D | H - E_0 | \psi_0 + \psi_D \rangle}{1 + \langle \psi_D | \psi_D \rangle} \approx F_c - F_c \langle \psi_D | \psi_D \rangle \tag{48}$$

In F_c one has deleted the normalization denominator or, alternatively, the second term on the right-hand side of (48). It is in fact well known that the term $F_c \langle \psi_D | \psi_D \rangle$ is cancelled to a large extent by the contribution of quadruples. The Davidson correction[41] of the CI(SD) correlation energy and related procedures[42,43] is actually designed to cancel the unphysical term $F_c \langle \psi_D | \psi_D \rangle$ in (48).

There is a problem concerning both CEPA-0 and the Davidson correction. The normalization denominator in (48) cannot be cancelled completely by effects of higher excitations, as is obvious for two-electron systems where the CI(SD) is exact. The same difficulty occurs in the derivation of (46) from (42). In order to get (46) we have performed the unconstrained summation over k, l, c and d, but antisymmetric entities U_{ijkl}^{abcd} vanish whenever two indices are identical and the contributions from the first term are cancelled by the remaining terms. In other words, CEPA-0 and the Davidson correction formally include contributions of quadruples $(ijkl) \rightarrow (abcd)$ which do not appear in (7) due to the exclusion principle, e.g. for $i = k$ and/or $a = c$.

The main result of the above discussion may be phrased in the following way. If one wants to truncate the hierarchy of CI equations (36)–(38) on the CI(SD) level, one has to account for the coupling to higher excitations, especially ψ_Q, in order to maintain size extensivity. The dominant effect of quadruples is expressed in (44): it cancels or almost cancels E_c occurring on the left-hand side of the CI equation (38) or (42), respectively. This is taken into account in the CEPA techniques, which are quite generally written in the following form:

$$T_P^{ab} = \langle \phi_P^{ab} | H - E_0 - A_P | \psi_0 + \psi_S + \psi_D \rangle = 0 \qquad (49)$$

$$e_P = \langle \psi_P | H | \psi_0 \rangle \qquad (50)$$

$$A_P = \sum_Q T_{PQ} e_P \qquad (51)$$

Equation (49) differs from the CI(SD) in the replacement of E_c by the shift A_P, which is defined in (51) with the aid of the topological factors T_{PQ}.[13] Depending on the actual choice made for T_{PQ} the following methods are covered by equations (49)–(51):

CEPA-0:
$$T_{PQ} = 0 \qquad (52)$$

CI(SD):
$$T_{PQ} = 1 \qquad (53)$$

CEPA-2:
$$T_{PQ} = \delta_{PQ} \qquad (54)$$

CEPA-1:
$$T_{PQ} = \begin{cases} 1 & P, Q \quad \text{joint} \\ \frac{1}{2} & P, Q \quad \text{semi-joint} \\ 0 & \text{else} \end{cases} \qquad (55)$$

The specification of T_{PQ} refers to a treatment on the spin–orbit level. P and Q are called joint if $P = Q = (ij)$ or $P = (i)$ and $Q = (ij)$; they are called semi-joint if $P = (ij)$ and $Q = (ik)$, $k \neq j$. The choice of shifts A_P for the singles appears to be different in existing CEPA implementations.[18]

Let us comment briefly on the above equations. The choices (52) and (53), which lead to CEPA-0 and the CI(SD), need no explanation. CEPA-2, Eq. (54), is related to the IEPA where one treats one electron pair at a time, but different from IEPA we have now properly included the coupling $\langle \psi_P | H | \psi_Q \rangle$ in our treatment.

It appears that most workers in this field now (slightly) prefer CEPA-1, which was first proposed by Meyer.[12] His reasoning—which led to the choice (55)—is interesting and typical of the development of pair theories. Let us again neglect singles, which contribute to the energy in fourth order only. A modification of the CI(D) to achieve size consistency requires the replacement of E_c occurring in Eq. (38) by a shift A_P, the absolute value of which is much smaller than E_c, while neglecting the right-hand side. Meyer then looked for an expression for A_P which reproduced the exact (size-extensive) correlation energy for the model case of n separated electron pairs. CEPA-2 is exact in this case if localized occupied MOs φ_i are used—all CEPA variants are invariant with respect to a unitary transformation of virtual MOs—but it is not if the φ_i are delocalized. The choice (55) for T_{PQ} was made because it leads to the exact correlation energy both for a localized and a completely delocalized description. Although this holds only for the model system of separated pairs, it may be safely expected that CEPA-1 is also a reliable approximation in the general case.

The above formulation refers to a treatment on the spin–orbit level. CI calculations are usually done by using properly spin-coupled configuration-state functions (CSFs) with external pairs coupled to singlets or triplets, since this appears to be computationally most efficient. A conversion of equations (49)–(55) then requires an averaging procedure—over functions differing in spin distributions only—which has been discussed in detail by Hurley.[44] The present authors recommend for CEPA-1 the T_{PQ} given below in connection with the CPF approach.[13,18]

B. Variational Coupled Electron-pair Approach

In recent work on gradient evaluations in coupled pair theories, Pulay[14] gave a succinct description of the properties which should ideally be incorporated in a fictitious pair theory:

1 Computational simplicity: the method should not require significantly more computational effort than the corresponding CI problem.
2 The energy should ideally be a variational upper bound.
3 The total energy should be size-consistent.
4 The method should be exact for two-electron systems.
5 The energy and the wave function should be invariant with respect to unitary transformations among the strongly occupied orbitals, and also among the virtual orbitals.
6 The method should be free of singularity even in the quasi-degenerate case.
7 It should be possible to efficiently evaluate energy gradients with respect to an external perturbation, primarily nuclear motion.

One is forced to drop some of these features in the design of a computational method since the desired properties are to some extent mutually inconsistent.[14]

The advantage of coupled pair methods results from combining computational simplicity and size consistency while incorporating partly— depending on the version—the features (4), (5) and (6). The variational coupled pair methods, which will be discussed further below, allow for an efficient evaluation of gradients and, therefore, combine the largest subset of the desired features. The functional formulation has two additional advantages: (i) The density matrices are easily defined by means of PT.[13] (ii) Using a functional formulation of a method usually improves its convergence characteristics.

We now will discuss the method of matching a functional to a given CEPA variant. It has already been pointed out by one of us (R.A.)[18] that the CEPA equation (49) can be obtained by a variation of the functional

$$F_c = \langle \psi_0 + \psi_c | H - E_0 | \psi_0 + \psi_c \rangle - \sum_P A_P \langle \psi_P | \psi_P \rangle \qquad (56)$$

if the A_P are considered as external parameters which are *not* varied. At the stationary point, i.e. if Eq. (49) holds, one gets

$$F_c = E_c = \langle \psi_0 | H | \psi_c \rangle \qquad (57)$$

Pulay reconsidered Eq. (56) in his paper on gradient evaluation in coupled pair theories. There he aimed for a thorough variation of (56) and named this approach variational CEPA (CEPA-V).

We shall demonstrate the underlying idea and follow the derivation of CEPA-2V—a CEPA-2 like method—by Pulay.[15] Pulay contends that a full variation of the functional (56) may yield equations resembling CEPA-2 if the shifts $A_P = e_P$ are properly scaled by a factor X. X is then determined by the condition that unwanted terms introduced by the variation should cancel as completely as possible. The variation of the modified functional

$$F_c = \langle \psi_0 + \psi_c | H - E_0 | \psi_0 + \psi_c \rangle - X \sum_P e_P \langle \psi_P | \psi_P \rangle \qquad (58)$$

results in the set of equations

$$\langle \phi_P^{ab} | H - E_0 - e_P | \psi_0 + \psi_c \rangle - \tfrac{1}{2} X \frac{\partial e_P}{\partial C_P^{ab}} \langle \psi_P | \psi_P \rangle + (1 - X) e_P C_P^{ab} = 0 \qquad (59)$$

Equation (59) deviates from CEPA-2 by the second and third terms.

Pulay's original reasoning[14,15] was that the wavefunction determined as a solution of (59) should be as close as possible to the CEPA-2 wavefunction. This led to the choice $X = \tfrac{2}{3}$. Later on the value $X = 1$ was favoured since the functional F_c (58) is then identical to the original one, (56). However, inserting the solution of (59) into (58) results in

$$E_c = F_c = \langle \psi_0 | H | \psi_c \rangle + (X/2) \sum_P e_P \langle \psi_P | \psi_P \rangle \qquad (60)$$

This expression for E_c deviates already in fourth order from the usual transition energy formula (57). Therefore, CEPA-2V is expected to overshoot CEPA-2 correlation energies for $X > 0$, in agreement with model calculations.[45]

In the opinion of the present authors, a minor flaw of the functional (58) is its dependence on the third power of ψ_P, which renders it unbound. F_c depends on ψ_c in third order and a complete variation yields $\|\psi_c\| = \infty$ and $E_c = -\infty$.[13]

C. The Coupled Pair Functional Method

We shall now demonstrate the second approach towards variational pair theory. A convenient starting point is the classification of the problems arising from the use of CI(SD). Again we shall rely on PT as a tool for investigations.

In order to motivate the following modifications of the CI(SD) energy expression, we reconsider the statement concerning Eq. (48). The CI(SD) correlation energy may be written

$$E_c = \langle \psi_0 + \psi_c | H - E_0 | \psi_0 + \psi_c \rangle / (1 + \langle \psi_c | \psi_c \rangle) \tag{61}$$

if the intermediate normalization of the wavefunction is chosen. The expansion of the denominator leads to

$$E_c = \langle \psi_0 + \psi_c | H - E_0 | \psi_0 + \psi_c \rangle (1 - \langle \psi_c | \psi_c \rangle) + O(V^6) \tag{62}$$

where $O(V^6)$ denotes a deviation of sixth order in V. As already mentioned, the second term on the right-hand side of Eq. (62) will be cancelled to a large extent by contributions of higher excitations.

This fact is exploited by Davidson's correction[41] which recommends omitting the second term and the sixth-order terms in (62). This prescription, however, suffers from two basic shortcomings. First, the formula tends to overshoot the contributions of higher excitations, which is obvious for the two-electron case where no correction should be made. Secondly, the Davidson corrected energy expression is evaluated with the CI(SD) wavefunction. One thereby underestimates the cluster corrections, as is easily verified for simple models such as He_n, where both the CI(SD) and the corrected correlation energies increase like $\sqrt{n}$.

The discussion so far reveals the CI(SD) denominator to be too large—which causes the size-consistency problem—which is rectified to some extent by the Davidson correction. A reasonable procedure to cure these shortcomings from the very beginning would be to modify the denominator in the CI(SD) energy expression in order to get rid of the size-consistency problem and then to treat the modified energy expression by means of a variational procedure. To be more specific, one would attempt to fix the denominator by forcing the whole procedure to give exact results for some representative but simple systems, for example He_n.

The approach just sketched was used in the derivation of the CPF method. We shall not go into all details but only give a short outline of our reasoning. The expression (61) for the CI(SD) correlating energy is rewritten in the form

$$E_c = \left(2 \sum_P e_P + \sum_{P,Q} \langle \psi_P | H - E_0 | \psi_Q \rangle \right) \Big/ \left(1 + \sum_P \langle \psi_P | \psi_P \rangle \right) \qquad (63)$$

stressing the contributions arising from each electron pair P. To be more precise, since singles are included, one should call P a portion,[23] which labels all kinds of excitations.

The modification of the CI(SD) denominator is performed in the most flexible way by introducing a topological matrix $\mathbf{T}$ which mediates the coupling of different electron pairs, thus allowing for individual denominators corresponding to different electron pairs. The CPF energy expression then reads

$$F_c = 2 \sum_P e_P / N_P + \sum_{P,Q} \langle \psi_P | H - E_0 | \psi_Q \rangle / (M_P M_Q) \qquad (64)$$

$$N_P = 1 + \sum_Q T_{PQ} \langle \psi_Q | \psi_Q \rangle \qquad (65)$$

$$T_{PQ} = T_{QP} \qquad (66)$$

$$M_P = \sqrt{N_P} \qquad (67)$$

The topological factors T_{PQ} were determined for the single-reference case considering all singles and doubles interacting with the reference. We need the following definitions (compare with nomenclature):

We write $P = (ijp)$ for doubles and $P = i$ for singles, where p labels the different spin couplings if necessary. This nomenclature is self-evident for closed-shell cases. For open-shell cases, it implies a special spin coupling of singles. $P = i$ is understood to label the CSFs which have vanishing interaction with ψ_0 by means of the Brillouin theorem. The singles—from i—which in addition involve a spin flip of a singly occupied MO j are indexed as $P = (ijp)$, since the corresponding CSFs are double replacements on the spin–orbit level. An analogous route is pursued for the fully internal excitations.

The matrix $\mathbf{T}$ was fixed by the requirement that the following simple systems should be described as exactly as possible by the functional (64) which clearly is size extensive:

1. a system of separated closed-shell electron pairs, i.e. two-electron systems at mutually infinite distances,
2. identical separated closed-shell pairs, e.g. He_n, without restriction to a localized representation (this implies unitary invariance with respect to a transformation of occupied MOs),
3. separated triplet pairs, and
4. identical separated triplet pairs.

CPF complies exactly with conditions 1. and 2., and up to $O(V^6)$[13] with conditions 3. and 4.

Using the definitions introduced above and denoting the occupation number of the orbital i in the reference configuration by n_i we obtain

$$T_{PQ} = \frac{\delta_{ik} + \delta_{il}}{2n_i} + \frac{\delta_{jk} + \delta_{jl}}{2n_j} \tag{68}$$

for $P = (ijp)$, $Q = (klq)$. This formula covers singles if we formally equate $P = i$ with $P = ii$ for this purpose.

We shall demonstrate our thought for the first case, a system of separated closed-shell electron pairs. Starting from localized orbitals, it is easy to write a functional of the form (64) which yields the correct correlation energy on the 'CI(SD) level'

$$F_c = 2 \sum_P e_P / N_P + \sum_P \langle \psi_P | H - E_0 | \psi_P \rangle / N_P \tag{69}$$

$$N_i = N_{ii} = 1 + \langle \psi_i | \psi_i \rangle + \langle \psi_{ii} | \psi_{ii} \rangle \tag{70}$$

Here one has to consider only the cases $P = i$ and $P = ii$ since the remaining ψ_P vanish. F_c from Eqs. (69) and (70) is clearly a special case of Eqs. (64)–(67). In other words, this trivial model leads to the requirement $T_{PQ} = 1$ for $P = Q = ii$ and $T_{PQ} = 0$ for $(P, Q) = (ii, jj)$, $i \neq j$, and analogously for the singles as indicated in connection with Eq. (68). A similar reasoning is used to squeeze further conditions on the **T** matrix out of the remaining cases 2. to 4.

Let us finally comment on the term $\langle \psi_P | H - E_0 | \psi_Q \rangle / (M_P M_Q)$ occurring in Eq. (64). For separated pairs, only the case $P = Q$ occurs, and all considerations sketched above do not fix the denominator in a unique way for $P \neq Q$. Any other averaging instead of $M_P M_Q$, e.g. $0.5(N_P + N_Q)$, would also have been possible. The actual choice of the denominator affects F_c only in fifth order in V for $P \neq Q$ (the corresponding contribution for $P = Q$ enters in fourth order), and we have, therefore, made in Eq. (64) what appeared to be a simple and reasonable choice to us.

Variation of the functional F_c, Eqs. (64)–(67), with respect to the coefficients C_P^{ab} is straightforward and yields

$$\left\langle \phi_P^{ab} \middle| H - E_0 - A_P \middle| \psi_0 + M_P \sum_Q \psi_Q / M_Q \right\rangle = 0 \tag{71}$$

where

$$A_P = \sum_Q (N_P / N_Q) T_{PQ} \tilde{e}_Q \tag{72}$$

$$\tilde{e}_P = \frac{2 \langle \psi_0 | H - E_0 | \psi_P \rangle}{N_P} + \sum_Q \frac{\langle \psi_Q | H - E_0 | \psi_P \rangle}{M_Q M_P} \tag{73}$$

The $\tilde{e}_P$ constitute a decomposition of the total correlation energy into contributions corresponding to the respective P, since

$$F_c = \sum_P \tilde{e}_P \qquad (74)$$

In order to discuss some general features of the present method we shall now have a closer look at $\tilde{e}_P$ and F_c in terms of the correlation function obtained as a solution of Eqs. (71)–(74).

Since $\tilde{e}_P = O(V^2)$, $N_P = 1 + O(V^2)$, $A_P = O(V^2)$, it follows immediately that the ψ_P depend on the choice of T_{PQ} in terms proportional to V^3:

$$\psi_c(T) - \psi_c(T') = O(|T - T'| * V^3) \qquad (75)$$

and consequently

$$\tilde{e}_P(T) - \tilde{e}_P(T') = O(|T - T'| * V^4) \qquad (76)$$

$$E_c(T) - E_c(T') = O(|T - T'| * V^4) \qquad (77)$$

This is expected, of course, since cluster corrections enter the energy in fourth order. Rewriting Eqs. (73) and (74) we get

$$\tilde{e}_P = \langle \psi_0 | H - E_0 | \psi_P \rangle + O(V^4) \qquad (78)$$

$$E_c = \langle \psi_0 | H - E_0 | \psi_c \rangle + O(V^6) \qquad (79)$$

The present method does not lead to the usual transition energy formula, but the deviation—the second term on the right-hand side of Eq. (79)—starts in sixth order in V. The individual terms furthermore cancel to some extent at least, and do so exactly for special cases, e.g. for separated pairs.

In connection with the variational equation, let us comment briefly on the variational principle. F_c is not bounded from below by the lowest eigenvalue of H, as was shown by a rather extreme application.[13] However, it is easily shown that F_c is at least bounded from below by some real number B:[13]

$$F_c[\psi_c] \geqslant B \qquad B > -\infty \qquad (80)$$

Although Eq. (80) is probably of little practical help, one may hope that the present functional is also quite stable in the critical case of near-degeneracies, since F_c cannot collapse to $-\infty$.

It is worth noting that the physical reasoning used to fix the coupled pair functional leads to variational equations which are closely connected to Meyer's CEPA-1. This is obvious if one writes the CEPA-1 equations in the present nomenclature

$$\left\langle \phi_P^{ab} \middle| H - E_0 - A_P \middle| \psi_0 + \sum_Q \psi_Q \right\rangle = 0 \qquad (81)$$

$$A_P = \sum_Q T_{PQ} e_Q \qquad e_P = \langle \psi_0 | H - E_0 | \psi_P \rangle \qquad (82)$$

$$E_c = \sum_P e_P$$

and compares them with Eqs. (71)–(74). For the present choice of the matrix **T** the quantities A_P have the same structure as the CEPA-1 energy shifts for the closed-shell case if only double substitutions are considered. Using the same matrix **T** in Eq. (82) and in Eqs. (65) and (72), one finds agreement up to and including third order in the correlation functions ψ_P and fourth order in E_c.

The functional formulation of the pair approach leads to an unambiguous way to evaluate density matrices required for the computation of properties. For this purpose one uses perturbation theory and considers a perturbation G:

$$H(\lambda) = H + \lambda G \tag{83}$$

The density matrix γ is then uniquely determined by the requirement

$$\left(\frac{\partial E}{\partial \lambda}\right)_{\lambda = 0} = \text{tr}(\gamma G) \tag{84}$$

For the detailed analysis concerning one-particle operators G, which leads to the one-particle density matrix, the reader is referred to the literature.[13]

IV. APPLICATIONS

A. Introductory Comments

In this section we will review some recent applications in order to assess the scope and limitations of pair approaches. This is an apparently simple task since theories have to be judged by a comparison with measurements. Unfortunately, matters are more complicated: electronic structure calculations are almost exclusively performed within the orbital approximation and it is difficult to distinguish between technical (basis set saturation) deficiencies and shortcomings of methods. A comparison with experiment may further suffer from uncertainties in the measurements or their evaluation and such usually subtle effects as relativistic corrections or zero-point vibrations (for R_e or D_e).

These difficulties can be avoided by the comparison of full CI results with those obtained by approximate methods—such as CEPA, CPF, CP-MET or many-body perturbation theory (MBPT)—using identical basis sets. Proceeding in this way, one exclusively establishes the errors introduced by the corresponding approximations. This approach will be pursued in Section IV.B where pair approaches will be compared with full CI and other methods aiming for size extensivity. Such a comparison is instructive but clearly not too conclusive since full CI calculations are available for relatively small systems and small basis sets only.

In the remaining subsections we therefore have to compare with experimental results. By virtue of the basis saturation problems of electronic structure calculations, this can only be meaningful if it is considered in connection with the basis set convergence of computed properties. We want to demonstrate the importance of this point by a typical example.

Single-reference CI(SD) calculations of the ground state of Cl_2 within an $(11, 7, 2, 1)/[7, 4, 2, 1]$ basis yield[46]

CI(SD):	$R_e = 199.0$ pm
CPF:	$R_e = 200.9$ pm
Expt:	$R_e = 198.8$ pm

The CI(SD) result is in perfect agreement with experiment.[47] This would prove the superiority of the CI(SD) over the CPF method of calculation if the basis set were sufficiently saturated. The latter point can be checked by increasing the basis set, e.g. to $(11, 7, 3, 2, 1)/[7, 4, 3, 2, 1]$, which leads to

CI(SD):	$R_e = 198.2$ pm
CPF:	$R_e = 199.9$ pm

The CI(SD) result is now 0.6 pm too short, the CPF value 1.1 pm too long! A further extension of the basis certainly leads to a further shortening of computed R_e, which then decreases the agreement of CI(SD) with experiment, whereas the CPF result becomes better.

The conclusion is clear cut: agreement with experiment does not necessarily prove the superiority of a computational procedure, since it may simply result from a fortuitous error cancellation. On the other hand, there would be no objection against utilizing error cancellations if only considerable regularity could be assumed. In fact, the whole secret of numerical treatments is, of course, to exploit error cancellation where it can be relied upon, e.g. in a series of molecules with comparable electronic structure. However, *ab initio* methods are typically required to treat exceptional cases and for these one cannot rely on a cancellation of errors. Under these circumstances, we are fortunate to have available a number of coupled pair calculations with sufficiently large basis sets.

B. Comparison of Methods

The comparison with full CI results provides the best test of approximate methods since the errors introduced by the approximations then become immediately obvious. Thanks to the remarkable progress achieved in the field of full CI treatments, such results are available, e.g. for BH, NH_3, H_2O (at R_e, $1.5R_e$ and $2R_e$) and HF.[48,49]

In Table I we compare the full CI correlation energies with those obtained from single-reference type treatments which aim for size consistency, such as the Davidson corrected CI(SD) (and the CI(SD) itself, of course), the MBPT(2) and MBPT(4),[50] the CCSD,[50] the symmetry-adapted cluster methods SAC-A and SAC-B[51] and the CPF methods. (We have not included MR-CI(SD)

TABLE I
Comparison of computed correlation energies E_c from full CI treatments with approximate methods.[a] (See text for details of methods.)

	$\overline{\Delta E}$	s.d.	$\overline{\Delta E}$	s.d.
CI(SD) + D.C.	1.6	0.8	1.8	0.6
SAC-B[b]	2.8	1.4	2.6	1.5
CPF	3.9	2.8	2.9	1.6
CCSD[c]	3.9	3.1	2.8	1.7
SAC-A[b]	4.2	5.2	2.7	1.8
SDTQ MBPT(4)[c]	5.2	6.0	2.9	2.5
SD MBPT(2)[c]	7.1	7.0	4.5	3.2
CI(SD)	18.7	21.4	10.3	6.9

[a] $\overline{\Delta E}$ is the mean absolute deviation, s.d. the corresponding standard deviation with respect to the full CI in mhartree. The first two columns refer to the sample BH, NH_3, HF, $H_2O(R_e)$, $H_2O(1.5R_e)$, $H_2O(2R_e)$. The case $H_2O(2R_e)$ has been left out in the last two columns.
[b] Ref. 51.
[c] Ref. 50.

calculations[52,53] since they are in a different class.) Since similar comparisons have already been discussed in detail elsewhere,[13] we have only listed the average absolute deviations for the sample BH, NH_3, $H_2O(R_e)$, $H_2O(1.5R_e)$, $H_2O(2R_e)$ and HF—with respect to full CI—and the corresponding standard deviations. In this way one gets a very concise description, although it has to be mentioned that some interesting details are suppressed. Of the methods included in Table I, the deviations from the full CI are largest for the CI(SD), as expected, and are smallest for the Davidson corrected CI(SD), which is not so expected. However, the Davidson corrected CI(SD) becomes markedly poorer if larger systems are considered, e.g. $(HF)_2$ or $(HF)_3$ where a comparison is easily done.[13] This simply reflects the fact that the Davidson correction does not achieve size extensivity.[32]

The comparison of techniques which aim for size extensivity by approximate inclusion of higher excitations—cluster corrections—reveals a slight advantage of SAC-B over the CPF and the CCSD methods, which are of comparable accuracy, whereas SAC-A is slightly poorer. The largest deviations, besides the CI(SD), are found for the second- and fourth-order perturbation expansions, MBPT(2) and MBPT(4). However, this results mainly from the $H_2O(2R_e)$ case, for which one would not normally apply single-reference procedures. If this case, $H_2O(2R_e)$, is left out of the sampling, then SAC-A, SAC-B, CCSD, MBPT(4) and CPF show comparable accuracy, with ΔE_c between 2.6 and 2.9 mhartree (and 1.8 mhartree for the Davidson corrected CI(SD), 4.5 mhartree for the MBPT(2)).

TABLE II

Energies (a.u.) of BeH_2 for three geometries[a] along the C_{2v} reaction path for the symmetric insertion reaction. (See text for details of methods.)

	Geometry 1	Geometry 2	Geometry 3
MR-LCCM[b]	−15.625 50	−15.605 32	−15.630 46
MR-MBPT(3)[b]	−15.627 8	−15.601 1	−15.612 0
CI(SD)[b]	−15.619 71	−15.591 47	−15.618 53
CI(SD) + D.C.[c]	−15.625 08	−15.609 16	−15.630 87
CCSD[b]	−15.621 71	−15.599 2	−15.624 19
CPF[c]	−15.622 67	−15.601 28	−15.626 87
Full CI[b]	−15.622 88	−15.602 92	−15.624 96

[a]Geometry 1: $R(Be–H_2) = 2.5a_0$, $R(H–H) = 2.78a_0$.
Geometry 2: $R(Be–H_2) = 2.75a_0$, $R(H–H) = 2.55a_0$.
Geometry 3: $R(Be–H_2) = 3.0a_0$, $R(H–H) = 2.32a_0$.
[b]Ref. 54.
[c]Ref. 55.

We want to be somewhat more explicit in the discussion of the insertion reaction of Be into H_2 for which full CI results[54] are available too. Recently, Jankowski[55] performed calculations using the CPF method for three points on the C_{2v} reaction path, as in previous work.[54] The results obtained are listed in Table II. They reveal a slight superiority of CPF over CCSD and the other methods for the geometries 1 and 2. CPF, however, is somewhat less accurate than CCSD for geometry 3, for which the distance between the Be and H_2 is largest. Nevertheless, CPF results are better than those of all other methods shown. The example demonstrates again that CPF is able to compete with other more complicated methods even in cases with near-degeneracies, although this reaction has to be looked upon with some care, since only three electron pairs are involved. The CPF energies differ from the full CI results by only 0.21, 1.64 and 1.91 mhartree for cases 1–3 and show the best overall performance of the approximate methods considered in Table II. Only for case 3 is the CCSD result better, deviation of 0.77 mhartree from full CI, than CPF. It should especially be noted that the (single reference) CPF energies are even more accurate than those obtained with the linearized multiple reference coupled cluster method (MR-LCCM),[54] which differ from full CI by 2.62, 2.40 and 5.50 mhartree for cases 1–3, respectively.

Any comparison of methods has to take into account the respective computational expenditures involved. Although no computation times have been published for the approximate treatments considered in Table I, the following comments can safely be made. The MBPT(2) certainly requires the least effort, whereas SAC-A, SAC-B, CCSD, CPF and CI(SD) are roughly comparable, probably with an advantage for the CI(SD) and the CPF (the

Davidson correction requires virtually no time). The MBPT(4) is the only method involving an $n^3 N^4$ step (n = number of orbitals correlated, N = number of virtual MOs), whereas the CI(SD) and CPF include $n^2 N^4$ and $n^3 N^3$ steps only. An MBPT(4) treatment can easily require one (or even two) orders of magnitude more CPU time than CI(SD) or CPF, depending on the size of the system and/or the basis set.

The results collected in Table I show a very close agreement between CPF, CCSD and the SAC methods. Available evidence also shows only small deviations between CPF and CEPA-1.[13] These methods provide an accurate description of cluster effects by approximating the coefficients of quadruples and higher excitations as products of doubles. The remaining deviation from full CI has been shown (for the CCSD) to be mainly due to contributions of linked triples.[56] The latter terms are not accounted for in CPF and the other methods.

A comparison of CP-MET with several versions of CEPA was performed by Koch and Kutzelnigg.[57] These authors employed approximate PNOs. The effect of this approximation on computed properties is hard to estimate. However, this uncertainty should not affect the conclusions drawn with respect to the comparison of methods.

The comparison is based on results for correlation energies of BH_3, H_2O and HCN at their equilibrium geometries and the potential curves of N_2 and F_2. Furthermore, some calculations on Be, Be_2 and BH are described. For the sake of consistency with previous work, the authors chose basis sets regarded nowadays as relatively small (only a single polarization set is used in most cases).

First of all, the authors found the CEPA schemes to deviate very little from CP-MET for the 'good' cases (deviations of around $\pm 2\%$ for correlation energies), that is for those which may be calculated reliably with either method. For these calculations they found the CEPA-1 version to be closest to CP-MET. Secondly, Koch and Kutzelnigg report that CEPA-2 agrees best with experiment. This is attributed to the simulation of triple excitations in the CEPA-2 version.[57,58] However, one may argue that this good agreement arises from error cancellation: basis set deficiencies reduce the usual overshoot of CEPA-2 correlation energies.

The cases studied by Koch and Kutzelnigg do not exhibit significant superiority of any of the schemes considered. Thus, their results purport the CEPA methods to be preferable if computational simplicity is taken into account.

Similar conclusions have been reached by Jankowski and Paldus[59] in a careful comparative study involving CP-MET and various CEPA-type versions. These investigations were concerned with H_4 as a model system and especially concentrated on the reliability of coupled pair approaches in the presence of quasi-degeneracy. It turned out that single-reference coupled pair methods except CEPA-0 and CEPA-2 performed unexpectedly well.

C. Diatomic Hydrides

The reliability of CEPA in molecular calculations was assessed in a series of landmark papers on first- and second-row diatomic hydrides by Meyer and Rosmus.[60] They reported the results of systematic investigations using appropriate basis sets and the PNO-CEPA method. They calculated bond lengths R_e, values for D_0, the dipole moment μ_e and spectroscopic constants ω_e, α_e and $\omega_e x_e$. The basis sets employed were designed to account for roughly 90% of the valence correlation for first-row hydrides and about 85–90% for second-row hydrides (2d1f polarization sets were used).

The small number of electrons involved in the binding of diatomic hydrides allows for a relatively good description even on the PNO-CI level. The consistently better results obtained with PNO-CEPA, however, show the superiority of this method for a series of very different molecules. The molecular constants derived from the CEPA potential curves exhibit high accuracy when compared with experiment.

For the molecules LiH–HF and NaH–HCl, the standard deviations (CEPA-1 vs. experiment) are, for the bond lengths,

$$\Delta R_e = 0.003 \text{ Å}$$

and for the vibrational frequencies,

$$\Delta \omega_e = 14 \text{ cm}^{-1}$$

For the dissociation energies a maximum error of

$$\Delta D_e(\text{max}) = 0.3 \text{ eV}$$

is calculated, which is not unusual for this very demanding quantity.

In their work, Meyer and Rosmus also investigated the influence of intershell correlation. They found non-negligible effects for the left side of the periodic table which stems from the softer cores of alkali and alkaline-earth metals compared to the atoms in the neighbourhood of halogens. The effect on R_e and ω_e decreases along the rows, as expected. The authors extrapolate negligible influence of intershell correlation right of BH and SiH.

The work on the diatomic hydrides is completed by two papers on the ionization energies and the spectroscopic constants of the resulting ions and the electron affinities of the hydrides and spectroscopic constants of the negative ions.[61]

D. Diatomic Molecules

In this subsection we will summarize results obtained with the CPF approach. First, we will elaborate on CO, N_2, NO, O_2 and F_2,[62] secondly on the molecules Cl_2 and P_2,[46] and, thirdly, we will present some results obtained for Cu_2.[63,64]

1. CO, N_2, NO, O_2 and F_2 [62]

The calculations were performed to assess the typical accuracy of CPF for diatomic molecules involving first-row atoms. We expected to get marked improvement of calculated dissociation energies D_e when using the size-consistent CPF instead of CI(SD). This was established for a standard sp basis, usually $(10,6)/[6,4]$, and 2d1f polarization sets. The mean deviation of calculated dissociation energies from experimental ones was

CI(SD): $\qquad \overline{\Delta D_e} = 1.1\,\text{eV}$

CPF: $\qquad \overline{\Delta D_e} = 0.53\,\text{eV}$

D_e is always computed too small for both methods. The error ΔD_e on the CPF level is roughly proportional to D_e and corresponds to a deviation of 12%; the CI(SD) behaves in a less regular way. An extension of the polarization set from 2d1f to 3d2f1g reduced the error with respect to experiment to

CPF: $\qquad \overline{\Delta D_e} = 0.39\,\text{eV}$

The computed D_e(CPF) are still too small throughout, the average deviation being 8%. Further basis set extension will probably reduce $\overline{\Delta D_e}$ to 0.2–0.3 eV, but computed D_e appear to stay below experiment by about 5%.

The results obtained confirm the importance of very large polarization sets in order to converge D_e values to within chemical accuracy, i.e. 0.1 eV. The convergence of D_e on basis set extension, as displayed by the CPF results, should be comparable to that of other methods, e.g. MR-CI(SD) or MP(4). The relative importance of individual polarization sets was investigated but will not be treated in detail here.

The calculated bond lengths for CO, N_2, NO, O_2 and F_2 show the following pattern. With a $(10,6)/[6,4]$ sp basis augmented by a 2d1f polarization set, one gets the following mean deviation $\overline{\Delta R_e}$ with respect to experiment

CI(SD): $\qquad \overline{\Delta R_e} = 2\,\text{pm}$

CPF: $\qquad \overline{\Delta R_e} = 0.2\,\text{pm}$

The CI(SD) always yields R_e too short; CPF results are usually slightly too short, except for F_2 where R_e is 0.2 pm too long. The maximum deviation of computed CPF distances occurs for O_2: $\Delta R_e = 0.5\,\text{pm}$.

The extension of the polarization set to 3d2f1g results in a reduction of computed R_e on the CPF level by about 0.2 pm on average (as compared to 2d1f). The CPF values are then always shorter than experiment and

CPF: $\qquad \overline{\Delta R_e} = 0.4\,\text{pm}$

The authors estimate that fully saturated basis sets will result in bond lengths which are about 0.5 pm too short on the CPF level and about 2.4 pm too short on the CI(SD) level for first-row diatomics.

2. Cl_2 and P_2 [46]

We shall now comment on the results calculated for the molecule Cl_2. Although extensive investigations of convergence characteristics for the dissociation energy were performed, we only want to cite the best result which was obtained with a standard sp basis and 3d2f1g polarization set. The calculated binding energy for Cl_2 is

CPF: $\qquad D_e = 2.40\,eV$

and has to be compared with the experimental result of $D_e = 2.51\,eV$.[47] The error is of the same magnitude as for first-row diatomics, e.g. 0.2 eV for F_2 with a 3d2f1g polarization set.[62,65]

A somewhat different state of affairs is found for the behaviour of CPF bond lengths. The best calculated value is obtained with a 3d2f1g polarization set

CPF: $\qquad R_e = 199.9\,pm$

which is 1 pm longer than the experimental value $R_e = 198.8\,pm$. In order to see if this state of affairs is typical for second-row diatomics, let us quote briefly some results for P_2. P_2 and Cl_2 constitute the extreme cases of strongly and weakly bonded second-row diatomics and molecules like S_2 and PS probably fit in between P_2 and Cl_2, as is the case for the corresponding first-row diatomics.

The results for the dissociation energy of P_2 match the already observed pattern. With the large 3d2f1g polarization basis a value of $D_e = 4.5\,eV$ is calculated (expt 5.1 eV). The corresponding bond length, $R_e = 189.5\,pm$ (expt 189.3 pm), is again slightly too large, similarly as for Cl_2. The discussion of CPF results indicates that second-row diatomics behave differently from first-row diatomics where bond lengths are concerned, whereas similar patterns may be assumed for dissociation energies.

Since deviations of CPF dissociation energies from experiment are anticipated even for saturated basis sets, a few comments are in order to explain the error sources—which are in all probability shared by other coupled pair approaches:

1. In CPF, cluster corrections are only approximately accounted for. These effects are difficult to estimate but could easily lead to errors of 0.1 eV.
2. Correlation involving inner-shell orbitals is expected to increase D_e and so to reduce the error. For N_2 we have computed this contribution (0.025 eV) with a 2d1f polarization set both for core and valence regions. A value of 0.1 eV should, therefore, be a safe upper limit for N_2. For other diatomics,

with larger bond distances and fewer bonding electron pairs interacting with the cores, smaller contributions are expected.

3. The CPF method—and other coupled pair theories developed to date—nowhere accounts for effects of linked triples, quadruples and higher terms. In view of available evidence[66] we expect these terms to increase both R_e and D_e, which would reduce discrepancies with experiment. The neglect of linked triples may be the major error source of the CPF method, as has already been mentioned in Section IV.B.

3. Cu_2 [63,64]

We shall now outline some work done on Cu_2, which is the simplest of the transition metal diatomics. In spite of its closed-shell structure, the calculation of bond length and dissociation energy for Cu_2 has long resisted a thorough *ab initio* treatment.

Only recently has the work of Bauschlicher, Walch and Siegbahn[67] showed the need to include d correlation, whereas the work of Werner and Martin[64] and of Scharf, Brode and Ahlrichs[63] stressed the importance of cluster corrections and relativistic corrections. The results of Werner *et al.*[64] were obtained using CEPA-1 to account for cluster contributions, while Scharf *et al.* used the CPF approach. Both groups accounted for relativistic corrections by employing first-order perturbation theory, i.e. by evaluating the Cowan–Griffin operator[68] which consists of the mass–velocity and the one-electron Darwin term of the Breit–Pauli Hamiltonian.

Both investigations used basis sets of similar size and led to very similar results, which again shows the closeness of CPF and CEPA-1. The CPF bond distance for the largest basis set used (16, 11, 6, 3)/[10, 7, 4, 3] is

$$R_e = 223.8 \text{ pm} \qquad (\text{expt } 222 \text{ pm})$$

and the value for the dissociation energy is

$$D_e = 1.84 \text{ eV} \qquad (\text{expt } 2.05 \text{ eV})$$

Inclusion of a g set had only a small effect.[63] Bauschlicher *et al.*[69] have recently looked again at the basis saturation problem for Cu_2. They found that even larger f sets than used previously are required—especially diffuse functions—for a proper description of the electronic structure. Their final result was $D_e = 1.9 \text{ eV}$ on the CPF level, which shows a deviation of only 0.15 eV from experiment.

A look at the effect of cluster corrections and relativistic contributions explains the delay of successful *ab initio* descriptions of Cu_2. Both corrections are of roughly the same size for the bond length (3–4 pm, depending on the method), whereas the inclusion of cluster corrections yields the main improvement for the dissociation energy—a calculation of this quantity on the CI(SD) level is clearly a hopeless venture.

E. Dipole Moments

In this section we will report results of CPF calculations of dipole moments for alkaline-earth halides performed by Langhoff et al.[70] and for SO_2, SF_2 and SCl_2 performed by Becherer.[71] Langhoff et al. calculated dipole moments of the $X^2\Sigma^+$ ground states of BeF, BeCl, MgF, MgCl, CaF, CaCl and SrF on SCF, CI(SD) and CPF levels. They found varying influence of electron correlation for these molecules. Electron correlation reduces computed dipole moments for the more covalent molecules BeF and BeCl, whereas it increases those of the more ionic halides of Ca and Sr. Langhoff et al. further point out that reliable values for the dipole moments of alkaline-earth halides may only be obtained if the polarization of the lone electron is described correctly. Since a thorough discussion of the calculated values is beyond the scope of this review, we only want to emphasize that CPF yields consistently better values than SCF and CI(SD) for the dipole moments of the molecules considered (see Table III). The calculated dipole moments of CaF and CaCl are in very good agreement with experiment, whereas for SrF a deviation of 0.269 D from the experimental value $\mu = 3.468$ D is observed, which is, however, far smaller than the deviations on SCF and CI(SD) levels.

We shall now comment on the results for the molecules SO_2, SF_2 and SCl_2 which were obtained by Becherer[71] using the CPF method. Although several calculations (on the SCF level) for either of these molecules are described in the literature, the results usually reproduce experimental values insufficiently.[75,76] This seems to be partly due to basis set deficiencies and partly due to neglect of electron correlation. Contrary to some of the alkaline-earth halides, there exist experimental values for the dipole moments of the SX_2 molecules. The calculations were performed using a basis including up to 2d1f polarization

TABLE III
Dipole moments μ (debye) of alkaline-earth halides.[70]

	CI(SD)	CPF	Expt
BeF	1.831	1.086	–
BeCl	0.88	0.796	–
MgF	3.048	3.077	–
MgCl	3.381	3.382	–
CaF	2.59	3.06	3.07[72]
CaCl	3.629	4.192	4.265[73]
SiF	2.523	3.199	3.468[74]

TABLE IV

Dipole moments μ (debye) of SX_2 molecules.[71]

	SCF	CI(SD)	CPF	Expt
SO_2	2.188	1.938	1.519	1.61[77]
SF_2	1.611	1.418	1.137	1.05[78]
SCl_2	0.503	0.441	0.33	0.36[79]

sets and, therefore, good results were expected. In Table IV we give the dipole moments calculated with SCF, CI(SD), CPF and the experimental values. Worth mentioning is the fact that for all three SX_2 molecules the inclusion of electron correlation seems mandatory since far too large values for the dipole moments are calculated on the SCF level.

The results discussed in this subsection again show that CPF rectifies the main drawbacks of CI(SD) and yields a more reliable description of the electronic structure. The authors are not aware of systematic large-scale applications of Davidson-type corrections, CCSD or CP-MET, or perturbation treatments for the calculation of properties.

F. Further Applications

In the preceding subsections we have discussed in some detail applications which dealt with a series of molecules and/or employed very extended basis sets. This was helpful to assess with some confidence the reliability of pair approaches. There are, of course, numerous further applications of pair methods and we will briefly review some recent articles.

Staemmler and coworkers have treated a number of open-shell systems in close connection with experimental problems, such as the photodissociation of H_2O[80] or the Penning ionization of water by metastable He.[81] Further work is concerned with Rydberg states of H_4 and H_5,[82,83] rotational barriers about $C\!=\!C$ double bonds[84] and ionization potentials.[85]

Dykstra and coworkers have applied CCSD and especially his ACCD (approximate CCSD).[86] Although the abbreviations CCSD and ACCD do not contain the letter P—the only pragmatical definition of pair approaches—these methods are closely related to SCEP. Recent applications include treatments of intermolecular interactions in $(HF)_2$,[87] of HF with N_2, CO and HCN[88] and of HF with Mg.[89]

Numerous applications of CEPA have been performed by Botschwina and coworkers in order to compute spectroscopic properties of small molecules and ions. It is beyond the scope of this chapter to review this work, and we only point to some recent papers concerned with SiO and $HOSi^+$,[90] HCN and CN^-,[91] HC_2Cl^+,[92] and CS and HCS^+.[93]

Last but not least we mention very recent work on infrared absorption intensities for the isoelectronic systems NH_3, H_3O^+ and CH_3^-,[94] and a detailed study of the water molecule.[95] The computations were performed on the CI and CPA (coupled pair approximation) level, which is closely related to CEPA.[44]

V. SUMMARY

Any CI truncated below the full excitation level necessarily suffers from size-consistency problems. The severity of this shortcoming depends very much on the case: it is certainly a problem of major concern for large-scale single-reference CI(SD) calculations which are now feasible (e.g. correlating 28 electrons in using a basis of 150 contracted Gauss type orbitals[96]); it is a minor problem for MR-CI(SD) treatments based on a sufficiently large reference space or CI(SDTQ) calculations as far as they are feasible at present.[48,49]

A variety of methods have been developed and tested which exploit features of the physical structure of wavefunctions in order to avoid the shortcomings of brute-force CI(SD), CI(SDTQ) or full CI calculations. We mention especially internally[28] or externally[97] contracted MR-CI(SD) techniques, full CI methods in a limited orbital space such as the CAS-SCF,[98,99] perturbation theory procedures like MP(4), MP(5), etc.,[100-102] and pair methods like CP-MET (and CCSD),[11] CEPA[12] and CPF.[13] All these methods certainly have their virtues as is obvious from the chapters in this volume.

Recent work has demonstrated that pair methods have achieved their main objective. We now have facts available—discussed in Section IV.B of this chapter—which show that single-reference CPF and CEPA-1 yield a reliable account of unlinked clusters and essentially miss the contributions of linked triples. This holds as long as ψ_0 is still a useful zeroth-order approximation.

CEPA and CPF have the advantage of structural simplicity. From a programming point of view, they differ only in minor details—which are easily implemented[12,13,21]—from CI(SD) procedures for which very efficient codes are available. Efficiency is an important aspect: this is demonstrated by applications of CEPA and CPF which are routinely performed for relatively large systems (i.e. number of basis functions and orbitals correlated).

Recent work has concentrated on the development of pair approaches based on a functional formulation, e.g. CPF[13] and CEPA-V.[14,15] This offers various advantages, of which we mention especially gradient calculations and a clearcut definition of density matrices required for first-order properties.

Besides size consistency, pair approaches have always exploited special features of two-electron functions as mentioned at the beginning of this review. However, a remarkable convergence of methods has taken place in this respect. Efficient matrix-oriented direct CI algorithms—which avoid logic in inner loops and are well suited for vector computers[28,29,34]—were first developed within pair approaches of SCEP type.[18,26] These features can be transferred rather easily to GUGA (graphical unitary group approach) based MR-CI(SD) procedures[13,35] since program structure and data flow are very similar to those required by matrix algorithms. These modifications not only facilitate vectorization, they also lead to much simpler and shorter codes.[35]

The extension of pair approaches to the multiple-reference case is still in its infancy. Some promising attempts have been published[36,103] but further systematic work is desirable to test the viability of such an approach.

Acknowledgement

Thanks are due to K. Jankowski who carefully read the manuscript and made valuable comments.

References

1. Hurley, A. C., Lennard-Jones, J., and Pople, J. A., *Proc. R. Soc. A*, **220**, 446 (1953).
2. Hylleraas, E. A., *Z. Phys.*, **54**, 347 (1929).
3. James, H. M., and Coolidge, A. S., *J. Chem. Phys.*, **1**, 825 (1933).
4. Bartlett, R. J., and Purvis, G. D., *Int. J. Quantum Chem.*, **14**, 561 (1978).
5. Clementi, E., *J. Chem. Phys.*, **39**, 487 (1963).
6. Jungen, M., and Ahlrichs, R., *Theor. Chim. Acta*, **17**, 339 (1970).
7. Mehler, E. L., Ruedenberg, K., and Silver, D. M., *J. Chem. Phys.*, **52**, 1181 (1970).
8. Kutzelnigg, W., *Top. Curr. Chem.*, **41**, 31 (1973).
9. Sinanoglu, O., *J. Chem. Phys.*, **36**, 706, 3198 (1962).
   Nesbet, R. K., *Phys. Rev.*, **109**, 1632 (1958); *Adv. Chem. Phys.*, **9**, 321 (1965).
10. Prime, S., and Robb, M. A., *Theor. Chim. Acta*, **42**, 181 (1976).
11. Bartlett, R. J., *Adv. Chem. Phys.* (1986).
12. Meyer, W., *Int. J. Quantum Chem. Symp.*, **5**, 341 (1971); *J. Chem. Phys.*, **58**, 1017 (1973).
13. Ahlrichs, R., Scharf, P., and Ehrhardt, C., *J. Chem. Phys.*, **82**, 890 (1985).
14. Pulay, P., *J. Mol. Struct.*, **103**, 57 (1983).
15. Pulay, P., *Int. J. Quantum Chem. Symp.*, **17**, 257 (1983).
16. Cizek, J., *J. Chem. Phys.*, **45**, 4256 (1966); *Adv. Chem. Phys.*, **14**, 35 (1969).
17. Paldus, J., in *New Horizons of Quantum Chemistry* (Eds. P.-O. Löwdin and B. Pullman), Reidel, Dordrecht, 1983.
18. Ahlrichs, R., *Comput. Phys. Commun.*, **17**, 31 (1979).
19. Löwdin, P.-O., and Shull, H., *Phys. Rev.*, **101**, 1730 (1956); Shull, H., *J. Chem. Phys.*, **30**, 1405 (1959).
20. Ahlrichs, R., and Driessler, F., *Theor. Chim. Acta*, **36**, 275 (1975).
21. Ahlrichs, R., Lischka, H., Staemmler, V., and Kutzelnigg, W., *J. Chem. Phys.*, **62**, 1225 (1975).
22. Meyer, W., in *Modern Theoretical Chemistry*, Vol. 3 (Ed. H. F. Schaefer), Plenum, New York, 1977.
23. Taylor, P. R., *J. Chem. Phys.*, **74**, 1256 (1981).
24. Roos, B. O., *Chem. Phys. Lett.*, **15**, 153 (1972).
25. Roos, B. O., and Siegbahn, P. E. M., in *Modern Theoretical Chemistry*, Vol. 3 (Ed. H. F. Schaefer), Plenum, New York, 1977.

26. Meyer, W., *J. Chem. Phys.*, **64**, 2901 (1976).
27. Pulay, P., Saebo, S., and Meyer, W., *J. Chem. Phys.*, **81**, 1901 (1984).
28. Werner, H.-J., and Reinsch, F.-A., *J. Chem. Phys.*, **76**, 3144 (1982).
29. Ahlrichs, R., in *Proceedings of the Fifth Seminar on Computational Methods in Quantum Chemistry* (Eds. P. T. van Duijnen and W. C. Nieuwpoort), MPI Garching, München, 1981.
30. Siegbahn, P. E. M., *J. Chem. Phys.*, **72**, 1647 (1980).
31. Siegbahn, P. E. M., *Int. J. Quantum Chem.*, **18**, 1229 (1980).
32. Ahlrichs, R., in *Methods in Computational Molecular Physics* (Eds. G. H. F. Diercksen and S. Wilson), Reidel, Dordrecht, 1983.
33. Meyer, W., Ahlrichs, R., and Dykstra, C. E., in *Advanced Theories and Computational Approaches to the Electronic Structure of Molecules* (Ed. C. E. Dykstra), Reidel, Dordrecht, 1984.
34. Saunders, V. R., and van Lenthe, J. H., *Mol. Phys.*, **48**, 923 (1983).
35. Ahlrichs, R., Böhm, H.-J., Ehrhardt, C., Scharf, P., Schiffer, H., Lischka, H., and Schindler, M., *J. Comput. Chem.*, **6**, 200 (1985).
36. Werner, H.-J., *Adv. Chem. Phys.*, (1986).
37. Paldus, J., Cizek, J., and Shavitt, I., *Phys. Rev. A*, **5**, 50 (1972).
38. Taylor, P. R., Bacskay, G. B., Hush, N. S., and Hurley, A. C., *Chem. Phys. Lett.*, **41**, 444 (1976).
39. Pople, J. A., Krishnan, R., Schlegel, H. B., and Binkley, J. S., *Int. J. Quantum Chem.*, **14**, 545 (1978).
40. Chiles, R. A., and Dykstra, C. E., *J. Chem. Phys.*, **74**, 4544 (1981).
41. Davidson, E. R., in *The World of Quantum Chemistry* (Eds. R. Daudel and B. Pullman), Reidel, Dordrecht, 1974.
42. Pople, J. A., Seeger, R., and Krishnan, R., *Int. J. Quantum Chem. Symp.*, **11**, 149 (1977).
43. Siegbahn, P. E. M., *Chem. Phys. Lett.*, **55**, 386 (1978).
44. Hurley, A. C., *Electron Correlation in Small Molecules*, Academic Press, New York, 1976.
45. Pulay, P., and Saebo, S., *Chem. Phys. Lett.*, **117**, 37 (1985).
46. Becherer, R., and Ahlrichs, R., *Chem. Phys.*, **99**, 389 (1985).
47. Huber, K. P., and Herzberg, G., *Molecular Spectra and Molecular Structure*, Vol. IV, *Constants of Diatomic Molecules*, Van Nostrand Reinhold, New York, 1979.
48. Saxe, P., Schaefer, H. F., and Handy, N. C., *Chem. Phys. Lett.*, **79**, 202 (1981).
49. Harrison, R. J., and Handy, N. C., *Chem. Phys. Lett.*, **95**, 386 (1983).
50. Bartlett, R. J., Sekino, H., and Purvis, G. D., III, *Chem. Phys. Lett.*, **98**, 66 (1983).
51. Hirao, K., and Hatano, Y., *Chem. Phys. Lett.*, **100**, 519 (1983).
52. Phillips, R. A., Buenker, R. J., Bruna, P. J., and Peyerimhoff, S. D., *Chem. Phys.*, **84**, 11 (1984).
53. Brown, F. B., Shavitt, I., and Shepard, R., *Chem. Phys. Lett.*, **105**, 363 (1984).
54. Laidig, W. D., and Bartlett, R. J., *Chem. Phys. Lett.*, **104**, 424 (1984).
55. Jankowski, K., private communication.
56. Bartlett, R. J., Dykstra, C. E., and Paldus, J., in *Advanced Theories and Computational Approaches to the Electronic Structure of Molecules* (Ed. C. E. Dykstra), Reidel, Dordrecht, 1984.
57. Koch, S., and Kutzelnigg, W., *Theor. Chim. Acta*, **59**, 387 (1981).
58. Meyer, W., *Theor. Chim. Acta*, **35**, 271 (1974).
59. Jankowski, K., and Paldus, J., *Int. Quantum Chem.*, **18**, 1243 (1980).
60. Meyer, W., and Rosmus, P., *J. Chem. Phys.*, **63**, 2356 (1975).

61. Rosmus, P., and Meyer, W., *J. Chem. Phys.*, **66**, 13 (1977); *J. Chem. Phys.*, **69**, 2745 (1978).
62. Ahlrichs, R., Scharf, P., and Jankowski, K., *Chem. Phys.*, **98**, 381 (1985).
63. Scharf, P., Brode, S., and Ahlrichs, R., *Chem. Phys. Lett.*, **113**, 447 (1985).
64. Werner, H.-J. and Martin, R. L., *Chem. Phys. Lett.*, **113**, 451 (1985).
65. Jankowski, K., Becherer, R., Scharf, P., Schiffer, H., and Ahlrichs, R., *J. Chem. Phys.*, **82**, 1413 (1985).
66. Urban, M., and Noga, J., *Theor. Chim. Acta*, **62**, 549 (1983).
67. Bauschlicher, C. W., Walch, S. P., and Siegbahn, P. E. M., *J. Chem. Phys.*, **76**, 6015 (1982); *J. Chem. Phys.*, **78**, 3347 (1983).
68. Cowan, R. D., and Griffin, D. C., *J. Opt. Soc. Am.*, **66**, 1010 (1976).
69. Bauschlicher, C. W., Walch, S. P., and Langhoff, S. R., to be published.
70. Langhoff, S. R., Bauschlicher, C. W., Partridge, H., and Ahlrichs, R., *J. Chem. Phys.*, **84**, 5025 (1986).
71. Becherer, R., Doctoral thesis, Karlsruhe, 1985.
72. Childs, W. J., Goodman, L. S., Nielsen, U., and Pfeufer, V., *J. Chem. Phys.*, **80**, 2283 (1984).
73. Ernst, W. E., Kindt, S., and Törring, T., *Phys. Rev. Lett.*, **51**, 979 (1983).
    Ernst, W. E., Kindt, S., Nair, K. P. R., and Törring, T., *Phys. Rev. A*, **29**, 1158 (1984).
74. Ernst, W. E., Kandler, J., Kindt, S., and Törring, T., *Chem. Phys. Lett.*, **113**, 351 (1985).
75. Burton, P. G., and Carlsen, N. R., *Chem. Phys. Lett.*, **46**, 48 (1977).
76. Solouki, B., Rosmus, P., and Bock, H., *Chem. Phys. Lett.*, **26**, 20 (1974).
77. McClellan, A. L., *Tables of Experimental Dipole Moments*, Freeman, San Francisco, 1963.
78. Grable, G. F., and Smith, W. V., *J. Chem. Phys.*, **19**, 502 (1951).
79. Murray, J. T., Little, W. A., Williams, Q., and Weatherly, T., *J. Chem. Phys.*, **65**, 985 (1976).
80. Staemmler, V., and Palma, A., *Chem. Phys.*, **93**, 63 (1985).
81. Haug, B., Morgner, H., and Staemmler, V., *J. Phys. B: At. Mol. Phys.*, **18**, 259 (1985).
82 Jungen, M., and Staemmler, V., *Chem. Phys. Lett.*, **103**, 191 (1983).
83. Kaufmann, K., Jungen, M., and Staemmler, V., *Chem. Phys.*, **79**, 111 (1983).
84. Staemmler, V., and Jaquet, R., in *Energy Storage and Redistribution in Molecules* (Ed. J. Hinze), Plenum, New York, 1983.
85. Staemmler, V., *Theor. Chim. Acta*, **64**, 205 (1983).
86. Chiles, R. A., and Dykstra, C. E., *Chem. Phys. Lett.*, **80**, 69 (1981).
87. Michael, D. W., Dykstra, C. E., and Lisy, J. M., *J. Chem. Phys.*, **81**, 5998 (1984).
88. Benzel, M. A., and Dykstra, C. E., *J. Chem. Phys.*, **78**, 4052 (1983).
89. Jasien, P. G., and Dykstra, C. E., *Chem. Phys. Lett.*, **106**, 276 (1984).
90. Botschwina, P., and Rosmus, P., *J. Chem. Phys.*, **82**, 1420 (1985).
91. Botschwina, P., *Chem. Phys. Lett.*, **114**, 58 (1985).
92. Botschwina, P., Sebald, P., and Maier, J. P., *Chem. Phys. Lett.*, **114**, 353 (1985).
93. Botschwina, P., and Sebald, P., *J. Mol. Spectrosc.*, **110**, 1 (1985).
94. Swanton, D. J., Bacskay, G. B., and Hush, N. S., *Chem. Phys.*, **107**, 9 (1986).
95. Swanton, D. J., Bacskay, G. B., and Hush, N. S., *J. Chem. Phys.*, **84**, 5715 (1986).
96. Scharf, P., and Ahlrichs, R., *Chem. Phys.*, **100**, 237 (1985).
97. Siegbahn, P. E. M., in *Proceedings of the Fifth Seminar on Computational Methods in Quantum Chemistry* (Eds. P. T. van Duijen and W. C. Nievwpoort), MPI Garching, München, 1981.

98. Ruedenberg, K., and Sundberg, K. R., in *Quantum Science, Methods and Structure* (Eds. J. L. Calais, O. Goscinski, J. Linderberg, and Y. Öhrn), Plenum, New York, 1976.
99. Roos, B. O., Taylor, P. R., and Siegbahn, P. E. M., *Chem. Phys.*, **48**, 157 (1980).
100. Krishnan, R., and Pople, J. A., *Int. J. Quantum Chem.*, **14**, (1978).
101. Krishnan, R., Frisch, M. J., and Pople, J. A., *J. Chem. Phys.*, **72**, 4244 (1980).
102. Laidig, W. D., Fitzgerald, G., and Bartlett, R. J., *Chem. Phys. Lett.*, **113**, 151 (1985).
103. Ruttink, P. J. A., in *Proceedings of the Sixth Seminar on Computational Methods in Quantum Chemistry* (Eds. W. P. Kraemer and R. Beardsworth), MPI Garching, München, 1984.

AUTHOR INDEX

SUBJECT INDEX